Lecture Notes in Physics

Springer
Berlin
Heidelberg
New York
Hong Kong
London
Milan
Paris
Tokyo

The Editorial Policy for Edited Volumes

The series *Lecture Notes in Physics* (LNP), founded in 1969, reports new developments in physics research and teaching - quickly, informally but with a high degree of quality. Manuscripts to be considered for publication are topical volumes consisting of a limited number of contributions, carefully edited and closely related to each other. Each contribution should contain at least partly original and previously unpublished material, be written in a clear, pedagogical style and aimed at a broader readership, especially graduate students and nonspecialist researchers wishing to familiarize themselves with the topic concerned. For this reason, traditional proceedings cannot be considered for this series though volumes to appear in this series are often based on material presented at conferences, workshops and schools.

Acceptance

A project can only be accepted tentatively for publication, by both the editorial board and the publisher, following thorough examination of the material submitted. The book proposal sent to the publisher should consist at least of a preliminary table of contents outlining the structure of the book together with abstracts of all contributions to be included. Final acceptance is issued by the series editor in charge, in consultation with the publisher, only after receiving the complete manuscript. Final acceptance, possibly requiring minor corrections, usually follows the tentative acceptance unless the final manuscript differs significantly from expectations (project outline). In particular, the series editors are entitled to reject individual contributions if they do not meet the high quality standards of this series. The final manuscript must be ready to print, and should include both an informative introduction and a sufficiently detailed subject index.

Contractual Aspects

Publication in LNP is free of charge. There is no formal contract, no royalties are paid, and no bulk orders are required, although special discounts are offered in this case. The volume editors receive jointly 30 free copies for their personal use and are entitled, as are the contributing authors, to purchase Springer books at a reduced rate. The publisher secures the copyright for each volume. As a rule, no reprints of individual contributions can be supplied.

Manuscript Submission

The manuscript in its final and approved version must be submitted in ready to print form. The corresponding electronic source files are also required for the production process, in particular the online version. Technical assistance in compiling the final manuscript can be provided by the publisher's production editor(s), especially with regard to the publisher's own LaTeX macro package which has been specially designed for this series.

LNP Homepage (springerlink.com)

On the LNP homepage you will find:

–The LNP online archive. It contains the full texts (PDF) of all volumes published since 2000. Abstracts, table of contents and prefaces are accessible free of charge to everyone. Information about the availability of printed volumes can be obtained.

–The subscription information. The online archive is free of charge to all subscribers of the printed volumes.

–The editorial contacts, with respect to both scientific and technical matters.

–The author's / editor's instructions.

D. Blaschke M.A. Ivanov T. Mannel (Eds.)

Heavy Quark Physics

Editors

David Blaschke
Universität Rostock
Fachbereich Physik
Universitätsplatz 3
18051 Rostock, Germany

Thomas Mannel
Universität Karlsruhe
Institut für Theoretische Teilchenphysik
76128 Karlsruhe, Germany

Mikhal A. Ivanov
Bogoliubov Laboratory
for Theoretical Physics
Joint Institute for Nuclear Research
6 Joliot-Curie Street
141980 Dubna, Russian Federation

D. Blaschke, M.A. Ivanov, T. Mannel (Eds.), *Heavy Quark Physics*, Lect. Notes Phys. **647** (Springer, Berlin Heidelberg 2004), DOI 10.1007/b97728

Library of Congress Control Number: 2004104709

Bibliographic information published by Die Deutsche Bibliothek Die Deutsche Bibliothek lists this publication in the Deutsche Nationalbibliografie; detailed bibliographic data is available in the Internet at <http://dnb.ddb.de>

ISSN 0075-8450
ISBN 3-540-21921-8 Springer-Verlag Berlin Heidelberg New York

Springer-Verlag is a part of Springer Science+Business Media

springeronline.com

Printed in Germany

Typesetting: Camera-ready by the authors/editor
Data conversion: PTP-Berlin Protago-TeX-Production GmbH
Cover design: *design & production*, Heidelberg

Printed on acid-free paper
54/3141/ts - 5 4 3 2 1 0

Preface

The International School was arranged as a continuation of a series of workshops on Heavy Quark Physics held in Dubna (1993, 1996, 2000), Bad Honnef (1994) and Rostock (1997). Starting with this event the workshops were transformed into schools with the lectures given by well-known experts in this area on a pedagogic level and original talks given by junior and senior scientists. While the first few workshops were mainly German-Russian, the school has turned into a truly international event with participation from all over world.

The school covered the main topics in heavy flavor physics. The comprehensive introduction to B-physics (heavy mass expansion, non-leptonic decays, QCD factorization) was given by M. Neubert (Cornell). CP-violation in the Standard Model and in the B-meson system was discussed in lectures of R. Fleischer (DESY). T. Mannel (Karlsruhe) gave lectures on rare B-meson decays. A.G. Grozin (Novosibirsk) presented methods of calculation of loop diagrams in Heavy Quark Effective Theory.

V.K. Mitrushkin (Dubna) gave an introduction to the Lattice QCD with fermions. C. McNeile (Liverpool) discussed application of the Lattice QCD to heavy quark physics. C.D. Roberts (Argonne) provided an introduction to the use of the Dyson–Schwinger equations in calculating both light and heavy hadron observables.

Recent ideas on physics beyond the Standard Model were discussed in lectures given by R. Rückl (Würzburg) (extra dimensions) and D.I. Kazakov (Dubna) (supersymmetry).

Model-independent calculations of one-loop corrections to polarization observables were presented in lectures of J.G. Körner (Mainz). The study of exclusive decays of heavy hadrons requires the model-dependent nonperturbative methods. Exclusive rare B and B_c decays and electromagnetic properties of heavy baryons within a relativistic quark model were discussed in lectures by M.A. Ivanov (Dubna) and V.E. Lyubovitskij (Tuebingen). I.M. Narodetskii (Moscow) discussed the properties of heavy baryons in the nonperturbative string approach to QCD. A description of exclusive nonleptonic B-decays from QCD light-cone sum rules was given by B. Melic (Würzburg). A.A. Pivovarov (Moscow) has shown a way of extracting the strong coupling constant and strange quark mass from semileptonic τ lepton decays. A lecture on quark–hadron duality sum rules and two-photon decays of mesons was given by one of the founder of this direction – S.B. Gerasimov (Dubna). The role of instantons in the explanation of the $\Delta = 1/2$ rule in the weak decays was presented by N.I. Kochelev

(Dubna). Multibaryons with heavy flavors in the Skyrme model were discussed by V.B. Kopeliovich (Moscow).

The modern experimental status of B-physics was presented by two collaborations: Belle (R. Chistov) and BaBar (G. De Nardo). Special emphasis was devoted to the first measurement of time dependent CP Asymmetries in $B \to \pi\pi$ decay.

Another aspect of heavy quark physivs presented in the School was the production of heavy quarks and quarkonia. Lectures on heavy flavor production in the nuclear environment were given by B.B. Kopeliovich (Heidelberg and Dubna). Charmonium in a hot and dense medium was discussed in lectures by D. Blaschke (Rostock and Dubna) and A. Polleri (Munich). The lectures of N.P. Zotov (MSU), V.A. Saleev (Samara) and A.V. Kotikov (Dubna) were devoted to photoproduction of charmonia and charmed mesons at high energy. O.V. Teryaev (Dubna) discussed the collinear and $k_\perp$-factorization in heavy quarks hadroproduction. Charm production in hadron–nucleon collisions within quark–gluon string models was discussed by G.I. Lykasov (Dubna). The experimental status of the charmonium production in proton–nucleus collisions at HERA-B was presented by R.V. Miziuk and A.V. Lanyov.

Finally the organizers thank the German Bundesministerium für Forschung und Bildung which provided the basic funding of the school. We also appreciate the Heisenberg–Landau program and the Russian Fund for Basic Research for their partial support. We want to thank Sergei Nedelko (Dubna) for his assistance in preparing the manuscript of this book.

Rostock, Dubna and Karlsruhe
June 2003

David Blaschke
Mikhail A. Ivanov
Thomas Mannel

Contents

Part I B–Physics

Part III Production of Heavy Flavors

List of Contributors

D. Blaschke
Universität Rostock,
Fachbereich Physik,
18051 Rostock, Germany
and
Joint Institute
for Nuclear Research,
Bogoliubov Laboratory for
Theoretical Physics,
141980 Dubna, Russia

G. De Nardo
Naples University and INFN, Naples
Dipartimento di Scienze Fisiche
Complesso Universitario
di Monte Sant'Angelo
via Cintia
80126 Napoli, Italy

R. Fleischer
CERN
Theory Division
1211 Geneva 23, Switzerland

M.A. Ivanov
Joint Institute
for Nuclear Research,
Bogoliubov Laboratory for
Theoretical Physics,
141980 Dubna, Russia

Yu.L. Kalinovsky
Joint Institute for Nuclear Research,
Laboratory for Information
Technologies,
141980 Dubna, Russia

B.Z. Kopeliovich
Max-Planck Institut für Kernphysik
Postfach 103980
69029 Heidelberg, Germany
and
Institut für Theoretische Physik
der Universität
93040 Regensburg, Germany
and
Joint Institute for Nuclear Research,
Dubna,
141980 Moscow Region, Russia

J.G. Körner
Johannes Gutenberg–Universität
Institut für Physik
Staudinger Weg 7
55099 Mainz, Germany

A. Lanyov
Particle Physics Laboratory
Joint Institute for Nuclear Research
141980 Dubna, Moscow region, Russia

A. Lipatov
M.V. Lomonosov Moscow
State University
Physical Department
119992 Moscow, Russia

V.E. Lyubovitskij
Universität Tübingen
Institut für Theoretische Physik
Auf der Morgenstelle 14
72076 Tübingen, Germany

T. Mannel
Universität Karlsruhe
Institut für Theoretische
Teilchenphysik
76128 Karlsruhe, Germany
and
Universität Siegen,
Fachbereich Physik,
57068 Siegen, Germany

M.C. Mauser
Johannes Gutenberg–Universität
Institut für Physik
Staudinger Weg 7
55099 Mainz, Germany

B. Melić
Julius-Maximilians-Universität
Würzburg
Institut für Theoretische Physik
und Astrophysik
97074 Würzburg, Germany
and
Johannes-Gutenberg Universität
Mainz
Institut für Physik
55099 Mainz, Germany

C. McNeile
University of Liverpool
Department of Mathematical Sciences,
L69 3BX, UK

R. Mizuk
Institute for Theoretical
and Experimental Physics
117259 Moscow, Russia

A. Mück
Universität Würzburg
Institut für Theoretische Physik
und Astrophysik
Am Hubland
97074 Würzburg, Germany

I.M. Narodetskii
Institute of Theoretical
and Experimental Physics
Moscow 117259, Russia

M. Neubert
Cornell University
Newman Laboratory
of Elementary-Particle Physics
Ithaca, NY 14853, USA

A. Pilaftsis
University of Manchester
Department of Physics and Astronomy
Manchester M13 9PL, UK

A.A. Pivovarov
Institute for Nuclear Research
of the Russian Academy of Science
117312 Moscow, Russia

J. Raufeisen
Los Alamos National Laboratory
MS H846, Los Alamos
NM 87545, USA

C.D. Roberts
Argonne National Laboratory
Physics Division
Bldg 203, Argonne,
Illinois 60439-4843, USA
and
Universität Rostock
Fachbereich Physik
18051 Rostock, Germany

R. Rückl
Universität Würzburg
Institut für Theoretische Physik
und Astrophysik
Am Hubland
97074 Würzburg, Germany

V.A. Saleev
Samara State University
Samara, 443011, Russia

M.A. Trusov
Institute of Theoretical
and Experimental Physics
Moscow 117259, Russia

D.V. Vasin
Samara Municipal
Nayanova University
Samara, 443001, Russia

V.L. Yudichev
Joint Institute for
Nuclear Research,
Bogoliubov Laboratory
for Theoretical Physics,
141980 Dubna, Russia

N. Zotov
M.V. Lomonosov
Moscow State University
D.V. Skobeltsyn Institute
of Nuclear Physics
119992 Moscow, Russia

Part I

B–Physics

Theory of Exclusive Hadronic B Decays

Matthias Neubert

Newman Laboratory of Elementary-Particle Physics, Cornell University, Ithaca, NY 14853, USA

Abstract. These notes provide a pedagogical introduction to the theory of non-leptonic heavy-meson decays recently proposed by Beneke, Buchalla, Sachrajda and myself. We provide a rigorous basis for factorization for a large class of non-leptonic two-body B-meson decays in the heavy-quark limit. The resulting factorization formula incorporates elements of the naive factorization approach and the hard-scattering approach, and allows us to compute systematically radiative ("non-factorizable") corrections to naive factorization for decays such as $B \to D\pi$ and $B \to \pi\pi$.

1 Introduction

Non-leptonic two-body decays of B mesons, although simple as far as the underlying weak decay of the b quark is concerned, are complicated on account of strong-interaction effects. If these effects could be computed, this would enhance tremendously our ability to uncover the origin of CP violation in weak interactions from data on a variety of such decays being collected at the B factories. In these lecture, I review recent progress towards a systematic analysis of weak heavy-meson decays into two energetic mesons based on the factorization properties of decay amplitudes in QCD [1,2]. My discussion will follow very closely the detailed account of this approach given in [2]. (We have worked so hard on this paper that any attempt to improve on it were bound to fail and leave the author in despair.) Much of the credit for these notes belongs to my collaborators Martin Beneke, Gerhard Buchalla, and Chris Sachrajda.

As in the classic analysis of semi-leptonic $B \to D$ transitions [3,4], our arguments make extensive use of the fact that the b quark is heavy compared to the intrinsic scale of strong interactions. This allows us to deduce that non-leptonic decay amplitudes in the heavy-quark limit have a simple structure. The arguments to reach this conclusion, however, are quite different from those used for semi-leptonic decays, since for non-leptonic decays a large momentum is transferred to at least one of the final-state mesons. The results of our work justify naive factorization of four fermion operators for many, but not all, non-leptonic decays and imply that corrections termed "non-factorizable", which up to now have been thought to be intractable, can be calculated rigorously if the mass of the decaying quark is large enough. This leads to a large number of predictions for CP-violating B decays in the heavy-quark limit, for which measurements will soon become available.

Weak decays of heavy mesons involve three fundamental scales, the weak-interaction scale M_W, the b-quark mass m_b, and the QCD scale $\Lambda_{\rm QCD}$, which

M. Neubert, Theory of Exclusive Hadronic B Decays, Lect. Notes Phys. **647**, 3–41 (2004)
http://www.springerlink.com/

are strongly ordered: $M_W \gg m_b \gg \Lambda_{\rm QCD}$. The underlying weak decay being computable, all theoretical work concerns strong-interaction corrections. QCD effects involving virtualities above the scale m_b are well understood. They renormalize the coefficients of local operators O_i in the effective weak Hamiltonian [5], so that the amplitude for the decay $B \to M_1 M_2$ is given by

$$\mathcal{A}(B \to M_1 M_2) = \frac{G_F}{\sqrt{2}} \sum_i \lambda_i \, C_i(\mu) \, \langle M_1 M_2 | O_i(\mu) | B \rangle \,, \tag{1}$$

where each term in the sum is the product of a Cabibbo–Kobayashi–Maskawa (CKM) factor λ_i, a coefficient function $C_i(\mu)$, which incorporates strong-interaction effects above the scale $\mu \sim m_b$, and a matrix element of an operator O_i. The difficult theoretical problem is to compute these matrix elements or, at least, to reduce them to simpler non-perturbative objects.

A variety of treatments of this problem exist, which rely on assumptions of some sort. Here we identify two somewhat contrary lines of approach. The first one, which we shall call "naive factorization", replaces the matrix element of a four-fermion operator in a heavy-quark decay by the product of the matrix elements of two currents [6,7], e.g.

$$\langle D^+ \pi^- | (\bar{c}b)_{V-A} (\bar{d}u)_{V-A} | \bar{B}_d \rangle \to \langle \pi^- | (\bar{d}u)_{V-A} | 0 \rangle \, \langle D^+ | (\bar{c}b)_{V-A} | \bar{B}_d \rangle \,. \tag{2}$$

This assumes that the exchange of "non-factorizable" gluons between the π^- and the $(\bar{B}_d \, D^+)$ system can be neglected if the virtuality of the gluons is below $\mu \sim m_b$. The non-leptonic decay amplitude then reduces to the product of a form factor and a decay constant. This assumption is in general not justified, except in the limit of a large number of colours in some cases. It deprives the amplitude of any physical mechanism that could account for rescattering in the final state. "Non-factorizable" radiative corrections must also exist, because the scale dependence of the two sides of (2) is different. Since such corrections at scales larger than μ are taken into account in deriving the effective weak Hamiltonian, it appears rather arbitrary to leave them out below the scale μ. Various generalizations of the naive factorization approach have been proposed, which include new parameters that account for non-factorizable corrections. In their most general form, these generalizations have nothing to do with the original "factorization" ansatz, but amount to a general parameterization of the matrix elements. Such general parameterizations are exact, but at the price of introducing many unknown parameters and eliminating any theoretical input on strong-interaction dynamics.

The second method used to study non-leptonic decays is the hard-scattering approach, which assumes the dominance of hard gluon exchange. The decay amplitude is then expressed as a convolution of a hard-scattering factor with light-cone wave functions of the participating mesons, in analogy with more familiar applications of this method to hard exclusive reactions involving only light hadrons [8,9]. In many cases, the hard-scattering contribution represents the leading term in an expansion in powers of $\Lambda_{\rm QCD}/Q$, where Q denotes the hard scale. However, the short-distance dominance of hard exclusive processes is not enforced kinematically and relies crucially on the properties of hadronic wave functions. There is an important difference between light mesons and heavy mesons in this regard, because the light quark in a heavy meson at rest naturally has a small momentum of order $\Lambda_{\rm QCD}$, while for fast light mesons a configuration with a soft quark is suppressed by the endpoint behaviour of the meson wave function. As a consequence, the soft (or Feynman) mechanism is power suppressed for hard exclusive processes involving light mesons, but it is of leading power for heavy-meson decays.

It is clear from this discussion that a satisfactory treatment should take into account soft contributions, but also allow us to compute corrections to naive factorization in a systematic way. It is not at all obvious that such a treatment would result in a predictive framework. We will show that this does indeed happen for most non-leptonic two-body B decays. Our main conclusion is that "non-factorizable" corrections are dominated by hard gluon exchange, while the soft effects that survive in the heavy-quark limit are confined to the (BM_1) system, where M_1 denotes the meson that picks up the spectator quark in the B meson. This result is expressed as a factorization formula, which is valid up to corrections suppressed by powers of $\Lambda_{\rm QCD}/m_b$. At leading power, non-perturbative contributions are parameterized by the physical form factors for the $B \to M_1$ transition and leading-twist light-cone distribution amplitudes of the mesons. Hard perturbative corrections can be computed systematically in a way similar to the hard-scattering approach. On the other hand, because the $B \to M_1$ transition is parameterized by a form factor, we recover the result of naive factorization at lowest order in α_s.

An important implication of the factorization formula is that strong rescattering phases are either perturbative or power suppressed in $\Lambda_{\rm QCD}/m_b$. It is worth emphasizing that the decoupling of M_2 occurs in the presence of soft interactions in the (BM_1) system. In other words, while strong-interaction effects in the $B \to M_1$ transition are not confined to small transverse distances, the other meson M_2 is predominantly produced as a compact object with small transverse extension. The decoupling of soft effects then follows from "colour transparency". The colour-transparency argument for exclusive B decays has already been noted in the literature [10,11], but it has never been developed into a factorization formula that could be used to obtain quantitative predictions.

The approach described in [1,2] is general and applies to decays into a heavy and a light meson (such as $B \to D\pi$) as well as to decays into two light mesons (such as $B \to \pi\pi$). Factorization does not hold, however, for decays such as $B \to \pi D$ and $B \to D\bar{D}$, in which the meson that does *not* pick up the spectator quark in the B meson is heavy. For the main part in these lectures, we will focus on the case of $B \to D^{(*)}L$ decays (with L a light meson), for which the factorization formula takes its simplest form, and power counting will be relatively straightforward. Occasionally, we will point out what changes when we consider more complicated decays such as $B \to \pi\pi$. A detailed treatment of these processes can be found in [12].

The outline of these notes is as follows: In Sect. 2 we state the factorization formula in its general form. In Sect. 3 we collect the physical arguments that lead to factorization and introduce our power-counting scheme. We show how light-cone distribution amplitudes enter, discuss the heavy-quark scaling of the $B \to D$ form factor, and explain the cancellation of soft and collinear contributions in "non-factorizable" vertex corrections to non-leptonic decay amplitudes. We also comment on the implications of our results for final-state interactions in hadronic B decays. The cancellation of long-distance singularities is demonstrated in more detail in Sect. 4, where we present the calculation of the hard-scattering functions at one-loop order for decays into a heavy and a light meson. Various sources of power-suppressed effects, which give corrections to the factorization formula, are discussed in Sect. 5. They include hard-scattering contributions, weak annihilation, and contributions from multi-particle Fock states. We then point out some limitations of the factorization approach. In Sect. 7 we consider the phenomenology of $B \to D^{(*)}L$ decays on the basis of the factorization formula and discuss various tests of our theoretical framework. We also examine to what extent a charm meson should be considered as heavy or light. Section 8 contains the conclusion.

2 Statement of the Factorization Formula

In this section we summarize the factorization formula for non-leptonic B decays. We introduce relevant terminology and provide definitions of the hadronic quantities that enter the factorization formula as input parameters.

2.1 The Idea of Factorization

In the context of non-leptonic decays the term "factorization" is usually applied to the approximation of the matrix element of a four-fermion operator by the product of a form factor and a decay constant, as illustrated in (2). Corrections to this approximation are called "non-factorizable". We will refer to this approximation as "naive factorization" and use quotes on "non-factorizable" to avoid confusion with the (much less trivial) meaning of factorization in the context of hard processes in QCD. In the latter case, factorization refers to the separation of long-distance contributions to the process from a short-distance part that depends only on the large scale m_b. The short-distance part can be computed in an expansion in the strong coupling $\alpha_s(m_b)$. The long-distance contributions must be computed non-perturbatively or determined experimentally. The advantage is that these non-perturbative parameters are often simpler in structure than the original quantity, or they are process independent. For example, factorization applied to hard processes in inclusive hadron–hadron collisions requires only parton distributions as non-perturbative inputs. Parton distributions are much simpler objects than the original matrix element with two hadrons in the initial state. On the other hand, factorization applied to the $B \to D$ form factor leads to a non-perturbative object (the "Isgur–Wise function"), which is still a function of the momentum transfer. However, the benefit here is that symmetries relate this function to other form factors. In the case of non-leptonic B decays, the simplification is primarily of the first kind (simpler structure). We call those effects non-factorizable (without quotes) which depend on the long-distance properties of the B meson and both final-state mesons combined.

The factorization properties of non-leptonic decay amplitudes depend on the two-meson final state. We call a meson "light" if its mass m remains finite in the heavy-quark limit. A meson is called "heavy" if its mass scales with m_b in the heavy-quark limit, such that m/m_b stays fixed. In principle, we could still have $m \gg \Lambda_{\rm QCD}$ for a light meson. Charm mesons could be considered as light in this sense. However, unless otherwise mentioned, we assume that m is of order $\Lambda_{\rm QCD}$ for a light meson, and we consider charm mesons as heavy. In evaluating the scaling behaviour of the decay amplitudes, we assume that the energies of both final-state mesons (in the B-meson rest frame) scale with m_b in the heavy-quark limit.

2.2 The Factorization Formula

We consider a generic weak decay $B \to M_1 M_2$ in the heavy-quark limit and differentiate between decays into final states containing a heavy and a light meson or two light mesons. Our goal is to show that, up to power corrections of order $\Lambda_{\rm QCD}/m_b$, the transition matrix element of an operator O_i in the effective weak Hamiltonian can be written as

$$\begin{aligned}
\langle M_1 M_2 | O_i | B \rangle &= \sum_j F_j^{B\to M_1}(m_2^2)\, f_{M_2} \int_0^1 du\, T_{ij}^I(u)\, \Phi_{M_2}(u) \\
&\quad \text{if } M_1 \text{ is heavy and } M_2 \text{ is light,} \\
\langle M_1 M_2 | O_i | B \rangle &= \sum_j F_j^{B\to M_1}(m_2^2)\, f_{M_2} \int_0^1 du\, T_{ij}^I(u)\, \Phi_{M_2}(u) \; + \; (M_1 \leftrightarrow M_2) \\
&\quad + f_B f_{M_1} f_{M_2} \int_0^1 d\xi\, du\, dv\, T_i^{II}(\xi,u,v)\, \Phi_B(\xi)\, \Phi_{M_1}(v)\, \Phi_{M_2}(u) \\
&\quad \text{if } M_1 \text{ and } M_2 \text{ are both light.}
\end{aligned} \tag{3}$$

Here $F_j^{B\to M}(m^2)$ denotes a $B \to M$ form factor evaluated at $q^2 = m^2$, $m_{1,2}$ are the light meson masses, and $\Phi_X(u)$ is the light-cone distribution amplitude for the quark–antiquark Fock state of the meson X. These non-perturbative quantities will be defined below. $T_{ij}^I(u)$ and $T_i^{II}(\xi,u,v)$ are hard-scattering functions, which are perturbatively calculable. The factorization formula in its general form is represented graphically in Fig. 1.

The second equation in (3) applies to decays into two light mesons, for which the spectator quark in the B meson (in the following simply referred to as the "spectator quark") can go to either of the final-state mesons. An example is the decay $B^- \to \pi^0 K^-$. If the spectator quark can go only to one of the final-state mesons, as for example in $\bar{B}_d \to \pi^+ K^-$, we call this meson M_1, and the second form-factor term on the right-hand side of (3) is absent. The formula simplifies when the spectator quark goes to a heavy meson (first equation in (3)), such as in $\bar{B}_d \to D^+ \pi^-$. Then the second term in Fig. 1, which accounts for

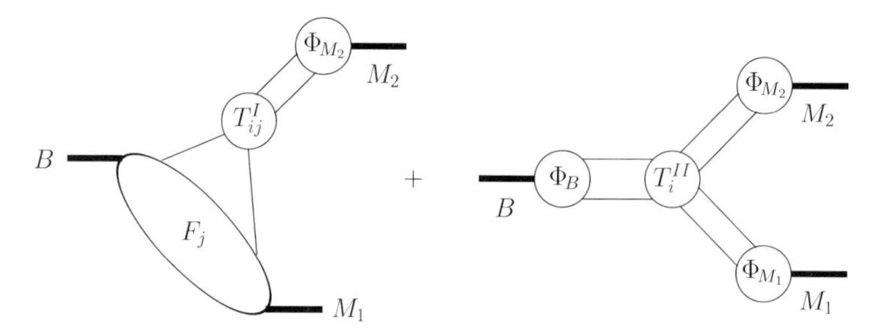

Fig. 1. Graphical representation of the factorization formula. Only one of the two form-factor terms in (3) is shown for simplicity.

hard interactions with the spectator quark, can be dropped because it is power suppressed in the heavy-quark limit. In the opposite situation that the spectator quark goes to a light meson but the other meson is heavy, factorization does not hold, because the heavy meson is neither fast nor small and cannot be factorized from the $B \to M_1$ transition. Finally, notice that annihilation topologies do not appear in the factorization formula, since they do not contribute at leading order in the heavy-quark expansion.

Any hard interaction costs a power of α_s. As a consequence, the hard-spectator term in the second formula in (3) is absent at order α_s^0. Since at this order the functions $T^I_{ij}(u)$ are independent of u, the convolution integral results in the normalization of the meson distribution amplitude, and (3) reproduces naive factorization. The factorization formula allows us to compute radiative corrections to this result to all orders in α_s. Further corrections are suppressed by powers of $\Lambda_{\mathrm{QCD}}/m_b$ in the heavy-quark limit.

The significance and usefulness of the factorization formula stems from the fact that the non-perturbative quantities appearing on the right-hand side of the two equations in (3) are much simpler than the original non-leptonic matrix elements on the left-hand side. This is because they either reflect universal properties of a single meson (light-cone distribution amplitudes) or refer only to a $B \to$ meson transition matrix element of a local current (form factors). While it is extremely difficult, if not impossible [13], to compute the original matrix element $\langle M_1 M_2 | O_i | B \rangle$ in lattice QCD, form factors and light-cone distribution amplitudes are already being computed in this way, although with significant systematic errors at present. Alternatively, form factors can be obtained using data on semi-leptonic decays, and light-cone distribution amplitudes by comparison with other hard exclusive processes.

After having presented the most general form of the factorization formula, we will from now on restrict ourselves to the case of heavy-light final states. Then the (simpler) first formula in (3) applies, and only the first term shown in Fig. 1 is present at leading power.

2.3 Definition of Non-perturbative Parameters

The form factors $F_j^{B\to M}(q^2)$ in (3) arise in the decomposition of current matrix elements of the form $\langle M(p')|\bar{q}\Gamma b|\bar{B}(p)\rangle$, where Γ can be any irreducible Dirac matrix that appears after contraction of the hard subgraph to a local vertex with respect to the $B \to M$ transition. We will often refer to the matrix element of the vector current evaluated between a B meson and a pseudoscalar meson P, which is conventionally parameterized as

$$\begin{aligned}\langle P(p')|\bar{q}\gamma^\mu b|\bar{B}(p)\rangle &= F_+^{B\to P}(q^2)\,(p^\mu + p'^\mu) \\ &\quad + \left[F_0^{B\to P}(q^2) - F_+^{B\to P}(q^2)\right] \frac{m_B^2 - m_P^2}{q^2}\, q^\mu\,, \end{aligned} \tag{4}$$

where $q = p - p'$, and $F_+^{B\to P}(0) = F_0^{B\to P}(0)$ at zero momentum transfer. Note that we write (3) in terms of physical form factors. In principle, Fig. 1 could be looked upon in two different ways. We could suppose that the region represented by F_j accounts only for the soft contributions to the $B \to M_1$ form factor. The hard contributions to the form factor would then be considered as part of T_{ij}^I (or as part of the second diagram). Performing this split-up would require that one understands the factorization of hard and soft contributions to the form factor. If M_1 is heavy, this amounts to matching the form factor onto a form factor defined in heavy-quark effective theory [14]. However, for a light meson M_1 the factorization of hard and soft contributions to the form factor is not yet completely understood. We bypass this problem by interpreting F_j as the physical form factor, including hard and soft contributions. This avoids the above problem, and in addition has the advantage that the physical form factors are directly related to measurable quantities.

Light-cone distribution amplitudes play the same role for hard exclusive processes that parton distributions play for inclusive processes. As in the latter case, the leading-twist distribution amplitudes, which are the ones we need at leading power in the $1/m_b$ expansion, are given by two-particle operators with a certain helicity structure. The helicity structure is determined by the angular momentum of the meson and the fact that the spinor of an energetic quark has only two large components. The leading-twist light-cone distribution amplitudes for pseudoscalar mesons (P) and longitudinally polarized vector mesons ($V_\parallel$) with flavour content ($\bar{q}q'$) are defined as

$$\begin{aligned}\langle P(q)|\bar{q}(y)_\alpha q'(x)_\beta|0\rangle &= \frac{if_P}{4}\,(\not{q}\gamma_5)_{\beta\alpha} \int_0^1 du\, e^{i(\bar{u}qx+uqy)}\,\Phi_P(u,\mu)\,, \\ \langle V_\parallel(q)|\bar{q}(y)_\alpha q'(x)_\beta|0\rangle &= -\frac{if_V}{4}\not{q}_{\beta\alpha} \int_0^1 du\, e^{i(\bar{u}qx+uqy)}\,\Phi_\parallel(u,\mu)\,, \end{aligned} \tag{5}$$

where $(x-y)^2 = 0$. We have suppressed the path-ordered exponentials that connect the two quark fields at different positions and make the light-cone operators gauge invariant. The equality sign is to be understood as "equal up to higher-twist terms". It is also understood that the operators on the left-hand side are colour singlets. When convenient, we use the "bar"-notation $\bar{u} \equiv 1-u$. The parameter μ is the renormalization scale of the light-cone operators on the left-hand side. The distribution amplitudes are normalized as $\int_0^1 du\, \Phi_X(u,\mu) = 1$ with $X = P, V_\parallel$. One defines the asymptotic distribution amplitude as the limit in which the renormalization scale is sent to infinity. In this case

$$\Phi_X(u,\mu) \stackrel{\mu\to\infty}{=} 6u(1-u)\,. \tag{6}$$

The use of light-cone distribution amplitudes in non-leptonic B decays requires justification, which we will provide in Sects. 3 and 4. The decay amplitude for a B decay into a heavy-light final state is then calculated by assigning momenta uq and $\bar{u}q$ to the quark and antiquark in the outgoing light meson (with momentum q), writing down the on-shell amplitude in momentum space, and performing the replacement

$$\bar{u}_{\alpha a}(uq)\,\Gamma(u,\ldots)_{\alpha\beta,ab}v_{\beta b}(\bar{u}q) \to \frac{if_P}{4N_c}\int_0^1 du\,\Phi_P(u)\,(\not{q}\gamma_5)_{\beta\alpha}\,\Gamma(u,\ldots)_{\alpha\beta,aa} \tag{7}$$

for pseudoscalars and, with obvious modifications, for vector mesons. (Even when working with light-cone distribution amplitudes it is not always justified to perform the collinear approximation on the external quark and antiquark lines right away. One may have to keep the transverse components of the quark and antiquark momenta until after some operations on the amplitude have been carried out. However, these subtleties do not concern calculations at leading-twist order.)

3 Arguments for Factorization

In this section we provide the basic power-counting arguments that lead to the factorized structure shown in (3). We do so by analyzing qualitatively the hard, soft and collinear contributions to the simplest Feynman diagrams.

3.1 Preliminaries and Power Counting

For concreteness, we label the charm meson which picks up the spectator quark by $M_1 = D^+$ and assign momentum p' to it. The light meson is labeled $M_2 = \pi^-$ and assigned momentum $q = E\,n_+$, where E is the pion energy in the B rest frame, and $n_\pm = (1,0,0,\pm1)$ are four-vectors on the light-cone. At leading power, we neglect the mass of the light meson.

The simplest diagrams that we can draw for a non-leptonic decay amplitude assign a quark and antiquark to each meson. We choose the quark and antiquark momenta in the pion as

$$l_q = uq + l_\perp + \frac{\boldsymbol{l}_\perp^2}{4uE}\,n_-\,, \qquad l_{\bar{q}} = \bar{u}q - l_\perp + \frac{\boldsymbol{l}_\perp^2}{4\bar{u}E}\,n_-\,. \tag{8}$$

Note that $q \neq l_q + l_{\bar{q}}$, but the off-shellness $(l_q+l_{\bar{q}})^2$ is of the same order as the light meson mass, which we can neglect at leading power. A similar decomposition (with longitudinal momentum fraction v and transverse momentum $l'_\perp$) is used for the charm meson.

To prove the factorization formula (3) for the case of heavy-light final states, one has to show that:

i) There is no leading (in powers of $\Lambda_{\rm QCD}/m_b$) contribution to the amplitude from the endpoint regions $u \sim \Lambda_{\rm QCD}/m_b$ and $\bar{u} \sim \Lambda_{\rm QCD}/m_b$.
ii) One can set $l_\perp = 0$ in the amplitude (more generally, expand the amplitude in powers of $l_\perp$) after collinear subtractions, which can be absorbed into the pion wave function. This, together with i), guarantees that the amplitude is legitimately expressed in terms of the light-cone distribution amplitudes of pion.
iii) The leading contribution comes from $\bar{v} \sim \Lambda_{\rm QCD}/m_b$ (the region where the spectator quark enters the charm meson as a soft parton), which guarantees the absence of a hard spectator interaction term.
iv) After subtraction of infrared contributions corresponding to the light-cone distribution amplitude and the form factor, the leading contributions to the amplitude come only from internal lines with virtuality that scales with m_b.
v) Non-valence Fock states are non-leading.

The requirement that after subtractions virtualities should be large is obvious to guarantee the infrared finiteness of the hard-scattering functions T^I_{ij}. Let us comment on setting transverse momenta in the wave functions to zero and on endpoint contributions. Neglecting transverse momenta requires that we count them as order $\Lambda_{\rm QCD}$ when comparing terms of different magnitude in the scattering amplitude. This conforms to our intuition and the assumption of the parton model, that intrinsic transverse momenta are limited to hadronic scales. However, in QCD transverse momenta are not limited, but logarithmically distributed up to the hard scale. The important point is that contributions that violate the starting assumption of limited transverse momentum can be absorbed into the universal light-cone distribution amplitudes. The statement that transverse momenta can be counted of order $\Lambda_{\rm QCD}$ is to be understood after these subtractions have been performed.

The second comment concerns endpoint contributions in the convolution integrals over longitudinal momentum fractions. These contributions are dangerous, because we may be able to demonstrate the infrared safety of the hard-scattering amplitude under assumption of generic u and independent of the shape of the meson distribution amplitude, but for $u \to 0$ or $u \to 1$ a propagator that was assumed to be off-shell approaches the mass-shell. If such a contribution were of leading power, we would not expect the perturbative calculation of the hard-scattering functions to be reliable.

Estimating endpoint contributions requires knowledge of the endpoint behaviour of the light-cone distribution amplitude. Since it enters the factorization formula at a renormalization scale of order m_b, we can use the asymptotic form (6) to estimate the endpoint contribution. (More generally, we only have to assume that the distribution amplitude at a given scale has the same endpoint behaviour as the asymptotic amplitude. This is generally the case, unless there is a conspiracy of terms in the Gegenbauer expansion of the distribution amplitude. If such a conspiracy existed at some scale, it would be destroyed by evolving the distribution amplitude to a different scale.) We count a light-meson distribution amplitude as order $\Lambda_{\rm QCD}/m_b$ in the endpoint region (defined as the region the quark or antiquark momentum is of order $\Lambda_{\rm QCD}$), and order 1 away from the endpoint, i.e. (for $X = P, V_\parallel$)

$$\Phi_X(u) \sim \begin{cases} 1\,; & \text{generic } u, \\ \Lambda_{\rm QCD}/m_b\,; & u,\,\bar{u} \sim \Lambda_{\rm QCD}/m_b. \end{cases} \tag{9}$$

Note that the endpoint region has a size of order $\Lambda_{\rm QCD}/m_b$, so that the endpoint suppression is $\sim (\Lambda_{\rm QCD}/m_b)^2$. This suppression has to be weighted against potential enhancements of the partonic amplitude when one of the propagators approaches the mass shell. The counting for B mesons, or heavy mesons in general, is different. Naturally, the heavy quark carries almost all of the meson momentum, and hence we count

$$\Phi_B(\xi) \sim \begin{cases} m_b/\Lambda_{\rm QCD}\,; & \xi \sim \Lambda_{\rm QCD}/m_b, \\ 0\,; & \xi \sim 1. \end{cases} \tag{10}$$

The zero probability for a light spectator with momentum of order m_b must be understood as a boundary condition for the wave function renormalized at a scale much below m_b. There is a small probability for hard fluctuations that transfer large momentum to the spectator. This "hard tail" is generated by evolution of the wave function from a hadronic scale to a scale of order m_b. If we assume that the initial distribution at the hadronic scale falls sufficiently rapidly for $\xi \gg \Lambda_{\rm QCD}/m_b$, this remains true after evolution. We shall assume a sufficiently fast fall-off, so that, for the purposes of power counting, the probability that the spectator-quark momentum is of order m_b can be set to zero. The same counting applies to the D meson. (Despite the fact that the charm meson has momentum of order m_b, we do not need to distinguish the rest frames of B and D for the purpose of power counting, because the two frames are not connected by a parametrically large boost. In other words, the components of the spectator quark in the D meson are still of order $\Lambda_{\rm QCD}$.)

3.2 The $B \to D$ Form Factor

We now demonstrate that the $B \to D$ form factor receives a leading contribution from soft gluon exchange. This implies that a non-leptonic decay cannot be treated completely in the hard-scattering picture, and so the form factor should enter the factorization formula as a non-perturbative quantity.

Consider the diagrams shown in Fig. 2. When the exchanged gluon is hard the spectator quark in the final state has momentum of order m_b. But according to the counting rule (10) this configuration has no overlap with the D-meson wave function. On the other hand, there is no suppression for soft gluons in

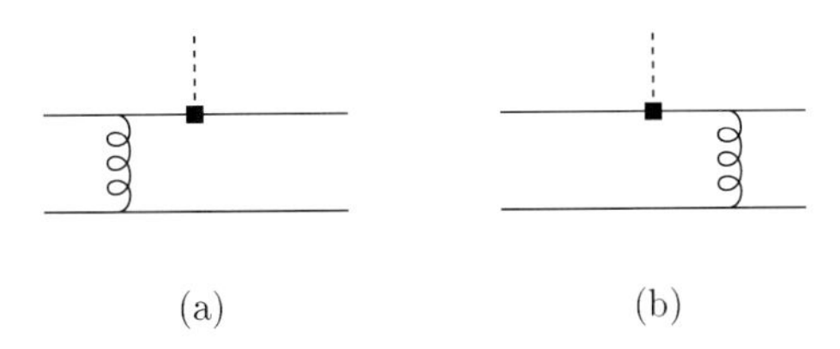

Fig. 2. Leading contributions to the $B \to D$ form factor in the hard-scattering approach. The dashed line represents the weak current. The two lines to the left belong to the B meson, the ones to the right to the recoiling charm meson.

Fig. 2. It follows that the dominant behaviour of the $B \to D$ form factor in the heavy-quark limit is given by soft processes.

Because of this argument, we can exploit the heavy-quark symmetries to determine how the form factor scales in the heavy-quark limit. The well-known result is that the form factor scales like a constant (modulo logarithms), since it is equal to one at zero velocity transfer and independent of m_b as long as the Lorentz boost that connects the B and D rest frames is of order 1. The same conclusion follows from the power-counting rules for light-cone wave functions. To see this, we represent the form factor by an overlap integral of wave functions (not integrated over transverse momentum),

$$F_{+,0}^{B \to D}(0) \sim \int \frac{d\xi d^2 k_\perp}{16\pi^3} \Psi_B(\xi, k_\perp) \Psi_D(\xi'(\xi), k_\perp) , \tag{11}$$

where $\xi'(\xi)$ is fixed by kinematics, and we have set $q^2 = 0$ for simplicity. The probability of finding the B meson in its valence Fock state is of order 1 in the heavy-quark limit, i.e.

$$\int \frac{d\xi d^2 k_\perp}{16\pi^3} |\Psi_{B,D}(\xi, k_\perp)|^2 \sim 1 . \tag{12}$$

Counting $k_\perp \sim \Lambda_{\rm QCD}$ and $d\xi \sim \Lambda_{\rm QCD}/m_b$, we deduce that $\Psi_B \sim m_b^{1/2}/\Lambda_{\rm QCD}^{3/2}$. From (11), we then obtain the scaling law $F_{+,0}^{B \to D}(0) \sim 1$, in agreement with the prediction of heavy-quark symmetry.

The representation (11) of the form factor as an overlap of wave functions for the two-particle Fock state of the heavy meson is not rigorous, because there is no reason to assume that the contribution from higher Fock states with additional soft gluons is suppressed. The consistency with the estimate based on heavy-quark symmetry shows that these additional contributions are not larger than the two-particle contribution.

3.3 Non-leptonic Decay Amplitudes

We now turn to a qualitative discussion of the lowest-order and one-gluon exchange diagrams that could contribute to the hard-scattering kernels $T_{ij}^I(u)$ in (3). In the figures below, the two lines directed upwards represent π^-, the lines on the left represent $\bar{B}_d$, and the lines on the right represent D^+.

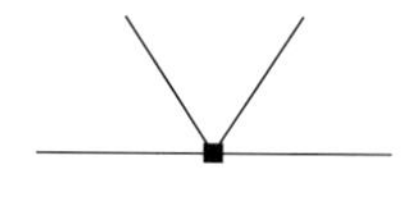

Fig. 3. Leading-order contribution to the hard-scattering kernels $T^I_{ij}(u)$. The weak decay of the b quark through a four-fermion operator is represented by the black square.

Lowest-Order Diagram

There is a single diagram with no hard gluon interactions shown in Fig. 3. According to (10) the spectator quark is soft, and since it does not undergo a hard interaction it is absorbed as a soft quark by the recoiling meson. This is evidently a contribution to the left-hand diagram of Fig. 1, involving the $B \to D$ form factor. The hard subprocess in Fig. 3 is just given by the insertion of a four-fermion operator, and hence it does not depend on the longitudinal momentum fraction u of the two quarks that form the emitted π^-. Consequently, the lowest-order contribution to $T^I_{ij}(u)$ in (3) is independent of u, and the u-integral reduces to the normalization condition for the pion distribution amplitude. The result is, not surprisingly, that the factorization formula reproduces the result of naive factorization if we neglect gluon exchange. Note that the physical picture underlying this lowest-order process is that the spectator quark (which is part of the $B \to D$ form factor) is soft. If this is the case, the hard-scattering approach misses the leading contribution to the non-leptonic decay amplitude.

Putting together all factors relevant to power counting, we find that in the heavy-quark limit the decay amplitude for a decay into a heavy-light final state (in which the spectator quark is absorbed by the heavy meson) scales as

$$\mathcal{A}(\bar{B}_d \to D^+\pi^-) \sim G_F m_b^2\, F^{B\to D}(0)\, f_\pi \sim G_F m_b^2\, \Lambda_{\rm QCD}\,. \tag{13}$$

Other contributions must be compared with this scaling rule.

Factorizable Diagrams

In order to justify naive factorization as the leading term in an expansion in α_s and $\Lambda_{\rm QCD}/m_b$, we must show that radiative corrections are either suppressed in one of these two parameters, or already contained in the definition of the form factor and the pion decay constant. Consider the graphs shown in Fig. 4. The first three diagrams are part of the form factor and do not contribute to the hard-scattering kernels. Since the first and third diagrams contain leading contributions from the region in which the gluon is soft, they should not be considered as corrections to Fig. 3. However, this is of no consequence since these soft contributions are absorbed into the physical form factor.

The fourth diagram in Fig. 4 is also factorizable. In general, this graph would split into a hard contribution and a contribution to the evolution of the pion distribution amplitude. However, as the leading-order diagram in Fig. 3 involves

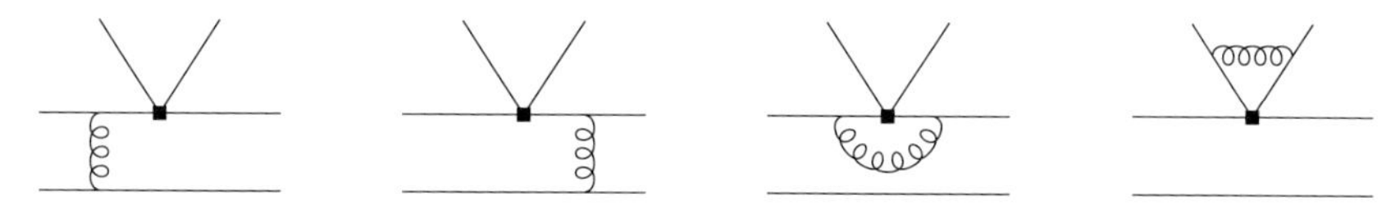

Fig. 4. Diagrams at order α_s that need not be calculated.

only the normalization integral of the pion distribution amplitude, the sum of the fourth diagram in Fig. 4 and the wave-function renormalization of the quarks in the emitted pion vanishes. In other words, these diagrams would renormalize the $(\bar{u}d)$ light-quark current, which however is conserved.

"Non-factorizable" Vertex Corrections

We now begin the analysis of "non-factorizable" diagrams, i.e. diagrams containing gluon exchanges that cannot be associated with the $B \to D$ form factor or the pion decay constant. At order α_s, these diagrams can be divided into three groups: vertex corrections, hard spectator interactions, and annihilation diagrams.

The vertex corrections shown in Fig. 5 violate the naive factorization ansatz (2). One of the key observations made in [1,2] is that these diagrams are calculable nonetheless. Let us summarize the argument here, postponing the explicit evaluation of these diagrams to Sect. 4. The statement is that the vertex-correction diagrams form an order-α_s contribution to the hard-scattering kernels $T_{ij}^I(u)$. To demonstrate this, we have to show that: i) The transverse momentum of the quarks that form the pion can be neglected at leading power, i.e. the two momenta in (8) can be approximated by uq and $\bar{u}q$, respectively. This guarantees that only a convolution in the longitudinal momentum fraction u appears in the factorization formula. ii) The contribution from the soft-gluon region and gluons collinear to the direction of the pion is power suppressed. In practice, this means that the sum of these diagrams cannot contain any infrared divergences at leading power in $\Lambda_{\mathrm{QCD}}/m_b$.

Neither of the two conditions holds true for any of the four diagrams individually, as each of them separately contains collinear and infrared divergences. As will be shown in detail later, the infrared divergences cancel when one sums over the gluon attachments to the two quarks comprising the emission pion ((a+b), (c+d) in Fig. 5). This cancellation is a technical manifestation of Bjorken's colour-transparency argument [10], stating that soft gluon interactions with the emitted colour-singlet $(\bar{u}d)$ pair are suppressed because they interact with the colour dipole moment of the compact light-quark pair. Collinear divergences cancel after summing over gluon attachments to the b and c quark lines ((a+c), (b+d) in Fig. 5). Thus the sum of the four diagrams (a–d) involves only hard gluon exchange at leading power. Because the hard gluons transfer large momentum to the quarks that form the emission pion, the hard-scattering factor now results in a non-trivial convolution with the pion distribution amplitude.

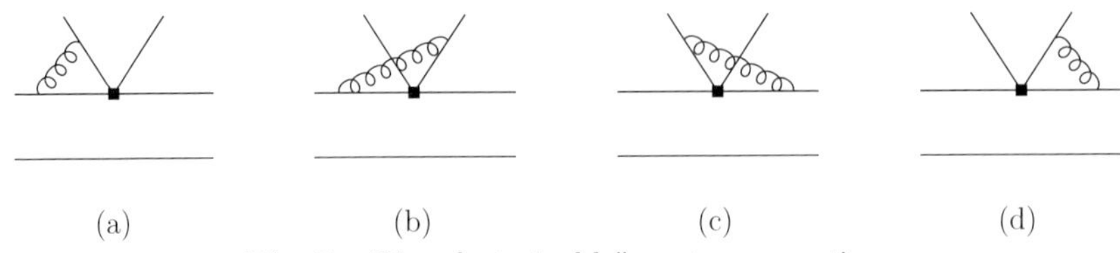

Fig. 5. "Non-factorizable" vertex corrections.

"Non-factorizable" contributions are therefore non-universal, i.e. they depend on the quantum numbers of the final-state mesons.

Note that the colour-transparency argument, and hence the cancellation of soft gluon effects, applies only if the $(\bar{u}d)$ pair is compact. This is not the case if the emitted pion is formed in a very asymmetric configuration, in which one of the quarks carries almost all of the pion momentum. Since the probability for forming a pion in such an endpoint configuration is of order $(\Lambda_{\rm QCD}/m_b)^2$, they could become important only if the hard-scattering amplitude favoured the production of these asymmetric pairs, i.e. if $T_{ij}^I \sim 1/u^2$ for $u \to 0$ (or $T_{ij}^I \sim 1/\bar{u}^2$ for $u \to 1$). However, we will see that such strong endpoint singularities in the hard-scattering amplitude do not occur.

To complete the argument, we have to show that all other types of contributions to the non-leptonic decay amplitudes are power suppressed in the heavy-quark limit. This includes interactions with the spectator quark, weak annihilation graphs, and contributions from higher Fock components of the meson wave functions. This will be done in Sect. 5. In summary, then, for hadronic B decays into a light emitted and a heavy recoiling meson the first factorization formula in (3) holds. At order α_s, the hard-scattering kernels $T_{ij}^I(u)$ are computed from the diagrams shown in Figs. 3 and 5. Naive factorization follows when one neglects all corrections of order $\Lambda_{\rm QCD}/m_b$ *and* α_s. The factorization formula allows us to compute systematically corrections to higher order in α_s, but still neglects power corrections.

3.4 Remarks on Final-State Interactions

Some of the loop diagrams entering the calculation of the hard-scattering kernels have imaginary parts, which contribute to the strong rescattering phases. It follows from our discussion that these imaginary parts are of order α_s or $\Lambda_{\rm QCD}/m_b$. This demonstrates that strong phases vanish in the heavy-quark limit (unless the real parts of the amplitudes are also suppressed). Since this statement goes against the folklore that prevails from the present understanding of this issue, and since the subject of final-state interactions (and of strong-interaction phases in particular) is of paramount importance for the interpretation of CP-violating observables, a few additional remarks are in order.

Final-state interactions are usually discussed in terms of intermediate hadronic states. This is suggested by the unitarity relation (taking $B \to \pi\pi$ for definiteness)

$$\mathrm{Im}\,\mathcal{A}_{B\to\pi\pi} \sim \sum_n \mathcal{A}_{B\to n}\,\mathcal{A}^*_{n\to\pi\pi}\,, \tag{14}$$

where n runs over all hadronic intermediate states. We can also interpret the sum in (14) as extending over intermediate states of partons. The partonic interpretation is justified by the dominance of hard rescattering in the heavy-quark limit. In this limit, the number of physical intermediate states is arbitrarily large. We may then argue on the grounds of parton–hadron duality that their average is described well enough (up to $\Lambda_{\rm QCD}/m_b$ corrections, say) by a partonic calculation. This is the picture implied by (3). The hadronic language is in principle exact. However, the large number of intermediate states makes it intractable to observe systematic cancellations, which usually occur in an inclusive sum over hadronic intermediate states.

A particular contribution to the right-hand side of (14) is elastic rescattering ($n = \pi\pi$). The energy dependence of the total elastic $\pi\pi$-scattering cross section is governed by soft pomeron behaviour. Hence the strong-interaction phase of the $B \to \pi\pi$ amplitude due to elastic rescattering alone increases slowly in the heavy-quark limit [15]. On general grounds, it is rather improbable that elastic rescattering gives an appropriate representation of the imaginary part of the decay amplitude in the heavy-quark limit. This expectation is also borne out in the framework of Regge behaviour, as discussed in [15], where the importance (in fact, dominance) of inelastic rescattering was emphasized. However, this discussion left open the possibility of soft rescattering phases that do not vanish in the heavy-quark limit, as well as the possibility of systematic cancellations, for which the Regge approach does not provide an appropriate theoretical framework.

Equation (3) implies that such systematic cancellations *do* occur in the sum over all intermediate states n. It is worth recalling that similar cancellations are not uncommon for hard processes. Consider the example of $e^+e^- \to$ hadrons at large energy q. While the production of any hadronic final state occurs on a time scale of order $1/\Lambda_{\rm QCD}$ (and would lead to infrared divergences if we attempted to describe it using perturbation theory), the inclusive cross section given by the sum over all hadronic final states is described very well by a $(q\bar{q})$ pair that lives over a short time scale of order $1/q$. In close analogy, while each particular hadronic intermediate state n in (14) cannot be described partonically, the sum over all intermediate states is accurately represented by a $(q\bar{q})$ fluctuation of small transverse size of order $1/m_b$. Because the $(q\bar{q})$ pair is small, the physical picture of rescattering is very different from elastic $\pi\pi$ scattering.

In perturbation theory, the pomeron is associated with two-gluon exchange. The analysis of two-loop contributions to the non-leptonic decay amplitude in [2] shows that the soft and collinear cancellations that guarantee the partonic interpretation of rescattering extend to two-gluon exchange. Hence, the soft final-state interactions are again subleading as required by the validity of (3). As far as the hard rescattering contributions are concerned, two-gluon exchange plus ladder graphs between a compact $(q\bar{q})$ pair with energy of order m_b and transverse size of order $1/m_b$ and the other pion does not lead to large logarithms, and hence there is no possibility to construct the (hard) pomeron. Note the difference with elastic vector-meson production through a virtual photon, which also involves a compact $(q\bar{q})$ pair. However, in this case one considers $s \gg Q^2$, where $\sqrt{s}$ is the photon–proton center-of-mass energy and Q the virtuality of the photon. This implies that the $(q\bar{q})$ fluctuation is born long before it hits the proton. It is this difference of time scales, non-existent in non-leptonic B decays, that permits pomeron exchange in elastic vector-meson production in $\gamma^* p$ collisions.

4 $B \to D\pi$: Factorization at One-Loop Order

We now present a more detailed treatment of the exclusive decays $\bar{B}_d \to D^{(*)+}L^-$, where L is a light meson. We illustrate explicitly how factorization emerges at one-loop order and compute the hard-scattering kernels $T_{ij}^I(u)$ in the factorization formula (3). For each final state f, we express the decay amplitudes in terms of parameters $a_1(f)$ defined in analogy with similar parameters used in the literature on naive factorization.

4.1 Effective Hamiltonian and Decay Topologies

The effective Hamiltonian for $B \to D\pi$ is

$$\mathcal{H}_{\text{eff}} = \frac{G_F}{\sqrt{2}}\, V_{ud}^* V_{cb} \left(C_0 O_0 + C_8 O_8 \right). \tag{15}$$

We choose to write the two independent four-quark operators in the singlet–octet basis

$$\begin{aligned} O_0 &= \bar{c}\gamma^\mu(1-\gamma_5)b\, \bar{d}\gamma_\mu(1-\gamma_5)u\,, \\ O_8 &= \bar{c}\gamma^\mu(1-\gamma_5)T^A b\, \bar{d}\gamma_\mu(1-\gamma_5)T^A u\,, \end{aligned} \tag{16}$$

rather than in the more conventional basis of O_1 and O_2. The Wilson coefficients C_0 and C_8 describe the exchange of hard gluons with virtualities between the high-energy matching scale M_W and a renormalization scale μ of order m_b. (These coefficients are related to the ones of the standard basis by $C_0 = C_1 + C_2/3$ and $C_8 = 2C_2$.) They are known at next-to-leading order in renormalization-group improved perturbation theory and are given by [5]

$$C_0 = \frac{N_c+1}{2N_c}\, C_+ + \frac{N_c-1}{2N_c}\, C_-\,, \qquad C_8 = C_+ - C_-\,, \tag{17}$$

where

$$C_\pm(\mu) = \left(1 + \frac{\alpha_s(\mu)}{4\pi}\, B_\pm \right) \bar{C}_\pm(\mu)\,, \qquad B_\pm = \pm\frac{N_c \mp 1}{2N_c}\, B\,, \tag{18}$$

and

$$\bar{C}_\pm(\mu) = \left[\frac{\alpha_s(M_W)}{\alpha_s(\mu)} \right]^{d_\pm} \left[1 + \frac{\alpha_s(M_W) - \alpha_s(\mu)}{4\pi}\, S_\pm \right]. \tag{19}$$

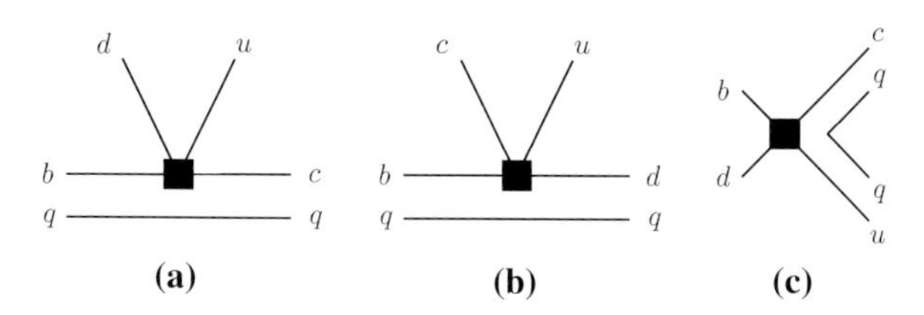

Fig. 6. Basic quark-level topologies for $B \to D\pi$ decays ($q = u,\, d$): (a) class-I, (b) class-II, (c) weak annihilation. $\bar{B}_d \to D^+\pi^-$ receives contributions from (a) and (c), $\bar{B}_d \to D^0\pi^0$ from (b) and (c), and $B^- \to D^0\pi^-$ from (a) and (b). Only (a) contributes in the heavy-quark limit.

For $N_c = 3$ and $f = 5$, we have $d_+ = \frac{6}{25}$ and $d_- = -\frac{12}{25}$, as well as $S_+ = \frac{6473}{3174}$ and $S_- = -\frac{9371}{1587}$. The scheme dependence of the Wilson coefficients at next-to-leading order is parameterized by the coefficient B in (18). We note that $B_{\rm NDR} = 11$ in the naive dimensional regularization (NDR) scheme with anticommuting γ_5, and $B_{\rm HV} = 7$ in the 't Hooft–Veltman (HV) scheme. We will demonstrate below that the scale and scheme dependence of the Wilson coefficients is canceled by a corresponding scale and scheme dependence of the hadronic matrix elements of the operators O_0 and O_8.

Before continuing with a discussion of these matrix elements, it is useful to consider the flavour structure for the various contributions to $B \to D\pi$ decays. The possible quark-level topologies are depicted in Fig. 6. In the terminology generally adopted for two-body non-leptonic decays, the decays $\bar{B}_d \to D^+\pi^-$, $\bar{B}_d \to D^0\pi^0$ and $B^- \to D^0\pi^-$ are referred to as class-I, class-II and class-III, respectively [16]. In $\bar{B}_d \to D^+\pi^-$ and $B^- \to D^0\pi^-$ decays the pion can be directly created from the weak current. We call this a class-I contribution, following the above terminology. In addition, in the case of $\bar{B}_d \to D^+\pi^-$ there is a contribution from weak annihilation, and a class-II amplitude contributes to $B^- \to D^0\pi^-$. The important point is that the spectator quark goes into the light meson in the case of the class-II amplitude. This amplitude is suppressed in the heavy-quark limit, as is the annihilation amplitude. It follows that the amplitude for $\bar{B}_d \to D^0\pi^0$, receiving only class-II and annihilation contributions, is subleading compared with $\bar{B}_d \to D^+\pi^-$ and $B^- \to D^0\pi^-$, which are dominated by the class-I topology.

We shall use the one-loop analysis for $\bar{B}_d \to D^+\pi^-$ as a concrete example to illustrate explicitly the various steps involved in establishing the factorization formula. Most of the arguments given below are standard from the theory of hard exclusive processes involving light hadrons [8]. However, it is instructive to repeat these arguments in the context of B decays.

4.2 Soft and Collinear Cancellations at One-Loop Order

In order to demonstrate the property of factorization for the decay $\bar{B}_d \to D^+\pi^-$, we now analyze the "non-factorizable" one-gluon exchange contributions shown in Fig. 5 in some detail. We consider the leading, valence Fock state of the emitted pion. This is justified since higher Fock components only give power-suppressed contributions to the decay amplitude in the heavy-quark limit (as demonstrated later). For the purpose of our discussion, the valence Fock state of the pion can be written as

$$|\pi(q)\rangle = \int \frac{du}{\sqrt{u\bar{u}}} \frac{d^2 l_\perp}{16\pi^3} \frac{1}{\sqrt{2N_c}} \left(a_\uparrow^\dagger(l_q)\, b_\downarrow^\dagger(l_{\bar{q}}) - a_\downarrow^\dagger(l_q)\, b_\uparrow^\dagger(l_{\bar{q}}) \right) |0\rangle\, \Psi(u, \boldsymbol{l}_\perp) , \quad (20)$$

where $a_s^\dagger$ ($b_s^\dagger$) denotes the creation operator for a quark (antiquark) in a state with spin $s =\uparrow$ or $s =\downarrow$, and we have suppressed colour indices. The wave function $\Psi(u, \boldsymbol{l}_\perp)$ is defined as the amplitude for the pion to be composed of two on-shell quarks, characterized by longitudinal momentum fraction u and transverse momentum $l_\perp$. The on-shell momenta of the quark and antiquark are chosen as in (8). For the purpose of power counting, $l_\perp \sim \Lambda_{\rm QCD} \ll E \sim m_b$. Note that the invariant mass of the valence state is $(l_q + l_{\bar{q}})^2 = \boldsymbol{l}_\perp^2/(u\bar{u})$, which is of order $\Lambda_{\rm QCD}^2$ and hence negligible in the heavy-quark limit unless u is in the vicinity of the endpoints $u = 0$ or 1. In this case, the invariant mass of the quark–antiquark pair becomes large, and the valence Fock state is no longer a valid representation of the pion. However, in the heavy-quark limit the dominant contributions to the decay amplitude come from configurations where both partons are hard (u and $\bar{u}$ both of order 1), and so the two-particle Fock state yields a consistent description. We will provide an explicit consistency check of this important feature later on.

As a next step, we write down the amplitude

$$\langle \pi(q) | u(0)_\alpha \bar{d}(y)_\beta | 0 \rangle = \int du\, \frac{d^2 l_\perp}{16\pi^3} \frac{1}{\sqrt{2N_c}}\, \Psi^*(u, \boldsymbol{l}_\perp)\, (\gamma_5\, \not{q})_{\alpha\beta}\, e^{i l_q \cdot y} , \quad (21)$$

which appears as an ingredient of the $B \to D\pi$ matrix element. It is now straightforward to obtain the one-gluon exchange contribution to the $B \to D\pi$ matrix element of the operator O_8. For the sum of the four diagrams in Fig. 5, we find

$$\langle D^+\pi^- | O_8 | \bar{B}_d \rangle_{\text{1-gluon}} = \quad (22)$$
$$i g_s^2 \frac{C_F}{2} \int \frac{d^4 k}{(2\pi)^4} \langle D^+ | \bar{c} A_1(k) b | \bar{B}_d \rangle \frac{1}{k^2} \int_0^1 du\, \frac{d^2 l_\perp}{16\pi^3} \frac{\Psi^*(u, \boldsymbol{l}_\perp)}{\sqrt{2N_c}}\, \text{tr}[\gamma_5\, \not{q} A_2(l_q, l_{\bar{q}}, k)] ,$$

where

$$\begin{aligned} A_1(k) &= \frac{\gamma^\lambda(\not{p}_c - \not{k} + m_c)\Gamma}{2p_c \cdot k - k^2} - \frac{\Gamma(\not{p}_b + \not{k} + m_b)\gamma^\lambda}{2p_b \cdot k + k^2} , \\ A_2(l_q, l_{\bar{q}}, k) &= \frac{\Gamma(\not{l}_{\bar{q}} + \not{k})\gamma_\lambda}{2l_{\bar{q}} \cdot k + k^2} - \frac{\gamma_\lambda(\not{l}_q + \not{k})\Gamma}{2l_q \cdot k + k^2} . \end{aligned} \quad (23)$$

Here $\Gamma = \gamma^\mu(1 - \gamma_5)$, and p_b, p_c are the momenta of the b- and c-quark, respectively. There is no correction to the matrix element of O_0 at order α_s, because in this case the $(d\bar{u})$ pair is necessarily in a colour-octet configuration and cannot form a pion.

In (22) the pion wave function $\Psi(u, l_\perp)$ appears separated from the $B \to D$ transition. This is merely a reflection of the fact that we have represented the pion state in the form shown in (20). It does not, by itself, imply factorization, since the right-hand side of (22) still involves non-trivial integrations over $\boldsymbol{l}_\perp$ and the gluon momentum k, and long- and short-distance contributions are not yet disentangled. In order to prove factorization, we need to show that the integral over k receives only subdominant contributions from the region of small k^2. This is equivalent to showing that the integral over k does not contain infrared divergences at leading power in $\Lambda_{\mathrm{QCD}}/m_b$.

To demonstrate infrared finiteness of the one-loop integral

$$J \equiv \int d^4k \, \frac{1}{k^2} \, A_1(k) \otimes A_2(l_q, l_{\bar{q}}, k) \tag{24}$$

at leading power, the heavy-quark limit and the corresponding large light-cone momentum of the pion are again essential. First note that when k is of order m_b, $J \sim 1$ by dimensional analysis. Potential infrared divergences could arise when k is soft or collinear to the pion momentum q. We need to show that the contributions from these regions are power suppressed. (Note that we do not need to show that J is infrared finite. It is enough that logarithmic divergences have coefficients that are power suppressed.)

We treat the soft region first. Here all components of k become small simultaneously, which we describe by scaling $k \sim \lambda$. Counting powers of λ ($d^4k \sim \lambda^4$, $1/k^2 \sim \lambda^{-2}$, $1/p \cdot k \sim \lambda^{-1}$) reveals that each of the four diagrams in Fig. 5, corresponding to the four terms in the product in (24), is logarithmically divergent. However, because k is small the integrand can be simplified. For instance, the second term in A_2 can be approximated as

$$\frac{\gamma_\lambda(\not{l}_q + \not{k})\Gamma}{2l_q \cdot k + k^2} = \frac{\gamma_\lambda(u \not{q} + \not{l}_\perp + \frac{\boldsymbol{l}_\perp^2}{4uE} \not{n}_- + \not{k})\Gamma}{2uq \cdot k + 2l_\perp \cdot k + \frac{\boldsymbol{l}_\perp^2}{2uE} n_- \cdot k + k^2} \simeq \frac{q_\lambda}{q \cdot k} \, \Gamma \,, \tag{25}$$

where we used that $\not{q}$ to the extreme left or right of an expression gives zero due to the on-shell condition for the external quark lines. We get exactly the same expression but with an opposite sign from the other term in A_2, and hence the soft divergence cancels out. More precisely, we find that the integral is infrared finite in the soft region when $l_\perp$ is neglected. When $l_\perp$ is not neglected, there is a divergence from soft k which is proportional to $l_\perp^2/m_b^2 \sim \Lambda_{\mathrm{QCD}}^2/m_b^2$. In either case, the soft contribution to J is of order $\Lambda_{\mathrm{QCD}}/m_b$ or smaller and hence suppressed relative to the hard contribution. This corresponds to the standard soft cancellation mechanism, which is a technical manifestation of colour transparency.

Each of the four terms in (24) is also divergent when k becomes collinear with the light-cone momentum q. This implies the scaling $k^+ \sim \lambda^0$, $k_\perp \sim \lambda$, and $k^- \sim \lambda^2$. Then $d^4k \sim dk^+ dk^- d^2k_\perp \sim \lambda^4$, and $q \cdot k = q^+ k^- \sim \lambda^2$, $k^2 = 2k^+k^- + k_\perp^2 \sim \lambda^2$. The divergence is again logarithmic, and it is thus sufficient to consider the leading behaviour in the collinear limit. Writing $k = \alpha q + \ldots$ we can now simplify the second term of A_2 as

$$\frac{\gamma_\lambda(\not{l}_q + \not{k})\Gamma}{2l_q \cdot k + k^2} \simeq q_\lambda \, \frac{2(u+\alpha)\Gamma}{2l_q \cdot k + k^2} \,. \tag{26}$$

No simplification occurs in the denominator (in particular, $l_\perp$ cannot be neglected), but the important point is that the leading contribution is proportional to q_λ. Therefore, substituting $k = \alpha q$ into A_1 and using $q^2 = 0$, we obtain

$$q_\lambda A_1 \simeq \frac{\not{q}(\not{p}_c + m_c)\Gamma}{2\alpha p_c \cdot q} - \frac{\Gamma(\not{p}_b + m_b)\not{q}}{2\alpha p_b \cdot q} = 0 \,, \tag{27}$$

employing the equations of motion for the heavy quarks. Hence the collinear divergence cancels by virtue of the standard Ward identity.

This completes the proof of the absence of infrared divergences at leading power in the hard-scattering kernel for $\bar{B}_d \to D^+\pi^-$ to one-loop order. Similar cancellations are observed at higher orders. A complete proof of factorization at two-loop order can be found in [2]. Having established that the "non-factorizable" diagrams of Fig. 5 are dominated by hard gluon exchange (i.e. that the leading contribution to J arises from k of order m_b), we may now use the fact that $|\boldsymbol{l}_\perp| \ll E$ to expand A_2 in powers of $|\boldsymbol{l}_\perp|/E$. To leading order the expansion simply reduces to neglecting $l_\perp$ altogether, which implies $l_q = uq$ and $l_{\bar{q}} = \bar{u}q$ in (8). As a consequence, we may perform the $l_\perp$ integration in (22) over the pion distribution amplitude. Defining

$$\int \frac{d^2 l_\perp}{16\pi^3} \, \frac{\Psi^*(u, \boldsymbol{l}_\perp)}{\sqrt{2N_c}} \equiv \frac{i f_\pi}{4N_c} \, \Phi_\pi(u) \,, \tag{28}$$

the matrix element of O_8 in (22) becomes

$$\begin{aligned} &\langle D^+\pi^-|O_8|\bar{B}_d\rangle_{\text{1-gluon}} = \\ &-g_s^2 \, \frac{C_F}{8N_c} \int \frac{d^4k}{(2\pi)^4} \, \langle D^+|\bar{c}A_1(k)b|\bar{B}_d\rangle \, \frac{1}{k^2} \, f_\pi \int_0^1 du \, \Phi_\pi(u) \, \mathrm{tr}[\gamma_5 \, \not{q} A_2(uq, \bar{u}q, k)] \,. \end{aligned} \tag{29}$$

On the other hand, putting y on the light-cone in (21) and comparing with (5), we see that the $l_\perp$-integrated wave function $\Phi_\pi(u)$ in (28) is precisely the light-cone distribution amplitude of the pion. This demonstrates the relevance of the light-cone wave function to the factorization formula. Note that the collinear approximation for the quark and antiquark momenta emerges automatically in the heavy-quark limit.

After the k integral is performed, the expression (29) can be cast into the form

$$\langle D^+\pi^-|O_8|\bar{B}_d\rangle_{\text{1-gluon}} \sim F^{B\to D}(0) \int_0^1 du \, T_8(u, z) \, \Phi_\pi(u) \,, \tag{30}$$

where $z = m_c/m_b$, $T_8(u, z)$ is the hard-scattering kernel, and $F^{B\to D}(0)$ the form factor that parameterizes the $\langle D^+|\bar{c}[\ldots]b|\bar{B}_d\rangle$ matrix element. Because of the absence of soft and collinear infrared divergences in the gluon exchange between the $(\bar{c}b)$ and $(\bar{d}u)$ currents, the hard-scattering kernel T_8 is calculable in QCD perturbation theory.

4.3 Matrix Elements at Next-to-Leading Order

We now compute these hard-scattering kernels explicitly to order α_s. The effective Hamiltonian (15) can be written as

$$\mathcal{H}_{\text{eff}} = \frac{G_F}{\sqrt{2}} V_{ud}^* V_{cb} \Bigg\{ \left[\frac{N_c+1}{2N_c} \bar{C}_+(\mu) + \frac{N_c-1}{2N_c} \bar{C}_-(\mu) + \frac{\alpha_s(\mu)}{4\pi} \frac{C_F}{2N_c} B C_8(\mu) \right] O_0 + C_8(\mu)\, O_8 \Bigg\}, \tag{31}$$

where the scheme-dependent term in the coefficient of the operator O_0 has been written explicitly. Because the light-quark pair has to be in a colour singlet to produce the pion in the leading Fock state, only O_0 gives a contribution to zeroth order in α_s. Similarly, to first order in α_s only O_8 can contribute. The result of evaluating the diagrams in Fig. 5 with an insertion of O_8 can be presented in a form that holds simultaneously for a heavy meson $H = D, D^*$ and a light meson $L = \pi, \rho$, using only that the $(\bar{u}d)$ pair is a colour singlet and that the external quarks can be taken on-shell. We obtain ($z = m_c/m_b$)

$$\begin{aligned} \langle H(p')L(q)|O_8|\bar{B}_d(p)\rangle &= \frac{\alpha_s}{4\pi}\frac{C_F}{2N_c}\, i f_L \int_0^1 du\, \Phi_L(u) \\ \times \Bigg[-\left(6\ln\frac{\mu^2}{m_b^2} + B \right) & (\langle J_V\rangle - \langle J_A\rangle) + F(u,z)\,\langle J_V\rangle - F(u,-z)\,\langle J_A\rangle \Bigg], \end{aligned} \tag{32}$$

where

$$\langle J_V\rangle = \langle H(p')|\bar{c}\,\not{q}\,b|\bar{B}_d(p)\rangle, \qquad \langle J_A\rangle = \langle H(p')|\bar{c}\,\not{q}\gamma_5 b\,|\bar{B}_d(p)\rangle\,. \tag{33}$$

It is worth noting that even after computing the one-loop correction the $(\bar{u}d)$ pair retains its $V-A$ structure. This, together with (5), implies that the form of (32) is identical for pions and longitudinally polarized ρ mesons. (The production of transversely polarized ρ mesons is power suppressed in Λ_{QCD}/m_b.) The function $F(u,z)$ appearing in (32) is given by

$$F(u,z) = \left(3 + 2\ln\frac{u}{\bar{u}}\right)\ln z^2 - 7 + f(u,z) + f(\bar{u},1/z)\,, \tag{34}$$

where

$$\begin{aligned} f(u,z) &= -\frac{u(1-z^2)[3(1-u(1-z^2))+z]}{[1-u(1-z^2)]^2}\ln[u(1-z^2)] - \frac{z}{1-u(1-z^2)} \\ &+ 2\left[\frac{\ln[u(1-z^2)]}{1-u(1-z^2)} - \ln^2[u(1-z^2)] - \text{Li}_2[1-u(1-z^2)] - \{u\to\bar{u}\}\right], \end{aligned} \tag{35}$$

and $\text{Li}_2(x)$ is the dilogarithm. The contribution of $f(u,z)$ in (34) comes from the first two diagrams in Fig. 5 with the gluon coupling to the b quark, whereas $f(\bar{u},1/z)$ arises from the last two diagrams with the gluon coupling to the charm quark. Note that the terms in the large square brackets in the definition of the function $f(u,z)$ vanish for a symmetric light-cone distribution amplitude. These terms can be dropped if the light final-state meson is a pion or a ρ meson, but they are relevant, e.g., for the discussion of Cabibbo-suppressed decays such as $\bar{B}_d \to D^{(*)+}K^-$ and $\bar{B}_d \to D^{(*)+}K^{*-}$.

The discontinuity of the amplitude, which is responsible for the occurrence of the strong rescattering phase, arises from $f(\bar{u},1/z)$ and can be obtained by recalling that z^2 is $z^2 - i\epsilon$ with $\epsilon > 0$ infinitesimal. We find

$$\begin{aligned} \frac{1}{\pi}\,\mathrm{Im}\,F(u,z) &= -\frac{(1-u)(1-z^2)[3(1-u(1-z^2))+z]}{[1-u(1-z^2)]^2} \\ &\quad - 2\left[\ln[1-u(1-z^2)] + 2\ln u + \frac{z^2}{1-u(1-z^2)} - \{u \to \bar{u}\}\right]. \end{aligned} \tag{36}$$

As mentioned above, (32) is applicable to all decays of the type $\bar{B}_d \to D^{(*)+}L^-$, where L is a light hadron such as a pion or a (longitudinally polarized) ρ meson. Only the operator J_V contributes to $\bar{B}_d \to D^+L^-$, and only J_A contributes to $\bar{B}_d \to D^{*+}L^-$. Our result can therefore be written as

$$\langle D^+L^-|O_{0,8}|\bar{B}_d\rangle = \langle D^+|\bar{c}\gamma^\mu(1-\gamma_5)b|\bar{B}_d\rangle \cdot i f_L q_\mu \int_0^1 du\, T_{0,8}(u,z)\,\Phi_L(u)\,, \tag{37}$$

where $L = \pi$, ρ, and the hard-scattering kernels are

$$\begin{aligned} T_0(u,z) &= 1 + O(\alpha_s^2)\,, \\ T_8(u,z) &= \frac{\alpha_s}{4\pi}\frac{C_F}{2N_c}\left[-6\ln\frac{\mu^2}{m_b^2} - B + F(u,z)\right] + O(\alpha_s^2)\,. \end{aligned} \tag{38}$$

When the D meson is replaced by a D^* meson, the result is identical except that $F(u,z)$ must be replaced with $F(u,-z)$. Since no order-α_s corrections exist for O_0, the matrix element retains its leading-order factorized form

$$\langle D^+L^-|O_0|\bar{B}_d\rangle = i f_L q_\mu\,\langle D^+|\bar{c}\gamma^\mu(1-\gamma_5)b|\bar{B}_d\rangle \tag{39}$$

to this accuracy. From (35) it follows that $T_8(u,z)$ tends to a constant as u approaches the endpoints ($u \to 0, 1$). (This is strictly true for the part of $T_8(u,z)$ that is symmetric in $u \leftrightarrow \bar{u}$; the asymmetric part diverges logarithmically as $u \to 0$, which however does not affect the power behaviour and the convergence properties in the endpoint region.) Therefore the contribution to (37) from the endpoint region is suppressed, both by phase space and by the endpoint suppression intrinsic to $\Phi_L(u)$. Consequently, the emitted light meson is indeed dominated by energetic constituents, as required for the self-consistency of the factorization formula.

The final result for the class-I, non-leptonic $\bar{B}_d \to D^{(*)+}L^-$ decay amplitudes, in the heavy-quark limit and at next-to-leading order in α_s, can be compactly expressed in terms of the matrix elements of a "transition operator"

$$\mathcal{T} = \frac{G_F}{\sqrt{2}} V_{ud}^* V_{cb} \Big[a_1(DL)\, Q_V - a_1(D^*L)\, Q_A \Big] \,, \tag{40}$$

where

$$Q_V = \bar{c}\gamma^\mu b \otimes \bar{d}\gamma_\mu(1-\gamma_5)u \,, \qquad Q_A = \bar{c}\gamma^\mu\gamma_5 b \otimes \bar{d}\gamma_\mu(1-\gamma_5)u \,, \tag{41}$$

and hadronic matrix elements of $Q_{V,A}$ are understood to be evaluated in factorized form, i.e.

$$\langle DL|j_1 \otimes j_2|\bar{B}\rangle \equiv \langle D|j_1|\bar{B}\rangle \, \langle L|j_2|0\rangle \,. \tag{42}$$

Equation (40) defines the quantities $a_1(D^{(*)}L)$, which include the leading "non-factorizable" corrections, in a renormalization-scale and -scheme independent way. To leading power in $\Lambda_{\rm QCD}/m_b$ these quantities should not be interpreted as phenomenological parameters (as is usually done), because they are dominated by hard gluon exchange and thus calculable in QCD. At next-to-leading order we get

$$\begin{aligned} a_1(DL) &= \frac{N_c+1}{2N_c}\bar{C}_+(\mu) + \frac{N_c-1}{2N_c}\bar{C}_-(\mu) \\ &\quad + \frac{\alpha_s}{4\pi}\frac{C_F}{2N_c}\, C_8(\mu) \left[-6\ln\frac{\mu^2}{m_b^2} + \int_0^1 du\, F(u,z)\, \Phi_L(u) \right] , \\[2ex] a_1(D^*L) &= \frac{N_c+1}{2N_c}\bar{C}_+(\mu) + \frac{N_c-1}{2N_c}\bar{C}_-(\mu) \\ &\quad + \frac{\alpha_s}{4\pi}\frac{C_F}{2N_c}\, C_8(\mu) \left[-6\ln\frac{\mu^2}{m_b^2} + \int_0^1 du\, F(u,-z)\, \Phi_L(u) \right] . \end{aligned} \tag{43}$$

We observe that the scheme-dependent terms parameterized by B have canceled between the coefficient of O_0 in (31) and the matrix element of O_8 in (37). Likewise, the μ dependence of the terms in brackets in (43) cancels against the scale dependence of the coefficients $\bar{C}_\pm(\mu)$, ensuring a consistent result at next-to-leading order. The coefficients $a_1(DL)$ and $a_1(D^*L)$ are seen to be non-universal, i.e. they depend explicitly on the nature of the final-state mesons. This dependence enters via the light-cone distribution amplitude of the light emission meson and via the analytic form of the hard-scattering kernel ($F(u,z)$ vs. $F(u,-z)$). However, the non-universality enters only at next-to-leading order.

Using the fact that violations of heavy-quark spin symmetry require hard gluon exchange, Politzer and Wise have computed the "non-factorizable" vertex corrections to the decay-rate ratio of the $D\pi$ and $D^*\pi$ final states many years ago [17]. In the context of our formalism, this calculation requires the symmetric part (with respect to $u \leftrightarrow \bar{u}$) of the difference $F(u,z) - F(u,-z)$. Explicitly,

$$\frac{\Gamma(\bar{B}_d \to D^+\pi^-)}{\Gamma(\bar{B}_d \to D^{*+}\pi^-)} = \left| \frac{\langle D^+|\bar{c}\, \not{q}(1-\gamma_5)b|\bar{B}_d\rangle}{\langle D^{*+}|\bar{c}\, \not{q}(1-\gamma_5)b|\bar{B}_d\rangle} \right|^2 \left| \frac{a_1(D\pi)}{a_1(D^*\pi)} \right|^2 , \tag{44}$$

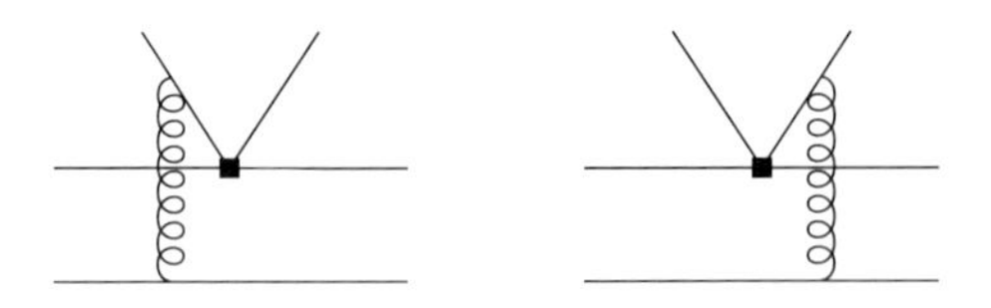

Fig. 7. "Non-factorizable" spectator interactions.

where for simplicity we neglect the light meson masses as well as the mass difference between D and D^* in the phase-space for the two decays. At next-to-leading order

$$\left|\frac{a_1(D\pi)}{a_1(D^*\pi)}\right|^2 = 1 + \frac{\alpha_s}{4\pi}\frac{C_F}{N_c}\frac{C_8}{C_0}\,\mathrm{Re}\int_0^1 du\,[F(u,z) - F(u,-z)]\,\Phi_\pi(u)\,. \qquad (45)$$

Our result for the symmetric part of the kernel agrees with that found in [17].

5 Power-Suppressed Contributions

Up to this point we have presented arguments in favour of factorization of non-leptonic B-decay amplitudes in the heavy-quark limit, and have explored in detail how the factorization formula works at one-loop order for the decays $\bar{B}_d \to D^{(*)+}L^-$. It is now time to show that other contributions not considered so far are indeed power suppressed. This is necessary to fully establish the factorization formula. Besides, it will also provide some numerical estimates of the corrections to the heavy-quark limit.

We start by discussing interactions involving the spectator quark and weak annihilation contributions, before turning to the more delicate question of the importance of non-valence Fock states.

5.1 Interactions with the Spectator Quark

Clearly, the diagrams shown in Fig. 7 cannot be associated with the form-factor term in the factorization formula (3). We will now show that for B decays into a heavy-light final state their contribution is power suppressed in the heavy-quark limit. (This suppression does *not* occur for decays into two light mesons, where hard spectator interactions contribute at leading power. In this case, they contribute to the kernels T_i^{II} in the factorization formula (second term in Fig. 1).)

In general, "non-factorizable" diagrams involving an interaction with the spectator quark would impede factorization if there existed a soft contribution at leading power. While such terms are present in each of the two diagrams separately, they cancel in the sum over the two gluon attachments to the $(\bar{u}d)$ pair by virtue of the same colour-transparency argument that was applied to the "non-factorizable" vertex corrections.

Focusing again on decays into a heavy and a light meson, such as $\bar{B}_d \to D^+\pi^-$, we still need to show that the contribution remaining after the soft

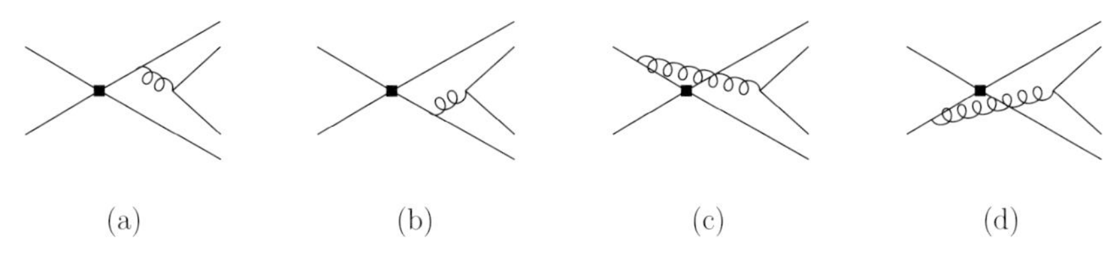

Fig. 8. Annihilation diagrams.

cancellation is power suppressed relative to the leading-order contribution (13). A straightforward calculation leads to the (simplified) result

$$\begin{aligned}\mathcal{A}(\bar{B}_d \to D^+\pi^-)_{\rm spec} &\sim G_F\, f_\pi f_D f_B\, \alpha_s \\ &\quad\times \int_0^1 \frac{d\xi}{\xi}\, \Phi_B(\xi) \int_0^1 \frac{d\eta}{\eta}\, \Phi_D(\eta) \int_0^1 \frac{du}{u}\, \Phi_\pi(u) \\ &\sim G_F\, \alpha_s\, m_b\, \Lambda_{\rm QCD}^2\,. \end{aligned} \tag{46}$$

This is indeed power suppressed relative to (13). Note that the gluon virtuality is of order $\xi\eta\, m_b^2 \sim \Lambda_{\rm QCD}^2$ and so, strictly speaking, the calculation in terms of light-cone distribution amplitudes cannot be justified. Nevertheless, we use (46) to deduce the scaling behaviour of the soft contribution, as we did for the heavy-light form factor in Sect. 3.2.

5.2 Annihilation Topologies

Our next concern are the annihilation diagrams shown in Fig. 8, which also contribute to the decay $\bar{B}_d \to D^+\pi^-$. The hard part of these diagrams could, in principle, be absorbed into hard-scattering kernels of the type T_i^{II}. The soft part, if unsuppressed, would violate factorization. However, we will see that the hard part as well as the soft part are suppressed by at least one power of $\Lambda_{\rm QCD}/m_b$.

The argument goes as follows. We write the annihilation amplitude as

$$\begin{aligned}\mathcal{A}(\bar{B}_d \to D^+\pi^-)_{\rm ann} &\sim G_F\, f_\pi f_D f_B\, \alpha_s \\ &\quad\times \int_0^1 d\xi\, d\eta\, du\, \Phi_B(\xi)\, \Phi_D(\eta)\, \Phi_\pi(u)\, T^{\rm ann}(\xi,\eta,u)\,, \end{aligned} \tag{47}$$

where the dimensionless function $T^{\rm ann}(\xi,\eta,u)$ is a product of propagators and vertices. The product of decay constants scales as $\Lambda_{\rm QCD}^4/m_b$. Since $d\xi\,\Phi_B(\xi)$ scales as 1 and so does $d\eta\,\Phi_D(\eta)$, while $du\,\Phi_\pi(u)$ is never larger than 1, the amplitude can only compete with the leading-order result (13) if $T^{\rm ann}(\xi,\eta,u)$ can be made of order $(m_b/\Lambda_{\rm QCD})^3$ or larger. Since $T^{\rm ann}(\xi,\eta,u)$ contains only two propagators, this can be achieved only if both quarks the gluon splits into are soft, in which case $T^{\rm ann}(\xi,\eta,u) \sim (m_b/\Lambda_{\rm QCD})^4$. But then $du\,\Phi_\pi(u) \sim (\Lambda_{\rm QCD}/m_b)^2$, so that this contribution is power suppressed.

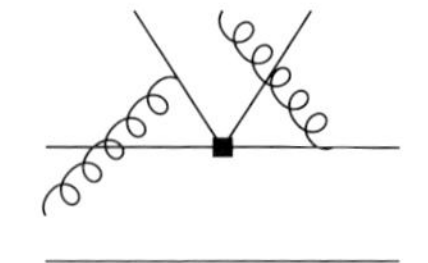

Fig. 9. Diagram that contributes to the hard-scattering kernel involving a quark–antiquark–gluon distribution amplitude of the B meson and the emitted light meson.

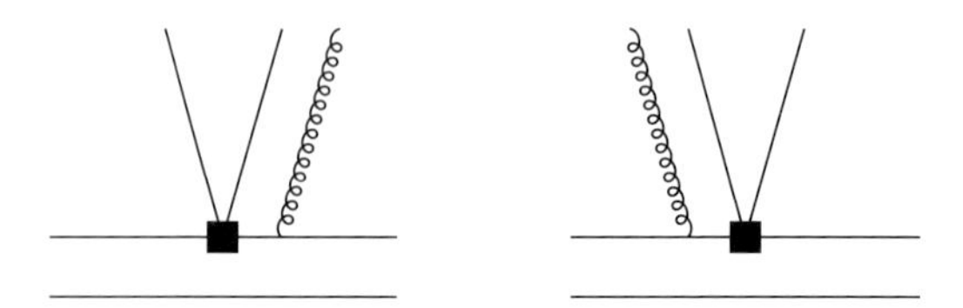

Fig. 10. The contribution of the $q\bar{q}g$ Fock state to the $\bar{B}_d \to D^+\pi^-$ decay amplitude.

5.3 Non-leading Fock States

Our discussion so far concentrated on contributions related to the quark–antiquark components of the meson wave functions. We now present qualitative arguments that justify this restriction to the valence-quark Fock components. Some of these arguments are standard [8,9]. We will argue that higher Fock states yield only subleading contributions in the heavy-quark limit.

Additional Hard Partons

An example of a diagram that would contribute to a hard-scattering function involving quark–antiquark–gluon components of the emitted meson and the B meson is shown in Fig. 9. For light mesons, higher Fock components are related to higher-order terms in the collinear expansion, including the effects of intrinsic transverse momentum and off-shellness of the partons by gauge invariance. The assumption is that the additional partons are collinear and carry a finite fraction of the meson momentum in the heavy-quark limit. Under this assumption, it is easy to see that adding additional partons to the Fock state increases the number of off-shell propagators in a given diagram (compare Fig. 9 to Fig. 3). This implies power suppression in the heavy-quark expansion. Additional partons in the B-meson wave function are always soft, as is the spectator quark. Nevertheless, when these partons are connected to the hard-scattering amplitudes the virtuality of the additional propagators is still of order $m_b\Lambda_{\mathrm{QCD}}$, which is sufficient to guarantee power suppression.

Let us study in more detail how the power suppression arises for the simplest non-trivial example, where the pion is composed of a quark, an antiquark, and an additional gluon. The contribution of this 3-particle Fock state to the $B \to D\pi$ decay amplitude is shown in Fig. 10. It is convenient to use the Fock–Schwinger gauge, which allows us to express the gluon field A_λ in terms of the field-strength tensor $G_{\rho\lambda}$ via

$$A_\lambda(x) = \int_0^1 dv\, v x^\rho\, G_{\rho\lambda}(vx)\,. \tag{48}$$

Up to twist-4 level, there are three quark–antiquark–gluon matrix elements that could potentially contribute to the diagrams shown in Fig. 10. Due to the $V-A$ structure of the weak-interaction vertex, the only relevant three-particle light-cone wave function has twist-4 and is given by [18,19]

$$\begin{aligned}
&\langle\pi(q)|\bar{d}(0)\gamma_\mu\gamma_5\, g_s G_{\alpha\beta}(vx)\, u(0)|0\rangle \\
&= f_\pi (q_\beta g_{\alpha\mu} - q_\alpha g_{\beta\mu}) \int \mathcal{D}u\, \phi_\perp(u_i)\, e^{ivu_3 q\cdot x} \\
&+ f_\pi \frac{q_\mu}{q\cdot x} (q_\alpha x_\beta - q_\beta x_\alpha) \int \mathcal{D}u\, \big(\phi_\perp(u_i) + \phi_\parallel(u_i)\big)\, e^{ivu_3 q\cdot x}\,.
\end{aligned} \tag{49}$$

Here $\int \mathcal{D}u \equiv \int_0^1 du_1\, du_2\, du_3\, \delta(1-u_1-u_2-u_3)$, with u_1, u_2 and u_3 the fractions of the pion momentum carried by the quark, antiquark and gluon, respectively. Evaluating the diagrams in Fig. 10, and neglecting the charm-quark mass for simplicity, we find

$$\langle D^+\pi^-|O_8|\bar{B}_d\rangle_{q\bar{q}g} = i f_\pi\, \langle D^+|\bar{c}\, \not{q}(1-\gamma_5) b|\bar{B}_d\rangle \int \mathcal{D}u\, \frac{2\phi_\parallel(u_i)}{u_3\, m_b^2}\,. \tag{50}$$

Since $\phi_\parallel \sim \Lambda_{\rm QCD}^2$, the suppression by two powers of $\Lambda_{\rm QCD}/m_b$ compared to the leading-order matrix element is obvious. Note that due to G-parity $\phi_\parallel$ is antisymmetric in $u_1 \leftrightarrow u_2$ for a pion, so that (50) vanishes in this case.

Additional Soft Partons

A more precarious situation may arise when the additional Fock components carry only a small fraction of the meson momentum, contrary to the assumption made above. It is usually argued [8,9] that these configurations are suppressed, because they occupy only a small fraction of the available phase space (since $\int du_i \sim \Lambda_{\rm QCD}/m_b$ when the parton that carries momentum fraction u_i is soft). This argument does not apply when the process involves heavy mesons. Consider, for example, the diagram shown in Fig. 11a for the decay $B \to D\pi$. Its contribution involves the overlap of the B-meson wave function involving additional soft gluons with the wave function of the D meson, also containing soft gluons. There is no reason to suppose that this overlap is suppressed relative to the soft overlap of the valence-quark wave functions. It represents (part of) the overlap of the "soft cloud" around the b quark with (part of) the "soft cloud" around the c quark after the weak decay. The partonic decomposition of this cloud is unrestricted up to global quantum numbers. (In the case where the B meson decays into two light mesons, there is a form-factor suppression $\sim (\Lambda_{\rm QCD}/m_b)^{3/2}$ for the overlap of the valence-quark wave functions, but once this price is paid there is again no reason for further suppression of additional soft gluons in the overlap of the B-meson wave function and the wave function of the recoiling meson.)

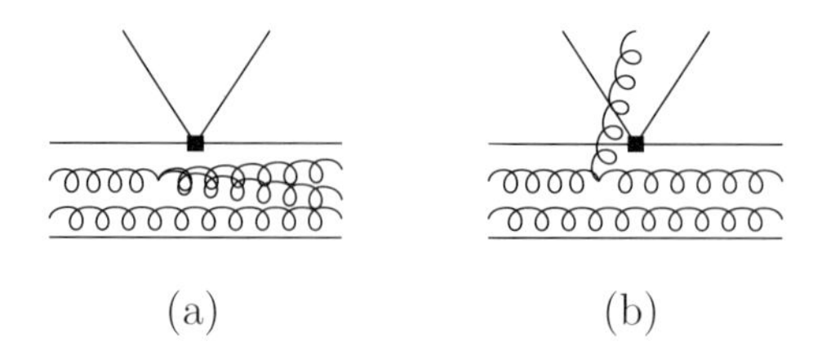

Fig. 11. (a) Soft overlap contribution which is part of the $B \to D$ form factor. (b) Soft overlap with the pion which would violate factorization, if it were unsuppressed.

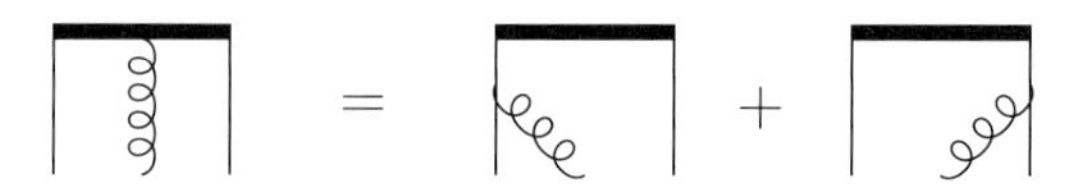

Fig. 12. Quark–antiquark–gluon distribution amplitude in the gluon endpoint region.

The previous paragraph essentially repeated our earlier argument against the hard-scattering approach, and in favour of using the $B \to D$ form factor as an input to the factorization formula. However, given the presence of additional soft partons in the $B \to D$ transition, we must now argue that it is unlikely that the emitted pion drags with it one of these soft partons, for instance a soft gluon that goes into the pion wave function, as shown in Fig. 11b. Notice that if the $(q\bar{q})$ pair is produced in a colour-octet state, at least one gluon (or a further $(q\bar{q})$ pair) must be pulled into the emitted meson if the decay is to result in a two-body final state. What suppresses the process shown in Fig. 11b relative to the one in Fig. 11a even if the emitted $(q\bar{q})$ pair is in a colour-octet state?

It is once more colour transparency that saves us. The dominant configuration has both quarks carry a large fraction of the pion momentum, and only the gluon might be soft. In this situation we can apply a non-local "operator product expansion" to determine the coupling of the soft gluon to the small $(q\bar{q})$ pair [2]. The gluon endpoint behaviour of the $q\bar{q}g$ wave function is then determined by the sum of the two diagrams shown on the right-hand side in Fig. 12. The leading term (for small gluon momentum) cancels in the sum of the two diagrams, because the meson (represented by the black bar) is a colour singlet. This cancellation, which is exactly the same cancellation needed to demonstrate that "non-factorizable" vertex corrections are dominated by hard gluons, provides one factor of $\Lambda_{\rm QCD}/m_b$ needed to show that Fig. 11b is power suppressed relative to Fig. 11a.

In summary, we have (qualitatively) covered all possibilities for non-valence contributions to the decay amplitude and find that they are all power suppressed in the heavy-quark limit.

6 Difficulties with Charm

The factorization formula (3) holds in the heavy-quark limit $m_b \to \infty$. Corrections to the asymptotic limit are power-suppressed in the ratio $\Lambda_{\rm QCD}/m_b$ and, generally speaking, do not assume a factorized form. Since m_b is fixed to about 5 GeV in the real world, one may worry about the magnitude of power corrections to hadronic B-decay amplitudes. Naive dimensional analysis would suggest that these corrections should be of order 10% or so. However, in some cases related to $B \to D\pi$ decays there are also corrections suppressed by $\Lambda_{\rm QCD}/m_c$, which are potentially much larger.

There are decay modes, such as $B^- \to D^0\pi^-$, in which the spectator quark can go to either of the two final-state mesons. The factorization formula (3) applies to the contribution that arises when the spectator quark goes to the D meson, but not when the spectator quark goes to the pion. However, even in the latter case we may use naive factorization to estimate the power behaviour of the decay amplitude. Adapting (13) to the decay $B^- \to D^0\pi^-$, we find that the non-factorizing (class-II) amplitude is suppressed compared to the factorizing (class-I) amplitude by

$$\frac{\mathcal{A}(B^- \to D^0\pi^-)_{\text{class-II}}}{\mathcal{A}(B^- \to D^0\pi^-)_{\text{class-I}}} \sim \frac{F^{B\to\pi}(m_D^2)\, f_D}{F^{B\to D}(m_\pi^2)\, f_\pi} \sim \left(\frac{\Lambda_{\rm QCD}}{m_b}\right)^2 . \tag{51}$$

Here we use that $F^{B\to\pi}(q^2) \sim (\Lambda_{\rm QCD}/m_b)^{3/2}$ even for $q^2 \sim m_b^2$, as long as $q_{max}^2 - q^2$ is also of order m_b^2. (It follows from our definition of heavy final-state mesons that these conditions are fulfilled.) As a consequence, strictly speaking factorization *does* hold for $B^- \to D^0\pi^-$ decays in the sense that the class-II contribution is power suppressed with respect to the class-I contribution.

Unfortunately, the scaling behaviour for real B and D mesons is far from the estimate (51) valid in the heavy-quark limit. Based on the dominance of the class-I amplitude we would expect that

$$R = \frac{\mathrm{Br}(B^- \to D^0\pi^-)}{\mathrm{Br}(\bar{B}_d \to D^+\pi^-)} \approx 1 \tag{52}$$

in the heavy-quark limit. This contradicts existing data which yield $R = 1.89 \pm 0.35$, despite the additional colour suppression of the class-II amplitude. One reason for the failure of power counting lies in the departure of the decay constants and form factors from naive power counting. The following compares the power counting to the actual numbers (square brackets):

$$\frac{f_D}{f_\pi} \sim \left(\frac{\Lambda_{\rm QCD}}{m_c}\right)^{1/2} \quad [\approx 1.5]\,, \qquad \frac{F_+^{B\to\pi}(m_D^2)}{F_+^{B\to D}(m_\pi^2)} \sim \left(\frac{\Lambda_{\rm QCD}}{m_b}\right)^{3/2} \quad [\approx 0.5]\,. \tag{53}$$

However, it is unclear whether the failure of power counting can be attributed to the form factors and decay constants alone.

Note that for the purposes of power counting we treated the charm quark as heavy, taking the heavy-quark limit for fixed m_c/m_b. This simplified the discussion, since we did not have to introduce m_c as a separate scale. However, in reality charm is somewhat intermediate between a heavy and a light quark, since m_c is not particularly large compared to $\Lambda_{\rm QCD}$. In this context it is worth noting that the first hard-scattering kernel in (3) cannot have $\Lambda_{\rm QCD}/m_c$ corrections, since there is a smooth transition to the case of two light mesons. The situation is different with the hard spectator interaction term, which we argued to be power suppressed for decays into a D meson and a light meson.

7 Phenomenology of $B \to D^{(*)}L$ Decays

The matrix elements we have computed in Sect. 4.3 provide the theoretical basis for a model-independent calculation of the class-I non-leptonic decay amplitudes for decays of the type $B \to D^{(*)}L$, where L is a light meson, to leading power in $\Lambda_{\rm QCD}/m_b$ and at next-to-leading order in renormalization-group improved perturbation theory. In this section we discuss phenomenological applications of this formalism and confront our numerical results with experiment. We also provide some numerical estimates of power-suppressed corrections to the factorization formula.

7.1 Non-leptonic Decay Amplitudes

The results for the class-I decay amplitudes for $B \to D^{(*)}L$ are obtained by evaluating the (factorized) hadronic matrix elements of the transition operator $\mathcal{T}$ defined in (40). They are written in terms of products of CKM matrix elements, light-meson decay constants, $B \to D^{(*)}$ transition form factors, and the QCD parameters $a_1(D^{(*)}L)$. The decay constants can be determined experimentally using data on the weak leptonic decays $P^- \to l^-\bar{\nu}_l(\gamma)$, hadronic $\tau^- \to M^-\nu_\tau$ decays, and the electromagnetic decays $V^0 \to e^+e^-$. Following [16], we use $f_\pi = 131\,\mathrm{MeV}$, $f_K = 160\,\mathrm{MeV}$, $f_\rho = 210\,\mathrm{MeV}$, $f_{K^*} = 214\,\mathrm{MeV}$, and $f_{a_1} = 229\,\mathrm{MeV}$. (Here a_1 is the pseudovector meson with mass $m_{a_1} \simeq 1230\,\mathrm{MeV}$.)

The non-leptonic $\bar{B}_d \to D^{(*)+}L^-$ decay amplitudes for $L = \pi,\ \rho$ can be expressed as

$$\begin{aligned}
\mathcal{A}(\bar{B}_d \to D^+\pi^-) &= i\frac{G_F}{\sqrt{2}}\, V_{ud}^* V_{cb}\, a_1(D\pi)\, f_\pi\, F_0(m_\pi^2)\,(m_B^2 - m_D^2)\,, \\
\mathcal{A}(\bar{B}_d \to D^{*+}\pi^-) &= -i\frac{G_F}{\sqrt{2}}\, V_{ud}^* V_{cb}\, a_1(D^*\pi)\, f_\pi A_0(m_\pi^2)\, 2m_{D^*}\, \varepsilon^*\cdot p\,, \\
\mathcal{A}(\bar{B}_d \to D^+\rho^-) &= -i\frac{G_F}{\sqrt{2}}\, V_{ud}^* V_{cb}\, a_1(D\rho)\, f_\rho\, F_+(m_\rho^2)\, 2m_\rho\, \eta^*\cdot p\,,
\end{aligned} \tag{54}$$

where p (p') is the momentum of the B (charm) meson, ε and η are polarization vectors, and the form factors F_0, F_+ and A_0 are defined in the usual way [16]. The decay mode $\bar{B}_d \to D^{*+}\rho^-$ has a richer structure than the decays with at least one pseudoscalar in the final state. The most general Lorentz-invariant decomposition of the corresponding decay amplitude can be written as

$$\mathcal{A}(\bar{B}_d \to D^{*+}\rho^-) = i\frac{G_F}{\sqrt{2}}\, V_{ud}^* V_{cb}\, \varepsilon^{*\mu}\eta^{*\nu}\left(S_1\, g_{\mu\nu} - S_2\, q_\mu p'_\nu + iS_3\, \epsilon_{\mu\nu\alpha\beta}\, p'^\alpha q^\beta \right), \tag{55}$$

where the quantities S_i can be expressed in terms of semi-leptonic form factors. To leading power in $\Lambda_{\rm QCD}/m_b$, we obtain

$$\begin{aligned} S_1 &= a_1(D^*\rho)\, m_\rho f_\rho\, (m_B + m_{D^*}) A_1(m_\rho^2)\,, \\ S_2 &= a_1(D^*\rho)\, m_\rho f_\rho\, \frac{2A_2(m_\rho^2)}{m_B + m_{D^*}}\,. \end{aligned} \tag{56}$$

The contribution proportional to S_3 in (55) is associated with transversely polarized ρ mesons and thus leads to power-suppressed effects, which we do not consider here.

The various $B \to D^{(*)}$ form factors entering the expressions for the decay amplitudes can be determined by combining experimental data on semi-leptonic decays with theoretical relations derived using heavy-quark effective theory [3,16]. Since we work to leading order in $\Lambda_{\rm QCD}/m_b$, it is consistent to set the light meson masses to zero and evaluate these form factors at $q^2 = 0$. In this case the kinematic relations

$$F_0(0) = F_+(0)\,, \qquad (m_B + m_{D^*})A_1(0) - (m_B - m_{D^*})A_2(0) = 2m_{D^*} A_0(0) \tag{57}$$

allow us to express the two $\bar{B}_d \to D^+ L^-$ rates in terms of $F_+(0)$, and the two $\bar{B}_d \to D^{*+} L^-$ rates in terms of $A_0(0)$. Heavy-quark symmetry implies that these two form factors are equal to within a few percent [14]. Below we adopt the common value $F_+(0) = A_0(0) = 0.6$. All our predictions for decay rates will be proportional to the square of this number.

7.2 Meson Distribution Amplitudes and Predictions for a_1

Let us now discuss in more detail the ingredients required for the numerical analysis of the coefficients $a_1(D^{(*)}L)$. The Wilson coefficients C_i in the effective weak Hamiltonian depend on the choice of the scale μ as well as on the value of the strong coupling α_s, for which we take $\alpha_s(m_Z) = 0.118$ and two-loop evolution down to a scale $\mu \sim m_b$. To study the residual scale dependence of the results, which remains because the perturbation series are truncated at next-to-leading order, we vary μ between $m_b/2$ and $2m_b$. The hard-scattering kernels depend on the ratio of the heavy-quark masses, for which we take $z = m_c/m_b = 0.30 \pm 0.05$.

Hadronic uncertainties enter the analysis also through the parameterizations used for the meson light-cone distribution amplitudes. It is convenient and conventional to expand the distribution amplitudes in Gegenbauer polynomials as

$$\Phi_L(u) = 6u(1-u)\left[1 + \sum_{n=1}^{\infty} \alpha_n^L(\mu)\, C_n^{(3/2)}(2u-1)\right], \tag{58}$$

Table 1. Numerical values for the integrals $\int_0^1 du\, F(u,z)\, \Phi_L(u)$ (upper portion) and $\int_0^1 du\, F(u,-z)\, \Phi_L(u)$ (lower portion) obtained including the first two Gegenbauer moments.

z	Leading term	Coefficient of α_1^L	Coefficient of α_2^L
0.25	$-8.41-9.51i$	$5.92-12.19i$	$-1.33+0.36i$
0.30	$-8.79-9.09i$	$5.78-12.71i$	$-1.19+0.58i$
0.35	$-9.13-8.59i$	$5.60-13.21i$	$-1.00+0.73i$
0.25	$-8.45-6.56i$	$6.72-10.73i$	$-0.38+0.93i$
0.30	$-8.37-5.99i$	$6.83-11.49i$	$-0.21+0.85i$
0.35	$-8.24-5.44i$	$6.81-12.29i$	$-0.08+0.75i$

where $C_1^{(3/2)}(x) = 3x$, $C_2^{(3/2)}(x) = \frac{3}{2}(5x^2-1)$, etc. The Gegenbauer moments $\alpha_n^L(\mu)$ are multiplicatively renormalized. The scale dependence of these quantities would, however, enter the results for the coefficients only at order α_s^2, which is beyond the accuracy of our calculation. We assume that the leading-twist distribution amplitudes are close to their asymptotic form and thus truncate the expansion at $n=2$. However, it would be straightforward to account for higher-order terms if desired. For the asymptotic form of the distribution amplitude, $\Phi_L(u) = 6u(1-u)$, the integral in (43) yields

$$\begin{aligned}\int_0^1 du\, F(u,z)\, \Phi_L(u) &= 3\ln z^2 - 7 \\ &+ \left[\frac{6z(1-2z)}{(1-z)^2(1+z)^3} \left(\frac{\pi^2}{6} - \mathrm{Li}_2(z^2) \right) - \frac{3(2-3z+2z^2+z^3)}{(1-z)(1+z)^2} \ln(1-z^2) \right. \\ &\quad \left. + \frac{4-17z+20z^2+5z^3}{2(1-z)(1+z)^2} + \{z \to 1/z\} \right], \end{aligned} \tag{59}$$

and the corresponding result with the function $F(u,-z)$ is obtained by replacing $z \to -z$. More generally, a numerical integration with a distribution amplitude expanded in Gegenbauer polynomials yields the results collected in Table 1. We observe that the first two Gegenbauer polynomials in the expansion of the light-cone distribution amplitudes give contributions of similar magnitude, whereas the second moment gives rise to much smaller effects. This tendency persists in higher orders. For our numerical discussion it is a safe approximation to truncate the expansion after the first non-trivial moment. The dependence of the results on the value of the quark mass ratio $z = m_c/m_b$ is mild and can be neglected for all practical purposes. We also note that the difference of the convolutions with the kernels for a pseudoscalar D and vector D^* meson are numerically very small. This observation is, however, specific to the case of $B \to D^{(*)}L$ decays and should not be generalized to other decays.

Next we evaluate the complete results for the parameters a_1 at next-to-leading order, and to leading power in $\Lambda_{\mathrm{QCD}}/m_b$. We set $z = m_c/m_b = 0.3$.

Table 2. The QCD coefficients $a_1(D^{(*)}L)$ at next-to-leading order for three different values of the renormalization scale μ. The leading-order values are shown for comparison.

	$\mu = m_b/2$	$\mu = m_b$	$\mu = 2m_b$
$a_1(DL)$	$1.074 + 0.037i$	$1.055 + 0.020i$	$1.038 + 0.011i$
	$-(0.024 - 0.052i)\,\alpha_1^L$	$-(0.013 - 0.028i)\,\alpha_1^L$	$-(0.007 - 0.015i)\,\alpha_1^L$
$a_1(D^*L)$	$1.072 + 0.024i$	$1.054 + 0.013i$	$1.037 + 0.007i$
	$-(0.028 - 0.047i)\,\alpha_1^L$	$-(0.015 - 0.025i)\,\alpha_1^L$	$-(0.008 - 0.014i)\,\alpha_1^L$
a_1^{LO}	1.049	1.025	1.011

Varying z between 0.25 and 0.35 would change the results by less than 0.5%. The results are shown in Table 2. The contributions proportional to the second Gegenbauer moment α_2^L have coefficients of order 0.2% or less and can safely be neglected. The contributions associated with α_1^L are present only for the strange mesons K and K^*, but not for π and ρ. Moreover, the imaginary parts of the coefficients contribute to their modulus only at order α_s^2, which is beyond the accuracy of our analysis. To summarize, we thus obtain

$$\begin{aligned} |a_1(DL)| &= 1.055^{+0.019}_{-0.017} - (0.013^{+0.011}_{-0.006})\alpha_1^L\,, \\ |a_1(D^*L)| &= 1.054^{+0.018}_{-0.017} - (0.015^{+0.013}_{-0.007})\alpha_1^L\,, \end{aligned} \tag{60}$$

where the quoted errors reflect the perturbative uncertainty due to the scale ambiguity (and the negligible dependence on the value of the ratio of quark masses and higher Gegenbauer moments), but not the effects of power-suppressed corrections. These will be estimated later. It is evident that within theoretical uncertainties there is no significant difference between the two a_1 parameters, and there is only a very small sensitivity to the differences between strange and non-strange mesons (assuming that $|\alpha_1^{K^{(*)}}| < 1$). In our numerical analysis below we thus take $|a_1| = 1.05$ for all decay modes.

7.3 Tests of Factorization

The main lesson from the previous discussion is that corrections to naive factorization in the class-I decays $\bar{B}_d \to D^{(*)+}L^-$ are very small. The reason is that these effects are governed by a small Wilson coefficient and, moreover, are colour suppressed by a factor $1/N_c^2$. For these decays, the most important implications of the QCD factorization formula are to restore the renormalization-group invariance of the theoretical predictions, and to provide a theoretical justification for why naive factorization works so well. On the other hand, given the theoretical uncertainties arising, e.g., from unknown power-suppressed corrections, there is little hope to confront the extremely small predictions for non-universal (process-dependent) "non-factorizable" corrections with experimental data. Rather, what we may do is ask whether data supports the prediction of a quasi-universal parameter $|a_1| \simeq 1.05$ in these decays. If this is indeed the case, it would support the usefulness of the heavy-quark limit in analyzing non-leptonic decay amplitudes. If, on the other hand, we were to find large non-universal effects, this would point towards the existence of sizeable power corrections to our predictions. We will see that within present experimental errors the data are in good agreement with our prediction of a quasi universal a_1 parameter. However, a reduction of the experimental uncertainties to the percent level would be very desirable for obtaining a more conclusive picture.

We start by considering ratios of non-leptonic decay rates that are related to each other by the replacement of a pseudoscalar meson by a vector meson. In the comparison of $B \to D\pi$ and $B \to D^*\pi$ decays one is sensitive to the difference of the values of the two a_1 parameters in (60) evaluated for $\alpha_1^L = 0$. This difference is at most few times 10^{-3}. Likewise, in the comparison of $B \to D\pi$ and $B \to D\rho$ decays one is sensitive to the difference in the light-cone distribution amplitudes of the pion and the ρ meson, which start at the second Gegenbauer moment α_2^L. These effects are suppressed even more strongly. From the explicit expressions for the decay amplitudes in (54) it follows that

$$\begin{aligned} \frac{\Gamma(\bar{B}_d \to D^+\pi^-)}{\Gamma(\bar{B}_d \to D^{*+}\pi^-)} &= \frac{(m_B^2 - m_D^2)^2 |\boldsymbol{q}\,|_{D\pi}}{4m_B^2 |\boldsymbol{q}\,|^3_{D^*\pi}} \left(\frac{F_0(m_\pi^2)}{A_0(m_\pi^2)}\right)^2 \left|\frac{a_1(D\pi)}{a_1(D^*\pi)}\right|^2, \\ \frac{\Gamma(\bar{B}_d \to D^+\rho^-)}{\Gamma(\bar{B}_d \to D^+\pi^-)} &= \frac{4m_B^2 |\boldsymbol{q}\,|^3_{D\rho}}{(m_B^2 - m_D^2)^2 |\boldsymbol{q}\,|_{D\pi}} \frac{f_\rho^2}{f_\pi^2} \left(\frac{F_+(m_\rho^2)}{F_0(m_\pi^2)}\right)^2 \left|\frac{a_1(D\rho)}{a_1(D\pi)}\right|^2. \end{aligned} \tag{61}$$

Using the experimental values for the branching ratios reported by the CLEO Collaboration [21] we find (taking into account a correlation between some systematic errors in the second case)

$$\begin{aligned} \left|\frac{a_1(D\pi)}{a_1(D^*\pi)}\right| \frac{F_0(m_\pi^2)}{A_0(m_\pi^2)} &= 1.00 \pm 0.11\,, \\ \left|\frac{a_1(D\rho)}{a_1(D\pi)}\right| \frac{F_+(m_\rho^2)}{F_0(m_\pi^2)} &= 1.16 \pm 0.11\,. \end{aligned} \tag{62}$$

Within errors, there is no evidence for any deviations from naive factorization.

Our next-to-leading order results for the quantities $a_1(D^{(*)}L)$ allow us to make theoretical predictions which are not restricted to ratios of hadronic decay rates. A particularly clean test of these predictions, which is essentially free of hadronic uncertainties, is obtained by relating the $\bar{B}_d \to D^{(*)+}L^-$ decay rates to the differential semi-leptonic $\bar{B}_d \to D^{(*)+}l^-\nu$ decay rate evaluated at $q^2 = m_L^2$. In this way the parameters $|a_1|$ can be measured directly [10]. One obtains

$$R_L^{(*)} = \frac{\Gamma(\bar{B}_d \to D^{(*)+}L^-)}{d\Gamma(\bar{B}_d \to D^{(*)+}l^-\bar{\nu})/dq^2\big|_{q^2=m_L^2}} = 6\pi^2 |V_{ud}|^2 f_L^2\, |a_1(D^{(*)}L)|^2\, X_L^{(*)}\,, \tag{63}$$

where $X_\rho = X_\rho^* = 1$ for a vector meson (because the production of the lepton pair via a $V - A$ current in semi-leptonic decays is kinematically equivalent to that of a vector meson with momentum q), whereas X_π and X_π^* deviate from

1 only by (calculable) terms of order m_π^2/m_B^2, which numerically are below the 1% level [16]. We emphasize that with our results for a_1 given in (43) the above relation becomes a prediction based on first principles of QCD. This is to be contrasted with the usual interpretation of this formula, where a_1 plays the role of a phenomenological parameter that is fitted from data.

The most accurate tests of factorization employ the class-I processes $\bar{B}_d \to D^{*+}L^-$, because the differential semi-leptonic decay rate in $B \to D^*$ transitions has been measured as a function of q^2 with good accuracy. The results of such an analysis, performed using CLEO data, have been reported in [23]. One finds

$$\begin{aligned} R_\pi^* &= (1.13 \pm 0.15)\,\mathrm{GeV}^2 & &\Rightarrow & |a_1(D^*\pi)| &= 1.08 \pm 0.07\,, \\ R_\rho^* &= (2.94 \pm 0.54)\,\mathrm{GeV}^2 & &\Rightarrow & |a_1(D^*\rho)| &= 1.09 \pm 0.10\,, \\ R_{a_1}^* &= (3.45 \pm 0.69)\,\mathrm{GeV}^2 & &\Rightarrow & |a_1(D^*a_1)| &= 1.08 \pm 0.11\,. \end{aligned} \tag{64}$$

This is consistent with our theoretical result in (43). In particular, the data show no evidence for large power corrections to our predictions obtained at leading order in $\Lambda_{\mathrm{QCD}}/m_b$. However, a further improvement in the experimental accuracy would be desirable in order to become sensitive to process-dependent, non-factorizable effects.

7.4 Predictions for Class-I Decay Amplitudes

We now consider a larger set of class-I decays of the form $\bar{B}_d \to D^{(*)+}L^-$, all of which are governed by the transition operator (40). In Table 3 we compare the QCD factorization predictions with experimental data. As previously we work in the heavy-quark limit, i.e. our predictions are model independent up to corrections suppressed by at least one power of $\Lambda_{\mathrm{QCD}}/m_b$. The results show good agreement with experiment within errors, which are still rather large. (Note that we have not attempted to adjust the semi-leptonic form factors $F_+(0)$ and $A_0(0)$ so as to obtain a best fit to the data.)

We take the observation that the experimental data on class-I decays into heavy-light final states show good agreement with our predictions obtained in the heavy-quark limit as evidence that in these decays there are no unexpectedly large power corrections. We will now address the important question of the size of power corrections theoretically. To this end we provide rough estimates of two sources of power-suppressed effects: weak annihilation and spectator interactions. We stress that, at present, a complete account of power corrections to the heavy-quark limit cannot be performed in a systematic way, since these effects are not dominated by hard gluon exchange. In other words, factorization breaks down beyond leading power, and there are other sources of power corrections, such as contributions from higher Fock states, which we will not address here. We believe that the estimates presented below are nevertheless instructive.

To obtain an estimate of power corrections we adopt the following, heuristic procedure. We treat the charm quark as *light* compared to the large scale

Table 3. Model-independent predictions for the branching ratios (in units of 10^{-3}) of class-I, non-leptonic $\bar{B}_d \to D^{(*)+}L^-$ decays in the heavy-quark limit. All predictions are in units of $(|a_1|/1.05)^2$. The last two columns show the experimental results reported by the CLEO Collaboration [21], and by the Particle Data Group [24].

Decay mode	Theory (HQL)	CLEO data	PDG98
$\bar{B}_d \to D^+\pi^-$	3.27	2.50 ± 0.40	3.0 ± 0.4
$\bar{B}_d \to D^+K^-$	0.25	—	—
$\bar{B}_d \to D^+\rho^-$	7.64	7.89 ± 1.39	7.9 ± 1.4
$\bar{B}_d \to D^+K^{*-}$	0.39	—	—
$\bar{B}_d \to D^+a_1^-$	7.76	8.34 ± 1.66	6.0 ± 3.3
	$\times[F_+(0)/0.6]^2$		
$\bar{B}_d \to D^{*+}\pi^-$	3.05	2.34 ± 0.32	2.8 ± 0.2
$\bar{B}_d \to D^{*+}K^-$	0.22	—	—
$\bar{B}_d \to D^{*+}\rho^-$	7.59	7.34 ± 1.00	6.7 ± 3.3
$\bar{B}_d \to D^{*+}K^{*-}$	0.40	—	—
$\bar{B}_d \to D^{*+}a_1^-$	8.53	11.57 ± 2.02	13.0 ± 2.7
	$\times[A_0(0)/0.6]^2$		

provided by the mass of the decaying b quark ($m_c \ll m_b$, and m_c *fixed* as $m_b \to \infty$) and use a light-cone projection similar to that of the pion also for the D meson. In addition, we assume that m_c is still large compared to Λ_{QCD}. We implement this by using a highly asymmetric D-meson wave function, which is strongly peaked at a light-quark momentum fraction of order Λ_{QCD}/m_D. This guarantees correct power counting for the heavy-light final states we are interested in. As discussed in Sect. 5.2, there are four annihilation diagrams with a single gluon exchange (see Fig. 8a–d). The first two diagrams are "factorizable" and their contributions vanish because of current conservation in the limit $m_c \to 0$. For non-zero m_c they therefore carry an additional suppression factor $m_D^2/m_B^2 \approx 0.1$. Moreover, their contributions to the decay amplitude are suppressed by small Wilson coefficients. Diagrams (a) and (b) can therefore safely be neglected. From the non-factorizable diagrams (c) and (d) in Fig. 8, the one with the gluon attached to the b quark turns out to be strongly suppressed numerically, giving a contribution of less than 1% of the leading class-I amplitude. We are thus left with diagram (d), in which the gluon couples to the light quark in the B meson. This mechanism gives the dominant annihilation contribution. (Note that by deforming the light spectator-quark line one can redraw this diagram in such a way that it can be interpreted as a final-state rescattering process.)

Adopting a common notation, we parameterize the annihilation contribution to the $\bar{B}_d \to D^+\pi^-$ decay amplitude in terms of a (power-suppressed) amplitude A such that $\mathcal{A}(\bar{B}_d \to D^+\pi^-) = T + A$, where T is the "tree topology", which contains the dominant factorizable contribution. A straightforward calculation using the approximations discussed above shows that the contribution of diagram (d) is (to leading order) independent of the momentum fraction ξ of the light quark inside the B meson:

$$A \simeq f_\pi f_D f_B \int du\, \frac{\Phi_\pi(u)}{u} \int dv\, \frac{\Phi_D(v)}{\bar{v}^2} \simeq 3 f_\pi f_D f_B \int dv\, \frac{\Phi_D(v)}{\bar{v}^2} \,. \tag{65}$$

The B-meson wave function simply integrates to f_B, and the integral over the pion distribution amplitude can be performed using the asymptotic form of the wave function. We take $\Phi_D(v)$ in the form of (58) with the coefficients $\alpha_1^D = 0.8$ and $\alpha_2^D = 0.4$ ($\alpha_i^D = 0$, $i > 2$). With this ansatz $\Phi_D(v)$ is strongly peaked at $\bar{v} \sim \Lambda_{\rm QCD}/m_D$. The integral over $\Phi_D(v)$ in (65) is divergent at $v = 1$, and we regulate it by introducing a cut-off such that $v \le 1 - \Lambda/m_B$ with $\Lambda \approx 0.3\,\text{GeV}$. Then $\int dv\, \Phi_D(v)/\bar{v}^2 \approx 34$. Evidently, the proper value of Λ is largely unknown, and our estimate will be correspondingly uncertain. Nevertheless, this exercise will give us an idea of the magnitude of the effect. For the ratio of the annihilation amplitude to the leading, factorizable contribution we obtain

$$\frac{A}{T} \simeq \frac{2\pi\alpha_s}{3} \, \frac{C_+ + C_-}{2C_+ + C_-} \, \frac{f_D f_B}{F_0(0)\, m_B^2} \int dv\, \frac{\Phi_D(v)}{\bar{v}^2} \approx 0.04 \,. \tag{66}$$

We have evaluated the Wilson coefficients at $\mu = m_b$ and used $f_D = 0.2\,\text{GeV}$, $f_B = 0.18\,\text{GeV}$, $F_0(0) = 0.6$, and $\alpha_s = 0.4$. This value of the strong coupling constant reflects that the typical virtuality of the gluon propagator in the annihilation graph is of order $\Lambda_{\rm QCD} m_B$. We conclude that the annihilation contribution is a correction of a few percent, which is what one would expect for a generic power correction to the heavy-quark limit. Taking into account that $f_B \sim \Lambda_{\rm QCD}(\Lambda_{\rm QCD}/m_B)^{1/2}$, $F_0(0) \sim (\Lambda_{\rm QCD}/m_B)^{3/2}$ and $f_D \sim \Lambda_{\rm QCD}$, we observe that in the heavy-quark limit the ratio A/T indeed scales as $\Lambda_{\rm QCD}/m_b$, exhibiting the expected linear power suppression. (Recall that we consider the D meson as a light meson for this heuristic analysis of power corrections.)

Using the same approach, we may derive a numerical estimate for the non-factorizable spectator interaction in $\bar{B}_d \to D^+\pi^-$ decays, discussed in Sect. 5.1. We find

$$\frac{T_{\rm spec}}{T_{\rm lead}} \simeq \frac{2\pi\alpha_s}{3} \, \frac{C_+ - C_-}{2C_+ + C_-} \, \frac{f_D f_B}{F_0(0)\, m_B^2} \, \frac{m_B}{\lambda_B} \int dv\, \frac{\Phi_D(v)}{\bar{v}} \approx -0.03 \,, \tag{67}$$

where the hadronic parameter $\lambda_B = O(\Lambda_{\rm QCD})$ is defined as $\int_0^1 (d\xi/\xi)\, \Phi_B(\xi) \equiv m_B/\lambda_B$. For the numerical estimate we have assumed that $\lambda_B \approx 0.3\,\text{GeV}$. With the same model for $\Phi_D(v)$ as above we have $\int dv\, \Phi_D(v)/\bar{v} \approx 6.6$, where the integral is now convergent. The result (67) exhibits again the expected power suppression in the heavy-quark limit, and the numerical size of the effect is at the few percent level.

We conclude from this discussion that the typical size of power corrections to the heavy-quark limit in class-I decays of B mesons into heavy-light final states is at the level of 10% or less, and thus our prediction for the near universality of the parameters a_1 governing these decay modes appears robust.

8 Conclusion

With the recent commissioning of the B factories and the planned emphasis on heavy-flavour physics in future collider experiments, the role of B decays in providing fundamental tests of the Standard Model and potential signatures of new physics will continue to grow. In many cases the principal source of systematic uncertainty is a theoretical one, namely our inability to quantify the non-perturbative QCD effects present in these decays. This is true, in particular, for almost all measurements of CP violation at the B factories.

In these lectures, I have reviewed a rigorous framework for the evaluation of strong-interaction effects for a large class of exclusive, two-body non-leptonic decays of B mesons. The main result is contained in the factorization formula (3), which expresses the amplitudes for these decays in terms of experimentally measurable semi-leptonic form factors, light-cone distribution amplitudes, and hard-scattering functions that are calculable in perturbative QCD. For the first time, therefore, we have a well founded field-theoretic basis for phenomenological studies of exclusive hadronic B decays, and a formal justification for the ideas of factorization. For simplicity, I have mainly focused on $B \to D\pi$ decays here. A detailed discussion of B decays into two light mesons will be presented in a forthcoming paper [12].

We hope that the factorization formula (3) will form the basis for future studies of non-leptonic two-body decays of B mesons. Before, however, a fair amount of conceptual work remains to be completed. In particular, it will be important to investigate better the limitations on the numerical precision of the factorization formula, which is valid in the formal heavy-quark limit. We have discussed some preliminary estimates of power-suppressed effects in the present work, but a more complete analysis would be desirable. In particular, for rare B decays into two light mesons it will be important to understand the role of chirally-enhanced power corrections and weak annihilation contributions [12,25]. For these decays, there are also still large uncertainties associated with the description of the hard spectator interactions.

Theoretical investigations along these lines should be pursued with vigor. We are confident that, ultimately, this research will result in a *theory* of non-leptonic B decays, which should be as useful for this area of heavy-flavour physics as the large-m_b limit and heavy-quark effective theory were for the phenomenology of semi-leptonic decays.

Acknowledgements

I would like to thank the organizers of the workshop, in particular David Blaschke and Mikhail Ivanov, for the invitation to present these lectures and for their hospitality. I am grateful to the students for attending the lectures and contributing with questions and discussions.

References

1. M. Beneke, G. Buchalla, M. Neubert and C.T. Sachrajda, *Phys. Rev. Lett.* **83**, 1914 (1999).
2. M. Beneke, G. Buchalla, M. Neubert and C.T. Sachrajda, *Nucl. Phys. B* **591**, 313 (2000).
3. N. Isgur and M.B. Wise, *Phys. Lett. B* **232**, 113 (1989); *ibid.* **237**, 527 (1990).
4. M.A. Shifman and M.B. Voloshin, *Sov. J. Nucl. Phys.* **45**, 292 (1987) [*Yad. Fiz.* **45**, 463 (1987)]; *ibid.* **47**, 511 (1988) [**47** (1988) 801].
5. For a review, see: G. Buchalla, A.J. Buras and M.E. Lautenbacher, *Rev. Mod. Phys.* **68**, 1125 (1996).
6. D. Fakirov and B. Stech, *Nucl. Phys. B* **133**, 315 (1978).
7. N. Cabibbo and L. Maiani, *Phys. Lett. B* **73**, 418 (1978); *ibid.* **76**, 663 (1978) (E).
8. G.P. Lepage and S.J. Brodsky, *Phys. Rev. D* **22**, 2157 (1980).
9. A.V. Efremov and A.V. Radyushkin, *Phys. Lett. B* **94**, 245 (1980).
10. J.D. Bjorken, *Nucl. Phys. (Proc. Suppl.) B* **11**, 325 (1989).
11. M.J. Dugan and B. Grinstein, *Phys. Lett. B* **255**, 583 (1991).
12. M. Beneke, G. Buchalla, M. Neubert and C.T. Sachrajda, *Nucl. Phys. B* **606**, 245 (2001).
13. L. Maiani and M. Testa, *Phys. Lett. B* **245**, 585 (1990).
14. For a review, see: M. Neubert, *Phys. Rep.* **245**, 259 (1994).
15. J.F. Donoghue, E. Golowich, A.A. Petrov and J.M. Soares, *Phys. Rev. Lett.* **77**, 2178 (1996).
16. For a review, see: M. Neubert and B. Stech, in: *Heavy Flavours II*, ed. A.J. Buras and M. Lindner (World Scientific, Singapore, 1998) pp. 294 [hep-ph/9705292]; M. Neubert, *Nucl. Phys. (Proc. Suppl.) B* **64**, 474 (1998).
17. H.D. Politzer and M.B. Wise, *Phys. Lett. B* **257**, 399 (1991).
18. A. Khodjamirian and R. Rückl, in: *Heavy Flavours II*, ed. A.J. Buras and M. Lindner (World Scientific, Singapore, 1998) pp. 345 [hep-ph/9801443].
19. V.M. Braun and I.E. Filyanov, *Z. Phys. C* **48**, 239 (1990).
20. B.V. Geshkenbein and M.V. Terentev, *Phys. Lett. B* **117**, 243 (1982); *Sov. J. Nucl. Phys.* **39**, 554 (1984) [*Yad. Fiz.* **39**, 873 (1984)].
21. B. Barish et al., CLEO Collaboration, Conference report CLEO CONF 97-01 (EPS 97-339).
22. B. Nemati et al., CLEO Collaboration, *Phys. Rev. D* **57**, 5363 (1998).
23. J.L. Rodriguez, in: Proceedings of the 2nd International Conference on *B Physics and CP Violation*, Honolulu, Hawaii, March 1997, ed. T.E. Browder et al. (World Scientific, Singapore, 1998) pp. 124 [hep-ex/9801028].
24. C. Caso et al., Particle Data Group, *Eur. Phys. J. C* **3**, 1 (1998).
25. Y.Y. Keum, H.-N. Li and A.I. Sanda, *Phys. Lett. B* **504**, 6 (2001).

B Physics and CP Violation

Robert Fleischer

Theory Division, CERN, 1211 Geneva 23, Switzerland

Abstract. After an introduction to the Standard-Model description of CP violation, we turn to the main focus of these lectures, the B-meson system. Since non-leptonic B decays play the key rôle for the exploration of CP violation, we have to discuss the tools to describe these transitions theoretically before classifying the main strategies to study CP violation. We will then have a closer look at the B-factory benchmark modes $B_d \to J/\psi K_S$, $B_d \to \phi K_S$ and $B_d \to \pi^+\pi^-$, and shall emphasize the importance of studies of B_s decays at hadron colliders. Finally, we focus on more recent developments related to $B \to \pi K$ modes and the $B_d \to \pi^+\pi^-$, $B_s \to K^+K^-$ system.

1 Introduction

The non-conservation of the CP symmetry, where C and P denote the charge-conjugation and parity transformation operators, respectively, is one of the most exciting phenomena in particle physics since its unexpected discovery through $K_L \to \pi^+\pi^-$ decays in 1964 [1]. At that time it was believed that – although weak interactions are neither invariant under P, nor invariant under C – the product CP was preserved. Consider, for instance, the process

$$\pi^+ \to e^+\nu_e \xrightarrow{C} \pi^- \to e^-\nu_e^C \xrightarrow{P} \pi^- \to e^-\overline{\nu}_e. \tag{1}$$

Here the left-handed ν_e^C state is not observed in nature; only after performing an additional P transformation do we obtain the right-handed electron antineutrino.

Before the start of the B factories, CP-violating effects could only be studied in the kaon system, where we distinguish between "indirect" CP violation, which is due to the fact that the mass eigenstates K_S and K_L of the neutral kaon system are not eigenstates of the CP operator, and "direct" CP violation, arising directly at the decay amplitude level of the neutral kaon system. The former kind of CP violation was already discovered in 1964 and is described by a complex parameter ε, whereas the latter one, described by the famous parameter $\mathrm{Re}(\varepsilon'/\varepsilon)$, could only be established in 1999 after tremendous efforts by the NA48 (CERN) [2] and KTeV (Fermilab) [3] collaborations, reporting the following results in 2002:

$$\mathrm{Re}(\varepsilon'/\varepsilon) = \begin{cases} (14.7 \pm 2.2) \times 10^{-4} & (\text{NA48 [4]}) \\ (20.7 \pm 2.8) \times 10^{-4} & (\text{KTeV [5]}). \end{cases} \tag{2}$$

Unfortunately, the theoretical interpretation of $\mathrm{Re}(\varepsilon'/\varepsilon)$ is still affected by large hadronic uncertainties and does not provide a stringent test of the Standard-Model description of CP violation, unless significant theoretical progress concerning the relevant hadronic matrix elements can be made [6–8].

R. Fleischer, B Physics and CP Violation, Lect. Notes Phys. **647**, 42–77 (2004)
http://www.springerlink.com/

In 2001, CP violation could also be established in B-meson decays by the BaBar (SLAC) [9] and Belle (KEK) [10] collaborations, representing the start of a new era in the exploration of CP violation. As we will discuss in these lecture notes, decays of neutral and charged B-mesons provide valuable insights into this phenomenon, offering in particular powerful tests of the Kobayashi–Maskawa (KM) mechanism [11], which allows us to accommodate CP violation in the Standard Model of electroweak interactions. In Sect. 2, we shall have a closer look at the Standard-Model description of CP violation, and shall introduce the Wolfenstein parametrization and the unitarity triangles of the Cabibbo–Kobayashi–Maskawa (CKM) matrix. Since non-leptonic decays of B mesons play the key rôle in the exploration of CP violation, we have to discuss the tools to deal with these transitions and the corresponding theoretical problems in Sect. 3. The main strategies to study CP violation are then classified in Sect. 4, before we focus on benchmark modes for the B factories in Sect. 5. The great physics potential of B_s-meson decays for experiments at hadron colliders is emphasized in Sect. 6, and will also be employed in Sect. 7, where we discuss interesting recent developments. Finally, we make a few comments on the "usual" rare B decays in Sect. 8, before we summarize our conclusions and give a brief outlook in Sect. 9.

A considerably more detailed presentation of CP violation in the B system can be found in [12], as well as in the textbooks listed in [13]. Another lecture on related topics was given by Neubert at this school [14].

2 CP Violation in the Standard Model

2.1 Charged-Current Interactions of Quarks

The CP-violating effects discussed in these lectures originate from the charged-current interactions of the quarks, described by the Lagrangian

$$\mathcal{L}_{\text{int}}^{\text{CC}} = -\frac{g_2}{\sqrt{2}} \left(\bar{u}_{\text{L}},\, \bar{c}_{\text{L}},\, \bar{t}_{\text{L}} \right) \gamma^\mu \, \hat{V}_{\text{CKM}} \begin{pmatrix} d_{\text{L}} \\ s_{\text{L}} \\ b_{\text{L}} \end{pmatrix} W_\mu^\dagger + \text{h.c.}, \tag{3}$$

where the gauge coupling g_2 is related to the gauge group $SU(2)_{\text{L}}$, the $W_\mu^{(\dagger)}$ field corresponds to the charged W bosons, and $\hat{V}_{\text{CKM}}$ denotes the CKM matrix, connecting the electroweak eigenstates of the down, strange and bottom quarks with their mass eigenstates through a unitary transformation.

Since the CKM matrix elements V_{UD} and V_{UD}^* enter in $D \to UW^-$ and the CP-conjugate process $\overline{D} \to \overline{U}W^+$, respectively, where $D \in \{d, s, b\}$ and $U \in \{u, c, t\}$, we observe that the phase structure of the CKM matrix is closely related to CP violation. It was pointed out by Kobayashi and Maskawa in 1973 that actually one complex phase is required – in addition to three generalized Euler angles – to parametrize the quark-mixing matrix in the case of three fermion generations, thereby allowing us to accommodate CP violation in the Standard Model [11].

More detailed investigations show that additional conditions have to be satisfied for CP violation. They can be summarized as follows:

$$\begin{aligned}&(m_t^2 - m_c^2)(m_t^2 - m_u^2)(m_c^2 - m_u^2)\\&\quad\times(m_b^2 - m_s^2)(m_b^2 - m_d^2)(m_s^2 - m_d^2) \times J_{\rm CP} \neq 0,\end{aligned} \tag{4}$$

where the Jarlskog parameter

$$J_{\rm CP} = \pm\,\mathrm{Im}\left(V_{i\alpha}V_{j\beta}V_{i\beta}^*V_{j\alpha}^*\right) \quad (i \neq j,\, \alpha \neq \beta) \tag{5}$$

represents a measure of the "strength" of CP violation within the Standard Model [15]. As data imply $J_{\rm CP} = \mathcal{O}(10^{-5})$, CP violation is a small effect in the Standard Model. In scenarios for physics beyond the Standard Model, typically also new sources of CP violation arise [16].

2.2 Wolfenstein Parametrization

The quark transitions caused by charged-current interactions exhibit an interesting hierarchy, which is made explicit in the Wolfenstein parametrization of the CKM matrix [17]:

$$\hat{V}_{\rm CKM} = \begin{pmatrix} 1-\frac{1}{2}\lambda^2 & \lambda & A\lambda^3(\rho - i\eta) \\ -\lambda & 1-\frac{1}{2}\lambda^2 & A\lambda^2 \\ A\lambda^3(1-\rho-i\eta) & -A\lambda^2 & 1 \end{pmatrix} + \mathcal{O}(\lambda^4). \tag{6}$$

This parametrization corresponds to an expansion in powers of the small quantity $\lambda = 0.22$, which can be fixed through semileptonic kaon decays. The other parameters are of order 1, where η leads to an imaginary part of the CKM matrix. The Wolfenstein parametrization is very useful for phenomenological applications, as we will see below. A detailed discussion of the next-to-leading order terms in λ can be found in [18].

2.3 Unitarity Triangles

The central targets for tests of the KM mechanism of CP violation are the unitarity triangles of the CKM matrix. As we have already noted, the CKM matrix is unitary. Consequently, it satisfies

$$\hat{V}_{\rm CKM}^\dagger \cdot \hat{V}_{\rm CKM} = \hat{1} = \hat{V}_{\rm CKM} \cdot \hat{V}_{\rm CKM}^\dagger, \tag{7}$$

implying a set of 12 equations, which consist of 6 normalization relations and 6 orthogonality relations. The latter can be represented as 6 triangles in the complex plane [19], all having the same area, $2A_\Delta = |J_{\rm CP}|$ [20]. However, in only two of them, all three sides are of comparable magnitude $\mathcal{O}(\lambda^3)$, while in the remaining ones, one side is suppressed with respect to the others by $\mathcal{O}(\lambda^2)$ or $\mathcal{O}(\lambda^4)$. The orthogonality relations describing the non-squashed triangles are given by

$$V_{ud}\,V_{ub}^* + V_{cd}\,V_{cb}^* + V_{td}\,V_{tb}^* = 0 \quad \text{[1st and 3rd column]} \tag{8}$$

$$V_{ub}^*\,V_{tb} + V_{us}^*\,V_{ts} + V_{ud}^*\,V_{td} = 0 \quad \text{[1st and 3rd row]}. \tag{9}$$

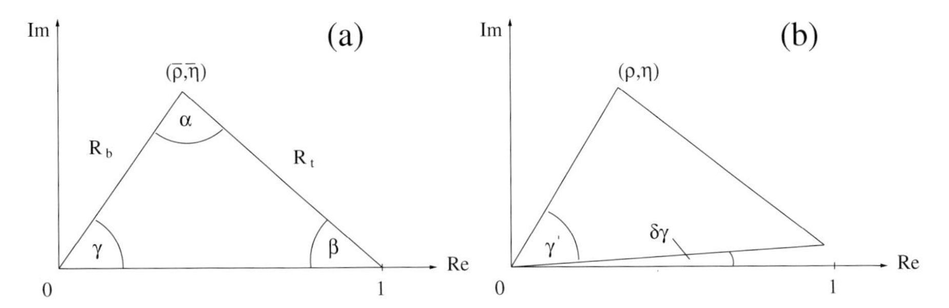

Fig. 1. The two non-squashed unitarity triangles of the CKM matrix: (a) and (b) correspond to the orthogonality relations (8) and (9), respectively.

At leading order in λ, these relations agree with each other, and yield

$$(\rho + i\eta)A\lambda^3 + (-A\lambda^3) + (1 - \rho - i\eta)A\lambda^3 = 0. \tag{10}$$

Consequently, they describe the same triangle, which is usually referred to as *the* unitarity triangle of the CKM matrix [20,21]. It is convenient to divide (10) by the overall normalization $A\lambda^3$. Then we obtain a triangle in the complex plane with a basis normalized to 1, and an apex given by (ρ, η).

In the future, the experimental accuracy will reach such an impressive level that we will have to distinguish between the unitarity triangles described by (8) and (9), which differ through $\mathcal{O}(\lambda^2)$ corrections. They are illustrated in Fig. 1, where $\overline{\rho}$ and $\overline{\eta}$ are related to ρ and η through [18]

$$\overline{\rho} \equiv \left(1 - \lambda^2/2\right)\rho, \quad \overline{\eta} \equiv \left(1 - \lambda^2/2\right)\eta, \tag{11}$$

and

$$\delta\gamma \equiv \gamma - \gamma' = \lambda^2\eta. \tag{12}$$

The sides R_b and R_t of the unitarity triangle shown in Fig. 1 (a) are given by

$$R_b = \left(1 - \frac{\lambda^2}{2}\right)\frac{1}{\lambda}\left|\frac{V_{ub}}{V_{cb}}\right| = \sqrt{\overline{\rho}^2 + \overline{\eta}^2} = 0.38 \pm 0.08 \tag{13}$$

$$R_t = \frac{1}{\lambda}\left|\frac{V_{td}}{V_{cb}}\right| = \sqrt{(1-\overline{\rho})^2 + \overline{\eta}^2} = \mathcal{O}(1), \tag{14}$$

and will show up at several places throughout these lectures. Whenever we refer to a unitarity triangle, we mean the one illustrated in Fig. 1 (a).

2.4 Standard Analysis of the Unitarity Triangle

There is a "standard analysis" to constrain the apex of the unitarity triangle in the $\overline{\rho}$–$\overline{\eta}$ plane, employing the following ingredients:

- Using heavy-quark arguments, exclusive and inclusive $b \to u, c\ell\overline{\nu}_\ell$ decays provide $|V_{ub}|$ and $|V_{cb}|$ [22], allowing us to fix the side R_b of the unitarity triangle, i.e. a circle in the $\overline{\rho}$–$\overline{\eta}$ plane around $(0,0)$ with radius R_b.

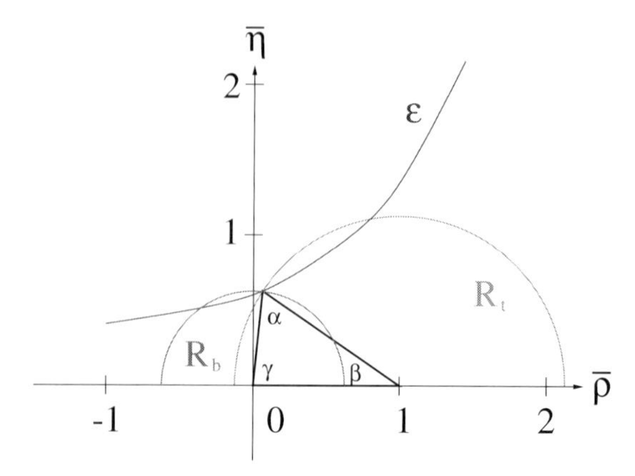

Fig. 2. Contours to determine the unitarity triangle in the $\overline{\rho}$–$\overline{\eta}$ plane.

- Using the top-quark mass m_t as an input, and taking into account certain QCD corrections and non-perturbative parameters, we may extract $|V_{td}|$ from B_d^0–$\overline{B_d^0}$ mixing (see below). The combination of $|V_{td}|$ with $|V_{cb}|$ allows us then to fix the side R_t of the unitarity triangle, i.e. a circle in the $\overline{\rho}$–$\overline{\eta}$ plane around $(1, 0)$ with radius R_t. Comparing B_d^0–$\overline{B_d^0}$ with B_s^0–$\overline{B_s^0}$ mixing, an $SU(3)$-breaking parameter ξ suffices to determine R_t.
- Using m_t and $|V_{cb}|$ as an input, and taking into account certain QCD corrections and non-perturbative parameters, the observable ε describing indirect CP violation in the kaon system allows us to fix a hyperbola in the $\overline{\rho}$–$\overline{\eta}$ plane.

These contours are sketched in Fig. 2; their intersection gives the apex of the unitarity triangle shown in Fig. 1 (a). Because of strong correlations between theoretical and experimental uncertainties, it is rather involved to convert the experimental information into an allowed range in the $\overline{\rho}$–$\overline{\eta}$ plane, and various analyses can be found in the literature: a simple scanning approach [7], a Gaussian approach [23], the "BaBar 95% scanning method" [24], a Bayesian approach [25], and a non-Bayesian statistical approach [26]. Other recent analyses can be found in [27,28]. A reasonable range for α, β and γ that is consistent with these approaches is given by

$$70^\circ \lesssim \alpha \lesssim 130^\circ, \quad 20^\circ \lesssim \beta \lesssim 30^\circ, \quad 50^\circ \lesssim \gamma \lesssim 70^\circ. \tag{15}$$

The question of how to combine the theoretical and experimental errors in an optimal way will certainly continue to be a hot topic in the future. This is also reflected by the Bayesian [25] vs. non-Bayesian [26] debate going on at present.

2.5 Quantitative Studies of CP Violation

As we have seen above, the neutral kaon system provides two different CP-violating parameters, ε and $\text{Re}(\varepsilon'/\varepsilon)$. The former is one of the ingredients of the "standard analysis" of the unitarity triangle, implying in particular $\overline{\eta} > 0$

if very plausible assumptions about a certain non-perturbative "bag" parameter are made. On the other hand, $\mathrm{Re}(\varepsilon'/\varepsilon)$ does not (yet) provide further stringent constraints on the unitarity triangle because of large hadronic uncertainties, although the experimental values are of the same order of magnitude as the range of theoretical estimates [6,7].

Considerably more promising in view of testing the Standard-Model description of CP violation are the rare kaon decays $K^+ \to \pi^+\nu\overline{\nu}$ and $K_{\mathrm{L}} \to \pi^0\nu\overline{\nu}$, which originate in the Standard Model from loop effects and are theoretically very clean since the relevant hadronic matrix elements can be fixed through semileptonic kaon decays [7,29]. In particular, they also allow an interesting determination of the unitarity triangle [30], and show interesting correlations with CP violation in the B sector [12,31]. Unfortunately, the $K \to \pi\nu\overline{\nu}$ branching ratios are at the 10^{-11} level in the Standard Model; two events of $K^+ \to \pi^+\nu\overline{\nu}$ have already been observed by the E787 Experiment at Brookhaven, yielding a branching ratio of $(1.57^{+1.75}_{-0.82}) \times 10^{-10}$ [32]. It is very important to measure $K^+ \to \pi^+\nu\overline{\nu}$ and $K_{\mathrm{L}} \to \pi^0\nu\overline{\nu}$ with reasonable statistics, and there are efforts under way to accomplish this challenging goal [33].

In the case of the B-meson system, consisting of charged mesons $B_u^+ \sim u\bar{b}$, $B_c^+ \sim c\bar{b}$, as well as neutral ones $B_d^0 \sim d\bar{b}$, $B_s^0 \sim s\bar{b}$, we have a "simplified" hadron dynamics, since the b quark is "heavy" with respect to the QCD scale parameter Λ_{QCD}. Moreover, hadronic uncertainties can be eliminated or cancel in appropriate CP-violating observables, thereby providing various tests of the KM mechanism of CP violation and direct determinations of the angles of the unitarity triangle. As we will see below, the Standard Model predicts large CP-violating asymmetries in certain decays, and large effects were actually observed recently in $B_d \to J/\psi K_{\mathrm{S}}$ [9,10]. The goal is now to overconstrain the unitarity triangle as much as possible and to test several Standard-Model predictions, with the hope to encounter discrepancies that could shed light on the physics lying beyond the Standard Model. In this decade, the asymmetric e^+e^- B factories operating at the $\Upsilon(4S)$ resonance with their detectors BaBar and Belle provide access to several benchmark decay modes of $B_u^\pm$ and B_d^0 mesons [34]. Moreover, experiments at hadron colliders allow us to study, in addition, large data samples of decays of B_s mesons, which are another very important element in the testing of the Standard-Model description of CP violation. Important first steps in this direction are already expected at run II of the Tevatron [35], whereas several strategies can only be fully exploited in the LHC era [36], in particular at LHCb (CERN) and BTeV (Fermilab).

In these lectures we shall focus on the B-meson system. For the exploration of CP violation, non-leptonic B decays play the central rôle, as CP-violating effects are due to certain interference effects that may show up in this decay class. Before turning to these modes, let us note that there are also other promising systems to obtain insights into CP violation, for example D mesons, where the Standard Model predicts very small CP violation, electric dipole moments or hyperon decays. These topics are, however, beyond the scope of this presentation.

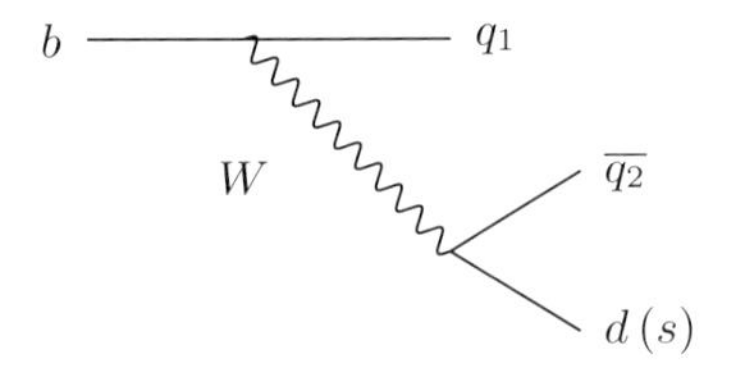

Fig. 3. Tree diagrams ($q_1, q_2 \in \{u, c\}$).

3 Non-leptonic B Decays

3.1 Classification

Non-leptonic $\overline{B}$ decays are mediated by $b \to q_1\,\overline{q_2}\,d\,(s)$ quark-level transitions, with $q_1, q_2 \in \{u, d, c, s\}$. There are two kinds of topologies contributing to non-leptonic B decays: tree-diagram-like and "penguin" topologies. The latter consist of gluonic (QCD) and electroweak (EW) penguins. In Figs. 3–5, the corresponding leading-order Feynman diagrams are shown. Depending on the flavour content of their final states, we may classify $b \to q_1\,\overline{q_2}\,d\,(s)$ decays as follows:

- $q_1 \neq q_2 \in \{u, c\}$: only tree diagrams contribute.
- $q_1 = q_2 \in \{u, c\}$: tree and penguin diagrams contribute.
- $q_1 = q_2 \in \{d, s\}$: only penguin diagrams contribute.

3.2 Low-Energy Effective Hamiltonians

In order to analyse non-leptonic B decays theoretically, one uses low-energy effective Hamiltonians, which are calculated by making use of the operator product expansion, yielding transition matrix elements of the following structure:

$$\langle f|\mathcal{H}_{\rm eff}|i\rangle = \frac{G_{\rm F}}{\sqrt{2}}\lambda_{\rm CKM}\sum_k C_k(\mu)\langle f|Q_k(\mu)|i\rangle\,. \tag{16}$$

The operator product expansion allows us to separate the short-distance contributions to this transition amplitude from the long-distance ones, which are described by perturbative Wilson coefficient functions $C_k(\mu)$ and non-perturbative

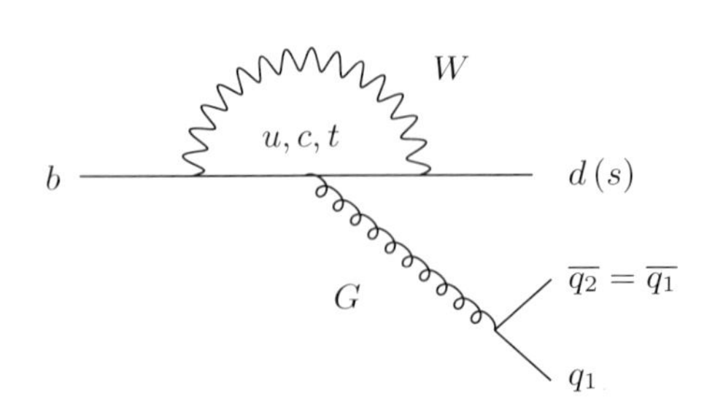

Fig. 4. QCD penguin diagrams ($q_1 = q_2 \in \{u, d, c, s\}$).

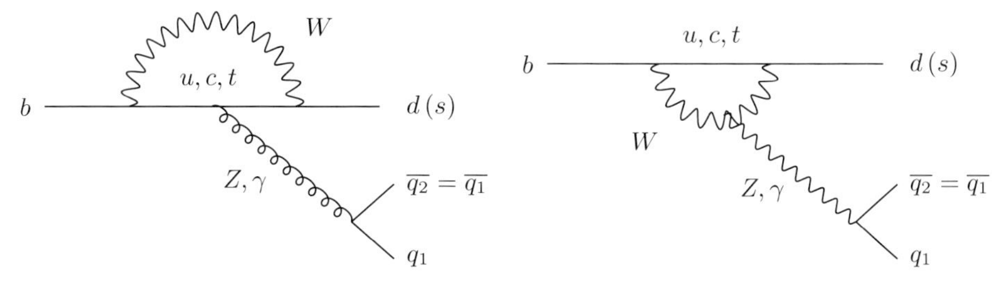

Fig. 5. Electroweak penguin diagrams ($q_1 = q_2 \in \{u, d, c, s\}$).

hadronic matrix elements $\langle f|Q_k(\mu)|i\rangle$, respectively. As usual, $G_{\rm F}$ is the Fermi constant, $\lambda_{\rm CKM}$ is a CKM factor, and μ denotes an appropriate renormalization scale. The Q_k are local operators, which are generated by electroweak interactions and QCD, and govern "effectively" the decay in question. The Wilson coefficients $C_k(\mu)$ can be considered as scale-dependent couplings related to the vertices described by the Q_k.

Let us consider $\overline{B^0_d} \to D^+K^-$, which is a pure "tree" decay, to discuss the evaluation of the corresponding low-energy effective Hamiltonian in more detail. At leading order, this decay originates from a $b \to c\overline{u}s$ quark-level transition, where the bc and $\overline{u}s$ quark currents are connected through the exchange of a W boson. Evaluating the corresponding Feynman diagram yields

$$-\frac{g_2^2}{8}V_{us}^*V_{cb}\left[\overline{s}\gamma^\nu(1-\gamma_5)u\right]\left[\frac{g_{\nu\mu}}{k^2-M_W^2}\right]\left[\overline{c}\gamma^\mu(1-\gamma_5)b\right]. \tag{17}$$

Because of $k^2 \approx m_b^2 \ll M_W^2$, we have

$$\frac{g_{\nu\mu}}{k^2-M_W^2} \quad \longrightarrow \quad -\frac{g_{\nu\mu}}{M_W^2} \equiv -\left(\frac{8G_{\rm F}}{\sqrt{2}g_2^2}\right)g_{\nu\mu}, \tag{18}$$

i.e. we may "integrate out" the W boson in (17), and arrive at

$$\begin{aligned}\mathcal{H}_{\rm eff} &= \frac{G_{\rm F}}{\sqrt{2}}V_{us}^*V_{cb}\left[\overline{s}_\alpha\gamma_\mu(1-\gamma_5)u_\alpha\right]\left[\overline{c}_\beta\gamma^\mu(1-\gamma_5)b_\beta\right] \\ &= \frac{G_{\rm F}}{\sqrt{2}}V_{us}^*V_{cb}(\overline{s}_\alpha u_\alpha)_{\rm V-A}(\overline{c}_\beta b_\beta)_{\rm V-A} \equiv \frac{G_{\rm F}}{\sqrt{2}}V_{us}^*V_{cb}O_2\,, \end{aligned} \tag{19}$$

where α and β denote $SU(3)_{\rm C}$ colour indices. Effectively, our decay process $b \to c\overline{u}s$ is now described by the "current–current" operator O_2.

If we take into account QCD corrections, operator mixing leads to a second "current–current" operator, which is given by

$$O_1 \equiv \left[\overline{s}_\alpha\gamma_\mu(1-\gamma_5)u_\beta\right]\left[\overline{c}_\beta\gamma^\mu(1-\gamma_5)b_\alpha\right]. \tag{20}$$

Consequently, we obtain a low-energy effective Hamiltonian of the following structure:

$$\mathcal{H}_{\rm eff} = \frac{G_{\rm F}}{\sqrt{2}}V_{us}^*V_{cb}\left[C_1(\mu)O_1 + C_2(\mu)O_2\right], \tag{21}$$

where $C_1(\mu) \neq 0$ and $C_2(\mu) \neq 1$ are due to QCD renormalization effects. In order to evaluate these coefficients, we have first to calculate QCD corrections to the decay processes both in the full theory, i.e. with W exchange, and in the effective theory, and have then to express the QCD-corrected transition amplitude in terms of QCD-corrected matrix elements and Wilson coefficients as in (16). This procedure is called "matching". The results for the $C_k(\mu)$ thus obtained contain terms of $\log(\mu/M_W)$, which become large for $\mu = \mathcal{O}(m_b)$, the scale governing the hadronic matrix elements of the O_k. Making use of the renormalization group, which exploits the fact that the transition amplitude (16) cannot depend on the chosen renormalization scale μ, we may sum up the following terms of the Wilson coefficients:

$$\alpha_s^n \left[\log\left(\frac{\mu}{M_W}\right)\right]^n \text{ (LO)}, \quad \alpha_s^n \left[\log\left(\frac{\mu}{M_W}\right)\right]^{n-1} \text{ (NLO)}, \quad ... \tag{22}$$

A very detailed discussion of these techniques can be found in [37].

In the case of decays receiving contributions both from tree and from penguin topologies, basically the only difference to (21) is that we encounter more operators:

$$\mathcal{H}_{\rm eff} = \frac{G_{\rm F}}{\sqrt{2}}\left[\sum_{j=u,c} V_{jr}^* V_{jb}\left\{\sum_{k=1}^{2} C_k(\mu)\, Q_k^{jr} + \sum_{k=3}^{10} C_k(\mu)\, Q_k^{r}\right\}\right]. \tag{23}$$

Here the current–current operators Q_1^{jr} and Q_2^{jr}, the QCD penguin operators Q_3^r–Q_6^r, and the EW penguin operators Q_7^r–Q_{10}^r are related to the tree, QCD and EW penguin processes shown in Figs. 3–5 (explicit expressions for these operators can be found in [12,37]). At a renormalization scale $\mu = \mathcal{O}(m_b)$, the Wilson coefficients of the current–current operators satisfy $C_1(\mu) = \mathcal{O}(10^{-1})$ and $C_2(\mu) = \mathcal{O}(1)$, whereas those of the penguin operators are $\mathcal{O}(10^{-2})$. Note that penguin topologies with internal charm- and up-quark exchanges are described in this framework by penguin-like matrix elements of the corresponding current–current operators [38], and may also have important phenomenological consequences [39,40].

Since the ratio $\alpha/\alpha_s = \mathcal{O}(10^{-2})$ of the QED and QCD couplings is very small, we would expect naïvely that EW penguins should play a minor rôle in comparison with QCD penguins. This would actually be the case if the top quark was not "heavy". However, since the Wilson coefficient C_9 increases strongly with m_t, we obtain interesting EW penguin effects in several B decays: $B^- \to K^-\phi$ is affected significantly by EW penguins, whereas $B \to \pi\phi$ and $B_s \to \pi^0\phi$ are even dominated by such topologies [41,42]. EW penguins also have an important impact on $B \to \pi K$ modes [43], as we will see in Sect. 7.

The low-energy effective Hamiltonians discussed in this section apply to all B decays that are caused by the same corresponding quark-level transition, i.e. they are "universal". Within this formalism, differences between various exclusive modes are only due to the hadronic matrix elements of the relevant four-quark operators. Unfortunately, the evaluation of such matrix elements is associated with large uncertainties and is a very challenging task. In this context, "factorization" is a widely used concept, which is our next topic.

3.3 Factorization of Hadronic Matrix Elements

In order to discuss "factorization", let us consider once more $\overline{B^0_d} \to D^+K^-$. Evaluating the corresponding transition amplitude, we encounter the hadronic matrix elements of the $O_{1,2}$ operators between the $\langle K^-D^+|$ final and $|\overline{B^0_d}\rangle$ initial states. If we use the well-known $SU(N_C)$ colour-algebra relation

$$T^a_{\alpha\beta}T^a_{\gamma\delta} = \frac{1}{2}\left(\delta_{\alpha\delta}\delta_{\beta\gamma} - \frac{1}{N_C}\delta_{\alpha\beta}\delta_{\gamma\delta}\right) \tag{24}$$

to rewrite the operator O_1, we obtain

$$\langle K^-D^+|\mathcal{H}_{\rm eff}|\overline{B^0_d}\rangle = \frac{G_F}{\sqrt{2}}V^*_{us}V_{cb}\Big[a_1\langle K^-D^+|(\overline{s}_\alpha u_\alpha)_{\rm V-A}(\overline{c}_\beta b_\beta)_{\rm V-A}|\overline{B^0_d}\rangle$$
$$+2\,C_1\langle K^-D^+|(\overline{s}_\alpha\, T^a_{\alpha\beta}\, u_\beta)_{\rm V-A}(\overline{c}_\gamma\, T^a_{\gamma\delta}\, b_\delta)_{\rm V-A}|\overline{B^0_d}\rangle\Big], \tag{25}$$

with

$$a_1 = \frac{C_1}{N_C} + C_2. \tag{26}$$

It is now straightforward to "factorize" the hadronic matrix elements:

$$\langle K^-D^+|(\overline{s}_\alpha u_\alpha)_{\rm V-A}(\overline{c}_\beta b_\beta)_{\rm V-A}|\overline{B^0_d}\rangle\Big|_{\rm fact}$$
$$= \langle K^-|\left[\overline{s}_\alpha\gamma_\mu(1-\gamma_5)u_\alpha\right]|0\rangle\langle D^+|\left[\overline{c}_\beta\gamma^\mu(1-\gamma_5)b_\beta\right]|\overline{B^0_d}\rangle$$
$$\propto f_K(\text{"decay constant"}) \times F_{BD}(\text{"form factor"}), \tag{27}$$

$$\langle K^-D^+|(\overline{s}_\alpha\, T^a_{\alpha\beta}\, u_\beta)_{\rm V-A}(\overline{c}_\gamma\, T^a_{\gamma\delta}\, b_\delta)_{\rm V-A}|\overline{B^0_d}\rangle\Big|_{\rm fact} = 0. \tag{28}$$

The quantity introduced in (26) is a phenomenological "colour factor", governing "colour-allowed" decays. In the case of "colour-suppressed" modes, for instance $\overline{B^0_d} \to \pi^0 D^0$, we have to deal with the combination

$$a_2 = C_1 + \frac{C_2}{N_C}. \tag{29}$$

The concept of the factorization of hadronic matrix elements has a long history [44], and can be justified, for example, in the large N_C limit [45]. Recently, the "QCD factorization" approach was developed [46–48], which may provide an important step towards a rigorous basis for factorization for a large class of non-leptonic two-body B-meson decays in the heavy-quark limit. The resulting formula for the transition amplitudes incorporates elements both of the naïve factorization approach sketched above and of the hard-scattering picture. Let us consider a decay $\overline{B} \to M_1M_2$, where M_1 picks up the spectator quark. If M_1 is either a heavy (D) or a light (π, K) meson, and M_2 a light (π, K) meson, QCD factorization gives a transition amplitude of the following structure:

$$A(\overline{B} \to M_1 M_2) = [\text{“naïve factorization”}] \times [1 + \mathcal{O}(\alpha_s) + \mathcal{O}(\Lambda_{\rm QCD}/m_b)] \,. \quad (30)$$

While the $\mathcal{O}(\alpha_s)$ terms, i.e. the radiative non-factorizable corrections to naïve factorization, can be calculated in a systematic way, the main limitation of the theoretical accuracy is due to the $\mathcal{O}(\Lambda_{\rm QCD}/m_b)$ terms. These issues are discussed in detail in [14]. Further interesting recent papers are listed in [49].

Another QCD approach to deal with non-leptonic B decays into charmless final states – the perturbative hard-scattering (or “PQCD”) approach – was developed independently in [50], and differs from the QCD factorization formalism in some technical aspects. An interesting avenue to deal with non-leptonic B decays is also provided by QCD light-cone sum-rule approaches [51].

4 Towards Studies of CP Violation in the B System

4.1 Amplitude Structure and Direct CP Violation

If we use the unitarity of the CKM matrix, it is an easy exercise to show that the amplitude for any given non-leptonic B decay can always be written is such a way that at most two weak CKM amplitudes contribute:

$$A(\overline{B} \to \overline{f}) = e^{+i\varphi_1}|A_1|e^{i\delta_1} + e^{+i\varphi_2}|A_2|e^{i\delta_2} \quad (31)$$

$$A(B \to f) = e^{-i\varphi_1}|A_1|e^{i\delta_1} + e^{-i\varphi_2}|A_2|e^{i\delta_2}. \quad (32)$$

Here $\varphi_{1,2}$ denote CP-violating weak phases, which are due to the CKM matrix, and the $|A_{1,2}|e^{i\delta_{1,2}}$ are CP-conserving “strong” amplitudes, containing the whole hadron dynamics of the decay at hand:

$$|A|e^{i\delta} \sim \sum_k \underbrace{C_k(\mu)}_{\text{pert. QCD}} \times \underbrace{\langle \overline{f}|Q_k(\mu)|\overline{B}\rangle}_{\text{non-pert.}} . \quad (33)$$

Employing (31) and (32), we obtain the following CP-violating rate asymmetry:

$$\mathcal{A}_{\rm CP} \equiv \frac{\Gamma(B \to f) - \Gamma(\overline{B} \to \overline{f})}{\Gamma(B \to f) + \Gamma(\overline{B} \to \overline{f})} = \frac{|A(B \to f)|^2 - |A(\overline{B} \to \overline{f})|^2}{|A(B \to f)|^2 + |A(\overline{B} \to \overline{f})|^2}$$

$$= \frac{2|A_1||A_2|\sin(\delta_1 - \delta_2)\sin(\varphi_1 - \varphi_2)}{|A_1|^2 + 2|A_1||A_2|\cos(\delta_1 - \delta_2)\cos(\varphi_1 - \varphi_2) + |A_2|^2} . \quad (34)$$

Consequently, a non-vanishing CP asymmetry $\mathcal{A}_{\rm CP}$ arises from interference effects between the two weak amplitudes, and requires both a non-trivial weak phase difference $\varphi_1 - \varphi_2$ and a non-trivial strong phase difference $\delta_1 - \delta_2$. This kind of CP violation is referred to as “direct” CP violation, as it originates directly at the amplitude level of the considered decay. It is the B-meson counterpart of the effects probed through $\mathrm{Re}(\varepsilon'/\varepsilon)$ in the neutral kaon system. Since $\varphi_1 - \varphi_2$ is in general given by one of the angles of the unitarity triangle – usually γ – the goal is to determine this quantity from the measured value of $\mathcal{A}_{\rm CP}$. Unfortunately, the extraction of $\varphi_1 - \varphi_2$ from $\mathcal{A}_{\rm CP}$ is affected by hadronic uncertainties, which are due to the strong amplitudes $|A_{1,2}|e^{i\delta_{1,2}}$ (see (34)).

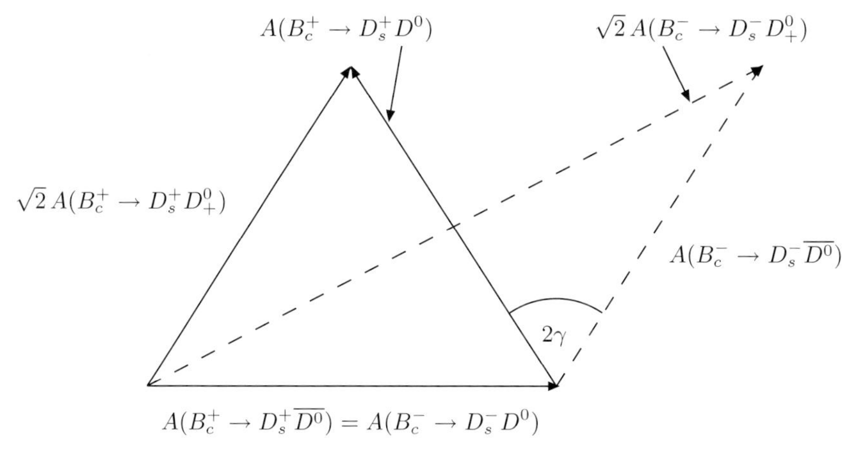

Fig. 6. The extraction of γ from $B_c^\pm \to D_s^\pm\{D^0, \overline{D^0}, D_+^0\}$ decays.

4.2 Classification of the Main Strategies

The most obvious – but also most challenging – strategy we may follow is to try to calculate the relevant hadronic matrix elements $\langle \overline{f}|Q_k(\mu)|\overline{B}\rangle$. As we have noted above, interesting progress has recently been made in this direction through the development of the QCD factorization [46–49], the PQCD [50], and the QCD light-cone sum-rule approaches [51].

Another avenue we may follow is to search for fortunate cases, where relations between decay amplitudes allow us to eliminate the hadronic uncertainties. This approach was pioneered by Gronau and Wyler [52], who proposed the extraction of γ from triangle relations between $B_u^\pm \to K^\pm\{D^0, \overline{D^0}, D_+^0\}$ amplitudes, where D_+^0 is the CP-even eigenstate of the neutral D-meson system. These modes receive only contributions from tree-diagram-like topologies. Unfortunately, this strategy, which is *theoretically clean*, is very difficult from an experimental point of view, since the corresponding triangles are very squashed ones (for other experimental problems and strategies to solve them, see [53]). As an alternative $B_d \to K^{*0}\{D^0, \overline{D^0}, D_+^0\}$ modes were proposed [54], where the triangles are more equilateral. Interestingly, from a theoretical point of view, the ideal realization of this "triangle" approach arises in the B_c-meson system. Here the $B_c^\pm \to D_s^\pm\{D^0, \overline{D^0}, D_+^0\}$ decays allow us to construct the amplitude triangles sketched in Fig. 6, where all sides are expected to be of the same order of magnitude [55]. The practical implementation of this strategy appears also to be challenging, but elaborate feasibility studies for experiments of the LHC era are strongly encouraged. Amplitude relations can also be derived with the help of the flavour symmetries of strong interactions, i.e. $SU(2)$ and $SU(3)$. Here we have to deal with $B_{(s)} \to \pi\pi, \pi K, KK$ decays, providing interesting determinations of weak phases and insights into hadronic physics. We shall have a closer look at these modes in Sect. 7.

The third avenue we may follow to deal with the problems arising from hadronic matrix elements is to employ decays of neutral B_d or B_s mesons. Here we

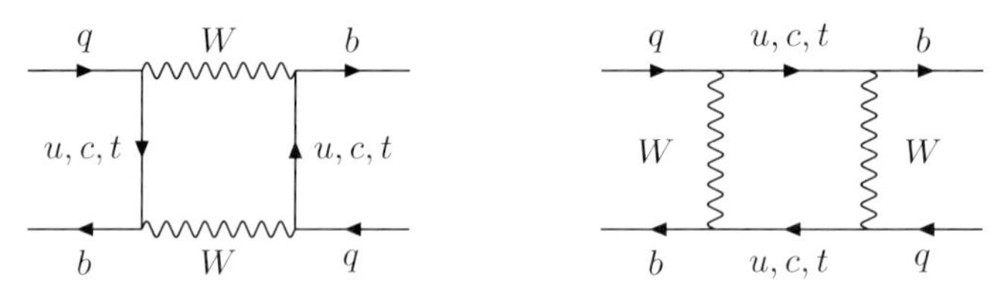

Fig. 7. Box diagrams contributing to B_q^0–$\overline{B_q^0}$ mixing ($q \in \{d, s\}$).

encounter a new kind of CP violation, which is due to interference effects between B_q^0–$\overline{B_q^0}$ mixing and decay processes, and is referred to as "mixing-induced" CP violation. Within the Standard Model, B_q^0–$\overline{B_q^0}$ mixing arises from the box diagrams shown in Fig. 7. Because of this phenomenon, an initially, i.e. at time $t = 0$, present B_q^0-meson state evolves into a time-dependent linear combination of B_q^0 and $\overline{B_q^0}$ states:

$$|B_q(t)\rangle = a(t)|B_q^0\rangle + b(t)|\overline{B_q^0}\rangle, \tag{35}$$

where $a(t)$ and $b(t)$ are governed by an appropriate Schrödinger equation. In order to solve it, mass eigenstates with mass and width differences

$$\Delta M_q \equiv M_{\rm H}^{(q)} - M_{\rm L}^{(q)} > 0 \quad \text{and} \quad \Delta\Gamma_q \equiv \Gamma_{\rm H}^{(q)} - \Gamma_{\rm L}^{(q)}, \tag{36}$$

respectively, are introduced. The decay rates $\Gamma(\overset{(-)}{B_q^0}(t) \to \overset{(-)}{f})$ then contain terms proportional to $\cos(\Delta M_q t)$ and $\sin(\Delta M_q t)$, describing the B_q^0–$\overline{B_q^0}$ oscillations. To be specific, let us consider the very important special case where the B_q^0 meson decays into a final CP eigenstate f, satisfying

$$(CP)|f\rangle = \pm|f\rangle. \tag{37}$$

The corresponding time-dependent CP asymmetry then takes the following form:

$$\begin{aligned} a_{\rm CP}(t) &\equiv \frac{\Gamma(B_q^0(t) \to f) - \Gamma(\overline{B_q^0}(t) \to f)}{\Gamma(B_q^0(t) \to f) + \Gamma(\overline{B_q^0}(t) \to f)} \qquad (38)\\ &= \left[\frac{\mathcal{A}_{\rm CP}^{\rm dir}(B_q \to f)\cos(\Delta M_q t) + \mathcal{A}_{\rm CP}^{\rm mix}(B_q \to f)\sin(\Delta M_q t)}{\cosh(\Delta\Gamma_q t/2) - \mathcal{A}_{\Delta\Gamma}(B_q \to f)\sinh(\Delta\Gamma_q t/2)}\right]. \end{aligned}$$

In order to calculate the CP-violating observables, it is convenient to introduce

$$\xi_f^{(q)} = \pm e^{-i\Theta_{\rm M}^{(q)}} \frac{A(\overline{B_q^0} \to \overline{f})}{A(B_q^0 \to f)}, \tag{39}$$

where $\pm$ refers to the CP eigenvalue of the final state f specified in (37), and

$$\Theta_{\rm M}^{(q)} - \pi = 2\arg(V_{tq}^* V_{tb}) \equiv \phi_q = \left\{ \begin{array}{ll} +2\beta = \mathcal{O}(50^\circ) & \text{for } q = d, \\ -2\delta\gamma = \mathcal{O}(-2^\circ) & \text{for } q = s \end{array} \right. \tag{40}$$

is the CP-violating weak B_q^0–$\overline{B_q^0}$ mixing phase. It should be noted that $\xi_f^{(q)}$ does not depend on the chosen CP or CKM phase conventions and is actually a physical observable (for a detailed discussion, see [12]). We then obtain

$$\mathcal{A}_{\rm CP}^{\rm dir}(B_q \to f) = \frac{1-\left|\xi_f^{(q)}\right|^2}{1+\left|\xi_f^{(q)}\right|^2} = \frac{|A(B\to f)|^2 - |A(\overline{B}\to \overline{f})|^2}{|A(B\to f)|^2 + |A(\overline{B}\to \overline{f})|^2}, \tag{41}$$

and conclude that this observable measures direct CP violation, which we have already encountered in (34). The interesting new aspect is "mixing-induced" CP violation, which is described by

$$\mathcal{A}_{\rm CP}^{\rm mix}(B_q \to f) = \frac{2{\rm Im}\xi_f^{(q)}}{1+\left|\xi_f^{(q)}\right|^2}, \tag{42}$$

and arises from interference effects between B_q^0–$\overline{B_q^0}$ mixing and decay processes. The width difference $\Delta\Gamma_q$, which may be sizeable in the B_s system, as we will see in Sect. 6.1, provides another observable,

$$\mathcal{A}_{\Delta\Gamma}(B_q \to f) \equiv \frac{2\,{\rm Re}\,\xi_f^{(q)}}{1+\left|\xi_f^{(q)}\right|^2}, \tag{43}$$

which is, however, not independent from $\mathcal{A}_{\rm CP}^{\rm dir}(B_q \to f)$ and $\mathcal{A}_{\rm CP}^{\rm mix}(B_q \to f)$:

$$\Big[\mathcal{A}_{\rm CP}^{\rm dir}(B_q \to f)\Big]^2 + \Big[\mathcal{A}_{\rm CP}^{\rm mix}(B_q \to f)\Big]^2 + \Big[\mathcal{A}_{\Delta\Gamma}(B_q \to f)\Big]^2 = 1. \tag{44}$$

Let us now have a closer look at $\xi_f^{(q)}$. Using (31) and (32), we obtain

$$\xi_f^{(q)} = \mp e^{-i\phi_q}\left[\frac{e^{+i\varphi_1}|A_1|e^{i\delta_1} + e^{+i\varphi_2}|A_2|e^{i\delta_2}}{e^{-i\varphi_1}|A_1|e^{i\delta_1} + e^{-i\varphi_2}|A_2|e^{i\delta_2}}\right], \tag{45}$$

and observe that the calculation of $\xi_f^{(q)}$ is in general affected by hadronic uncertainties. However, if one CKM amplitude plays the dominant rôle, the corresponding hadronic matrix element cancels:

$$\xi_f^{(q)} = \mp e^{-i\phi_q}\left[\frac{e^{+i\phi_f/2}|M_f|e^{i\delta_f}}{e^{-i\phi_f/2}|M_f|e^{i\delta_f}}\right] = \mp e^{-i(\phi_q-\phi_f)}. \tag{46}$$

In this special case, direct CP violation vanishes, i.e. $\mathcal{A}_{\rm CP}^{\rm dir}(B_q \to f) = 0$. However, we still have mixing-induced CP violation, measuring the CP-violating weak phase difference $\phi \equiv \phi_q - \phi_f$ *without* hadronic uncertainties:

$$\mathcal{A}_{\rm CP}^{\rm mix}(B_q \to f) = \pm \sin\phi. \tag{47}$$

The corresponding time-dependent CP asymmetry now takes the following simple form:

$$\left.\frac{\Gamma(B_q^0(t)\to f) - \Gamma(\overline{B_q^0}(t)\to \overline{f})}{\Gamma(B_q^0(t)\to f) + \Gamma(\overline{B_q^0}(t)\to \overline{f})}\right|_{\Delta\Gamma_q=0} = \pm\sin\phi\,\sin(\Delta M_q t), \tag{48}$$

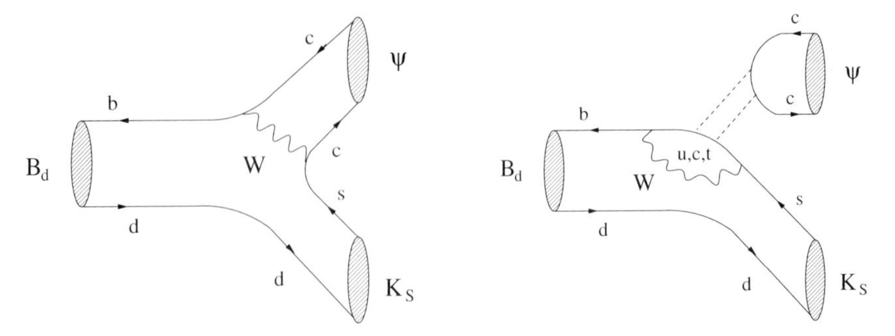

Fig. 8. Feynman diagrams contributing to $B_d^0 \to J/\psi K_S$. The dashed lines in the penguin topology represent a colour-singlet exchange.

and allows an elegant determination of $\sin\phi$. Let us apply this formalism, in the next section, to important benchmark modes for the B factories.

5 Benchmark Modes for the B Factories

5.1 The "Gold-Plated" Mode $B_d \to J/\psi K_S$

The decay $B_d^0 \to J/\psi\, K_S$ is a transition into a CP eigenstate with eigenvalue -1, and originates from $\overline{b} \to \overline{c}c\overline{s}$ quark-level decays. As can be seen in Fig. 8, we have to deal both with tree-diagram-like and with penguin topologies. The corresponding amplitude can be written as [56]

$$A(B_d^0 \to J/\psi K_S) = \lambda_c^{(s)}\left(A_{\rm CC}^{c'} + A_{\rm pen}^{c'}\right) + \lambda_u^{(s)} A_{\rm pen}^{u'} + \lambda_t^{(s)} A_{\rm pen}^{t'}\,, \tag{49}$$

where $A_{\rm CC}^{c'}$ denotes the current–current contributions, i.e. the "tree" processes in Fig. 8, and the strong amplitudes $A_{\rm pen}^{q'}$ describe the contributions from penguin topologies with internal q quarks ($q \in \{u,c,t\}$). These penguin amplitudes take into account both QCD and EW penguin contributions. The primes in (49) remind us that we are dealing with a $\overline{b} \to \overline{s}$ transition, and the

$$\lambda_q^{(s)} \equiv V_{qs}V_{qb}^* \tag{50}$$

are CKM factors. If we employ the unitarity of the CKM matrix to eliminate $\lambda_t^{(s)}$ through $\lambda_t^{(s)} = -\lambda_u^{(s)} - \lambda_c^{(s)}$, and the Wolfenstein parametrization, we may write

$$A(B_d^0 \to J/\psi K_S) \propto \left[1 + \lambda^2 a e^{i\theta} e^{i\gamma}\right], \tag{51}$$

where the hadronic parameter $ae^{i\theta}$ measures, sloppily speaking, the ratio of penguin- to tree-diagram-like contributions to $B_d^0 \to J/\psi\, K_S$. Since this parameter enters in a doubly Cabibbo-suppressed way, the formalism discussed in Sect. 4.2 gives, to a very good approximation [57]:

$$\mathcal{A}_{\rm CP}^{\rm dir}(B_d \to J/\psi K_S) = 0, \quad \mathcal{A}_{\rm CP}^{\rm mix}(B_d \to J/\psi K_S) = -\sin\phi_d. \tag{52}$$

After important first steps by the OPAL, CDF and ALEPH collaborations, the $B_d \to J/\psi K_S$ mode (and similar decays) led eventually, in 2001, to the observation of CP violation in the B system [9,10]. The present status of $\sin 2\beta$ is given as follows:

$$\sin 2\beta = \begin{cases} 0.741 \pm 0.067 \pm 0.033 \text{ (BaBar [58])} \\ 0.719 \pm 0.074 \pm 0.035 \text{ (Belle [59])}, \end{cases} \tag{53}$$

yielding the world average [60]

$$\sin 2\beta = 0.734 \pm 0.054, \tag{54}$$

which agrees well with the results of the "standard analysis" of the unitarity triangle (15), implying $0.6 \lesssim \sin 2\beta \lesssim 0.9$.

In the LHC era, the experimental accuracy of the measurement of $\sin 2\beta$ may be increased by one order of magnitude [36]. In view of such a tremendous accuracy, it will then be important to obtain deeper insights into the theoretical uncertainties affecting (52), which are due to penguin contributions. A possibility to control them is provided by the $B_s \to J/\psi K_S$ channel [56]. Moreover, also direct CP violation in $B \to J/\psi K$ modes allows us to probe such penguin effects [42,61]. So far, there are no experimental indications for non-vanishing CP asymmetries of this kind.

Although the agreement between (54) and the results of the CKM fits is striking, it should not be forgotten that new physics may nevertheless hide in $\mathcal{A}_{\rm CP}^{\rm mix}(B_d \to J/\psi K_S)$. The point is that the key quantity is actually ϕ_d, which is fixed through $\sin \phi_d = 0.734 \pm 0.054$ up to a twofold ambiguity,

$$\phi_d = \left(47^{+5}_{-4}\right)^\circ \vee \left(133^{+4}_{-5}\right)^\circ . \tag{55}$$

Here the former solution would be in perfect agreement with the range implied by the CKM fits, $40^\circ \lesssim \phi_d \lesssim 60^\circ$, whereas the latter would correspond to new physics. The two solutions can be distinguished through a measurement of the sign of $\cos \phi_d$: in the case of $\cos \phi_d = +0.7 > 0$, we would conclude $\phi_d = 47^\circ$, whereas $\cos \phi_d = -0.7 < 0$ would point towards $\phi_d = 133^\circ$, i.e. new physics. There are several strategies on the market to resolve the twofold ambiguity in the extraction of ϕ_d [62]. Unfortunately, they are rather challenging from a practical point of view. In the $B \to J/\psi K$ system, $\cos \phi_d$ can be extracted from the time-dependent angular distribution of the decay products of $B_d \to J/\psi[\to \ell^+\ell^-]K^*[\to \pi^0 K_S]$, if the sign of a hadronic parameter $\cos \delta$ involving a strong phase δ is fixed through factorization [63,64]. Let us note that analyses of this kind are already in progress at the B factories [65].

The preferred mechanism for new physics to manifest itself in CP-violating effects in $B_d \to J/\psi K_S$ is through B^0_d–$\overline{B^0_d}$ mixing, which arises in the Standard Model from the box diagrams shown in Fig. 7. However, new physics may also enter at the $B \to J/\psi K$ amplitude level. Employing estimates borrowed from effective field theory suggests that the effects are at most $\mathcal{O}(10\%)$ for a generic new-physics scale $\Lambda_{\rm NP}$ in the TeV regime. In order to obtain the whole picture, a set of appropriate observables can be introduced, using $B_d \to J/\psi K_S$ and its

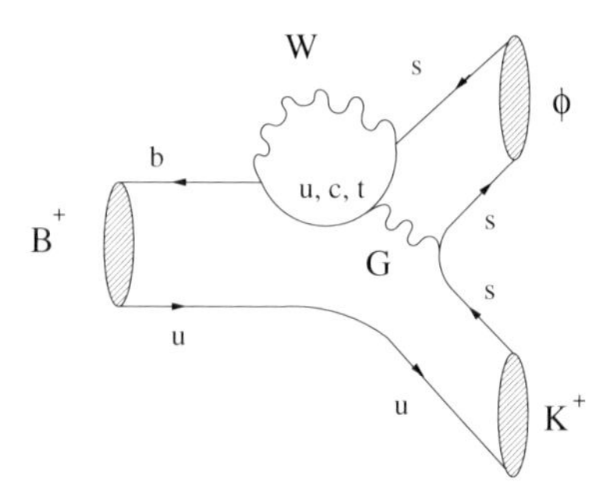

Fig. 9. QCD penguin contributions to $B^+ \to \phi K^+$.

charged counterpart $B^\pm \to J/\psi K^\pm$ [61]. So far, these observables do not yet indicate any deviation from the Standard Model.

In the context of new-physics effects in the $B \to J/\psi K$ system, it is interesting to note that an upper bound on ϕ_d is implied by an upper bound on $R_b \propto |V_{ub}/V_{cb}|$, as can be seen in Fig. 2. To be specific, we have

$$\sin \beta_{\max} = R_b^{\max}, \tag{56}$$

yielding $(\phi_d)_{\max}^{\rm SM} \sim 55°$ for $R_b^{\max} \sim 0.46$. As the determination of R_b from semi-leptonic tree-level decays is very robust concerning the impact of new physics, $\phi_d \sim 133°$ would require new-physics contributions to B_d^0–$\overline{B_d^0}$ mixing. As we will see in Sect. 7.2, an interesting connection between the two solutions for ϕ_d and constraints on γ is provided by CP violation in $B_d \to \pi^+\pi^-$ [66].

5.2 The $B \to \phi K$ System

An important testing ground for the Standard-Model description of CP violation is also provided by $B \to \phi K$ decays. As can be seen in Fig. 9, these modes are governed by QCD penguin processes [67], but also EW penguins are sizeable [41,68]. Consequently, $B \to \phi K$ modes represent a sensitive probe for new physics. In the Standard Model, we have the following relations [42,69–71]:

$$\mathcal{A}_{\rm CP}^{\rm dir}(B_d \to \phi K_{\rm S}) \ = 0 + \mathcal{O}(\lambda^2) \tag{57}$$

$$\mathcal{A}_{\rm CP}^{\rm mix}(B_d \to \phi K_{\rm S}) \ = \mathcal{A}_{\rm CP}^{\rm mix}(B_d \to J/\psi K_{\rm S}) + \mathcal{O}(\lambda^2). \tag{58}$$

As in the case of the $B \to J/\psi K$ system, a combined analysis of $B_d \to \phi K_{\rm S}$, $B^\pm \to \phi K^\pm$ modes should be performed in order to obtain the whole picture [71]. There is also the possibility of an unfortunate case, where new physics cannot be distinguished from the Standard Model, as discussed in detail in [12,71].

In the summer of 2002, the experimental status can be summarized as follows:

$$\mathcal{A}_{\rm CP}^{\rm dir}(B_d \to \phi K_{\rm S}) = \begin{cases} \text{n.a.} & \text{(BaBar [72])} \\ 0.56 \pm 0.41 \pm 0.12 & \text{(Belle [73])} \end{cases} \tag{59}$$

$$\mathcal{A}_{\rm CP}^{\rm mix}(B_d \to \phi K_{\rm S}) = \begin{cases} 0.19^{+0.50}_{-0.52} \pm 0.09 & \text{(BaBar [72])} \\ 0.73 \pm 0.64 \pm 0.18 & \text{(Belle [73]).} \end{cases} \tag{60}$$

Unfortunately, the experimental uncertainties are still very large. Because of $\mathcal{A}_{\rm CP}^{\rm mix}(B_d \to J/\psi K_{\rm S}) = -0.734 \pm 0.054$ (see (52) and (54)), there were already speculations about new-physics effects in $B_d \to \phi K_{\rm S}$ [74]. In this context, it is interesting to note that there are more data available from Belle:

$$\mathcal{A}_{\rm CP}^{\rm dir}(B_d \to \eta' K_{\rm S}) = -0.26 \pm 0.22 \pm 0.03 \tag{61}$$

$$\mathcal{A}_{\rm CP}^{\rm mix}(B_d \to \eta' K_{\rm S}) = -0.76 \pm 0.36^{+0.06}_{-0.05} \tag{62}$$

$$\mathcal{A}_{\rm CP}^{\rm dir}(B_d \to K^+K^-K_{\rm S}) = 0.42 \pm 0.36 \pm 0.09^{+0.22}_{-0.03} \tag{63}$$

$$\mathcal{A}_{\rm CP}^{\rm mix}(B_d \to K^+K^-K_{\rm S}) = -0.52 \pm 0.46 \pm 0.11^{+0.03}_{-0.27}. \tag{64}$$

The corresponding modes are governed by the same quark-level transitions as $B_d \to \phi K_{\rm S}$. Consequently, it is probably too early to be excited too much by the possibility of signals of new physics in $B_d \to \phi K_{\rm S}$ [60]. However, the experimental situation should improve significantly in the future.

5.3 The Decay $B_d \to \pi^+\pi^-$

Another benchmark mode for the B factories is $B_d^0 \to \pi^+\pi^-$, which is a decay into a CP eigenstate with eigenvalue $+1$, and originates from $\overline{b} \to \overline{u}u\overline{d}$ quark-level transitions, as can be seen in Fig. 10. In analogy to (49), the corresponding decay amplitude can be written in the following form [75]:

$$A(B_d^0 \to \pi^+\pi^-) = \lambda_u^{(d)}\left(A_{\rm CC}^u + A_{\rm pen}^u\right) + \lambda_c^{(d)} A_{\rm pen}^c + \lambda_t^{(d)} A_{\rm pen}^t. \tag{65}$$

If we use again the unitarity of the CKM matrix, yielding $\lambda_t^{(d)} = -\lambda_u^{(d)} - \lambda_c^{(d)}$, as well as the Wolfenstein parametrization, we obtain

$$A(B_d^0 \to \pi^+\pi^-) \propto \left[e^{i\gamma} - de^{i\theta}\right], \tag{66}$$

where

$$de^{i\theta} \equiv \frac{1}{R_b}\left(\frac{A_{\rm pen}^c - A_{\rm pen}^t}{A_{\rm CC}^u + A_{\rm pen}^u - A_{\rm pen}^t}\right) \tag{67}$$

measures, sloppily speaking, the ratio of penguin to tree contributions in $B_d \to \pi^+\pi^-$. In contrast to the $B_d^0 \to J/\psi K_{\rm S}$ amplitude (51), this parameter does *not* enter in (66) in a doubly Cabibbo-suppressed way, thereby leading to the well-known "penguin problem" in $B_d \to \pi^+\pi^-$. If we had negligible penguin contributions, i.e. $d = 0$, the corresponding CP-violating observables were given as follows:

$$\mathcal{A}_{\rm CP}^{\rm dir}(B_d \to \pi^+\pi^-) = 0, \quad \mathcal{A}_{\rm CP}^{\rm mix}(B_d \to \pi^+\pi^-) = \sin(2\beta + 2\gamma) = -\sin 2\alpha, \tag{68}$$

where we have also used the unitarity relation $2\beta + 2\gamma = 2\pi - 2\alpha$. We observe that actually the phases $2\beta = \phi_d$ and γ enter directly in the $B_d \to \pi^+\pi^-$

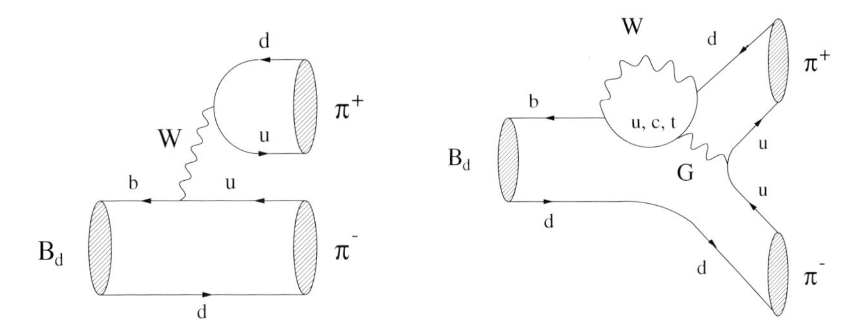

Fig. 10. Feynman diagrams contributing to $B_d^0 \to \pi^+\pi^-$.

observables, and not α. Consequently, since ϕ_d can be fixed straightforwardly through $B_d \to J/\psi K_S$, we may use $B_d \to \pi^+\pi^-$ to probe γ. This is advantageous to deal with penguins and possible new-physics effects, as we will see in Sect. 7.2.

Measurements of the $B_d \to \pi^+\pi^-$ CP asymmetries are already available:

$$\mathcal{A}_{\rm CP}^{\rm dir}(B_d \to \pi^+\pi^-) = \begin{cases} -0.30 \pm 0.25 \pm 0.04 & \text{(BaBar [76])} \\ -0.94^{+0.31}_{-0.25} \pm 0.09 & \text{(Belle [77])} \end{cases} \tag{69}$$

$$\mathcal{A}_{\rm CP}^{\rm mix}(B_d \to \pi^+\pi^-) = \begin{cases} -0.02 \pm 0.34 \pm 0.05 & \text{(BaBar [76])} \\ 1.21^{+0.27+0.13}_{-0.38-0.16} & \text{(Belle [77]).} \end{cases} \tag{70}$$

Unfortunately, the BaBar and Belle results are not fully consistent with each other; the experimental picture will hopefully be clarified soon. Forming nevertheless the weighted averages of (69) and (70), using the rules of the Particle Data Group (PDG), yields

$$\mathcal{A}_{\rm CP}^{\rm dir}(B_d \to \pi^+\pi^-) = -0.57 \pm 0.19\ (0.32) \tag{71}$$

$$\mathcal{A}_{\rm CP}^{\rm mix}(B_d \to \pi^+\pi^-) = 0.57 \pm 0.25\ (0.61), \tag{72}$$

where the errors in brackets are the ones increased by the PDG scaling-factor procedure [78]. Direct CP violation at this level would require large penguin contributions with large CP-conserving strong phases. A significant impact of penguins on $B_d \to \pi^+\pi^-$ is also indicated by data on $B \to \pi K, \pi\pi$ decays, as well as by theoretical considerations [40,48,79,80] (see Sect. 7.2). Consequently, it is already evident that the penguin contributions to $B_d \to \pi^+\pi^-$ *cannot* be neglected.

Many approaches to deal with the penguin problem in the extraction of weak phases from the CP-violating $B_d \to \pi^+\pi^-$ observables were developed; the best known is an isospin analysis of the $B \to \pi\pi$ system [81], yielding α. Unfortunately, this approach is very difficult in practice, as it requires a measurement of the $B_d^0 \to \pi^0\pi^0$ and $\overline{B_d^0} \to \pi^0\pi^0$ branching ratios. However, useful bounds may already be obtained from experimental constraints on the CP-averaged $B_d \to \pi^0\pi^0$ branching ratio [82,83]. Alternatively, we may employ the CKM unitarity to express $\mathcal{A}_{\rm CP}^{\rm mix}(B_d \to \pi^+\pi^-)$ in terms of α and hadronic parameters. Using $\mathcal{A}_{\rm CP}^{\rm dir}(B_d \to \pi^+\pi^-)$, a strong phase can be eliminated, allowing us to determine α as a function of a hadronic parameter $|p/t|$, which is, however, problematic to be determined reliably [40,46,48,83–86]. A different parametrization of the $B_d \to \pi^+\pi^-$ observables, involving a hadronic parameter P/T and $\phi_d = 2\beta$, is employed in [87], where, moreover, $\alpha + \beta + \gamma = 180°$ is used to eliminate γ, and β is fixed through the Standard-Model solution $\sim 26°$ implied by $\mathcal{A}_{\rm CP}^{\rm mix}(B_d \to J/\psi K_{\rm S})$. Provided $|P/T|$ is known, α can be extracted. To this end, $SU(3)$ flavour-symmetry arguments and plausible dynamical assumptions are used to fix $|P|$ through the CP-averaged $B^\pm \to \pi^\pm K$ branching ratio. On the other hand, $|T|$ is estimated with the help of factorization and data on $B \to \pi\ell\nu$. Refinements of this approach were presented in [88]. Another strategy to deal with penguins in $B_d \to \pi^+\pi^-$ is offered by $B_s \to K^+K^-$. Using the U-spin flavour symmetry of strong interactions, ϕ_d and γ can be extracted from the corresponding CP-violating observables [75]. Before coming back to this approach in more detail in Sect. 7.2, let us first have a closer look at the B_s-meson system.

6 The B_s-Meson System

6.1 General Features

At the e^+e^- B factories operating at the $\Upsilon(4S)$ resonance, no B_s mesons are accessible, since $\Upsilon(4S)$ states decay only to $B_{u,d}$-mesons, but not to B_s. On the other hand, the physics potential of the B_s system is very promising for hadron machines, where plenty of B_s mesons are produced. Consequently, B_s physics is in some sense the "El Dorado" for B experiments at hadron colliders. There are important differences between the B_d and B_s systems:

- Within the Standard Model, the B_s^0–$\overline{B_s^0}$ mixing phase probes the tiny angle $\delta\gamma$ in the unitarity triangle shown in Fig. 1 (b), and is hence negligibly small:
$$\phi_s = -2\delta\gamma = -2\lambda^2\eta = \mathcal{O}(-2°), \tag{73}$$
whereas $\phi_d = 2\beta = \mathcal{O}(50°)$.
- A large $x_s \equiv \Delta M_s/\Gamma_s = \mathcal{O}(20)$, where $\Gamma_s \equiv (\Gamma_{\rm H}^{(q)} + \Gamma_{\rm L}^{(q)})/2$, is expected in the Standard Model, whereas $x_d = 0.775 \pm 0.012$. The present lower bound on ΔM_s is given as follows [89]:
$$\Delta M_s > 14.4\,{\rm ps}^{-1}\ (95\%\ {\rm C.L.}). \tag{74}$$
- There may be a sizeable width difference $\Delta\Gamma_s/\Gamma_s = \mathcal{O}(-10\%)$ between the mass eigenstates of the B_s system that is due to CKM-favoured $b \to c\bar{c}s$ quark-level transitions into final states common to $\overline{B_s^0}$ and B_s^0, whereas $\Delta\Gamma_d$ is negligibly small [90]. The present CDF and LEP results imply [89]
$$|\Delta\Gamma_s|/\Gamma_s < 0.31\ (95\%\ {\rm C.L.}). \tag{75}$$

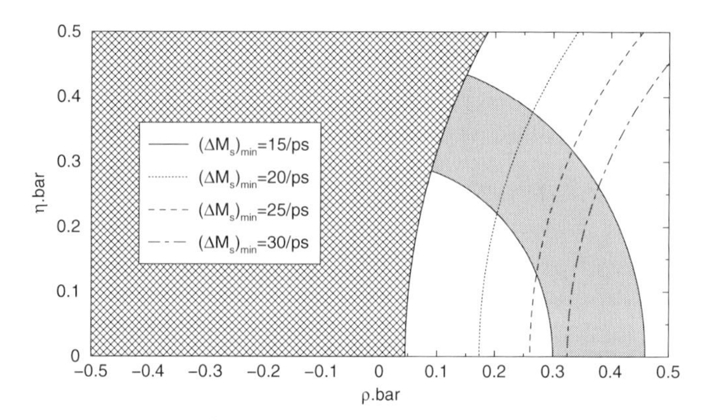

Fig. 11. The impact of the upper limit $(R_t)_{\rm max}$ on the allowed range in the $\overline{\rho}$–$\overline{\eta}$ plane for $\xi = 1.15$. The shaded region corresponds to $R_b = 0.38 \pm 0.08$.

Interesting applications of $\Delta\Gamma_s$ are extractions of weak phases from "untagged" B_s data samples, where we do not distinguish between initially present B_s^0 or $\overline{B_s^0}$ mesons, as discussed in [91].

Let us now discuss the rôle of ΔM_s for the determination of the unitarity triangle in more detail. As we have already noted in Sect. 2.4, the comparison of ΔM_d with ΔM_s allows an interesting determination of the side R_t of the unitarity triangle. To this end, only a single $SU(3)$-breaking parameter

$$\xi \equiv \frac{\sqrt{\hat{B}_{B_s}} f_{B_s}}{\sqrt{\hat{B}_{B_d}} f_{B_d}} = 1.15 \pm 0.06 \tag{76}$$

is required, which measures $SU(3)$-breaking effects in non-perturbative mixing and decay parameters. It can be determined through lattice or QCD sum-rule calculations. The mass difference ΔM_s has not yet been measured. However, lower bounds on ΔM_s can be converted into upper bounds on R_t through [92]

$$(R_t)_{\rm max} = 0.83 \times \xi \times \sqrt{\frac{15.0\,{\rm ps}^{-1}}{(\Delta M_s)_{\rm min}}}, \tag{77}$$

excluding already a large part in the $\overline{\rho}$–$\overline{\eta}$ plane, as can be seen in Fig. 11. In particular, $\gamma < 90^\circ$ is implied. In a recent paper [93], it is argued that ξ may actually be significantly larger than the conventional range given in (76), $\xi = 1.32 \pm 0.10$ (see also [27]). In this case, the excluded range in the $\overline{\rho}$–$\overline{\eta}$ plane would be reduced, shifting the upper limit for γ closer to 90°. Hopefully, the status of ξ will be clarified soon. In the near future, run II of the Tevatron should provide a measurement of ΔM_s, thereby constraining the unitarity triangle and γ in a much more stringent way.

6.2 Benchmark B_s Decays

An interesting class of B_s decays is due to $b(\bar{b}) \to c\bar{u}s(\bar{s})$ quark-level transitions. Here we have to deal with pure "tree" decays, where both B_s^0 and $\overline{B_s^0}$ mesons may decay into the same final state f. The resulting interference effects between decay and mixing processes allow a *theoretically clean* extraction of $\phi_s + \gamma$ from

$$\xi_f^{(s)} \times \xi_{\overline{f}}^{(s)} = e^{-2i(\phi_s+\gamma)}. \tag{78}$$

There are several well-known strategies on the market employing these features: we may consider the colour-allowed decays $B_s \to D_s^{\pm} K^{\mp}$ [94], or the colour-suppressed modes $B_s \to D^0 \phi$ [95]. In the case of $B_s \to D_s^{*\pm} K^{*\mp}$ or $B_s \to D^{*0}\phi$, the observables of the corresponding angular distributions provide sufficient information to extract $\phi_s + \gamma$ from "untagged" analyses [96], requiring a sizeable $\Delta\Gamma_s$. A "tagged" strategy involving $B_s \to D_s^{*\pm} K^{*\mp}$ modes was proposed in [97]. Recently, strategies making use of "CP-tagged" B_s decays were proposed [98], which require a symmetric e^+e^- collider operated at the $\Upsilon(5S)$ resonance. In this approach, initially present CP eigenstates B_s^{CP} are employed, which can be tagged through the fact that the $B_s^0/\overline{B_s^0}$ mixtures have anticorrelated CP eigenvalues at $\Upsilon(5S)$. Here $B_s \to D_s^{\pm} K^{\mp}, D_s^{\pm} K^{*\mp}, D_s^{*\pm} K^{*\mp}$ modes may be used. Let us note that there is also an interesting counterpart of (78) in the B_d system [99], which employs $B_d \to D^{(*)\pm}\pi^{\mp}$ decays, and allows a determination of $\phi_d + \gamma$.

The extraction of γ from the phase $\phi_s + \gamma$ provided by the B_s approaches sketched in the previous paragraph requires ϕ_s as an additional input, which is negligibly small in the Standard Model. Whereas it appears to be quite unlikely that the pure tree decays listed above are affected significantly by new physics, as they involve no flavour-changing neutral-current processes, this is not the case for the B_s^0–$\overline{B_s^0}$ mixing phase ϕ_s. In order to probe this quantity, $B_s \to J/\psi\,\phi$ offers interesting strategies [64,100]. Since this decay is the B_s counterpart of $B_d \to J/\psi K_{\mathrm{S}}$, the corresponding Feynman diagrams are analogous to those shown in Fig. 8. However, in contrast to $B_d \to J/\psi K_{\mathrm{S}}$, the final state of $B_s \to J/\psi\phi$ is an admixture of different CP eigenstates. In order to disentangle them, we have to use the angular distribution of the $J/\psi \to \ell^+\ell^-$ and $\phi \to K^+K^-$ decay products [101]. The corresponding observables are governed by [36]

$$\xi_{\psi\phi}^{(s)} \propto e^{-i\phi_s}\left[1 - 2\,i\,\sin\gamma \times \mathcal{O}(10^{-3})\right]. \tag{79}$$

Since we have $\phi_s = \mathcal{O}(-2^\circ)$ in the Standard Model, the extraction of ϕ_s from the $B_s \to J/\psi[\to \ell^+\ell^-]\phi[\to K^+K^-]$ angular distribution may well be affected by hadronic uncertainties at the 10% level. These hadronic uncertainties, which may become an important issue in the LHC era [36], can be controlled through $B_d \to J/\psi\,\rho^0$, exhibiting some other interesting features [102]. Since $B_s \to J/\psi\phi$ shows small CP-violating effects in the Standard Model because of (79), this mode represents a sensitive probe to search for new-physics contributions to B_s^0–$\overline{B_s^0}$ mixing [103]. Note that new-physics effects entering at the $B_s \to J/\psi\phi$ amplitude level are expected to play a minor rôle and can already be probed in the $B \to J/\psi K$ system [61]. For a detailed discussion of "smoking-gun" signals

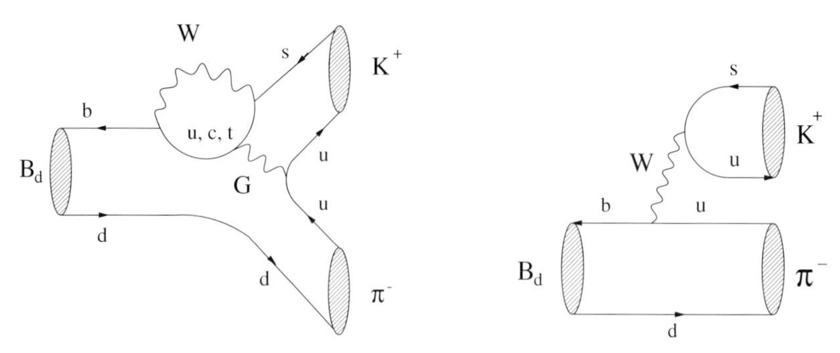

Fig. 12. Feynman diagrams contributing to $B_d^0 \to \pi^- K^+$.

of a sizeable value of ϕ_s, see [64]. There, also methods to determine this phase *unambiguously* are proposed.

7 Recent Developments

7.1 Status of $B \to \pi K$ Decays

If we employ flavour-symmetry arguments and make plausible dynamical assumptions, $B \to \pi K$ decays allow determinations of γ and hadronic parameters with a "minimal" theoretical input [104]–[116]. Alternative strategies, relying on a more extensive use of theory, are provided by the QCD factorization [46–48] and PQCD [50,79] approaches, which furthermore allow a reduction of the theoretical uncertainties of the flavour-symmetry strategies. These topics are discussed in detail in [14]. Let us here focus on the former kind of strategies.

To get more familiar with $B \to \pi K$ modes, let us consider $B_d^0 \to \pi^- K^+$. As can be seen in Fig. 12, this channel receives contributions from penguin and colour-allowed tree-diagram-like topologies, where the latter bring γ into the game. Because of the small ratio $|V_{us}V_{ub}^*/(V_{ts}V_{tb}^*)| \approx 0.02$, the QCD penguin topologies dominate this decay, despite their loop suppression. This interesting feature applies to all $B \to \pi K$ modes. Because of the large top-quark mass, we also have to care about EW penguins. However, in the case of $B_d^0 \to \pi^- K^+$ and $B^+ \to \pi^+ K^0$, these topologies contribute only in colour-suppressed form and are hence expected to play a minor rôle. On the other hand, EW penguins contribute also in colour-allowed form to $B^+ \to \pi^0 K^+$ and $B_d^0 \to \pi^0 K^0$, and may here even compete with tree-diagram-like topologies. Because of the dominance of penguin topologies, $B \to \pi K$ modes are sensitive probes for new-physics effects [117].

Relations between the $B \to \pi K$ amplitudes that are implied by the $SU(2)$ isospin flavour symmetry of strong interactions suggest the following combinations to probe γ: the "mixed" $B^\pm \to \pi^\pm K$, $B_d \to \pi^\mp K^\pm$ system [105]–[108], the "charged" $B^\pm \to \pi^\pm K$, $B^\pm \to \pi^0 K^\pm$ system [109]–[111], and the "neutral" $B_d \to \pi^0 K$, $B_d \to \pi^\mp K^\pm$ system [111,112]. Correspondingly, we may introduce

Table 1. CP-conserving $B \to \pi K$ observables as defined in (80)–(82). For the evaluation of R, we have used $\tau_{B^+}/\tau_{B^0_d} = 1.060 \pm 0.029$.

Observable	CLEO [118]	BaBar [76,119]	Belle [120]	Average
R	1.00 ± 0.30	1.08 ± 0.15	1.22 ± 0.26	1.10 ± 0.12
$R_{\rm c}$	1.27 ± 0.47	1.46 ± 0.25	1.34 ± 0.37	1.40 ± 0.19
$R_{\rm n}$	0.59 ± 0.27	0.86 ± 0.15	1.41 ± 0.65	0.82 ± 0.13

Table 2. CP-violating $B \to \pi K$ observables as defined in (80)–(82). For the evaluation of A_0, we have used $\tau_{B^+}/\tau_{B^0_d} = 1.060 \pm 0.029$.

Observable	CLEO [121]	BaBar [76,119]	Belle [120]	Average
A_0	0.04 ± 0.16	0.11 ± 0.06	0.07 ± 0.11	0.09 ± 0.05
$A_0^{\rm c}$	0.37 ± 0.32	0.13 ± 0.13	0.03 ± 0.26	0.14 ± 0.11
$A_0^{\rm n}$	0.02 ± 0.10	0.09 ± 0.05	0.08 ± 0.13	0.08 ± 0.04

the following sets of observables [111]:

$$\left\{ \begin{array}{c} R \\ A_0 \end{array} \right\} \equiv \left[\frac{\mathrm{BR}(B^0_d \to \pi^- K^+) \pm \mathrm{BR}(\overline{B^0_d} \to \pi^+ K^-)}{\mathrm{BR}(B^+ \to \pi^+ K^0) + \mathrm{BR}(B^- \to \pi^- \overline{K^0})} \right] \frac{\tau_{B^+}}{\tau_{B^0_d}} \tag{80}$$

$$\left\{ \begin{array}{c} R_{\rm c} \\ A_0^{\rm c} \end{array} \right\} \equiv 2 \left[\frac{\mathrm{BR}(B^+ \to \pi^0 K^+) \pm \mathrm{BR}(B^- \to \pi^0 K^-)}{\mathrm{BR}(B^+ \to \pi^+ K^0) + \mathrm{BR}(B^- \to \pi^- \overline{K^0})} \right] \tag{81}$$

$$\left\{ \begin{array}{c} R_{\rm n} \\ A_0^{\rm n} \end{array} \right\} \equiv \frac{1}{2} \left[\frac{\mathrm{BR}(B^0_d \to \pi^- K^+) \pm \mathrm{BR}(\overline{B^0_d} \to \pi^+ K^-)}{\mathrm{BR}(B^0_d \to \pi^0 K^0) + \mathrm{BR}(\overline{B^0_d} \to \pi^0 \overline{K^0})} \right]. \tag{82}$$

The experimental status of these observables is summarized in Tables 1 and 2. Moreover, there are stringent constraints on CP violation in $B^\pm \to \pi^\pm K$:

$$\mathcal{A}_{\rm CP}(B^\pm \to \pi^\pm K) = \left\{ \begin{array}{ll} 0.17 \pm 0.10 \pm 0.02 & \text{(BaBar [119])} \\ -0.46 \pm 0.15 \pm 0.02 & \text{(Belle [120])}. \end{array} \right. \tag{83}$$

Let us note that a very recent preliminary study of Belle indicates that the large asymmetry in (83) is due to a 3σ fluctuation [122]. Within the Standard Model, a sizeable value of $\mathcal{A}_{\rm CP}(B^\pm \to \pi^\pm K)$ could be induced by large rescattering effects. Other important indicators for such processes are branching ratios for $B \to KK$ decays, which are already strongly constrained by the B factories, and would allow us to take into account rescattering effects in the extraction of γ from $B \to \pi K$ modes [108,110,111,123]. Let us note that also the QCD factorization approach [14,46–48] is not in favour of large rescattering processes. For simplicity, we shall neglect such effects in the discussion given below. Interestingly, already CP-averaged $B \to \pi K$ branching ratios may lead to non-trivial constraints on γ [106,109,111], provided the corresponding $R_{\rm (c,n)}$ observables are found to be sufficiently different from 1. The final goal is, however, to determine γ.

Let us first turn to the charged and neutral $B \to \pi K$ systems in more detail. The starting point of our considerations are relations between the charged and

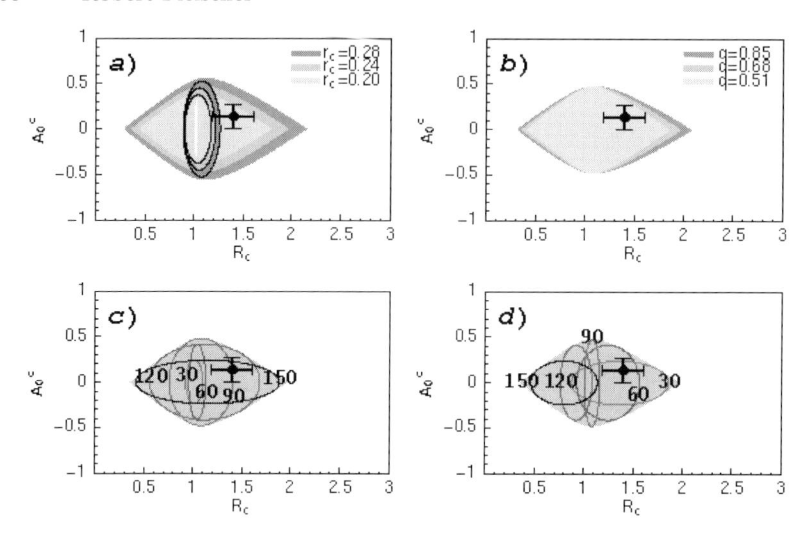

Fig. 13. The allowed regions in the R_c–A_0^c plane: (a) corresponds to $0.20 \leq r_c \leq 0.28$ for $q = 0.68$, and (b) to $0.51 \leq q \leq 0.85$ for $r_c = 0.24$; the elliptical regions arise if we restrict γ to the Standard-Model range specified in (86). In (c) and (d), we show also the contours for fixed values of γ and $|\delta_c|$, respectively ($r_c = 0.24$, $q = 0.68$).

neutral $B \to \pi K$ amplitudes that follow from the $SU(2)$ isospin symmetry of strong interactions. Assuming moreover that the rescattering effects discussed above are small, we arrive at a parametrization of the following structure [111] (for an alternative one, see [110]):

$$R_{\rm c,n} = 1 - 2r_{\rm c,n}\,(\cos\gamma - q)\cos\delta_{\rm c,n} + \left(1 - 2q\cos\gamma + q^2\right) r_{\rm c,n}^2 \quad (84)$$

$$A_0^{\rm c,n} = 2r_{\rm c,n}\sin\delta_{\rm c,n}\sin\gamma . \quad (85)$$

Here $r_{\rm c,n}$ measures – simply speaking – the ratio of tree to penguin topologies. Using $SU(3)$ flavour-symmetry arguments and data on the CP-averaged $B^\pm \to \pi^\pm\pi^0$ branching ratio [104], we obtain $r_{\rm c,n} \sim 0.2$. The parameter q describes the ratio of EW penguin to tree contributions, and can be fixed through $SU(3)$ flavour-symmetry arguments, yielding $q \sim 0.7$ [109]. In order to simplify (84) and (85), we have assumed that q is a real parameter, as is the case in the strict $SU(3)$ limit; for generalizations, see [111]. Finally, $\delta_{\rm c,n}$ is the CP-conserving strong phase between trees and penguins. Consequently, the observables $R_{\rm c,n}$ and $A_0^{\rm c,n}$ depend on the two "unknowns" $\delta_{\rm c,n}$ and γ. If we vary them within their allowed ranges, i.e. $-180^\circ \leq \delta_{\rm c,n} \leq +180^\circ$ and $0^\circ \leq \gamma \leq 180^\circ$, we obtain an allowed region in the $R_{\rm c,n}$–$A_0^{\rm c,n}$ plane [66,113]. Should the measured values of $R_{\rm c,n}$ and $A_0^{\rm c,n}$ lie outside this region, we would have an immediate signal for new physics. On the other hand, should the measurements fall into the allowed range, γ and $\delta_{\rm c,n}$ could be extracted. In this case, γ could be compared with the results of alternative strategies and with the values implied by the "standard analysis" of the unitarity triangle discussed in Sect. 2.4, whereas $\delta_{\rm c,n}$ provides valuable insights into hadron dynamics, thereby allowing tests of theoretical predictions.

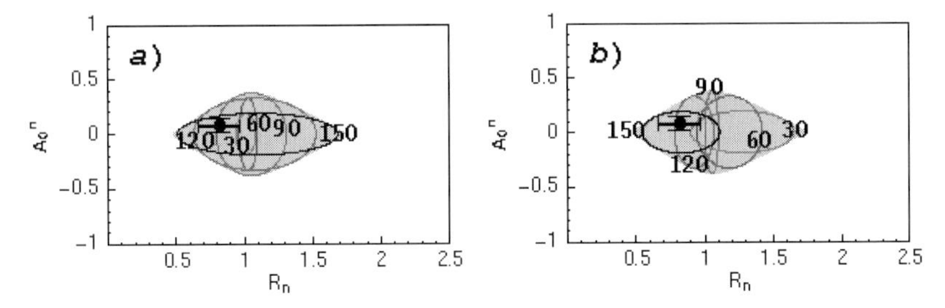

Fig. 14. The allowed regions in the R_n–A_0^n plane for $q = 0.68$ and $r_n = 0.19$. In (a) and (b), we show also the contours for fixed values of γ and $|\delta_n|$, respectively.

In Fig. 13, we show the allowed regions in the R_c–A_0^c plane for various parameter sets [66]. The crosses represent the averages of the experimental results given in Tables 1 and 2. If γ is constrained to the range implied by the "standard analysis" of the unitarity triangle,

$$50^\circ \lesssim \gamma \lesssim 70^\circ, \tag{86}$$

a much more restricted region arises in the R_c–A_0^c plane. The contours in Figs. 13c and d allow us to read off easily the preferred values for γ and δ_c, respectively, from the measured observables [66]. Interestingly, the present data seem to favour $\gamma \gtrsim 90^\circ$, which would be in conflict with (86). Moreover, they point towards $|\delta_c| \lesssim 90^\circ$; factorization predicts δ_c to be close to 0° [48]. The situation for the neutral $B \to \pi K$ system is illustrated in Fig. 14. Interestingly, here the data point to $\gamma \gtrsim 90^\circ$ as well, but favour also $|\delta_n| \gtrsim 90^\circ$ because of the average of R_n being smaller than 1 [66,112]. However, as can be seen in Table 1, the present data are unfortunately rather unsatisfactory in this respect.

If future, more accurate data really yield a value for γ in the second quadrant, the discrepancy with (86) may be due to new-physics contributions to B_q^0–$\overline{B_q^0}$ mixing ($q \in \{d, s\}$), or to the $B \to \pi K$ decay amplitudes. In the former case, the constraints implied by (77), which rely on the Standard-Model interpretation of B_q^0–$\overline{B_q^0}$ mixing, would no longer hold, so that γ may actually be larger than 90°. In the latter case, the Standard-Model expressions (84) and (85) would receive corrections due to the presence of new physics, so that also the extracted value for γ would not correspond to the Standard-Model result. In such a scenario – an example would be given by new-physics contributions to the EW penguin sector – also the extracted values for δ_c and δ_n may actually no longer satisfy $\delta_c \approx \delta_n$ [112].

An analysis similar to the one discussed above can also be performed for the mixed $B \to \pi K$ system, consisting of $B^\pm \to \pi^\pm K$, $B_d \to \pi^\mp K^\pm$ modes. To this end, only straightforward replacements of variables have to be made. The present data fall well into the Standard-Model region in observable space, but do not yet allow us to draw further definite conclusions [66]. At present, the situation in the charged and neutral $B \to \pi K$ systems appears to be more exciting.

There are also many other recent analyses of $B \to \pi K$ modes. For example, a study complementary to the one in $B \to \pi K$ observable space was performed in [115], where the allowed regions in the γ–$\delta_{\rm c,n}$ planes implied by $B \to \pi K$ data were explored. Another recent $B \to \pi K$ analysis can be found in [116], where the $R_{(\rm c)}$ were calculated for given values of $A_0^{(\rm c)}$ as functions of γ, and were compared with the B-factory data. Making more extensive use of theory than in the flavour-symmetry strategies discussed above, several different avenues to extract γ from $B \to \pi K$ modes are provided by the QCD factorization approach [14,48], which allows also a reduction of the theoretical uncertainties of the flavour-symmetry approaches discussed above, in particular a better control of $SU(3)$-breaking effects. In order to analyse $B \to \pi K$ data, also sum rules relating CP-averaged branching ratios and CP asymmetries of $B \to \pi K$ modes may be useful [114].

7.2 The $B_d \to \pi^+\pi^-$, $B_s \to K^+K^-$ System

As can be seen from Fig. 10, $B_d \to \pi^+\pi^-$ is related to $B_s \to K^+K^-$ through an interchange of all down and strange quarks. The corresponding decay amplitudes can be expressed as follows [75]:

$$A(B_d^0 \to \pi^+\pi^-) = \mathcal{C}\left[e^{i\gamma} - de^{i\theta}\right] \tag{87}$$

$$A(B_s^0 \to K^+K^-) = \left(\frac{\lambda}{1-\lambda^2/2}\right)\mathcal{C}'\left[e^{i\gamma} + \left(\frac{1-\lambda^2}{\lambda^2}\right)d'e^{i\theta'}\right], \tag{88}$$

where $de^{i\theta}$ was already introduced in (67), $d'e^{i\theta'}$ is its $B_s \to K^+K^-$ counterpart, and $\mathcal{C}$, $\mathcal{C}'$ are CP-conserving strong amplitudes. Using these general parametrizations, we obtain

$$\mathcal{A}_{\rm CP}^{\rm dir}(B_d \to \pi^+\pi^-) = \text{fct}(d, \theta, \gamma), \quad \mathcal{A}_{\rm CP}^{\rm mix}(B_d \to \pi^+\pi^-) = \text{fct}(d, \theta, \gamma, \phi_d) \tag{89}$$

$$\mathcal{A}_{\rm CP}^{\rm dir}(B_s \to K^+K^-) = \text{fct}(d', \theta', \gamma), \quad \mathcal{A}_{\rm CP}^{\rm mix}(B_s \to K^+K^-) = \text{fct}(d', \theta', \gamma, \phi_s), \tag{90}$$

where ϕ_s is negligibly small in the Standard Model, or can be fixed through $B_s \to J/\psi\phi$. We have hence four observables at our disposal, depending on six "unknowns". However, since $B_d \to \pi^+\pi^-$ and $B_s \to K^+K^-$ are related to each other by interchanging all down and strange quarks, the U-spin flavour symmetry of strong interactions implies

$$d'e^{i\theta'} = d\,e^{i\theta}. \tag{91}$$

Using this relation, the four observables in (89) and (90) depend on the four quantities d, θ, ϕ_d and γ, which can hence be determined [75]. The theoretical accuracy is only limited by the U-spin symmetry, as no dynamical assumptions about rescattering processes have to be made. Theoretical considerations give us confidence in (91), since this relation does not receive U-spin-breaking corrections within the factorization approach [75]. Moreover, we may also obtain experimental insights into U-spin breaking [75,124]. The U-spin arguments can be minimized, if the B_d^0–$\overline{B_d^0}$ mixing phase ϕ_d, which can be fixed through $B_d \to J/\psi K_{\rm S}$, is used as an input. We may then determine γ, as well as the

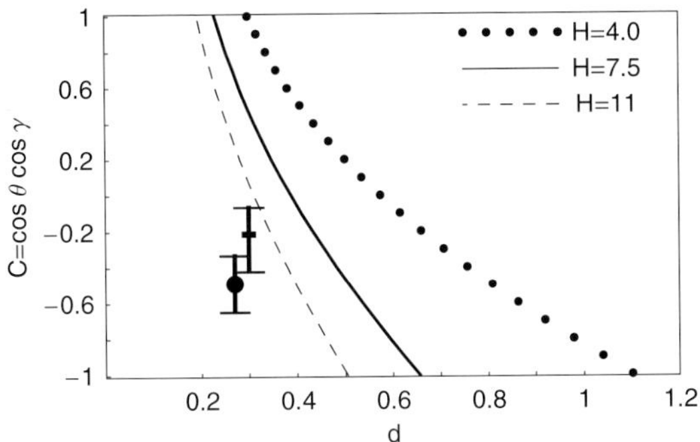

Fig. 15. Dependence of $C \equiv \cos\theta\cos\gamma$ on d for values of H consistent with (95). The "circle" and "square" with error bars represent the predictions of QCD factorization [48] and PQCD [79], respectively, for the Standard-Model range (86) of γ.

hadronic quantities d, θ, θ', by using only the U-spin relation $d' = d$; for a detailed illustration, see [75]. This approach is very promising for run II of the Tevatron and the experiments of the LHC era, where experimental accuracies for γ of $\mathcal{O}(10^\circ)$ [35] and $\mathcal{O}(1^\circ)$ [36] may be achieved, respectively. For other recently developed U-spin strategies, the reader is referred to [56,102,125,126].

Since $B_s \to K^+K^-$ is not accessible at the e^+e^- B factories operating at $\Upsilon(4S)$, data are not yet available. However, as can be seen by looking at the corresponding Feynman diagrams, $B_s \to K^+K^-$ is related to $B_d \to \pi^\mp K^\pm$ through an interchange of spectator quarks. Consequently, we have

$$\mathcal{A}_{\rm CP}^{\rm dir}(B_s \to K^+K^-) \approx \mathcal{A}_{\rm CP}^{\rm dir}(B_d \to \pi^\mp K^\pm) \tag{92}$$

$$\mathrm{BR}(B_s \to K^+K^-) \approx \mathrm{BR}(B_d \to \pi^\mp K^\pm)\,\frac{\tau_{B_s}}{\tau_{B_d}}. \tag{93}$$

For the following considerations, the quantity

$$H \equiv \frac{1}{\epsilon}\left|\frac{\mathcal{C}'}{\mathcal{C}}\right|^2 \left[\frac{\mathrm{BR}(B_d \to \pi^+\pi^-)}{\mathrm{BR}(B_s \to K^+K^-)}\right] \tag{94}$$

is particularly useful [127], where $\epsilon \equiv \lambda^2/(1-\lambda^2)$. Using (93), as well as factorization to estimate U-spin-breaking corrections to $|\mathcal{C}'| = |\mathcal{C}|$, H can be determined from the B-factory data as follows:

$$H \approx \frac{1}{\epsilon}\left(\frac{f_K}{f_\pi}\right)^2 \left[\frac{\mathrm{BR}(B_d \to \pi^+\pi^-)}{\mathrm{BR}(B_d \to \pi^\mp K^\pm)}\right] = \begin{cases} 7.3 \pm 2.9 \text{ (CLEO [118])} \\ 7.6 \pm 1.2 \text{ (BaBar [119])} \\ 7.1 \pm 1.9 \text{ (Belle [120])}. \end{cases} \tag{95}$$

If we employ the U-spin relation (91) and the amplitude parametrizations in (87) and (88), we obtain

$$H = \frac{1 - 2d\cos\theta\cos\gamma + d^2}{\epsilon^2 + 2\epsilon d\cos\theta\cos\gamma + d^2}. \tag{96}$$

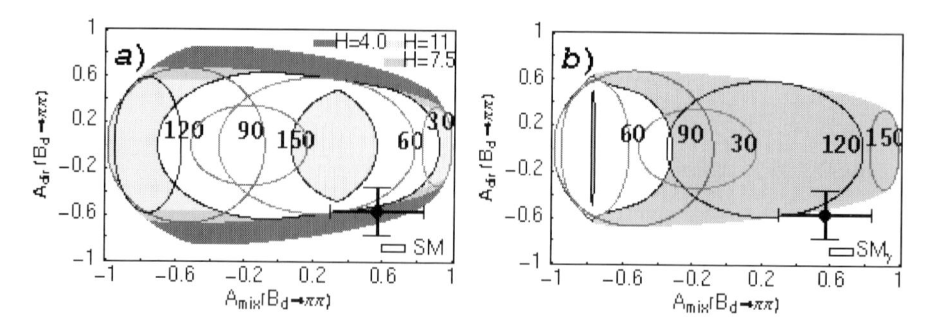

Fig. 16. Allowed regions in the $\mathcal{A}_{\rm CP}^{\rm mix}(B_d \to \pi^+\pi^-)$–$\mathcal{A}_{\rm CP}^{\rm dir}(B_d \to \pi^+\pi^-)$ plane for (a) $\phi_d = 47^\circ$ and various values of H, and (b) $\phi_d = 133^\circ$ ($H = 7.5$). The SM regions arise if we restrict γ to (86). Contours representing fixed values of γ are also included.

Consequently, H allows us to determine $C \equiv \cos\theta\cos\gamma$ as a function of d, as shown in Fig. 15. We observe that the data imply the rather restricted range $0.2 \lesssim d \lesssim 1$, thereby indicating that penguins cannot be neglected in $B_d \to \pi^+\pi^-$ analyses. Moreover, the experimental curves are not in favour of a Standard-Model interpretation of the theoretical predictions for $de^{i\theta}$ obtained within the QCD factorization [48] and PQCD [79] approaches. Interestingly, agreement could easily be achieved for $\gamma > 90^\circ$, as the circle and square in Fig. 15, calculated for $\gamma = 60^\circ$, would then move to positive values of C [66,127].

Let us now come back to the decay $B_d \to \pi^+\pi^-$ and its CP-violating observables, as parametrized in (89). As we have already noted, ϕ_d entering $\mathcal{A}_{\rm CP}^{\rm mix}(B_d \to \pi^+\pi^-)$ can be fixed through $\mathcal{A}_{\rm CP}^{\rm mix}(B_d \to J/\psi K_{\rm S})$, yielding the two-fold solution in (55). In order to deal with the penguins, we may employ H as an additional observable. Applying (91), we obtain $H = \mathrm{fct}(d, \theta, \gamma)$ (see (96)). We may then eliminate d in (89) through H. If we vary the remaining parameters θ and γ within their physical ranges, i.e. $-180^\circ \leq \theta \leq +180^\circ$ and $0^\circ \leq \gamma \leq 180^\circ$, we obtain an allowed region in the $\mathcal{A}_{\rm CP}^{\rm dir}(B_d \to \pi^+\pi^-)$–$\mathcal{A}_{\rm CP}^{\rm mix}(B_d \to \pi^+\pi^-)$ plane.

In Fig. 16, we show the corresponding results for the two solutions of ϕ_d and for various values of H, as well as the contours arising for fixed values of γ [66]. We observe that the experimental averages, represented by the crosses, overlap nicely with the SM region for $\phi_d = 47^\circ$, and point towards $\gamma \sim 55^\circ$. In this case, not only γ would be in accordance with the results of the CKM fits described in Sect. 1, but also the B_d^0–$\overline{B_d^0}$ mixing phase ϕ_d. On the other hand, for $\phi_d = 133^\circ$, the experimental values favour $\gamma \sim 125^\circ$, and have essentially no overlap with the SM region. Since a value of $\phi_d = 133^\circ$ would require CP-violating new-physics contributions to B_d^0–$\overline{B_d^0}$ mixing, also the γ range in (86) may no longer hold in this case, as it relies on a Standard-Model interpretation of the experimental information on $B_{d,s}^0$–$\overline{B_{d,s}^0}$ mixing. In particular, also values for γ larger than 90° could then in principle be accommodated. In order to put these observations on a more quantitative basis, we show in Fig. 17 the dependence of $|\mathcal{A}_{\rm CP}^{\rm dir}(B_d \to \pi^+\pi^-)|$ on γ for given values of $\mathcal{A}_{\rm CP}^{\rm mix}(B_d \to \pi^+\pi^-)$ [66]. If we vary $\mathcal{A}_{\rm CP}^{\rm mix}(B_d \to \pi^+\pi^-)$ within its whole positive range $[0, +1]$, the shaded "hills" in

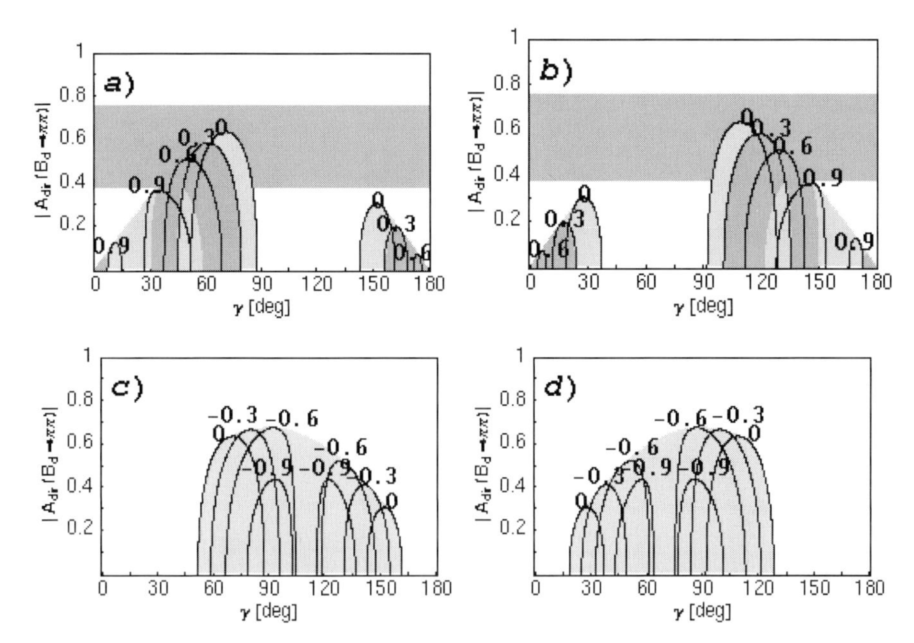

Fig. 17. $|\mathcal{A}_{\rm CP}^{\rm dir}(B_d \to \pi^+\pi^-)|$ as a function of γ in the case of $H = 7.5$ for various values of $\mathcal{A}_{\rm CP}^{\rm mix}(B_d \to \pi^+\pi^-)$. In (a) and (b), $\phi_d = 47°$ and $\phi_d = 133°$ were chosen, respectively. The shaded "hills" arise from a variation of $\mathcal{A}_{\rm CP}^{\rm mix}(B_d \to \pi^+\pi^-)$ within $[0, +1]$. The corresponding plots for negative $\mathcal{A}_{\rm CP}^{\rm mix}(B_d \to \pi^+\pi^-)$ are shown in (c) and (d) for $\phi_d = 47°$ and $\phi_d = 133°$, respectively. The bands arising from the experimental averages given in (71) and (72) are also included.

Figs. 17a and b arise. In the case of $\phi_d = 47°$, which is in agreement with the CKM fits, we may conveniently accommodate the Standard-Model range (86). On the other hand, we obtain a gap around $\gamma \sim 60°$ for $\phi_d = 133°$. Taking into account the experimental averages given in (71) and (72), we obtain

$$34° \lesssim \gamma \lesssim 75° \; (\phi_d = 47°), \quad 105° \lesssim \gamma \lesssim 146° \; (\phi_d = 133°). \tag{97}$$

If we vary $\mathcal{A}_{\rm CP}^{\rm mix}(B_d \to \pi^+\pi^-)$ within its whole negative range, both solutions for ϕ_d could accommodate (86), as can be seen in Figs. 17c and d, so that the situation would not be as exciting as for a positive value of $\mathcal{A}_{\rm CP}^{\rm mix}(B_d \to \pi^+\pi^-)$. In the future, the experimental uncertainties will be reduced considerably, i.e. the experimental bands in Fig. 17 will become much more narrow, thereby providing significantly more stringent results for γ, as well as the hadronic parameters. For a detailed discussion of the corresponding theoretical uncertainties, as well as simplifications that could be made through factorization, see [66].

In analogy to the analysis of the $B_d \to \pi^+\pi^-$ mode discussed above, we may also use H to eliminate d' in $\mathcal{A}_{\rm CP}^{\rm dir}(B_s \to K^+K^-)$ and $\mathcal{A}_{\rm CP}^{\rm mix}(B_s \to K^+K^-)$. If we then vary θ' and γ within their physical ranges, i.e. $-180° \leq \theta' \leq +180°$ and $0° \leq \gamma \leq 180°$, we obtain an allowed region in the $\mathcal{A}_{\rm CP}^{\rm mix}(B_s \to K^+K^-)$–$\mathcal{A}_{\rm CP}^{\rm dir}(B_s \to K^+K^-)$ plane [66], as shown in Fig. 18. There, also the impact of a

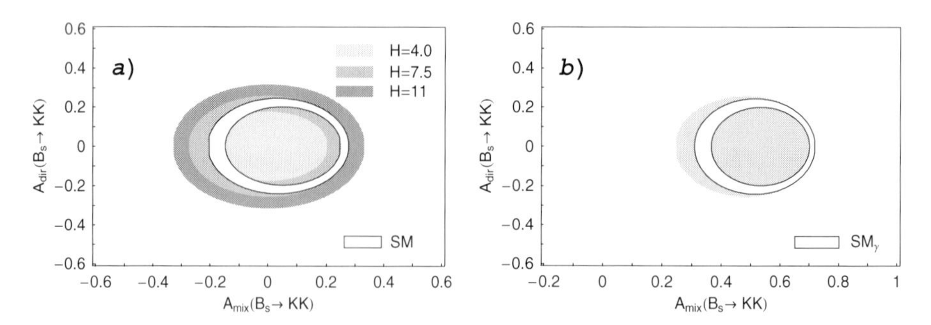

Fig. 18. Allowed regions in the $\mathcal{A}_{\rm CP}^{\rm mix}(B_s \to K^+K^-)$–$\mathcal{A}_{\rm CP}^{\rm dir}(B_s \to K^+K^-)$ plane for (a) $\phi_s = 0^\circ$ and various values of H, and (b) $\phi_s^{\rm NP} = 30^\circ$ ($H = 7.5$). The SM regions arise if γ is restricted to (86).

non-vanishing value of ϕ_s, which may be due to new-physics contributions to B_s^0–$\overline{B_s^0}$ mixing, is illustrated. If we constrain γ to (86), even more restricted regions arise. The allowed regions are remarkably stable with respect to variations of parameters characterizing U-spin-breaking effects [66], and represent a narrow target range for run II of the Tevatron and the experiments of the LHC era, in particular LHCb and BTeV. These experiments will allow us to exploit the whole physics potential of the $B_d \to \pi^+\pi^-$, $B_s \to K^+K^-$ system [75].

8 Remarks on the "Usual" Rare B Decays

Let us finally comment briefly on other "rare" B decays, which occur only at the one-loop level in the Standard Model, and involve $\bar{b} \to \bar{s}$ or $\bar{b} \to \bar{d}$ flavour-changing neutral-current transitions. Prominent examples are the following decay modes: $B \to K^*\gamma$, $B \to \rho\gamma$, $B \to K^*\mu^+\mu^-$ and $B_{s,d} \to \mu^+\mu^-$. The corresponding inclusive decays, for example $B \to X_s\gamma$, are also of particular interest, suffering from smaller theoretical uncertainties. Within the Standard Model, these transitions exhibit small branching ratios at the 10^{-4}–10^{-10} level, do not – apart from $B \to \rho\gamma$ – show sizeable CP-violating effects, and depend on $|V_{ts}|$ or $|V_{td}|$. A measurement of these CKM factors through such decays would be complementary to the one from $B_{s,d}^0$–$\overline{B_{s,d}^0}$ mixing. Since rare B decays are absent at the tree level in the Standard Model, they represent interesting probes to search for new physics. For detailed discussions of the many interesting aspects of rare B decays, the reader is referred to the lecture given by Mannel at this school [128], and to the overview articles listed in [29,129].

9 Conclusions and Outlook

The phenomenology of the B system is very rich and represents an exciting field of research. Thanks to the efforts of the BaBar and Belle collaborations, CP violation could recently be established in the B system with the help of the "gold-plated" mode $B_d \to J/\psi K_S$, thereby opening a new era in the exploration of CP-violating phenomena. The world average $\sin 2\beta = 0.734 \pm 0.054$ agrees now well with the Standard Model, but leaves a twofold solution for ϕ_d, given by $\phi_d = \left(47^{+5}_{-4}\right)^\circ \vee \left(133^{+4}_{-5}\right)^\circ$. The former solution is in accordance with the picture of the Standard Model, whereas the latter would point towards CP-violating new-physics contributions to B^0_d–$\overline{B^0_d}$ mixing. As we have seen, it is an important issue to resolve this ambiguity directly.

The physics potential of the B factories goes far beyond the famous $B_d \to J/\psi K_S$ decay, allowing us now to confront many more strategies to explore CP violation with data. Here the main goal is to overconstrain the unitarity triangle as much as possible, thereby performing a stringent test of the KM mechanism of CP violation. In this respect, important benchmark modes are given by $B \to \pi\pi$, $B \to \phi K$ and $B \to \pi K$ decays. First exciting data on these channels are already available from the B factories, but do not yet allow us to draw definite conclusions. In the future, the picture should, however, improve significantly.

Another important element in the testing of the Standard-Model description of CP violation is the B_s-meson system, which is not accessible at the e^+e^- B factories operating at the $\Upsilon(4S)$ resonance, BaBar and Belle, but can be studied nicely at hadron collider experiments. Already, run II of the Tevatron is expected to provide interesting results on B_s physics, and should discover B^0_s–$\overline{B^0_s}$ mixing soon, which is an important ingredient for the "standard" analysis of the unitarity triangle. Important B_s decays are $B_s \to J/\psi\phi$, $B_s \to K^+K^-$ and $B_s \to D^\pm_s K^\mp$. Although the Tevatron will provide first insights into these modes, they can only be fully exploited at the experiments of the LHC era, in particular LHCb and BTeV.

Apart from issues related to CP violation, several B-decay strategies allow also the determination of hadronic parameters, which can then be compared with theoretical predictions and may help us to control the corresponding hadronic uncertainties in a better way. Moreover, there are many other exciting aspects of B physics, for instance studies of certain rare B decays, which represent also sensitive probes for new physics. Hopefully, the future will bring many surprising results!

References

1. J.H. Christenson *et al.*, Phys. Rev. Lett. **13** 138 (1964)
2. V. Fanti *et al.* (NA48 Collaboration), Phys. Lett. B **465** 335 (1999)
3. A. Alavi-Harati *et al.* (KTeV Collaboration), Phys. Rev. Lett. **83** 22 (1999)
4. J.R. Batley *et al.* (NA48 Collaboration), Phys. Lett. B **544** 97 (2002)
5. A. Alavi-Harati *et al.* (KTeV Collaboration), hep-ex/0208007
6. S. Bertolini, hep-ph/0206095
7. A.J. Buras, TUM-HEP-435-01 [hep-ph/0109197]
8. M. Ciuchini and G. Martinelli, Nucl. Phys. Proc. Suppl. B **99** 27 (2001)
9. B. Aubert *et al.* (BaBar Collaboration), Phys. Rev. Lett. **87** 091801 (2001)
10. K. Abe *et al.* (Belle Collaboration), Phys. Rev. Lett. **87** 091802 (2001)

11. M. Kobayashi and T. Maskawa, Prog. Theor. Phys. **49** 652 (1973)
12. R. Fleischer, Phys. Rep. **370** 531 (2002)
13. G. Branco, L. Lavoura and J. Silva, *CP Violation* (Oxford Science Publications, Clarendon Press, Oxford 1999);
I.I. Bigi and A.I. Sanda, *CP Violation* (Cambridge Monographs on Particle Physics, Nuclear Physics and Cosmology, Cambridge University Press, Cambridge 2000)
14. M. Neubert, Theory of Exclusive Hadronic B Decays, Lect. Notes Phys. **647**, 3–41 (2004)
15. C. Jarlskog, Phys. Rev. Lett. **55** 1039 (1985); Z. Phys. C **29** 491 (1985)
16. For reviews, see A. Masiero and O. Vives, Annu. Rev. Nucl. Part. Sci. **51** 161 (2001);
Y. Grossman, Y. Nir and R. Rattazzi, Adv. Ser. Direct. High Energy Phys. **15** 755 (1998); M. Gronau and D. London, Phys. Rev. D **55** 2845 (1997)
17. L. Wolfenstein, Phys. Rev. Lett. **51** 1945 (1983)
18. A.J. Buras, M.E. Lautenbacher and G. Ostermaier, Phys. Rev. D **50** 3433 (1994)
19. R. Aleksan, B. Kayser and D. London, Phys. Rev. Lett. **73** 18 (1994)
20. G.C. Branco and L. Lavoura, Phys. Lett. B **208** 123 (1988);
C. Jarlskog and R. Stora, Phys. Lett. B **208** 268 (1988)
21. L.L. Chau and W.Y. Keung, Phys. Rev. Lett. **53** 1802 (1984)
22. For a recent review, see Z. Ligeti, LBNL-49214 [hep-ph/0112089]
23. A. Ali and D. London, Eur. Phys. J. C **18** 665 (2001)
24. S. Plaszczynski and M.H. Schune, LAL-99-67 [hep-ph/9911280]; Y. Grossman, Y. Nir, S. Plaszczynski and M.H. Schune, Nucl. Phys. B **511** 69 (1998)
25. M. Ciuchini *et al.*, JHEP **0107** 013 (2001)
26. A. Höcker, H. Lacker, S. Laplace and F. Le Diberder, Eur. Phys. J. C **21** 225 (2001)
27. D. Atwood and A. Soni, Phys. Lett. B **508** 17 (2001)
28. A.J. Buras, F. Parodi and A. Stocchi, TUM-HEP-465-02 [hep-ph/0207101];
A.J. Buras, TUM-HEP-489-02 [hep-ph/0210291]
29. G. Isidori, CERN-TH-2001-284 [hep-ph/0110255]
30. G. Buchalla and A.J. Buras, Phys. Lett. B **333** 221 (1994); Phys. Rev. D **54** 6782 (1996); Y. Grossman and Y. Nir, Phys. Lett. B **398** 163 (1997)
31. A.J. Buras and R. Fleischer, Phys. Rev. D **64** 115010 (2001)
32. S. Adler *et al.* (E787 Collaboration), Phys. Rev. Lett. **88** 041803 (2002)
33. L. Littenberg, hep-ex/0201026; M.V. Diwan, hep-ex/0205089
34. *The BaBar Physics Book*, eds. P. Harrison and H.R. Quinn, SLAC-R-504 (1998)
35. K. Anikeev *et al.*, FERMILAB-Pub-01/197 [hep-ph/0201071]
36. P. Ball *et al.*, CERN-TH-2000-101 [hep-ph/0003238]
37. G. Buchalla, A.J. Buras and M.E. Lautenbacher, Rev. Mod. Phys. **68** 1125 (1996)
38. R. Fleischer, Z. Phys. C **58** 483 (1993)
39. A.J. Buras and R. Fleischer, Phys. Lett. B **341** 379 (1995)
40. M. Ciuchini *et al.*, Phys. Lett. B **515** 33 (2001)
41. R. Fleischer, Z. Phys. C **62** 81; Phys. Lett. B **321** 259 and **332** 419 (1994).
42. R. Fleischer, Int. J. Mod. Phys. A **12** 2459 (1997)
43. N.G. Deshpande and X.-G. He, Phys. Rev. Lett. **74** 26 (1995) [E: ibid., p. 4099];
M. Gronau *et al.*, Phys. Rev. D **52** 6374 (1995)
44. M. Neubert and B. Stech, Adv. Ser. Direct. High Energy Phys. **15** 294 (1998), and references therein
45. A.J. Buras and J.-M. Gérard, Nucl. Phys. B **264** 371 (1986);
A.J. Buras, J.-M. Gérard and R. Rückl, Nucl. Phys. B **268** 16 (1986)

46. M. Beneke *et al.*, Phys. Rev. Lett. **83** 1914 (1999)
47. M. Beneke *et al.*, Nucl. Phys. B **591** 313 (2000)
48. M. Beneke *et al.*, Nucl. Phys. B **606** 245 (2001)
49. C.W. Bauer, D. Pirjol and I.W. Stewart, Phys. Rev. Lett. **87** 201806 (2001); C.W. Bauer, B. Grinstein, D. Pirjol and I.W. Stewart, hep-ph/0208034
50. H.-n. Li and H.L. Yu, Phys. Rev. D **53** 2480 (1996); Y.Y. Keum, H.-n. Li and A.I. Sanda, Phys. Lett. B **504** 6 (2001); Phys. Rev. D **63** 054008 (2001); Y.Y. Keum and H.-n. Li, Phys. Rev. D **63** 074006 (2001)
51. A. Khodjamirian, Nucl. Phys. B **605** 558 (2001); B. Melić: *in this Volume* [hep-ph/0209265]
52. M. Gronau and D. Wyler, Phys. Lett. B **265** 172 (1991)
53. D. Atwood, I. Dunietz and A. Soni, Phys. Rev. D **63** 036005 (2001); Phys. Rev. Lett. **78** 3257 (1997)
54. I. Dunietz, Phys. Lett. B **270** 75 (1991).
55. R. Fleischer and D. Wyler, Phys. Rev. D **62** 057503 (2000)
56. R. Fleischer, Eur. Phys. J. C **10** 299 (1999)
57. A.B. Carter and A.I. Sanda, Phys. Rev. Lett. **45** 952 (1980), Phys. Rev. D **23** 1567 (1981); I.I. Bigi and A.I. Sanda, Nucl. Phys. B **193** 85 (1981)
58. B. Aubert *et al.* (BaBar Collaboration), BABAR-PUB-02-008 [hep-ex/0207042]
59. K. Abe *et al.* (Belle Collaboration), BELLE-PREPRINT-2002-30 [hep-ex/0208025]
60. Y. Nir, WIS-35-02-DPP [hep-ph/0208080]
61. R. Fleischer and T. Mannel, Phys. Lett. B **506** 311 (2001)
62. Ya.I. Azimov, V.L. Rappoport and V.V. Sarantsev, Z. Phys. A **356** 437 (1997); Y. Grossman and H.R. Quinn, Phys. Rev. D **56** 7259 (1997); J. Charles *et al.*, Phys. Lett. B **425** 375 (1998); B. Kayser and D. London, Phys. Rev. D **61** 116012 (2000); H.R. Quinn *et al.*, Phys. Rev. Lett. **85** 5284 (2000)
63. A.S. Dighe, I. Dunietz and R. Fleischer, Phys. Lett. B **433** 147 (1998)
64. I. Dunietz, R. Fleischer and U. Nierste, Phys. Rev. D **63** 114015 (2001)
65. R. Itoh, KEK-PREPRINT-2002-106 [hep-ex/0210025]
66. R. Fleischer and J. Matias, Phys. Rev. D **66** 054009 (2002)
67. D. London and R.D. Peccei, Phys. Lett. B **223** 257 (1989); N.G. Deshpande and J. Trampetic, Phys. Rev. D **41** 895 and 2926 (1990); J.-M. Gérard and W.-S. Hou, Phys. Rev. D **43** 2909 (1991); Phys. Lett. B **253** 478 (1991)
68. N.G. Deshpande and X.-G. He, Phys. Lett. B **336** 471 (1994)
69. Y. Grossman and M.P. Worah, Phys. Lett. B **395** 241 (1997)
70. D. London and A. Soni, Phys. Lett. B **407** 61 (1997)
71. R. Fleischer and T. Mannel, Phys. Lett. B **511** 240 (2001)
72. B. Aubert *et al.* (BaBar Collaboration), BABAR-CONF-02/016 [hep-ex/0207070]
73. K. Abe *et al.* (Belle Collaboration), BELLE-CONF-0201 [hep-ex/0207098]
74. G. Hiller, SLAC-PUB-9326 [hep-ph/0207356]; A. Datta, UDEM-GPP-TH-02-103 [hep-ph/0208016]; M. Ciuchini and L. Silvestrini, hep-ph/0208087; M. Raidal, CERN-TH/2002-182 [hep-ph/0208091]; B. Dutta, C.S. Kim and S. Oh, hep-ph/0208226; Jong-Phil Lee and Kang Young Lee, KIAS-P02054 [hep-ph/0209290]
75. R. Fleischer, Phys. Lett. B **459** 306 (1999)

76. B. Aubert *et al.* (BaBar Collaboration), BABAR-PUB-02-009 [hep-ex/0207055]
77. K. Abe *et al.* (Belle Collaboration), Phys. Rev. Lett. **89** 071801 (2002)
78. K. Hagiwara *et al.* (Particle Data Group), Phys. Rev. D **66** 010001 (2002)
79. A.I. Sanda and K. Ukai, Prog. Theor. Phys. **107** 421 (2002)
80. Y.-Y. Keum, DPNU-02-30 [hep-ph/0209208]
81. M. Gronau and D. London, Phys. Rev. Lett. **65** 3381 (1990)
82. Y. Grossman and H.R. Quinn, Phys. Rev. D **58** 017504 (1998);
M. Gronau, D. London, N. Sinha and R. Sinha, Phys. Lett. B **514** 315 (2001)
83. J. Charles, Phys. Rev. D **59** 054007 (1999)
84. R. Fleischer and T. Mannel, Phys. Lett. B **397** 269 (1997)
85. D. London, N. Sinha and R. Sinha, Phys. Rev. D **63** 054015 (2001)
86. C.-D. Lü and Z.-j. Xiao, BIHEP-TH-2002-22 [hep-ph/0205134]
87. M. Gronau and J.L. Rosner, Phys. Rev. D **65** 093012 (2002)
88. M. Gronau and J.L. Rosner, Phys. Rev. D **66** 053003 (2002)
89. Working group on B oscillations, see
`http://lepbosc.web.cern.ch/LEPBOSC/`
90. For a recent review, see M. Beneke and A. Lenz, J. Phys. G **27** 1219 (2001)
91. I. Dunietz, Phys. Rev. D **52** 3048 (1995);
R. Fleischer and I. Dunietz, Phys. Rev. D **55** 259 (1997)
92. A.J. Buras, TUM-HEP-259-96 [hep-ph/9610461]
93. A.S. Kronfeld and S.M. Ryan, Phys. Lett. B **543** 59 (2002)
94. R. Aleksan, I. Dunietz and B. Kayser, Z. Phys. C **54** 653 (1992)
95. M. Gronau and D. London, Phys. Lett. B **253** 483 (1991)
96. R. Fleischer and I. Dunietz, Phys. Lett. B **387** 361 (1996)
97. D. London, N. Sinha and R. Sinha, Phys. Rev. Lett. **85** 1807 (2000)
98. A.F. Falk and A.A. Petrov, Phys. Rev. Lett. **85** 252 (2000);
D. Atwood and A. Soni, Phys. Lett. B **533** 37 (2002)
99. I. Dunietz, Phys. Lett. B **427** 179 (1998)
100. A.S. Dighe, I. Dunietz and R. Fleischer, Eur. Phys. J. C **6** 647 (1999)
101. A.S. Dighe, I. Dunietz, H.J. Lipkin and J.L. Rosner, Phys. Lett. B **369** 144 (1996)
102. R. Fleischer, Phys. Rev. D **60** 073008 (1999)
103. Y. Nir and D.J. Silverman, Nucl. Phys. B **345** 301 (1990)
104. M. Gronau, J.L. Rosner and D. London, Phys. Rev. Lett. **73** 21 (1994)
105. R. Fleischer, Phys. Lett. B **365** 399 (1996)
106. R. Fleischer and T. Mannel, Phys. Rev. D **57** 2752 (1998)
107. M. Gronau and J.L. Rosner, Phys. Rev. D **57** 6843 (1998)
108. R. Fleischer, Eur. Phys. J. C **6** 451 (1999); Phys. Lett. B **435** 221 (1998)
109. M. Neubert and J.L. Rosner, Phys. Lett. B **441** 403 (1998);
Phys. Rev. Lett. **81** 5076 (1998)
110. M. Neubert, JHEP **9902** 014 (1999)
111. A.J. Buras and R. Fleischer, Eur. Phys. J. C **11** 93 (1999)
112. A.J. Buras and R. Fleischer, Eur. Phys. J. C **16** 97 (2000)
113. R. Fleischer and J. Matias, Phys. Rev. D **61** 074004 (2000)
114. J. Matias, Phys. Lett. B **520** 131 (2001)
115. M. Bargiotti *et al.*, Eur. Phys. J. C **24** 361 (2002)
116. M. Gronau and J.L. Rosner, Phys. Rev. D **65** 013004 [E: D **65** 079901] (2002)
117. R. Fleischer and T. Mannel, TTP-97-22 [hep-ph/9706261];
D. Choudhury, B. Dutta and A. Kundu, Phys. Lett. B **456** 185 (1999);
X.-G. He, C.-L. Hsueh and J.-Q. Shi, Phys. Rev. Lett. **84** 18 (2000);
Y. Grossman, M. Neubert and A.L. Kagan, JHEP **9910** 029 (1999)

118. D. Cronin-Hennessy *et al.* (CLEO Collaboration), Phys. Rev. Lett. **85** 515 (2000)
119. B. Aubert *et al.* (BaBar Collaboration), hep-ex/0207065; hep-ex/0206053
120. B.C. Casey *et al.* (Belle Collaboration), hep-ex/0207090
121. S. Chen *et al.* (CLEO Collaboration), Phys. Rev. Lett. **85** 535 (2000)
122. K. Suzuki, talk at ICHEP02, Amsterdam, The Netherlands, 24–31 July 2002
123. A.J. Buras, R. Fleischer and T. Mannel, Nucl. Phys. B **533** 3 (1998);
A.F. Falk, A.L. Kagan, Y. Nir and A.A. Petrov, Phys. Rev. D **57** 4290 (1998);
D. Atwood and A. Soni, Phys. Rev. D **58** 036005 (1998)
124. M. Gronau, Phys. Lett. B **492** 297 (2000)
125. M. Gronau and J.L. Rosner, Phys. Lett. B **482** 71 (2000)
126. P.Z. Skands, JHEP **0101** 008 (2001)
127. R. Fleischer, Eur. Phys. J. C **16** 87 (2000)
128. T. Mannel, Theory of Rare B Decays: $b \to s\gamma$ and $b \to s\ell^+\ell^-$, Lect. Notes Phys. **647**, 78–99 (2004)
129. A. Ali, CERN-TH-2002-284 [hep-ph/0210183];
A.J. Buras and M. Misiak, TUM-HEP-468-02 [hep-ph/0207131]

Theory of Rare B Decays: $b \to s\gamma$ and $b \to s\ell^+\ell^-$

Thomas Mannel

Institut für Theoretische Teilchenphysik, Universität Karlsruhe, 76128 Karlsruhe, Germany

Abstract. In this lectures some of the theoretical aspects of rare B decays are discussed. The focus is on inclusive decays, since these can be computed more reliably. Topics covered are (1) short distance effects, (2) long distance QCD effects and (3) effects of "new physics" in these decays.

1 Introduction: $\mathcal{H}_{eff}$ for $b \to s\gamma$ and $b \to s\ell^+\ell^-$

These lectures are devoted to a discussion of rare B decays, focussing on so called flavour changing neutral current (FCNC) decays of the type $b \to s\gamma$ and $b \to s\ell^+\ell^-$. In the standard model these processes cannot appear at tree level and hence are loop-induced transitons. Thus they constitute an important check of the standard model and may open a window on physics beyond this model.

Radiative rare B decays have attracted considerable attention in the last few years. After the first observation in 1994, by the CLEO collaboration [1], data have become quite precise [2] so that even a measurement of the CP asymmetry in these decays [3] became possible. As far as data are concerned, the situation clearly will improve further, after the excellent start of both B factories at KEK and at SLAC.

$B \to X_s\gamma$ as well as $B \to X_s\ell^+\ell^-$ test the Standard Model (SM) in a particular way. The GIM cancellation, which is present in all the FCNC processes, is lifted in this case by the large top-quark mass; if the top quark were as light as the b quark, these decays would be too rare to be observable.

Since the SM contribution is small, these decays have a good sensitivity to "new physics", e.g. to new (heavy) particles contributing to the loop. In fact, already the first CLEO data could constrain some models for "new physics" in a stringent way[1].

The most general effective Hamiltonian describing the decays of the type $b \to s\gamma$ is given by

$$H_{eff} = \sum_i c_i O_i \,, \tag{1}$$

where the O_i are local operators

$$\begin{aligned} O_{1\cdots 6} &= \text{four-fermion operators} \\ O_7 &= m_b \bar{s}\sigma_{\mu\nu}(1+\gamma_5)b\, F^{\mu\nu} \\ O_7' &= m_s \bar{s}\sigma_{\mu\nu}(1-\gamma_5)b\, F^{\mu\nu} \end{aligned}$$

T. Mannel, Theory of Rare B Decays: $b \to s\gamma$ and $b \to s\ell^+\ell^-$, Lect. Notes Phys. **647**, 78–99 (2004)
http://www.springerlink.com/

$$\begin{aligned} O_8 &= m_b \bar{s}\sigma^{\mu\nu}T^a(1+\gamma_5)b\, G^a_{\mu\nu} \\ O'_8 &= m_s \bar{s}\sigma^{\mu\nu}T^a(1-\gamma_5)b\, G^a_{\mu\nu} \end{aligned} \tag{2}$$

and c_i are pertubatively calculable coefficients.

In any new physics analysis of B decays only the coefficients c_i are tested [4]. The decay $B \to X_s\gamma$ (and the corresponding exclusive decays) are practically determined by the two operators O_7 and O'_7, and hence these decays are mainly testing c_7 and c'_7. In the SM these two coefficients are

$$c_7 = -\frac{G_F^2 e}{32\sqrt{2}\pi^2} V_{tb}V^*_{ts} C_7 m_b, \tag{3}$$

$$c'_7 = -\frac{G_F^2 e}{32\sqrt{2}\pi^2} V_{tb}V^*_{ts} C_7 m_s \tag{4}$$

where C_7 is a function of $(m_t/M_W)^2$, which we shall discuss later.

Furthermore, the two operators differ by the handedness of the quarks; in order to disentangle these two contributions there has to be a handle on the polarization of the quarks or of the photon, which is impossile at a B factory. Consequently, from $b \to s\gamma$ alone only the combination $|c_7|^2 + |c'_7|^2$ can be determined in the near future.

Once the effective interaction for the quark transition is fixed, one has to calculate from this the actual hadronic process. This step is only for the inclusive decays under reasonable theoretical control; for exclusive decays, form factors are needed, which either need to be modelled or will finally come from the lattice.

For inclusive decays the machinery used is the heavy mass expansion[1]. Using this framework for the total rate one can establish that (1) the leading term as $m_b \to \infty$ is the free quark decay, (2) there are no subleading corrections of order $1/m_b$, (2) the first non-vanising corrections are of order $1/m_b^2$ and are given in terms of two parameters. This will be discussed in Sect. 3. Additional non-perturbative uncertainties are induced by a cut on the photon energy, which is necessary from the exprimental point of view to suppress backgrounds.

One may discuss the decays $b \to s\ell^+\ell^-$ in a similar way. Here we have to extend the operator basis (2) by two more operators involving leptons

$$\mathcal{O}_9 = \frac{\alpha}{\alpha_s}(\bar{s}_{L\alpha}\gamma_\mu b_{L\alpha})(\bar{\ell}\gamma^\mu \ell) \tag{5}$$

$$\mathcal{O}_{10} = (\bar{s}_{L\alpha}\gamma_\mu b_{L\alpha})(\bar{\ell}\gamma^\mu\gamma_5 \ell) \tag{6}$$

This involves two more wilson coefficients C_9 and C_{10} and the contribution of these two operators have tp be added to the effective Hamiltonian (1).

The QCD corrections to the effective hamiltonian (1) have been calculated already to next-to-leading order. A detailed review on this subject is given in [10]. The value of the coefficients at subldeading order involves subtelties such as the question of how to deal with γ_5 in dimensional regularization. We do

[1] A non-exhaustive selection of revies is [5–9].

not want to discuss these points in this lecture, although some of the results for $B \to X_s\gamma$ are quoted at NLLO precision. In order to give some feeling for the size of the Wilson coefficients we quote in Table 1 their values at leading logarithmic acuracy, where the abovementioned subtelties do not matter.

Table 1. Values for the Wilson coefficients $C_i(\mu)$ at the scale $\mu = m_W$ ("matching conditions") and at three other scales, $\mu = 10.0$ GeV, $\mu = 5.0$ GeV and $\mu = 2.5$ GeV, evaluated with one-loop β-function and the leading-order anomalous-dimension matrix, with $m_t = 174$ GeV and $\Lambda_{\rm QCD} = 225$ MeV. Note that the relation between c_i appearing in (1) and the tabulated coefficients is $c_i = 4G_F V_{ts}^* V_{tb} C_i/\sqrt{2}$; furthermore, $C_9'(\mu) \equiv C_9(\mu)/\alpha_s(\mu)$ (see secction 6.1).

$C_i(\mu)$	$\mu = m_W$	$\mu = 10.0$	$\mu = 5.0$	$\mu = 2.5$
C_1	0.0	0.182	0.275	0.40
C_2	−1.0	−1.074	−1.121	−1.193
C_3	0.0	−0.008	−0.013	−0.019
C_4	0.0	0.019	0.028	0.040
C_5	0.0	−0.006	−0.008	−0.011
C_6	0.0	0.022	0.035	0.055
C_7	0.195	0.286	0.325	0.371
C_8	0.097	0.138	0.153	0.172
C_9'	−2.08	−2.31	−2.36	−2.38
C_{10}	4.54	4.54	4.54	4.54

2 Perturbative Corrections to $b \to s\gamma$

The main perturbative corrections are the QCD corrections, which are substantial. These corrections are calculated using an effective field-theory framework and yield the QCD corrections to the Wilson coefficents.

To set this up, we have to write down first the relevant effective Hamiltonian as in (1). The operators appearing in (1) mix under renormalization as we evolve down from the M_W mass scale to the relevant scale, which is the mass of the b quark. The cofficient functions are calculated at the scale $\mu = M_W$ as a power series in the strong coupling

$$c_i(M_W) = c_i^{(0)}(M_W) + \frac{\alpha_s(M_W)}{\pi} c_i^{(1)}(M_W) + \cdots \tag{7}$$

Changing the scale μ results in a change of the coefficient functions and in the matrix elements, such that the matrix element of the effective Hamiltonian remains μ-independent. This change can be computed perturbatively for sufficiently large μ, using the standard machinery of renormalization group, which involves a calculation of the anomalous-dimension matrix that describes the mixing of the operators (2).

The solution of the renormalizaton group equation yields the coefficient functions at some lower scale μ, which take the form (schematically)

$$c_i(\mu) = c_i^{(0)}(M_W) \sum_{n=0} b_n^{(0)} \left(\frac{\alpha_s}{\pi} \ln \left(\frac{M_W^2}{\mu^2} \right) \right)^n + \frac{\alpha_s}{\pi} c_i^{(1)}(M_W) \sum_{n=0} b_n^{(1)} \left(\frac{\alpha_s}{\pi} \ln \left(\frac{M_W^2}{\mu^2} \right) \right)^n + \cdots \tag{8}$$

where the b_n are obtained from the solution of the renormalization group equation.

The last step is to compute the matrix elements of the operators at a scale $\mu \approx m_b$. This can be done for the inclusive case using the $1/m_b$ expansion. For the exclusive case, one would need the form factor in the corresponding approximation, which cannot be done with present theoretical techniques.

At present, the leading and the subleading terms of the coefficients have been calculated [11–14], including electroweak contributions[15], the main part of which is due to the correct scale setting in α_{em}.

A complete and up-to-date compilation can be found in [16]. Without going into any more detail we only quote the result from [16]

$$Br(B \to X_s\gamma) = (3.29 \pm 0.33) \times 10^{-4} \,. \tag{9}$$

where this result includes a cut on the photon energy at $E_{\gamma,min} = 0.05\, m_b$.

The QCD corrections are in fact dramatic; they increase the rate for $b \to s\gamma$ by about a factor of two. For example, already at the leading-log level we have $c_7(m_b)/c_7(M_W) = 1.63$, see Table 1. Another indication of this fact is a substantial dependence of the leading-order result on the choice of the renormalization scale μ. This is usually estimated by varying the scale μ between $m_b/2$ and $2m_b$. In this way one obtains a variation of $\delta_\mu = {}^{+27.4\%}_{-20.4\%}$ for the leading-order result.

Taking into account the subleading terms reduces the scale dependence substantially. In fact, one has at subleading order [16] $\delta_\mu = {}^{+0.1\%}_{-3.2\%}$, which is smaller than naively expected [17]. It has been argued that the smallness of δ_μ is accidental [16]. However, arguments have been given recently [18] that these cancellations are not accidental. In fact, most of the large radiative corrections may be assigned to the running of the b quark mass appearing in the operator O_7.

3 Non-perturbative Corrections to $b \to s\gamma$

Non-perturbative corrections arise from different sources. We shall consider here

- Long-distance effects from intermediate vector mesons $B \to J/\Psi X_s \to X_s\gamma$,
- Subleading terms in the heavy mass expansion: $1/m_b$ and $1/m_c$ corrections,
- Non-perturbative contributions to the photon spectrum ("shape function").

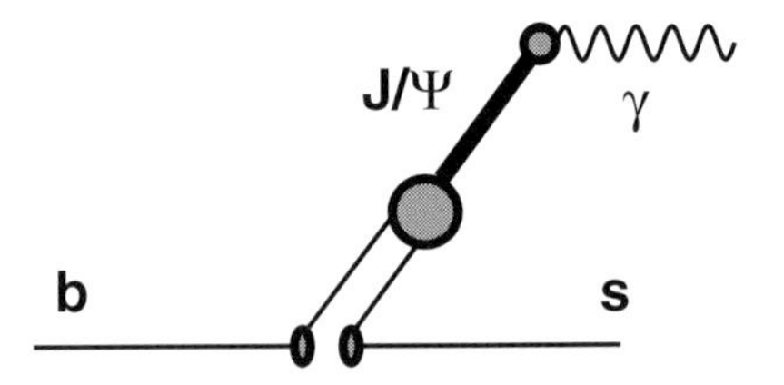

Fig. 1. Long distance contribution $B \to J/\Psi X_s \to X_s\gamma$

3.1 $B \to J/\Psi X_s \to X_s\gamma$

One long-distance contribution comes from the process $B \to X_s J/\Psi$ and the subsequent decay of the (off-shell) J/Ψ into a photon, see Fig. 1. The first process $B \to X_s J/\Psi$ has a branching ratio of order 1%, at least for an on-shell J/Ψ. Assuming that this is similar for the off-shell case, we have to multiply it with another factor $1/m_c^2$ for the propagation of the J/Ψ and a factor $f_{J/\Psi}^2$, since the J/Ψ has to annihilate into a photon. This leads us to the conclusion that this contribution is indeed negligibly small. However, one has to keep in mind that some extrapolation from $q^2 = m_{J/\Psi}^2$ to $q^2 = 0$ is involved, assuming that this will not lead to a strong enhancement.

3.2 $1/m_b$ and $1/m_c$ Corrections

A set of "standard" non-perturbative corrections arises from the heavy mass expansion [5–9]. As far as the total rate is concerned, we have the subleading corrections of order $1/m_b^2$, which are parametrized in terms of the kinetic energy λ_1 and the chromomagnetic moment λ_2 defined by the matrix elements

$$2M_H\lambda_1 = \langle H(v)|\bar{h}_v(iD)^2 h_v|H(v)\rangle \tag{10}$$
$$6M_H\lambda_2 = \langle H(v)|\bar{h}_v\sigma_{\mu\nu}iD^\mu iD^\nu h_v|H(v)\rangle\,. \tag{11}$$

In terms of these two matrix elements the total rate reads at tree level up to order $1/m_b^2$

$$\Gamma = \frac{G_F^2\alpha m_b^5}{32\pi^4}|V_{ts}V_{tb^*}|^2|C_7|^2\left(1+\frac{\lambda_1-9\lambda_2}{2m_b^2}+\cdots\right)\,. \tag{12}$$

This result is fully integrated over the photon energy spectrum. One can also compute the energy spectrum of the photon within the $1/m_b$ expansion, which is given, again at tree level, by

$$\frac{d\Gamma}{dx} = \frac{G_F^2\alpha m_b^5}{32\pi^4}|V_{ts}V_{tb^*}|^2|C_7|^2 \tag{13}$$
$$\left(\delta(1-x)-\frac{\lambda_1+3\lambda_2}{2m_b^2}\delta'(1-x)+\frac{\lambda_1}{6m_b^2}\delta''(1-x)+\cdots\right)$$

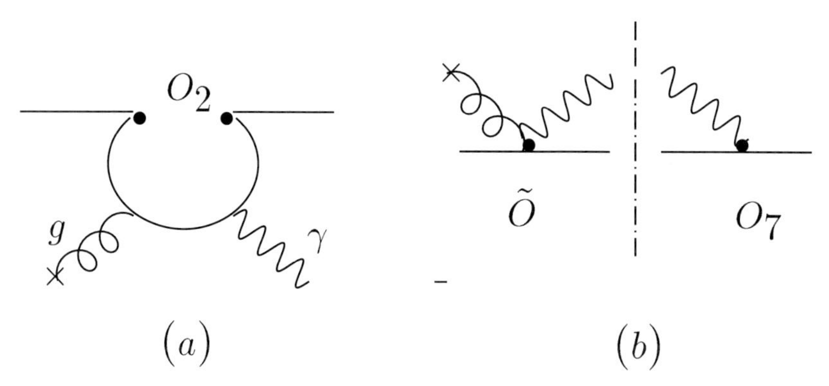

Fig. 2. Interference between O_7 and one of the four-fermion operators (O_2), leading to a contribution of order $1/m_c^2$

which can only be interpreted in terms of moments of the spectrum. We shall return to this point in the next subsection.

If the charm quark is also assumed to be heavy, one may discuss the charm-mass dependence in terms of a $1/m_c$ expansion [19]. The relevant contribution originates from the four-fermion operators (e.g. the operator O_2) involving the charm quark, see Fig. 2. Expanding the matrix element of O_2 in powers of $1/m_c$ we obtain a local operator of the form

$$O_{1/m_c^2} = \frac{1}{m_c^2} \bar{s}\gamma_\mu(1-\gamma_5)T^a b\, G^a_{\nu\lambda}\epsilon^{\mu\nu\rho\sigma}\partial^\lambda F_{\rho\sigma} \tag{14}$$

which can interfere with the leading term O_7 (see Fig. 2).

The detailed calculation [20,21] reveals that this contribution is rather small

$$\frac{\delta\Gamma_{1/m_c^2}}{\Gamma} = -\frac{C_2}{9C_7}\frac{\lambda_2}{m_c^2} \approx 0.03 \tag{15}$$

3.3 Non-perturbative Corrections in the Photon Energy Spectrum

The non-perturbative corrections for the total rate are thus quite small and can safely be neglected against the perturbative ones. However, to extract the process $B \to X_s\gamma$ there has to be a lower cut on the photon energy to get rid of the uninteresting processes such as ordinary bremsstrahlung. Clearly it is desirable to have this cut as high as possible, but this makes the process "less inclusive" and hence more sensitive to non-perturbative contributions to the photon-energy spectrum.

Since we are dealing at tree level with a two-body decay, the naive calculation of the photon spectrum yields a δ function at partonic level and the $1/m_b^n$ corrections are again distributions located at the partonic energy $E_\gamma = m_b/2$, see (13). Clearly (13) cannot be used to implement a cut on the photon energy spectrum, since this is not a smooth function.

The perturbative contributions have been calculated and yield a spectrum that is mainly determined by the bremsstrahlung of a radiated gluon. This part of the calculation is fully perturbative and enters the next-to-leading order analysis described in the previous section. In particular, the partonic δ function smoothens and turns into "plus distributions" of the form

$$\frac{d\Gamma}{dx} = \cdots + \frac{\alpha_s}{\pi}\left[\left(\frac{\ln(1-x)}{1-x}\right)_+ , \left(\frac{1}{1-x}\right)_+\right], \tag{16}$$

where the ellipses denote terms that are regular as $x \to 1$ and contributions proportional to $\delta(1-x)$, which are determined by virtual gluons.

Here we shall focus on the non-perturbative contributions close to the endpoint. The general structure of the terms in the $1/m_b$ expansion is

$$\frac{d\Gamma}{dx} = \Gamma_0 \left[\sum_i a_i \left(\frac{1}{m_b}\right)^i \delta^{(i)}(1-x) \right. \\ \left. + \mathcal{O}((1/m_b)^{i+1}\delta^{(i)}(1-x))\right], \tag{17}$$

where $\delta^{(i)}$ is the ith derivative of the δ function.

It has been shown [22,23] that the terms with $\delta^{(i)}(1-x)/m_b^i$ can be resummed into a non-perturbative function such that the photon energy spectrum becomes

$$\frac{d\Gamma}{dx} = \frac{G_F^2 \alpha m_b^5}{32\pi^4}|V_{ts}V_{tb^*}|^2|C_7|^2 f(m_b[1-x])\,, \tag{18}$$

where the non-perturbative fuction f is formally defined by the matrix element

$$2M_B f(\omega) = \langle B|\bar{Q}_v \delta(\omega + iD_+)Q_v|B\rangle\,. \tag{19}$$

Here D_+ is the light-cone component of the covariant derivative, acting on Q_v, which denotes a heavy-quark field in the static approximation.

The shape function is in fact a universal function, which appears for any heavy-to-light transition in the corresponding kinematical region. In general these transitions should be written as a convolution of a (perturbatively calculable) Wilson coefficient and the non-perturbative matrix element

$$d\Gamma = \int d\omega\, C_0(\omega)\langle B|O_0(\omega)|B\rangle \tag{20}$$

with

$$O_0(\omega) = \bar{Q}_v \delta(\omega + iD_+)Q_v \tag{21}$$

At tree level this leads to a simple and intuitive formula in which the mass m_b is replaced by an "effective mass" $m_b^* = m_b - \omega$ such that

$$d\Gamma = \int d\omega\, d\Gamma_{tree}(m_b \to m_b^*) f(\omega) \tag{22}$$

Since this function is universal, it appears in the semileptonic $b \to u\ell\bar{\nu}$ transitions as well as in the $b \to s\gamma$ decays. At leading twist, this leads to a model-independent relation between these inclusive decays, which may be used to obtain $|V_{ts}/V_{ub}|$ [25,24].

Moments of the shape function can be related to the parameters describing the subleading effects in the $1/m_b$ expansion. One has

$$\int d\omega\, f(\omega) = 1\,, \quad \int d\omega\, \omega\, f(\omega) = 0\,, \tag{23}$$
$$\int d\omega\, \omega^2\, f(\omega) = -\frac{\lambda_1}{3m_b^2}\,, \quad \int d\omega\, \omega^2\, f(\omega) = -\frac{\rho_1}{3m_b^2}\,.$$

Radiative corrections can be included using the machinery of effective field theory. However, here some ambiguity arises from the appearance of a double logarithm (see (16)), which makes the matching ambiguous. Various authors [25,26] have used the mass convolution formula (22), although this has not yet been proven to be correct beyond the tree level.

Finally one may also try to resum the subleading terms in $1/m_b$, i.e. the terms of order $\delta^{(i)}(1-x)/m_b^{i+1}$ in (17). This has been discussed in [27], where it has been shown that the relevant operators are

$$O_1^\mu(\omega) = \bar{Q}_v\,\{iD^\mu, \delta(iD_+ + \omega)\}\,Q_v \tag{24}$$
$$O_2^\mu(\omega) = i\bar{Q}_v\,[iD^\mu, \delta(iD_+ + \omega)]\,Q_v$$
$$O_3^{\mu\nu}(\omega_1,\omega_2) = \tag{25}$$
$$\bar{Q}_v\delta(iD_+ + \omega_2)\,\{iD_\perp^\mu, iD_\perp^\nu\}\,\delta(iD_+ + \omega_1)Q_v$$
$$O_4^{\mu\nu}(\omega_1,\omega_2) =$$
$$g_s\bar{Q}_v\delta(iD_+ + \omega_2)G_\perp^{\mu\nu}\delta(iD_+ + \omega_1)Q_v\,,$$

plus the corresponding ones where a Pauli spin matrix appears between the quark spinors.

The effect of the subleading terms can be parametrized by four universal functions, which appear again in both $b \to u\ell\bar{\nu}$ and $b \to s\gamma$. Using a simple but realistic model the effects of the subleading terms may be estimated as a function of the lower photon energy cut. In Fig. 3 we plot the rate integrated from a lower cut as a function of this cut for various values of the parameters. As expected, the subleading terms at cut values of 2.3 GeV are of order 10% and negligibly small below 2.1 GeV.

4 "New Physics" in $b \to s\gamma$

In the Standard Model, $b \to s\gamma$ is a loop-induced process; it thus has considerable sensitivity to new physics effects. However, as already pointed out in the introduction, any B physics experiment tests the coefficients c_i appearing in the effective Hamiltonian (1) and thus all the information on new effects is encoded in combinations of the low-energy parameters c_i, which have to be computed

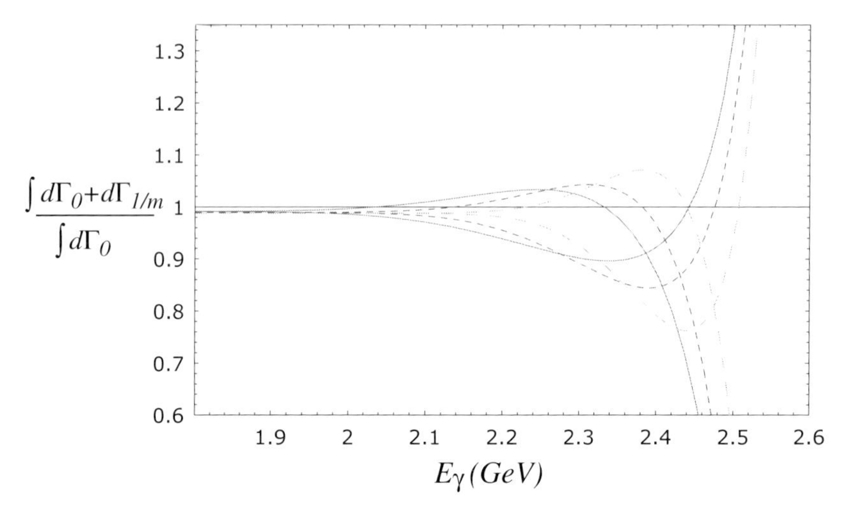

Fig. 3. Partially integrated rate normalized to the leading twist result. The three lines with a peak correspond to $\rho_2 = (500 \text{ MeV})^3$, and $\bar{\Lambda} = 570$ MeV (solid line), $\bar{\Lambda} = 470$ MeV (short-dashed line) $\bar{\Lambda} = 370$ MeV (long-dashed line). The two lines with a dip have $\rho_2 = -(500 \text{ MeV})^3$ and $\bar{\Lambda} = 470$ MeV (dashed line), $\bar{\Lambda} = 370$ MeV (dotted line).

in the Standard Model with the best possible accuracy. Comparing this to B decay data, it will clearly be impossible to find clean evidence for some specific scenario of new physics.

At present, no significant deviation from the Standard Model has been observed in $B \to X_s\gamma$ nor in any other B decay. Given that there are processes that are sensitive to new effects, B physics (and $b \to s\gamma$ in particular) can contribute to constrain new physics scenarios.

Keeping this in mind one may try various scenarios of new physics and calculate the effects on $b \to s\gamma$, i.e. calculate the coefficients of the low energy effective Hamiltonian (1) in specific scenarios. There is an enormous variety of models for physics beyond the Standard Model on the market, and is is impossible to cover all these ideas.

For that reason I shall only consider two examples, which are instructive and demonstrate the kind of sensitivity one may expect. In the next subsection I shall consider the Type-II two-Higgs doublet model and in Sect. 4.2 I shall discuss a few recent papers on supersymmetry with large values of $\tan\beta$.

4.1 Two-Higgs-Doublet Model (Type II)

One popular and consistent way to extend the Standard Model is to add one or more Higgs doublets. This can be done in various ways, but one well motivated way is to have two Higgs doublets where one doublet gives the mass to the up quarks, the other doublet to the down quarks.

Out of the eight degrees of freedom of the Higgs sector, three are needed to give mass to the heavy weak bosons, while the other five become physical states.

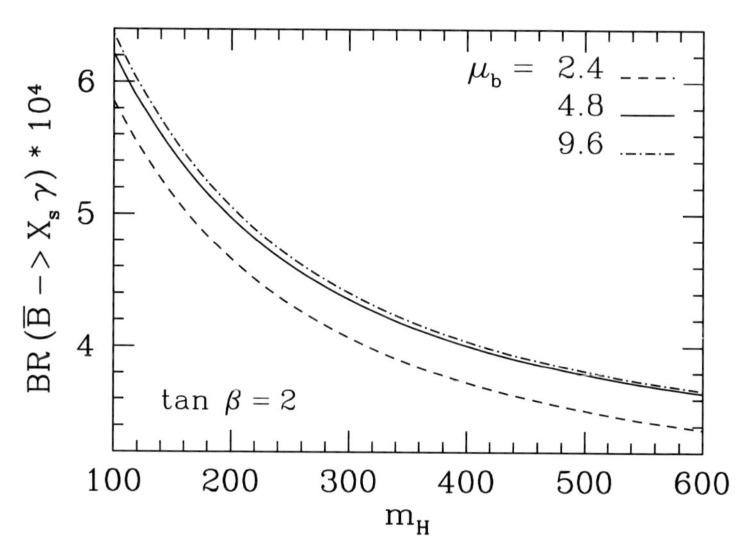

Fig. 4. The branching ratio for $B \to X_s\gamma$ as a function of the charged Higgs mass; figure taken from [29].

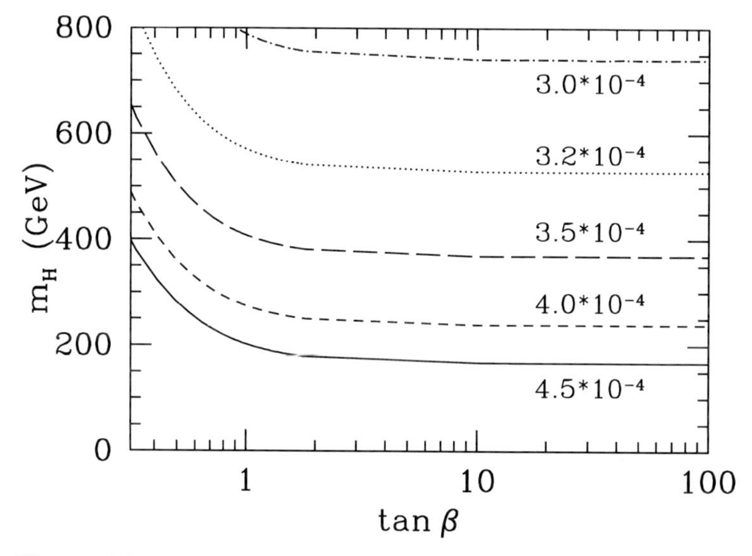

Fig. 5. The $\tan\beta$-M_{H^+} plane; contours indicate different experimental values for $B \to X_s\gamma$; figure taken from [29].

In particular, the spectrum contains a charged Higgs boson, which appears in the loops relevant to $b \to s\gamma$. The first analysis of this decay in this type of two-Higgs doublet model has been performed in [28].

The parameters of this model are the ratio of the two vacuum expectation values (usually expressed as $\tan\beta = v_1/v_2$), the mass M_{H^+} of the charged Higgs boson, all other parameters are irrelevant for our discussion.

In Fig. 4 (taken from [29]) the branching ratio of $b \to s\gamma$ is plotted as a function of the charged Higgs mass, for three different values of the renormalization scale μ.

In Fig. 5 (taken from [29]) we plot contours in the $\tan\beta$–M_{H^+} plane for different values of the $B \to X_s\gamma$ branching ratio. From this figure it becomes clear that there is no large effect induced by enlarging $\tan\beta$. One may derive bounds on the charged Higgs mass independently of $\tan\beta$; the current bound is $M_{H^+} > 314$ GeV at 95% CL [30].

4.2 Supersymmetry with Large $\tan\beta$

If supersymmetry were an exact symmetry, $b \to s\gamma$ would vanish, owing to the cancellations between particles and sparticles [31]. This means that $b \to s\gamma$ tests the breaking of supersymmetry. Clearly many different scenarios for this symmetry breaking can be invented, having complicated flavour structure.

Again I shall pick an example from a recent analysis [32,33]. In these papers it has been pointed out that $B \to X_s\gamma$ can indeed be enhanced in scenarios whith large $\tan\beta$. Working in the MSSM with a flavour-diagonal supersymmetry-breaking sector, the relevant parameters are the charged Higgs mass M_{H^+}, the light stop mass $m_{\tilde{t}_1}$, the supersymmetric μ parameter, and the parameter A_t from the sector of soft-supersymmetry breaking.

For large $\tan\beta$, renormalization group methods may be used to resum these terms [33] and one may confront these results with the recent data. In Fig. 6 (taken from [33]) we plot the rate for $B \to X_s\gamma$ as a function of $\tan\beta$ for $\mu = \pm 500$ GeV; the values of the other parameters are $M_{H^+} = 200$ GeV, $m_{\tilde{t}_1} = 250$ GeV, all other sparticle masses being at 800 GeV.

A similar plot can be made for negative A_t, but for the parameters chosen here this scenario is already practically excluded.

Given such a scenario, one may also scan over some range for the parameters and identify regions that are still allowed by the experimental constraints. In Fig. 7 such a scan was performed with $m_{\tilde{t}_2} \leq m_{\tilde{t}_1} \leq 1$ TeV, $m_{\tilde{\chi}_2^+} \leq m_{\tilde{\chi}_1^+} \leq 1$ TeV, $|A_t| \leq 500$ GeV, all other sparticle masses being 1 TeV.

Clearly $B \to X_s\gamma$ places significant constraints on the parameter space of certain supersymmetric scenarios; however, these studies have been performed with a flavour diagonal supersymmetry-breaking sector. An analysis witout this constraint can be found in [34]

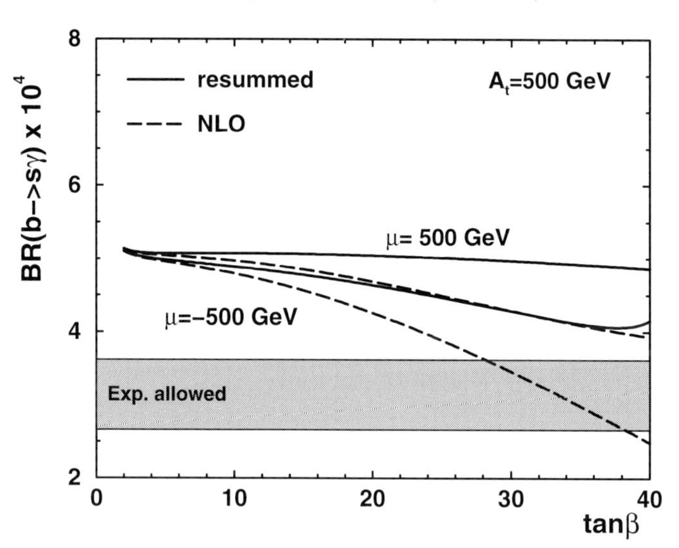

Fig. 6. The rate for $B \to X_s\gamma$ versus $\tan\beta$; for the values of the parameters see text. Figure taken from [33].

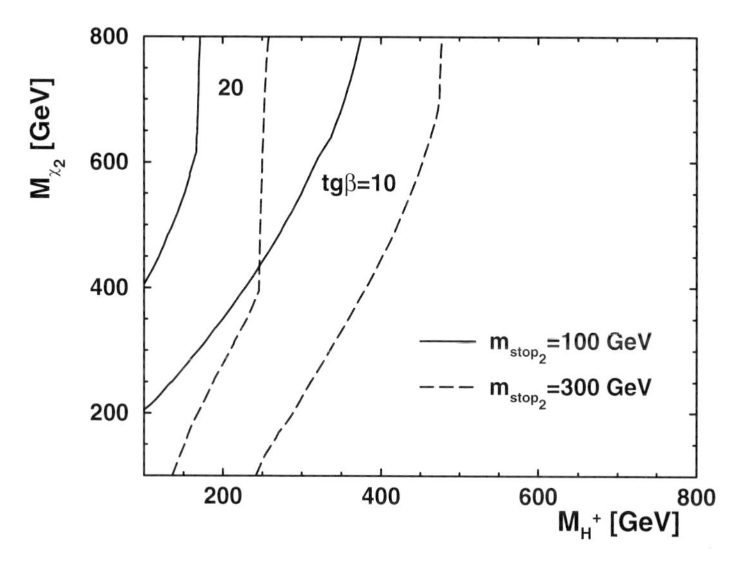

Fig. 7. Allowed range in the M_{H^+}–$M_{\tilde{\chi}2}$ plane for different values of $\tan\beta$ for two values of $m_{\tilde{t}_2}$. Figure taken from [33].

5 Summary on $b \to s\gamma$

The inclusive radiative rare decay $B \to X_s\gamma$ is under reasonable theoretical control; the latest theoretical prediction [18] is slightly higher than (9)

$$Br_{th}(B \to X_s\gamma) = (3.71 \pm 0.30) \times 10^{-4} ,$$

where the difference originates from a different value for the ratio m_c/m_b; while [16] use the ratio of pole masses, in [18] $m_c^{\overline{MS}}/m_b^{Pole}$ is is used as an "educated guess" of NNLO corrections.

The latest (combined) experimental result is [18]

$$Br_{exp} = (B \to X_s\gamma) = (2.96 \pm 0.35) \times 10^{-4} , \tag{26}$$

which is in agreement with theory within 1.6σ.

The theoretical uncertainty is mainly determined by our ignorance of some of the input parameters (quark masses, mixing angles) and to some extent also by the uncertainty of higher-order radiative corrections. Improving the current theoretical uncertainty will be very difficult with current theoretical tools.

6 The Inclusive Decay $B \to X_s\ell^+\ell^-$

6.1 Differential Rate and Forward-Backward Asymmetry

We make again use of the heavy mass expansion in which case the differential rates for the inclusive process $B \to X_s\ell^+\ell^-$ to leading order is given by the partonic rate. The quantities we are going to consider in some detail are the invariant mass spectrum ($\hat{s} = s/m_b^2$) of the leptons and the forward backward asymmetry defined by

$$\mathcal{A}(\hat{s}) \equiv \int_0^1 dz \frac{d^2\mathcal{B}}{dz\, d\hat{s}}(B \to X_s\ell^+\ell^-) - \int_{-1}^0 dz \frac{d^2\mathcal{B}}{dz\, d\hat{s}}(B \to X_s\ell^+\ell^-). \tag{27}$$

where $d^2\mathcal{B}/(dz\, d\hat{s})$ is the doubly differential rate with $z \equiv \cos\theta$, where θ is the angle of the ℓ^+ with respect to the b-quark direction in the centre-of-mass system of the dilepton pair.

Defining the kinematic variables as

$$\begin{aligned} u &= (p_b - p_1)^2 - (p_b - p_2)^2, \\ s &= (p_1 + p_2)^2, \\ \hat{s} &= \frac{s}{\mathrm{mb}^2}, \\ w(s) &= \sqrt{(s - (\mathrm{mb} + m_s)^2)(s - (\mathrm{mb} - m_s)^2)}. \end{aligned} \tag{28}$$

where p_b, p_1 and p_2 denote, respectively, the momenta of the b quark (= B hadron), ℓ^+ and ℓ^-, we obtain for the leptonic invariant mass spectrum

$$\frac{d\mathcal{B}}{d\hat{s}} = \mathcal{B}_{sl}\frac{\alpha^2}{4\pi^2}\frac{|V_{tb}V_{ts}^*|^2}{|V_{cb}|^2}\frac{1}{f(m_c^2/m_b^2)}w(\hat{s})\left[\left(|C_9+Y(\hat{s})|^2+C_{10}^2\right)\alpha_1(\hat{s},\hat{m}_s) \right. \tag{29}$$
$$\left. +\frac{4}{\hat{s}}C_7^2\alpha_2(\hat{s},\hat{m}_s)+12\alpha_3(\hat{s},\hat{m}_s)C_7(C_9+\text{ Re }Y(s))\right],$$

where f is the usual phase space function appearing in the calculation of the inclusive semileptonic rate

$$f(x) = 1-8x-8x^3-x^4-12x^2\ln x \tag{30}$$

the auxiliary functions are defined as follows:

$$\alpha_1(\hat{s},\hat{m}_s) = -2\hat{s}^2+\hat{s}(1+\hat{m}_s^2)+(1-\hat{m}_s^2)^2, \tag{31}$$
$$\alpha_2(\hat{s},\hat{m}_s) = -(1+\hat{m}_s^2)\hat{s}^2-(1+14\hat{m}_s^2+\hat{m}_s^4)\hat{s}+2(1+\hat{m}_s^2)(1-\hat{m}_s^2)^2, \tag{32}$$
$$\alpha_3(\hat{s},\hat{m}_s) = (1-\hat{m}_s^2)^2-(1+\hat{m}_s^2)\hat{s}, \tag{33}$$
$$Y(\hat{s}) = g(m_c/m_b,\hat{s})(3C_1+C_2+3C_3+C_4+3C_5+C_6) \tag{34}$$
$$-\frac{1}{2}g(1,\hat{s})(4C_3+4C_4+3C_5+C_6)$$
$$-\frac{1}{2}g(0,\hat{s})(C_3+3C_4)+\Delta C_9,$$

and $g(z,\hat{s})$ is the one-loop function given by

$$\text{Re }g(z,s) = -\frac{4}{9}\ln z^2+\frac{8}{27}+\frac{16z^2}{9s} \tag{35}$$
$$-\frac{2}{9}\sqrt{1-\frac{4z^2}{s}}\left(2+\frac{4z^2}{s}\right)\ln\left|\frac{1+\sqrt{1-\frac{4z^2}{s}}}{1-\sqrt{1-\frac{4z^2}{s}}}\right| \quad \text{for } s>4z^2$$

$$\text{Re }g(z,s) = -\frac{4}{9}\ln z^2+\frac{8}{27}+\frac{16z^2}{9s} \tag{36}$$
$$-\frac{2}{9}\sqrt{1-\frac{4z^2}{s}}\left(2+\frac{4z^2}{s}\right)\text{ atan }\left(\frac{1}{\sqrt{\frac{4z^2}{s}-1}}\right) \quad \text{for } s<4z^2$$

$$\text{Im }g(z,s) = -\frac{2\pi}{9}\sqrt{1-\frac{4z^2}{s}}\left(2+\frac{4z^2}{s}\right)\Theta(s-4z^2). \tag{37}$$

The constant ΔC_9 is given by

$$\Delta C_9(\mu) = \frac{C_9(\mu)}{\alpha_s(\mu)}-\frac{C_9(M_W)}{\alpha_s(M_W)}, \tag{38}$$

and takes into account the fact that the one-loop matrix elements of the operators $\mathcal{O}_1\cdots\mathcal{O}_6$ contains a large logarithm of the form $\ln(M_W^2/m_b^2)$, which is not due to

QCD effects. The one-loop function $g(z, \hat{s})$ is a finite piece and the choice of the renormalization scheme will effect the value of the one-loop function as well as the corresponding value of C_9. The term ΔC_9 cancels this scheme dependence,

In the same way the differential asymmetry as defined in (27) is [35]

$$\mathcal{A}(\hat{s}) = -\mathcal{B}_{sl} \frac{3\alpha^2}{8\pi^2} \frac{C_{10}}{f(m_c/m_b)} w^2(\hat{s}) \left[\hat{s}(C_9 + \text{ Re } Y(\hat{s})) + 4C_7(1 + \hat{m}_s^2)\right] . \quad (39)$$

Using these two observables we may perform an analysis of new physics effects by taking the Wilson coefficiencts to be free parameters to be determined from experiment.

6.2 Model Independent Analysis of New Physics Effects in $B \to X_s \ell^+ \ell^-$ and $B \to X_s \gamma$

In this section we shall discuss how the Wilson coefficients appearing in the effective Hamiltonian may be extracted from the experimental information. The argumentation follows closely the one in [4] We shall assume that all the matrix elements are normalized at the scale $\mu \sim m_b$, the mass of the b quark and hence the decay distributions are given in terms of the Wilson coefficients at the scale m_b. The SM makes specific predictions for these coefficients, but if there is physics beyond the SM, these coefficients will in general be modified.

We will somewhat elaborate on this point. A specific model provides the set of Wilson coefficients at high scales, which we shall choose to be the scale of the weak bosons $\mu = M_W$. Furthermore, we shall integrate out heavy degrees of freedom at the same scale $\mu = M_W$; this procedure introduces an uncertainty due to the difference in the masses of the heavy degrees of freedom. However, since the QCD coupling constant is small at these very high scales and does not appreciably change between these thresholds, it is a reasonably accurate approximation to neglect QCD corrections for scales above $\mu = M_W$. Starting from this scale, the Wilson coefficients are obtained from the solution of the renormalization group equations at the scale $\mu \sim m_b$, where we use the one-loop result for the anomalous dimensions and the beta function (see appendix).

In order to determine the sign of C_7 and the other two coefficients C_9 and C_{10}, one has to study the decay distributions and rates in $B \to X_s \ell^+ \ell^-$, where ℓ is either electron or muon. As already discussed, these decays are sensitive to the sign of C_7, and to C_9 and C_{10}. The first experimental information available in the decay $B \to X_s \ell^+ \ell^-$ will be a measurement of the branching fraction in a certain kinematic region of the invariant mass s of the lepton pair. In order to minimize long-distance effects we shall consider the kinematic regime for s below the J/ψ mass (low invariant mass) and for s above the mass of the ψ' (high invariant mass). Integrating (20) over these regions for the invariant mass one finds[2]

$$\mathcal{B}(\Delta s) = A(\Delta s) \left(C_9^2 + C_{10}^2\right) + B(\Delta s) C_9 + C(\Delta s), \quad (40)$$

[2] In performing the integrations over Δs we have set the resolution parameter δ to zero, since we do not consider any long-distance contribution. The long-distance

where A, B and C are fixed in terms of the Wilson coefficients $C_1 \cdots C_6$ and C_7. We derive from (29):

$$A(\Delta s) = \mathcal{B}_{sl} \frac{\alpha^2}{4\pi^2} \frac{1}{f(m_c/m_b)} \int\limits_{\Delta s} d\hat{s}\, \hat{u}(\hat{s}) \alpha_1(\hat{s}, \hat{m}_s) \tag{41}$$

$$B(\Delta s) = \mathcal{B}_{sl} \frac{\alpha^2}{4\pi^2} \frac{1}{f(m_c/m_b)} \int\limits_{\Delta s} d\hat{s}\, \hat{u}(\hat{s}) \left[2\alpha_1(\hat{s}, \hat{m}_s) \text{ Re } Y(\hat{s}) + 12 C_7 \alpha_3(\hat{s}, \hat{m}_s)\right] \tag{42}$$

$$C(\Delta s) = \mathcal{B}_{sl} \frac{\alpha^2}{4\pi^2} \frac{1}{f(m_c/m_b)} \int\limits_{\Delta s} d\hat{s}\, \hat{u}(\hat{s}) \left[\alpha_1(s, \hat{m}_s) \left\{(\text{ Re } Y(\hat{s}))^2 + (\text{ Im } Y(\hat{s}))^2\right\}\right.$$

$$\left. + \frac{4}{s} |C_7|^2 \alpha_2(s, \hat{m}_s) + 12 \alpha_3(s, \hat{m}_s) C_7 \text{ Re } Y(\hat{s})\right], \tag{43}$$

where the auxiliary functions α_i, $i = 1, 2, 3$, are as given above.

In our analysis we keep the values for $C_1 \cdots C_6$ and the modulus of C_7 fixed and hence $A(\Delta s)$, $B(\Delta s)$ and $C(\Delta s)$ may be calculated for the two invariant-mass ranges of interest. For the numerical analysis we use $m_b = 4.7$ GeV, $m_c = 1.5$ GeV, $m_s = 0.5$ GeV. The resulting coefficients A, B, and C are listed in Table 2.

Table 2. Values for the coefficients $A(\Delta s)$, $B(\Delta s)$ and $C(\Delta s)$ for the decay $B \to X_s \ell^+ \ell^-$.

Δs	C_7	$A(\Delta s)/10^{-8}$	$B(\Delta s)/10^{-8}$	$C(\Delta s)/10^{-8}$ $\ell = e$	$C(\Delta s)/10^{-8}$ $\ell = \mu$
$4m_\ell^2 < s < m_{J/\psi}^2$	+0.3	2.86	−5.76	84.1	76.6
$4m_\ell^2 < s < m_{J/\psi}^2$	−0.3	2.86	−20.8	124	116
$m_{\psi'}^2 < s < (1 - m_s^2)$	+0.3	0.224	−0.715	0.654	0.654
$m_{\psi'}^2 < s < (1 - m_s^2)$	−0.3	0.224	−1.34	2.32	2.32

Inspection of (43) shows that the integrand for $C(\Delta s)$ behaves as $1/s$ for small values of s, leading to a logarithmic dependence of $C(\Delta s)$ on the lepton mass for the case of the low invariant mass region. In fact, this is the only point where the masses of the leptons enter our analysis, and from this one may obtain the corresponding coefficients for $\ell = \mu$:

$$C(4m_\mu^2 < s < m_{J/\psi}^2) = C(4m_e^2 < s < m_{J/\psi}^2) - 8|C_7|^2 (1 + \hat{m}_s^2)(1 - \hat{m}_s^2)^3 \ln\left(\frac{m_\mu^2}{m_e^2}\right). \tag{44}$$

Of course one may apply (44) also to obtain C for any other lower cut s_0 on the lepton invariant mass, as long as $s_0 \ll m_{J/\psi}^2$.

contribution peaks strongly at the J/ψ and ψ' and δ has to be several times the width of these resonances in order to avoid large long-distance effects. However, calculating only the short-distance part one may safely neglect δ, since the short-distance contribution is flat in this region.

For a measured branching fraction $\mathcal{B}(\Delta s)$, one can solve the above equation for $\mathcal{B}(\Delta s)$, obtaining concentric circles in the C_9-C_{10} plane, with their centre lying at $C_9^* = B(\Delta s)/(2A(\Delta s))$ and $C_{10}^* = 0$. The radius R of these circles is proportional to

$$R = \sqrt{\mathcal{B}(\Delta s) - \mathcal{B}_{min}(\Delta s)}, \tag{45}$$

where the minimum branching fraction

$$\mathcal{B}_{min}(\Delta s) = C(\Delta s) - \frac{B^2(\Delta s)}{4A(\Delta s)} \tag{46}$$

is determined mainly by the present data on $B \to X_s\gamma$, i.e. by $|C_7|$. For the cases of interest one obtains, with the help of Table 2:

$$\mathcal{B}_{min}(4m_e^2 < s < m_{J/\psi}^2) = \begin{cases} 8.1 \times 10^{-7} \text{ for } C_7 = 0.3 \\ 8.6 \times 10^{-7} \text{ for } C_7 = -0.3, \end{cases} \tag{47}$$

$$\mathcal{B}_{min}(m_{\psi'}^2 < s < (1 - m_s^2)) = \begin{cases} 8.5 \times 10^{-10} \text{ for } C_7 = 0.3 \\ 3.0 \times 10^{-9} \text{ for } C_7 = -0.3. \end{cases} \tag{48}$$

Note that $\mathcal{B}(\Delta s)$ is an even function of C_{10}, so one is not able to fix the sign of C_{10} from a measurement of $\mathcal{B}(\Delta s)$ alone.

To further pin down the Wilson coefficients, one could perform a measurement of the forward-backward asymmetry $\mathcal{A}$, which has been defined above. The asymmetry is an odd function of C_{10}, and for a fixed value of the total branching ratio in this kinematic region one obtains, from integrating over a range (Δs):

$$\mathcal{A}(\Delta s) = C_{10}\left(\alpha(\Delta s)C_9 + \beta(\Delta s)\right), \tag{49}$$

where

$$\alpha(\Delta s) = -\mathcal{B}_{sl}\frac{3\alpha^2}{8\pi^2}\frac{1}{f(m_c/m_b)}\int_{\Delta s} d\hat{s}\,\hat{u}^2(\hat{s})\hat{s} \tag{50}$$

$$\beta(\Delta s) = -\mathcal{B}_{sl}\frac{3\alpha^2}{8\pi^2}\frac{1}{f(m_c/m_b)}\int_{\hat{s}} d\hat{s}\,\hat{u}^2(\hat{s})\left[\hat{s}\,\mathrm{Re}\,Y(\hat{s}) + 4C_7(1+\hat{m}_s^2)\right]. \tag{51}$$

For a fixed value of $\mathcal{A}(\Delta s)$, one obtains hyperbolic curves in the C_9-C_{10} plane; like the coefficients A, B and C, the parameters α and β are given in terms of the Wilson coefficients $C_1 \cdots C_6$ and C_7, and the kinematic region of s considered; their values are presented in Table 3.

Given the two experimental inputs, the branching fraction $\mathcal{B}(\Delta s)$ and the corresponding asymmetry $\mathcal{A}(\Delta s)$, one obtains a fourth-order equation for the Wilson coefficients C_9 and C_{10}, which admits in general four solutions. In Figs. 8–10 we plot the contours for a fixed value for the branching fraction $\mathcal{B}(\Delta s)$ and the FB asymmetry $\mathcal{A}(\Delta s)$. Since $\mathcal{B}(\Delta s)$ is an even function of C_{10} and $\mathcal{A}(\Delta s)$ is an odd one, we only plot positive values for C_{10}. The asymmetry vanishes

Table 3. Values for the coefficients $\alpha(\Delta s)$and $\beta(\Delta s)$.

Δs	C_7	$\alpha(\Delta s)/10^{-9}$	$\beta(\Delta s)/10^{-9}$
$4m_\ell^2 < s < m_{J/\psi}^2$	+0.3	−6.08	−24.0
$4m_\ell^2 < s < m_{J/\psi}^2$	−0.3	−6.08	55.4
$m_{\psi'}^2 < s < (1 - m_s^2)$	+0.3	−0.391	0.276
$m_{\psi'}^2 < s < (1 - m_s^2)$	−0.3	−0.391	1.37

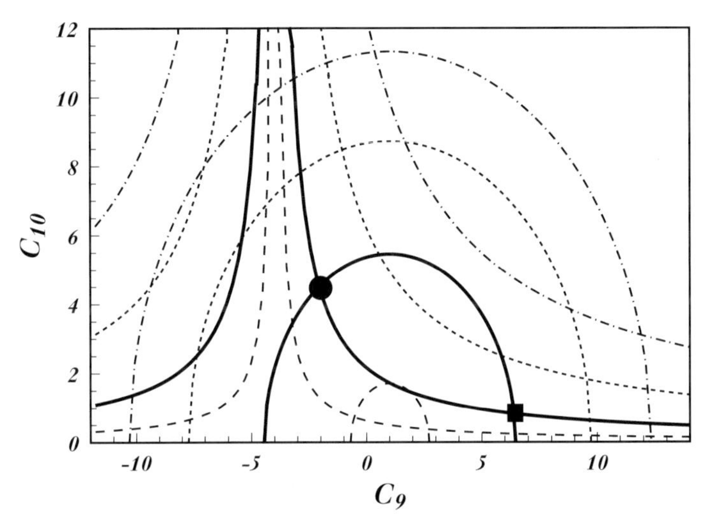

Fig. 8. Contour plots of $\mathcal{B}(\Delta s)$ and $\mathcal{A}(\Delta s)$ in the C_9-C_{10} plane for the low-invariant-mass region $4m_\ell^2 < s < m_{J/\psi}^2$ and $C_7 = 0.3$. The circles correspond to fixed values of $\mathcal{B}$: $\mathcal{B} = 5.6\times 10^{-6}$ (solid curve), $\mathcal{B} = 3.0\times 10^{-6}$ (long-dashed curve), $\mathcal{B} = 1.0\times 10^{-5}$ (short-dashed curve), $\mathcal{B} = 1.5\times 10^{-5}$ (dash-dotted curve). The left branches of the hyperbolae correspond to positive values of $\mathcal{A}$: $\mathcal{A} = 1.7 \times 10^{-7}$ (solid curve), $\mathcal{A} = 5.0 \times 10^{-8}$ (long-dashed curve), $\mathcal{A} = 5.0 \times 10^{-7}$ (short-dashed curve), $\mathcal{A} = 1.0 \times 10^{-6}$ (dash-dotted curve). The right branches of the hyperbolae correspond to negative values of $\mathcal{A}$: $\mathcal{A} = -1.41\cdot 10^{-8}$ (solid curve), $\mathcal{A} = -5.0\cdot 10^{-9}$ (long-dashed curve), $\mathcal{A} = -3.0\cdot 10^{-8}$ (short-dashed curve), $\mathcal{A} = -6.0\cdot 10^{-8}$ (dash-dotted curve). For negative values of C_{10}, the figure is simply reflected with $\mathcal{A} \to -\mathcal{A}$. The solid dot indicates the SM values for C_9 and C_{10}. The solid square is another allowed solution resulting from the SM values of $\mathcal{B}$ and $\mathcal{A}$.

for $C_{10} = 0$, but also for $C_9 = -\beta(\Delta s)/\alpha(\Delta s)$. The two lines $C_{10} = 0$ and $C_9 = -\beta(\Delta s)/\alpha(\Delta s)$ divide the C_9-C_{10} plane into four quadrants, in which the asymmetry has a definite sign. Reflecting the hyperbolae on the line $C_{10} = 0$ or $C_9 = -\beta(\Delta s)/\alpha(\Delta s)$ results in a sign change of the asymmetry.

Figures 8 and 9 show the contours in the $C_9(\mu)$-$C_{10}(\mu)$ plane for the low-invariant-mass region $4m_e^2 < s < m_{J/\psi}^2$, and Fig. 10 is for the high-invariant-mass region $m_{\psi'}^2 < s < (1-\hat{m}_s)^2$. Figures 8 and 10 are obtained for $C_7(\mu) = 0.3$, while Fig. 9 is for $C_7(\mu) = -0.3$. The possible solutions for C_9 and C_{10} are given by the intersections of the circle corresponding to the measured branching fraction

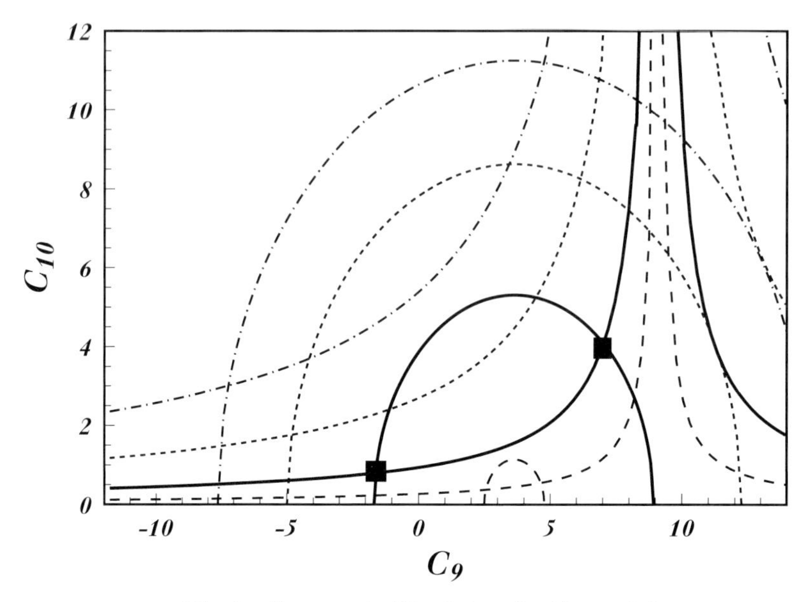

Fig. 9. Same as in Fig. 1, but for $C_7 = -0.3$.

and the hyperbola, corresponding to the measured asymmetry. Assuming the SM values for both $\mathcal{B}(\Delta s)$ and $\mathcal{A}(\Delta s)$, one obtains the solid lines in Figs. 8–10. The possible solutions in this case are represented by solid dots (SM solutions) and solid squares (other non-SM possible solutions).

From the figures one reads off that for the SM values of $\mathcal{B}$ and $\mathcal{A}$ one has more than one solution for the coefficients C_9 and C_{10}, but the ambiguity may in general be resolved by measuring both the low and the high invariant mass regions.

However, there is in principle also the possibility that the equations do not have a solution for C_9 and C_{10}. This is the case, for example, when the asymmetry is large and the branching fraction small, in which case the hyperbola may not intersect with the corresponding circle any more. If this happens one has to conclude that the present analysis is not complete; in other words, the operator basis we started from is not complete and physics beyond the SM will be present in the form of additional operators such as right-handed currents.

7 Conclusions

Rare decays mediated through flavour changing neutral current processes provide an important testing ground of the flavour structure of the standard model. In particular, the inclusive decays of B mesons – due to the large mass of the b quark – can be treated from the theoretical side in a systematic way. The main tool at our disposal is the expansion in inverse powers of the heavy quark mass, which allows us to make clean theoretical predictions including estimates of uncertainties. Such an estimate is indispensable to pin down any effect of physics

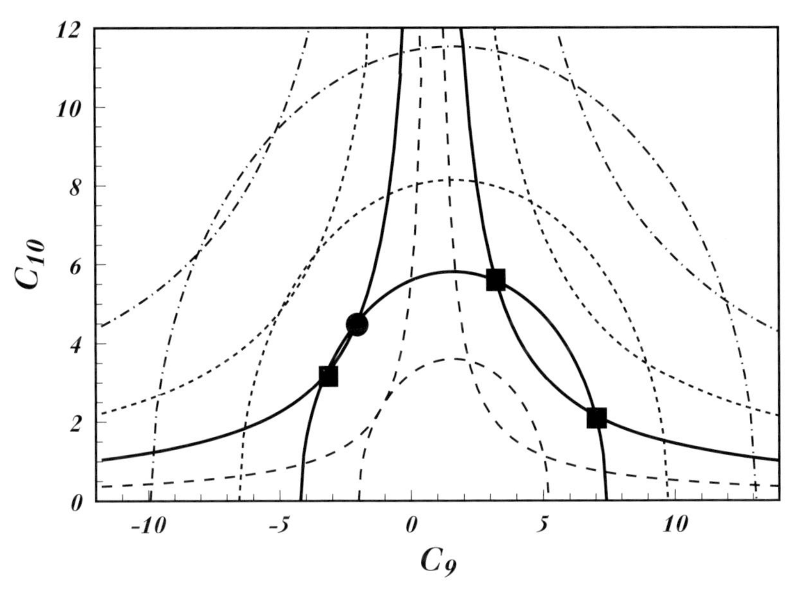

Fig. 10. Contour plots of $\mathcal{B}(\Delta s)$ and $\mathcal{A}(\Delta s)$ in the C_9-C_{10} plane for the high-invariant-mass region $m_{\psi'}^2 < s < (1-m_s)^2$ and for $C_7 = 0.3$. The circles correspond to fixed values of $\mathcal{B}$: $\mathcal{B} = 2.56 \times 10^{-7}$ (solid curve), $\mathcal{B} = 1.0 \times 10^{-7}$ (long-dashed curve), $\mathcal{B} = 5.0 \times 10^{-7}$ (short-dashed curve), $\mathcal{B} = 1.0 \times 10^{-6}$ (dash-dotted curve). The left branches of the hyperbolae correspond to positive values of $\mathcal{A}$: $\mathcal{A} = 1.41 \times 10^{-8}$ (solid curve), $\mathcal{A} = 5.0 \times 10^{-9}$ (long-dashed curve), $\mathcal{A} = 3.0 \times 10^{-8}$ (short-dashed curve), $\mathcal{A} = 6.0 \times 10^{-8}$ (dash-dotted curve). The right branches of the hyperbolae correspond to negative values of $\mathcal{A}$: $\mathcal{A} = -1.41 \times 10^{-8}$ (solid curve), $\mathcal{A} = -5.0 \times 10^{-9}$ (long-dashed curve), $\mathcal{A} = -3.0 \times 10^{-8}$ (short-dashed curve), $\mathcal{A} = -6.0 \times 10^{-8}$ (dash-dotted curve). For negative values of C_{10}, the figure is simply reflected with $\mathcal{A} \to -\mathcal{A}$. The solid dot indicates the SM Values for C_9 and C_{10}. The solid squares are other allowed solutions resulting from the SM values of $\mathcal{B}$ and $\mathcal{A}$.

beyond the Standard model in a unique way. In the meantime, many perturbative and nonperturbative contributions to rare decays have been identified and calculated in the standard model, and in particular the inclusive rare decays are under reasonable theoretical control.

Many scenarios of new physics have been invented. However, in B meson decays we can only obtain a limited ammount of information, which cannot finally settle the question of the details of physics beyond the standard model. The information that can be extracted is encoded in the Wilson coefficients of the effective Hamiltonian mediating the decays under study. In turn, this may as well be used as a model independent way to analyse B decays. Focussing on the Wilson coefficients which correspond to loop processes (i.e. $C_7, C_9 and C_{10}$), one may calculate certain observables as a function of these coeffients and – by measuring the values of these coefficients – try to extract the value of the coefficients.

In the near future the B factories as well as the hadron colliders will produce a large amount of data, including information on FCNC rare decays. Although at present there is no hint to physics beyond the standard model, there is still ample of room for a discovery of first hints at "new physics" at the B factories.

References

1. M. S. Alam *et al.* [CLEO Collaboration], Phys. Rev. Lett. **74**, 2885 (1995).
2. K. Hagiwara *et al.* [Particle Data Group Collaboration], *Review of particle Properties* Phys. Rev. D **66**, 010001 (2002).
3. T. E. Coan *et al.* [CLEO Collaboration], hep-ex/0010075.
4. A. Ali, G. F. Giudice and T. Mannel, Z. Phys. C **67**, 417 (1995) [hep-ph/9408213].
5. A. V. Manohar and M. B. Wise, *Cambridge Monographs on Particle Physics, Nuclear Physics, and Cosmology, Vol. 10.*
6. N. Isgur and M. B. Wise, CEBAF-TH-92-10, appeared in: Stone, S. (ed.): *B physics*, World Scientific 1994, p. 158-209.
7. M. Neubert, Phys. Rept. **245**, 259 (1994) [hep-ph/9306320].
8. I. Bigi, M. Shifman and N. Uraltsev, Ann. Rev. Nucl. Part. Sci. **47**, 591 (1997) [hep-ph/9703290].
9. T. Mannel, Rept. Prog. Phys. **60**, 1113 (1997).
10. G. Buchalla, A. J. Buras and M. E. Lautenbacher, Rev. Mod. Phys. **68**, 1125 (1996) [arXiv:hep-ph/9512380].
11. K. Chetyrkin, M. Misiak and M. Munz, Phys. Lett. B **400**, 206 (1997) [hep-ph/9612313].
12. A. Ali and C. Greub, Z. Phys. C **49**, 431 (1991).
13. K. Adel and Y. Yao, Phys. Rev. D **49**, 4945 (1994) [hep-ph/9308349].
14. C. Greub, T. Hurth and D. Wyler, hys. Lett. B **380**, 385 (1996) [hep-ph/9602281], Phys. Rev. D **54**, 3350 (1996) [hep-ph/9603404].
15. A. Czarnecki and W. J. Marciano, Phys. Rev. Lett. **81**, 277 (1998) [hep-ph/9804252].
16. A. L. Kagan and M. Neubert, Eur. Phys. J. C **7**, 5 (1999) [hep-ph/9805303].
17. A. J. Buras, M. Misiak, M. Munz and S. Pokorski, Nucl. Phys. B **424**, 374 (1994) [hep-ph/9311345].
18. M. Misiak, talk given at the "XXVIII Rencontres de Moriond", 10 - 17 March 2001, Les Arcs, France.
19. M. B. Voloshin, Phys. Lett. B **397**, 275 (1997) [hep-ph/9612483], A. K. Grant, A. G. Morgan, S. Nussinov and R. D. Peccei, Phys. Rev. D **56**, 3151 (1997) [hep-ph/9702380].
20. G. Buchalla, G. Isidori and S. J. Rey, Nucl. Phys. B **511**, 594 (1998) [hep-ph/9705253].
21. Z. Ligeti, L. Randall and M. B. Wise, Phys. Lett. B **402**, 178 (1997) [hep-ph/9702322].
22. M. Neubert, Phys. Rev. D **49**, 4623 (1994) [hep-ph/9312311].
23. I. I. Bigi, M. A. Shifman, N. G. Uraltsev and A. I. Vainshtein, Int. J. Mod. Phys. A **9**, 2467 (1994) [hep-ph/9312359].
24. A. K. Leibovich, I. Low and I. Z. Rothstein, Phys. Lett. B **486**, 86 (2000) [hep-ph/0005124].
25. T. Mannel and S. Recksiegel, Phys. Rev. D **60**, 114040 (1999) [hep-ph/9904475].
26. F. De Fazio and M. Neubert, JHEP**9906**, 017 (1999) [hep-ph/9905351].

27. C. W. Bauer, M. Luke and T. Mannel, hep-ph/0102089.
28. M. Ciuchini, G. Degrassi, P. Gambino and G. F. Giudice, two-Higgs doublet model," Nucl. Phys. B **527** (1998) 21 [hep-ph/9710335].
29. F. M. Borzumati and C. Greub, Phys. Rev. D **58**, 074004 (1998) [hep-ph/9802391].
30. P. Gambino, as in [18]
31. S. Ferrara and E. Remiddi, Phys. Lett. B **53**, 347 (1974).
32. G. Degrassi, P. Gambino and G. F. Giudice, JHEP**0012**, 009 (2000) [hep-ph/0009337].
33. M. Carena, D. Garcia, U. Nierste and C. E. Wagner, Phys. Lett. B **499**, 141 (2001) [hep-ph/0010003].
34. F. Gabbiani, E. Gabrielli, A. Masiero and L. Silvestrini, Nucl. Phys. B **477**, 321 (1996) [hep-ph/9604387], F. Borzumati, C. Greub, T. Hurth and D. Wyler, Phys. Rev. D **62**, 075005 (2000) [hep-ph/9911245].
35. A. Ali, T. Mannel and T. Morozumi, Phys. Lett. B **273**, 505 (1991).

Heavy Quarks on the Lattice

Craig McNeile

Department of Mathematical Sciences, University of Liverpool, L69 3BX, UK

Abstract. I review the basic ideas behind lattice QCD calculations that involve charm and bottom quarks. I report on the progress in getting the correct hyperfine splitting in charmonium from lattice QCD. Some of the basic technology behind numerical lattice QCD calculations is explained by studying some specific examples: computation of the charm quark mass, and the calculation of f_B.

1 Introduction

The B factories at KEK and SLAC are producing a wealth of new data on the decays of the B meson. One of the main goals of the current heavy flavor program is to check the CKM matrix formalism by measuring the matrix elements with sufficient accuracy. To convert the experimental data into information about the quarks requires the accurate computation of hadronic matrix elements. The best (and some would say only) way of computing the required matrix elements is to use lattice QCD.

In this paper I will explain the generic features of lattice QCD calculations that involve heavy quarks. Heavy quark lattice calculations share many common features to continuum calculations, such as matching to effective field theories. However, the more general formalism of lattice QCD allows a richer set of tools beyond just using perturbation theory. As most lattice QCD calculations share generic features, I will work through an example of computing the charm quark mass to show the important parts of the calculation. I will then discuss the calculation of the f_B decay constant. I assume the reader is already familiar with the basic ideas of heavy quark effective field theory [1,2].

The latest results on heavy quark physics from the lattice are reported in the reviews at the annual lattice conference [3–5]. The contents of the proceedings of the lattice conference have been put on hep-lat for the past couple of years [6,7] The longer reviews by Kronfeld [8], Davies [9,10] and Flynn and Sachrajda [11] contain other perspectives on heavy quarks on the lattice. Gupta gives a general overview of lattice QCD [12]. There have been recent (political) developments to set up a working group on producing a "particle data table" for lattice QCD results [13].

C. McNeile, Heavy Quarks on the Lattice, Lect. Notes Phys. **647**, 100–128 (2004)
http://www.springerlink.com/

2 A Brief Introduction to Numerical Lattice QCD

Most lattice QCD calculations start from the calculation of the correlator $c_{ij}(t)$ defined in terms of the path integral as:

$$c_{ij}(t) = \frac{1}{Z} \int dU \int d\psi d\overline{\psi} O(t)_i O(0)_j^\dagger e^{-S_F - S_G} \tag{1}$$

where S_F is the action of the fermions and S_G is the action of pure gauge theory.

The path integral is regulated by the introduction of a four dimensional space-time lattice. A typical lattice volume would be 24^3 48. The path integral is evaluated in Euclidean space for convergence. The fermion action is

$$S_F = \overline{\psi} M \psi \tag{2}$$

where M is called the fermion operator, a lattice approximation to the Dirac operator. The quadratic structure of the fermion action in (2) allows the integration over the fermion fields to be done explicitly.

$$c_{ij}(t) = \frac{1}{Z} \int dU O(t)_i O(0)_j^\dagger det(M) e^{-S_G} \tag{3}$$

The $det(M)$ term controls the dynamics of the sea quarks. The $O(t)$ operator controls the valence content of the state. For example an operator ($O(t)$) for a B meson would be:

$$O(t) = \overline{b}(t)\gamma_5 q(t) \tag{4}$$

where b and q are operators for the bottom and light quarks respectively. The operators in the path integral are Wick contracted to form a combination of quark propagators inside the path integral over the gauge fields.

$$c(t) = \frac{1}{Z} \int dU \overline{b}(t)\gamma_5 q(t) \overline{q}(0)\gamma_5 b(0) det(M) e^{-S_G} \tag{5}$$

The physical picture for the expression in 5 is a B meson created at time 0 and propagating to time t where it is destroyed. Figure 1 shows the propagation of the light and heavy quark in the vacuum.

The path integral expression for the correlator in (1) is calculated using Monte Carlo techniques on the computer. The ideas are sophisticated variants of the Monte Carlo method used to compute integrals.

The algorithms, usually based on importance sampling, produce N samples of the gauge fields on the lattice. Each gauge field is a snapshot of the vacuum. The QCD vacuum is a complicated structure. There is a community of people who are trying to describe the QCD vacuum in terms of objects such as a liquid of instantons (for example [14]). The correlator $c(t)$ is a function of the bottom ($M(U(i))_b^{-1}$) and light quark $M(U(i))_q^{-1}$ propagators averaged over the samples of the gauge fields.

$$c(t) \sim \frac{1}{N} \sum_{i}^{N} f(M(U(i))_b^{-1}, M(U(i))_q^{-1}) \tag{6}$$

The quark propagator is the inverse of the fermion operator. In perturbative calculations the quark propagator can be computed analytically from the fermion operator. In lattice QCD calculations the gauge fields have complicated space-time dependence so the quark propagator is computed numerically using variants of conjugate gradient algorithms.

The physics is extracted from the correlators by fitting the correlator to a functional form such as 7.

$$c_{ij}(t) = c_{ij}^{0} e^{-m_0 t} + c_{ij}^{1} e^{-m_1 t} + \ldots \tag{7}$$

To visually judge the quality of the data, the correlators are often displayed as effective mass plots

$$m_{eff}(t) = log(\frac{c(t)}{c(t+1)}) \tag{8}$$

An example of an effective mass plot (using the data generated for [15]) is in Fig. 2.

The computationally expensive part of lattice QCD calculations is generating the samples of gauge fields. The most expensive part of a lattice calculation is incorporating the determinant in (3). The SESAM collaboration [16] estimated that the number of floating point operations (N_{flop}) needed for n_f =2 full QCD calculations as:

$$N_{flop} \propto (\frac{L}{a})^5 (\frac{1}{am_{pi}})^{2.8} \tag{9}$$

A flop is a floating point operation such as a multiplication or addition. The value of N_{flop} represents amount of calculation required on the computer and even more importantly the cost of the computer required.

In some sense (9) (or some variant of it) is the most important equation in numerical lattice QCD. To half the size of the pion mass used in the calculations requires essentially a computer that is seven times faster. Equation 9 is not a hard physical limit. Improved algorithms or techniques may be cheaper. In

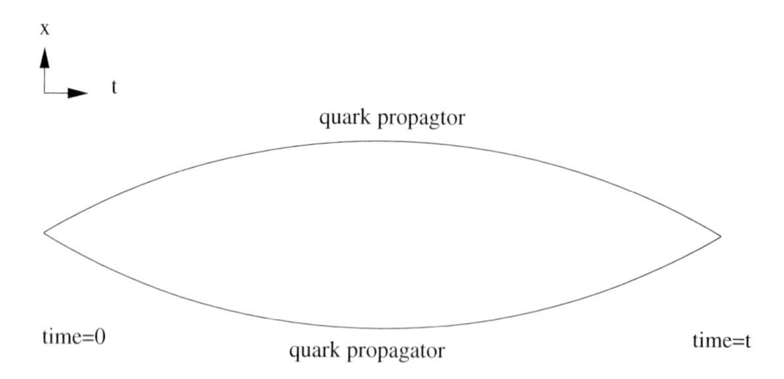

Fig. 1. Two point correlator

Table 1. Typical parameters in recent unquenched lattice QCD calculations.

Collaboration	n_f	a fm	L fm	$\frac{M_{PS}}{M_V}$
MILC [21]	2+1	0.09	2.5	0.4
CP-PACS [22]	2	0.11	2.5	0.6
UKQCD [23]	2	0.1	1.6	0.58
SESAM [24]	2	0.08	2.0	0.56

fact the "Asqtad" fermion action designed by the MILC collaboration is already computationally cheaper [17] than the cost estimates in (9).

The cost formula in (9) is for the generation of the gauge configurations. Once the gauge configurations have been generated, correlators for many different processes can be computed using some generalization of (6). This class of calculation can be carried out on a farm of workstations. The lattice QCD community are starting to create publicly available source code [18] and gauge configurations [19,20].

Table 1 shows the parameters of some recent large scale unquenched calculations. It is not considered necessary to do lattice calculations with physical light masses $M_{PS}/M_V = M_\pi/M_\rho \sim 0.18$. The aim is calculate with light enough quarks so that chiral perturbation theory can be used to make contact with the experiment. Sharpe [25] estimates that going as light as $M_{PS}/M_V \sim 0.3$ may be necessary.

The high computational cost of the fermion determinant led to development of quenched QCD, where the dynamics of the determinant is not included in (5),

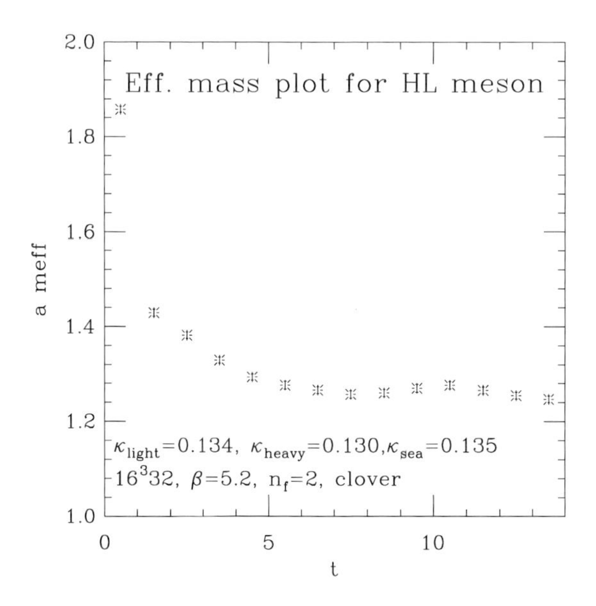

Fig. 2. Effective mass plot

hence the dynamics of the sea quarks is omitted. Until recently the majority of lattice QCD calculations were done in quenched QCD. When the dynamics of the sea quarks are included I will call the calculation unquenched.

Lattice QCD calculations produce results in units of the lattice spacing. One experimental number must be used to calculate the lattice spacing from:

$$a = am_{latt}^{X}/m_{expt}^{X} \tag{10}$$

As the lattice spacing goes to zero any choice of m_{expt}^{X} should produce the same lattice spacing – this is known scaling. Unfortunately, no calculations are in this regime yet. The recent unquenched calculations by the MILC collaboration [26] may be close.

Popular choices to set the scale are the mass of the rho, mass splitting between the S and P wave mesons in charmonium, and a quantity defined from the lattice potential called r_0. The quantity r_0 is defined by r_0 [27].

$$r_0^2 \frac{dV}{dr} |_{r_0} = 1.65 \tag{11}$$

Many potential [27] models predict $r_0 \sim 0.5$ fm. There is no perfect way to compute the lattice spacing. Although it may seem a little strange to use r_0 to calculate the lattice spacing, when it is not directly known from experiment, there are problems with all methods to set the lattice spacing. For example, to set the scale from the mass of the rho meson requires a long extrapolation in light quark mass. Also it is not clear how to deal with the decay width of the rho in Euclidean space.

2.1 Fermion Actions for Light Quarks

The lattice QCD formalism has the quark fields on the nodes of the lattice. The gauge fields are $SU(3)$ matrices and lie between the nodes of the lattice. There are a variety of different ways of writing a lattice approximation to the Dirac operator on the lattice. Gupta [12] reviews the problems and possibilities of fermion actions on the lattice. Discussions between lattice gauge theorists over which lattice fermion action is best must seem to outsiders to have the flavour of fanatical religious discussions, with the lattice "community" breaking into various sects, and accusations of "idolatry" being flung around. In the continuum limit all the fermion actions should produce the same results. This is clearly a good check on the results.

As a starting point I will consider the Wilson fermion action.

$$S_f^W = \sum_x (\kappa \sum_\mu \{\overline{\psi}_x(\gamma_\mu - 1)U_\mu(x)\psi_{x+\mu} - \overline{\psi}_{x+\mu}(\gamma_\mu + 1)U_\mu^\dagger(x)\psi_x\} + \overline{\psi}_x\psi_x) \tag{12}$$

The tree level relation between κ and the quark mass m is $\kappa = \frac{1}{2(4+m)}$.

There is a lot of effort in the lattice gauge community on designing new fermion actions for light quarks (with masses lighter than the strange quark mass). There has been a long standing concern about fermion doubling on the lattice [12]. There are a number of pragmatic solutions to the doubling problem. For example the action in (12) breaks chiral symmetry with an $O(a)$ term. Chiral symmetry will be restored as the continuum limit ($a \rightarrow 0$) is taken, but lack of chiral symmetry at finite lattice spacing causes problems, such as the difficulty of reaching light quark masses.

Our understanding of chiral symmetry on the lattice has increased by the rediscovery of the Ginsparg-Wilson relation [28]:

$$M\gamma_5 + \gamma_5 M = aM\gamma_5 M \tag{13}$$

where M is the fermion operator in (2) at zero mass.

Lattice fermion operators that obey the Ginsparg-Wilson relation (13) have a form of lattice chiral symmetry [29]. Explicit solutions, such as overlap-Dirac [30] or perfect actions [31], to (13) are known. Actions that obey the Ginsparg-Wilson relation are increasingly being used for quenched QCD calculations [32]. This class of actions have not been used for heavy quark calculations (see [33] for some speculations). Domain Wall actions, that can loosely be thought of as being approximate solutions to the Ginsparg-Wilson relation are being used in calculations [34,35] of the matrix elements for the ϵ'/ϵ. Unfortunately, solutions of the Ginsparg-Wilson relation are too expensive computationally to be used for unquenched calculations [36].

A more pragmatic development in the design of light fermion actions is the development of improved staggered fermion actions [37,21]. This class of action is being used for unquenched lattice QCD calculations with very light quarks (see Table 1) by the MILC collaboration. The improved staggered quark formalism is quite ugly compared to actions that are solutions of the Ginsparg-Wilson relation. It is not understood why calculations using improved staggered quarks are much faster [17] than calculations using Wilson fermions [16].

The largest systematic study of light hadron spectroscopy in quenched QCD has been carried out by the CP-PACS collaboration [38]. CP-PACS controlled the systematic errors by using $a \approx 0.1$ - 0.05 fm, $m_\pi/m_\rho \approx 0.75$ - 0.4, and box sizes greater than 3 fm. A summary of CP-PACS's results is in Fig. 3 CP-PACS [38] summarize their calculation of the light hadron spectrum in quenched QCD by the masses showed a deviation from experiment of less than 11%.

Although an agreement between experiment and the results of quenched QCD at the 11% level might seem impressive, many heavy-light matrix elements need to computed to an accuracy of under 5% to have an impact on tests of the CKM matrix.

There are now indications of the effects of the sea quarks. For example, the CP-PACS collaboration [22] have used $n_f = 2$ lattice QCD to calculate the strange quark mass. CP-PACS's results are in Table 2. The results show a sizable reduction in the mass of the strange quark between quenched and two flavour QCD. See the review by Lubicz [39] for a discussion of the results of CP-PACS in comparison to those from other groups. The results from CP-PACS for the strange quark mass need to be checked by other calculations with lighter quarks and finer lattice spacings,

Table 2. Mass of the strange quark from CP-PACS [22] in the $\overline{MS}$ scheme at a scale of 2 GeV.

n_f	input	m_s MeV
2	ϕ	90^{+5}_{-11}
2	K	88^{+4}_{-6}
0	ϕ	132^{+4}_{-6}
0	K	110^{+3}_{-4}

3 The Different Ways of Treating Heavy Quarks on the Lattice

In principle all the above formalism can be used to do calculations that include charm and bottom quarks. Unfortunately, in practice there is a restriction that the quark mass should be much less than the lattice spacing.

$$aM_Q << 1 \tag{14}$$

As the lattice spacings accessible to current calculations are $\frac{1}{a} \sim 2$ Gev (see Table 1) and 4 GeV [40] for unquenched and quenched QCD respectively. Hence calculations using traditional techniques will just about work for the charm mass (m_c ~1.3 GeV), but will not work for the bottom quark (m_b ~5 GeV). It is computationally expensive (see (9)) to reduce the lattice spacing, so that a b quark will be resolved by the lattice. There are a variety of special techniques

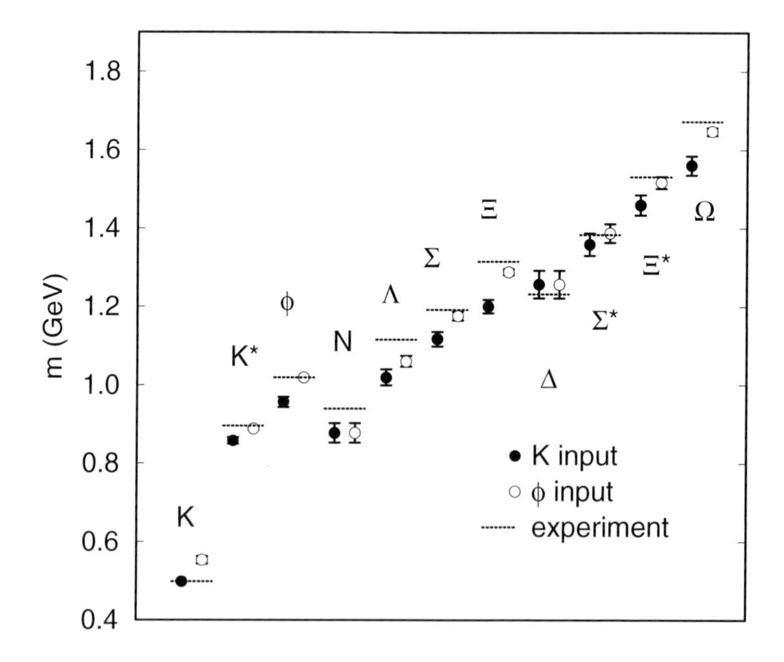

Fig. 3. Spectrum of light hadrons from CP-PACS [38]

for including the bottom quark in lattice QCD calculations, all of them are based on heavy quark effective field theory.

I do not discuss the method of using the potentials measured in lattice QCD with Schrödinger's equation to compute the mass spectrum of heavy-heavy mesons. This subject is reviewed by Bali [41]. The potential based approach is not applicable to computing matrix elements, so is not useful for checks of the CKM matrix.

Various subsets of the lattice QCD community have strong opinions on the right (and wrong) approach to including heavy quarks in lattice calculations. Obtaining consistent results from calculations that use different heavy quark actions is a good check on the systematic errors. In the next sections, I describe some of the more popular techniques used for heavy quarks on the lattice. I report on the results from this class of methods in Sect. 3.5.

3.1 The Improvement View

There are concerns that the results from lattice calculations have large lattice spacing errors because $aM_Q \sim 1$ even for charm quarks. The Wilson action in (12) has $O(a)$ lattice spacing errors. If the $O(a)$ term is removed then perhaps larger quark masses could be used in the lattice QCD calculations.

A standard technique from numerical analysis is to use derivatives that are closer approximations to the continuum derivatives. For example the lattice derivative in (16) should be more accurate with a larger lattice spacing than derivative in (15).

$$\frac{f(x+a) - f(x-a)}{2a} = \frac{df}{dx} + O(a^2) \tag{15}$$

$$\frac{4}{3}\{\frac{f(x+a) - f(x-a)}{2a} - \frac{f(x+2a) - f(x-2a)}{16a}\} = \frac{df}{dx} + O(a^4) \tag{16}$$

However in a quantum field theory there are additional complications, such as operators mixing under renormalization.

There is a formalism due to Symanzik [42,43] called improvement where new terms are added to the lattice action that cancel $O(a)$ terms in a way that is consistent with quantum field theory. The "simplest" improvement [44] to the Wilson action is to add the clover term 17 to remove tree level lattice spacing errors:

$$S_f^{clover} = S_f^W + c_{SW} \frac{ia\kappa}{2} \sum_x (\overline{\psi_x} \sigma_{\nu\mu} F_{\nu\mu} \psi_x) \tag{17}$$

where $F_{\nu\mu}$ is the lattice field strength tensor.

If the c_{SW} coefficient is computed in perturbation theory is used then the errors are $O(ag^4)$. The ALPHA collaboration [45] have computed c_{SW} to all orders in g^2 using a numerical technique. The result for c_{SW} from ALPHA is:

$$c_{SW} = \frac{1 - 0.656g^2 - 0.152g^4 - 0.054g^6}{1 - 0.922g^2} \tag{18}$$

for $0 < g < 1$, where g is the coupling. The estimate of c_{SW}, by ALPHA collaboration, agrees with the one loop perturbation theory for $g < 1/2$.

Some groups tried to use the results from lattice QCD calculations with quark masses around charm with the scaling laws from heavy quark effective field theory to compute matrix elements for the b quark. An example tried by the UKQCD collaboration [46] was to extrapolate the f_B decay constant from masses around charm, where the clover action can be legitimately used, to the bottom mass, using a functional form based on HQET [2].

$$\Phi(M_P) \equiv (\frac{\alpha(M_P)}{\alpha(M_B)})^{2/\beta_0} f_P \sqrt{M_P} = \gamma_P (1 + \frac{\delta_P}{M_P} + \frac{\eta_P}{M_P^2} + \ldots) \tag{19}$$

The review by Bernard [4] describes the potential problems with this approach. An extrapolation in mass from 2 GeV to around the bottom quark mass at 5 GeV is problematic. Note that UKQCD [46] did address some of Bernard's criticism [4]

Computers are fast enough to directly include quark masses close to the bottom mass in quenched calculations. However, this approach will not work for unquenched calculations for some time.

3.2 The Static Limit of QCD

It would clearly be better to interpolate in the heavy quark mass rather than use extrapolations. The static theory of QCD can be used to compute the properties mesons with light ante-quark and static (infinitely heavy) quarks. A combined analysis of data from static-light and heavy-light calculations can be used to interpolate to the b quark mass. This was the approach taken by the MILC collaboration [47] in one of the largest calculations of the f_B decay constant.

The lattice static theory of Eichten and Hill [48]

$$S_{static} = i a^3 \sum_x b^\dagger(x)(b(x) - U_0(x - \hat{t}) b(x - \hat{t})) \tag{20}$$

has been used for B meson physics.

One of the reasons that static quarks have not been included in many calculations is that it can be difficult to extract masses and amplitudes using (7) because the signal to noise ratio is poor. However there are numerical techniques [49] that are better, but not in wide spread use in matrix element determinations.

To extract matrix elements from static-light calculations requires the static-light operators to be matched to QCD. The ALPHA collaboration have started [50] a program to compute the matching and renormalization factors numerically. The one loop matching factors are available.

3.3 Nonrelativistic QCD

It would be better to actually do lattice calculations at the physical bottom or charm quark masses, rather than extrapolate or interpolate to the physical points.

The formalism called nonrelativistic QCD(NRQCD) allows this [51,52]. NRQCD is an effective field approximation to QCD for heavy quarks. The operators in the NRQCD Lagrangian are ordered by the velocity v.

A low order Lagrangian for NRQCD is

$$\mathcal{L}_0 = \overline{Q}(\Delta_t - \sum_{i=1}^{3} \frac{\Delta_i \Delta_{-i}}{2M_Q a} - c_{NR} \frac{\sigma.B}{2M_Q a} + ...)Q \tag{21}$$

where Δ_μ are the covariant derivatives on the lattice and B is the chrmomagnetic field.

The NRQCD formalism works both for both heavy-light (B) and heavy-heavy systems (Υ). Estimates from potential models [53] suggest that the $v^2 \sim 0.1$ in Υ and $v^2 \sim 0.3$ in charmonium. In Sect. 3.5, I review the evidence that shows that the NRQCD is not as convergent in charmonium as the naive power counting arguments suggest.

The main theoretical disadvantage of NRQCD is that the continuum limit can not be taken because of the $\frac{1}{M_Q a}$ terms in the Lagrangian. In practice improvement techniques can be used.

The coefficients, such as c_{NR} in (21), in the Lagrangian are fixed by matching to QCD. The matching calculations involve lattice perturbation theory that is harder than continuum perturbation theory because the Feynman rules are more complicated [54].

A physically motivated (but not rigorous) way of improving the convergence of lattice perturbation theory is to use tadpole improvement [53]. Tadpole perturbation theory can be used to produce “reasonable” tree level estimates for coefficients in the Lagrangian.

To do perturbation theory the gauge links are expanded in terms of the gauge potential:

$$U_\mu(x) = e^{iagA_\mu(x)} \rightarrow 1 + iagA_\mu(x) + \dots \tag{22}$$

Equation 22 suggests that the $\langle U \rangle \sim 1$, however this is not seen in numerical lattice calculations. Also the complicated vacuum structure of QCD would make $\langle U \rangle \sim 1$ unlikely. Lepage and Mackenzie [53] suggest:

$$U_\mu(x) = e^{iagA_\mu(x)} \rightarrow u_0(1 + iagA_\mu(x)) \tag{23}$$

where u_0 is the “mean gauge link”. Unfortunately there is no unique way of defining the mean gauge link. The expectation value of the mean link is zero because it is not gauge invariant. Some estimates are based on taking the quartic root of the plaquette

$$u_{0,P} = \langle \frac{1}{3} ReTr(U_{plaq}) \rangle^{1/4} \tag{24}$$

or computing the mean link in Landau gauge.

There are projects [55] under way that attempt to estimate the coefficients such as c_{NR} to order α^2. The basic idea [56] is to try to use weak coupling numerical lattice QCD calculations to obtain information on perturbative quantities.

A similar approach to NRQCD is taken by the Fermilab group [57,58], except that they match to a relativistic fermion action, essentially the clover action with mass dependent coefficients.

3.4 Anisotropic Lattices

A technique that has been used for heavy quarks [59–61] is to use a lattice spacing that is smaller in the time direction than in the space direction to circumvent the restriction $am_Q << 1$. A finer lattice spacing is used in the time direction such that $a_t m_q \leq 1$ but a larger lattice spacing a_s is used the spatial direction to keep the cost down and stop any problems with finite size effects. This approach assumes that the discretization error is only weakly dependent on $a_s m_q$.

The anisotropic clover operator [61] is

$$M = m_0 + \nu_0 \widehat{W}_0 \gamma_0 + \frac{\nu}{\xi_0} \sum_i \widehat{W}_i \gamma_i + \frac{i}{2}(w_0 \sum_{x,i} \sigma_{0i} \widehat{F_{0i}}(x) + \frac{w}{\xi_0} \sum_{x,i<j} \sigma_{ij} F_{ij}(x)) \quad (25)$$

The clover action ((12) plus (17)) is reproduced by the conditions: $\xi_0 = 1$, $\nu = \nu_0$, and $w = w_0$.

The parameters: w_0, w_i, ν_0, and ν need to be correctly chosen. For example Klassen [59] proposed to tune ν, by computing the pseudoscalar meson mass at nonzero momentum, and to choose the value of ν that gave $c(\underline{p}) = 1$.

$$E(\underline{p})^2 = E(0)^2 + c(\underline{p})^2 \underline{p}^2 \quad (26)$$

Klassen's [59] original motivation for using this class of action was that is was potentially easier to tune the unknown parameters using the techniques developed by the ALPHA collaboration [45] than for the Fermilab heavy quark action [57].

The pure gauge action is also modified [61]

$$S_g = \beta(\frac{1}{\xi_0} \sum_{x,s>s'} [1 - P_{ss'}(x)] + \xi_0 \sum_{x,s} [1 - P_{ss}(x)]) \quad (27)$$

where P_{ss} are purely spatial plaquettes and $P_{ss'}$ are plaquettes in space and time. The renormalized anisotropy $\xi_0 = a_s/a_t$ (ratio of lattice spacings in time and space) can be measured by comparing the lattice potential in space and time [62].

A practical problem in lattice calculations is that the signal in (7) is lost in the noise for large times. The smaller lattice spacing in the time direction from anisotropic lattices means that the region in lattice units, where there is a signal, is longer, thus it is easier to fit (7) to the data. Collins at al. [63] used this feature of anisotropic lattices to get improved signals for form factors. Although the fit region in lattice units is longer, the actual fit region in physical units may be smaller, this may cause problems.

3.5 The Hyperfine Splitting in Charmonium

It is obviously important to test the methods used to solve lattice QCD by comparing the results against experiment. This validation procedure ensures that the various errors in the calculations are under control. Figure 4 shows the charmonium spectrum from lattice QCD calculations by the CP-PACS collaboration [61]. The overall agreement with experiment is quite good.

A particularly good test of lattice QCD techniques is to compute the mass splitting between the J/ψ and η_c. In the section I will use Δm_H to denote the mass difference between J/ψ and η_c. The masses of these two meson can usually be computed with the smallest statistical error bars. Also, as these masses are independent of light valence quarks, this splitting does not depend on a chiral extrapolation of the valence quarks. Hein et al. [64] discuss the various systematic errors in the mass splittings between other heavy hadrons.

I will start by discussing the results from quenched QCD. I will use Δm_H as the mass splitting between the J/ψ and η_c. The experimental value for the mass splitting between the J/ψ and η_c is 116 MeV. It was therefore disappointing that some of the lattice QCD calculations in the early 90's were: $\Delta m_H =$ 28(2) MeV and 52(4) MeV from the Wilson action and tree level clover action respectively [65] ($a^{-1} = 2.73$ GeV from the string tension).

Using the clover action with the a value of the clover coefficient c_{SW} (see (17)) motivated by tadpole perturbation theory, the Fermilab group [66] obtained $\Delta m_H = 93 \pm 10$ MeV.

The hyperfine splitting is sensitive c_{SW} at nonzero lattice spacing, but the hyperfine splitting should be independent of the c_{SW} as the continuum limit is taken, because the clover term is an irrelevant operator.

Recently the QCD-TARO collaboration [67] have studied the charmonium spectrum using the clover action at a smaller lattice spacing ($a^{-1} \sim 5$ GeV)

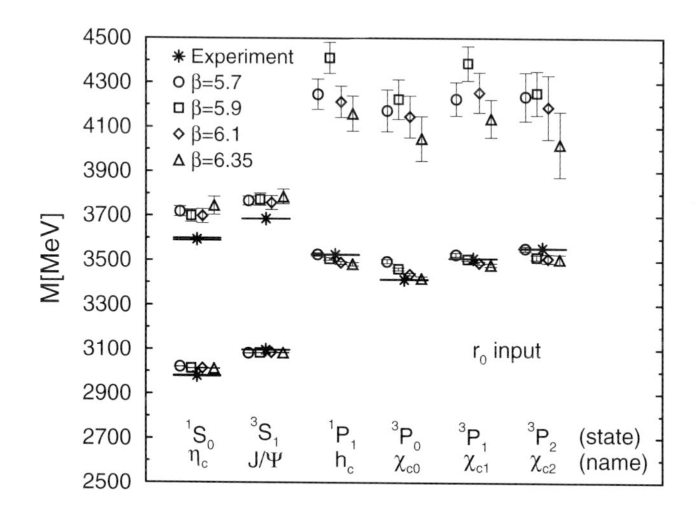

Fig. 4. Spectrum of charmonium from CP-PACS [61]

than previously used. Using the clover action QCD-TARO collaboration [67] obtained $\Delta m_H = 99 \pm 7$ MeV, $\Delta M = 87 \pm 2$ MeV, for quadratic and linear extrapolations in the lattice spacing to the continuum.

The NRQCD collaboration calculated Δm_H to be 96(2) MeV [68] using an NRQCD action that included terms of $O(m_c v^4)$. The lattice spacing was a^{-1} = 1.23(4) GeV. Using power accounting arguments the size of the next order in the NRQCD expansion was estimated to be 30 to 40 MeV. Unfortunately, when Trottier [69] included the $O(m_c v^6)$ relativistic corrections, he obtained $\Delta m_H = 55(5)$ MeV. The hyperfine splitting is sensitive to the c_{NR} coefficient (see (21)). Further work by Shakespeare and Trottier [70] showed that the hyperfine splitting was sensitive to the tadpole prescription used to estimate c_{NR}, so the final word on the utility of NRQCD for charm quarks may have to wait for c_{NR} to be computed beyond one loop. The caveat is that the current estimates for c_{NR} seem to produce good agreement with the hyperfine splittings in the baryon sector [71].

Although NRQCD is clearly not the technique to use to compute the hyperfine splitting in charmonium, NRQCD may be valid for hadrons with charm quarks, such as the mass splittings between the S-wave states and the speculated 1^{-+} state [72] and D mesons [64].

There was a preliminary attempt [73] to use an action motivated by renormalisation group arguments (perfect action [74]) to compute the charmonium spectrum. This class of action should produce results with reduced lattice spacing dependence. Unfortunately the action used in [73] did not produce a result for the hyperfine splitting closer to experiment than any other approach.

Klassen [59] first used anisotropic lattices to study the charmonium system. Using spatial lattice spacings in the range 0.17 to 0.3 fm (the scale was set using $r_0 = 0.5fm$) and two anisotropies 2 and 3, Klassen obtained a hyperfine splitting of just over 90 MeV in the charmonium system. Chen [60] obtained $\Delta m_H = 71.8(2.0)$ MeV with anisotropy $\xi_0 = a_s/a_t$ =2.

The definitive study of the anisotropic lattice technique for charmonium spectroscopy was carried out by CP-PACS [61]. They fixed the anisotropy at 3 and used spatial lattice spacings between 0.07 and 2 fm (finer then both Chen [60] and Klassen [59]). The results from CP-PACS [61] were: $\Delta m_H = = 73(4)$ MeV using r_0 to set the scale and 85(8) MeV using the P-S splitting for the lattice spacing. I have combined the different errors using quadrature. CP-PACS concluded that $a_s m_q < 1$ is still required for a reliable continuum extrapolation.

A qualitative explanation for the low value of the hyperfine splitting in charmonium from quenched QCD was given by El-khadra [66] using potential model ideas. In El-khadra's model the Richardson potential [75] is used

$$V(q^2) = C_F \frac{4\pi}{\beta_0^{n_f}} \frac{1}{q^2 \log(1 + q^2/\lambda^2)} \tag{28}$$

with

$$\beta_0^{n_f} = 11 - 2n_f/3 \tag{29}$$

to solve for the wave function of the charm quark. The wave function depends on the number of flavours n_f. El-khadra obtained the result

$$\frac{\Psi^0(0)}{\Psi^3(0)} = 0.86 \tag{30}$$

In this model the hyperfine splitting is related to the wave function and coupling (α_s) as

$$\Delta m_H \sim \frac{\alpha_s(m_c)}{m_c^2} \mid \Psi(0) \mid^2 \tag{31}$$

Including the suppression of the coupling in the quenched theory, El-khadra estimated

$$\Delta m_H^{quenched} \sim 70MeV. \tag{32}$$

There have been a number of unquenched lattice QCD calculations [24,76,23,77] that have seen evidence for the n_f dependence of the heavy quark potential at small distances. The MILC collaboration [77] have systematically studied the wave functions from the measured heavy quark potential from quenched and unquenched calculations.

There has not been much work on the charmonium spectrum from unquenched lattice QCD calculations. El-Khadra et al. [78] did look at the charmonium spectrum on (unimproved) staggered gauge configurations from the MILC collaboration. No significant increase in the hyperfine splitting was reported. The m_π / m_ρ was 0.6 and the lattice spacing was $a^{-1} \sim 0.99(4)$ GeV.

Stewart and Koniuk [79] studied the charmonium spectrum using NRQCD on unquenched (unimproved) staggered gauge configurations ($m_\pi / m_\rho \sim 0.45$ and $a \sim 0.16$ fm). Any signal for the effect of unquenching was hidden beneath the other systematic uncertainties in using NRQCD for charmonium.

Although the potential model argument of El-Khadra for the effect of quenching on the hyperfine splitting gives some insight, it does not explain the sea quark mass dependence of the splitting. Grinstein and Rothstein [80] have developed a formalism based on Chiral Lagrangian for the dependence of quarkonium mass splittings between 1P-1S and 2S-1S on the sea quark mass. Up to chiral logs they predict for the splitting δm

$$\delta m \sim A + B m_\pi^2 \tag{33}$$

where A and B are unknown parameters and m_π is the mass of the pion made out of light sea quarks.

In my opinion a decade's worth of lattice QCD calculations of the J/ψ -η_c mass splitting can be summarized as waiting for unquenched gauge configurations with light sea quark masses. Recently, progress has been made in the Upsilon system using the gauge configurations generated by the MILC collaboration. The preliminary work by Gray et al [26] found that the correct ratio was produced for the (P-S)/(2S-1S) mass splittings in Upsilon. It will be interesting to see the charmonium spectrum on these lattices, particularly if relations such as 33 can be tested and used.

There is another possible reason that the hyperfine mass splitting between the J/ψ and η_c is smaller than experiment in current simulations. All lattice calculations have computed the non-singlet correlator (see Fig. 1). However, charmonium interpolating operators are actually singlet, so the Wick contractions contain bubble diagrams (see Fig. 5). The bubble diagrams are OZI suppressed so should be small. However, this argument will fail if there is additional nonperturbative physics. For light mesons [81], it has been found that the effects of the bubbles can be large for the pseudoscalar and scalar mesons where the additional physics is the anomaly and the 0^{++} glueball, but not for other channels.

Morningstar and Peardon [82] have computed the glueball spectrum in quenched QCD. They obtained masses of 2590(40)(130) MeV and 3640(60)(180) MeV for the ground and first excited states of the 0^{-+} glueball respectively. Morningstar and Peardon computed the mass of the 1^{--} glueball to be 3850(50)(190) MeV. So it is not inconceivable that the η_c mass (2980 MeV) is effected more by glueball states than the J/ψ state. The above comments are speculations and can be checked by explicit lattice calculations. As the effect of the bubble diagrams is almost certainly less than 50 MeV, hence this will be a very hard mass splitting to estimate.

The mass spectrum of heavy-light mesons introduces the additional complication of the light valence quark. Lattice QCD calculations can be done with quark masses around charm, but for computational reasons the light quarks have masses that are typically greater than half the strange quark mass. The lattice data is extrapolated in the light quark mass to the physical points using a fit model based on chiral symmetry.

In Table 3 I have collected some results for the $D^\star - D$ mass splitting. Currently there is a lot of effort in the lattice gauge theory community to study the chiral extrapolations of quantities with the light quark mass. The Adelaide group [86,87] have developed various phenomenological forms for the light quark mass dependence of hadron masses loosely motivated by effective field theories. The fit models have had some empirical success with extrapolating the masses of the rho and nucleon [86,87], with the caveat that lattice spacing errors were not

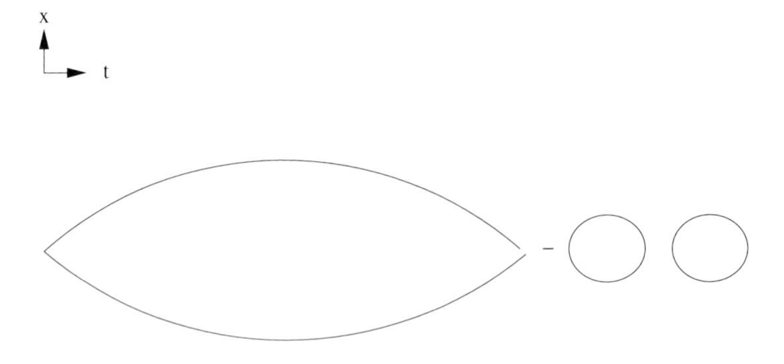

Fig. 5. Two point correlator

Table 3. Collection of hyperfine splittings between the D and $D^\star$.

Group	Method	$M_{D^\star} - M_D$ MeV
Boyle [83]	clover	124(5)(15)
Boyle [84]	β=6.0 tadpole clover	106(8)
Hein et al. [64]	NRQCD $\beta = 5.7$	$103^{+3+22}_{-0-0}(3)(6)(5)$
UKQCD [46]	NP clover $\beta = 6.2$	130^{+6+15}_{-6-35}
PDG [85]	Experiment	140.64(10)

taken into account. Similar techniques were applied to the $B^\star - B$ and $D^\star - D$ mass splittings by Guo and Thomas [88]. No improvement with the agreement between experiment and the lattice data was seen.

4 Case Study: Calculating the Charm Mass from Lattice QCD

The general steps involved in many lattice QCD calculations are fairly similar. To explain the component parts of a lattice QCD calculation, I will explain the use of lattice QCD data to extract the charm quark mass from experimental data. Full details of the calculation of the charm quark mass can be found in [89,90]. I will not discuss the approach [91,92] to computing the charm quark mass based on Fermilab's heavy quark action [57]) and the pole mass. There is a useful review by El-Khadra and Luke [93] on computing the mass of bottom quark.

The error for the mass of the charm quark quoted in the particle data table is 8% - an unbelievably large error for a basic parameter of the standard model that was discovered in 1974. The experimental mass of the D meson is 1869.3 ± 0.5, hence the error on the charm quark mass is predominantly from theory. There are many places in particle physics where a more accurate value of the charm mass would be useful. Some models of quark matrices predict relationships between quark masses and CKM matrix elements. For example Fritzsch and Xing [94] predict

$$\frac{\mid V_{ub} \mid}{\mid V_{cb} \mid} = \sqrt{\frac{m_u}{m_c}} \tag{34}$$

To test such relations we need to accurately determine all the component parameters of the standard model.

The starting point, I shall take is the masses of the heavy-light mesons as a function of lattice parameters. The hadron masses come from a fit of the correlator in (7). For example, Table 4 contains the masses of a heavy-light meson in lattice units $\beta = 6.2$ from UKQCD [46]. The κ value is defined in the action in (12). After this point no supercomputers are required, just a nonlinear χ^2 fitting program, physical insight and theoretical physics.

To start the journey from the lattice to the real world, the lattice parameters need to be converted to more physical parameters. The first job is to convert from the kappa value into the quark mass. There are a number of different expressions for the quark mass in terms of the lattice parameters.

Table 4. Heavy-light meson mass from UKQCD [46] as a function of κ value. M_{PS} and M_V are the pseudoscalar and vector meson masses

κ_H	κ_L	aM_P	aM_V
0.1200	0.1346	0.841^{+1}_{-1}	0.871^{+2}_{-2}
0.1200	0.1351	0.823^{+2}_{-1}	0.856^{+2}_{-2}
0.1233	0.1346	0.739^{+1}_{-1}	0.775^{+2}_{-2}

One definition of the quark mass is based on the vector ward identity.

$$m_V = \frac{1}{2}(\frac{1}{\kappa} - \frac{1}{\kappa_{crit}}) \tag{35}$$

The m_V quark mass suffers from an additive renormalisation for Wilson like fermions. The value of $\frac{1}{\kappa_{crit}}$ is obtained by the value of κ that gives zero pion mass. To remove $O(a)$ corrections the improvement formalism requires that

$$\hat{m} = m(1 + b_m ma) \tag{36}$$

The perturbative value of b_m is $-\frac{1}{2} - 0.096g^2$ [95].

The quark mass can also be defined in terms of the PCAC relation.

$$m_{AW} = \frac{\langle \partial_4 A_4(t) P(0) \rangle}{2\langle P(t) P(0) \rangle} \tag{37}$$

There are also O(a) corrections to (37), see [90] for details.

In principle the masses m_V and m_{AW} should agree, however at finite lattice spacing, where the calculations are actually done they disagree. For quark masses below strange, it has been shown that the two definitions agree as the lattice spacing is taken to zero (see [96] for a review).

The masses must be converted from lattice units into MeV. As explained in Sect. 2, one quantity must be sacrificed to find the lattice spacing. For example at $\beta = 6.2$ in quenched QCD, UKQCD find $a^{-1} = 2.66^{+7}_{-7}$, 2.91^{-1}_{+1}, 2.54^{-9}_{+4}, GeV from f_π, r_0, and m_ρ respectively. The spread in different choices should reduce as the continuum limit is taken in an unquenched lattice QCD calculation. If there are different choices of lattice spacing, this is usually included in the systematic error.

Now we have a table of data heavy-light meson masses in GeV versus the lattice quark masses in GeV. The meson masses must be interpolated and extrapolated to the physical meson masses. The theory behind the extrapolations is an effective Lagrangian for mesons with heavy quark and chiral symmetry [2].

The value of κ corresponding to the strange quark mass is usually determined from light quark spectroscopy by interpolating to the mass of kaon or phi meson. Becirevic et al. [89] investigated using three different fit models to extrapolate the meson masses in the heavy quark mass.

$$M^1_{HL}(m_Q) = A_1 + B_1(m_Q) + C_1 m_Q^2 \tag{38}$$

$$M_{HL}^2(m_Q) = A_2 + B_2(\frac{1}{m_Q}) + C_2(\frac{1}{m_Q^2}) \quad (39)$$

$$M_{HL}^3(m_Q) = A_3 + B_3(\frac{1}{m_Q}) + C_3 m_Q \quad (40)$$

The $1/M_Q$ terms are motivated by heavy quark symmetry [2]. Kronfeld and Simone [97] have used the fit model in (39) to estimate λ_1 and λ_2 parameters of HQET.

4.1 Quark Mass Renormalization Factors

In the last section I showed how to find the charm quark mass in the lattice scheme. However, normally quark masses are used in application in the $\overline{MS}$ scheme. Also, a consistent scheme is also required to compare the results from different calculations.

The quark mass in the lattice scheme ($m_L(a)$) is matched to the $\overline{MS}$ scheme $m_{\overline{MS}}(\mu)$.

$$m_{\overline{MS}}(\mu) = Z_m(a\mu) m_L(a) \quad (41)$$

where $Z_m(a\mu)$ is the matching factor.

The matching factor has been computed in perturbation theory to one loop order.

$$Z_m(a\mu) = 1 - \frac{\alpha(\mu)}{4\pi}(8\ln(\mu a) - C_M) \quad (42)$$

The value of C_M depends on the fermion action. For the clover action, $C_M = 25.8$ [98].

As usual with one loop calculations, there is an ambiguity as to what scale (μ) to evaluate the matching at The "best guess scale" for the μ (called $q\star$) can in principle be computed using the formalism described by Lepage and Mackenzie [53]. Most people include the effect of varying μ in some range from $1/a$ to π/a in the systematic errors.

The accuracy of the quark mass determination would improve if the matching could be done to higher order than one loop. The Feynman rules on the lattice are more complicated than in the continuum, hence calculations beyond one loop are very hard. Some groups are starting to try to automate lattice perturbation theory [99].

The general framework of lattice QCD allows other approaches to computing matching factors without using Feynman diagrams on the lattice. Sint [100] and Sommer [50] review some of the ways that matching factors are computed on the lattice. For example the α^3 term of the residual mass (important for the extraction of the bottom quark mass) of static theory in quenched QCD was computed using a numerical technique [101].

A general technique [102] for matching between the lattice and $\overline{MS}$ schemes has been developed by the Rome and Southampton groups. The basic idea is to use the quark propagator calculated in lattice QCD to do the lattice part of the matching. The gauge has to be fixed in this approach. Usually Landau gauge is chosen.

Table 5. The charm quark mass from quenched lattice QCD.

Group	$m_c^{\overline{MS}}(m_c) GeV$
Becirevic et al. [89]	1.26(4)(12)
Rolf and Sint [90]	1.301(34)
Juge [92]	1.27(5)
Kronfeld [91]	1.33(8)

4.2 Evolving the Quark Mass to a Reference Scale

The matching procedure produces the charm mass at the matching scale. To compare different mass determinations, the quark mass has to be evolved to a standard reference scale, essentially the same as that used by the particle data table. The reference scale chosen for the charm quark is the charm quark mass itself.

The running quark mass equation is used to evolve the quark mass to the standard reference scale of the charm quark mass.

$$\mu^2 \frac{d}{d\mu^2} m^{n_f}(\mu) = m^{n_f}(\mu) \gamma_m^{n_f}(\alpha_s^{n_f}) = - \sum_{i \geq 0} \gamma_{m,i}^{n_f} (\frac{\alpha_s^{n_f}(\mu)}{\pi})^{i+1} \tag{43}$$

The coefficients $\gamma_{m,i}^{n_f}$ are known to four loop order. The required equations are conveniently packaged in the RunDec Mathematica package [103]. The ALPHA collaboration have a method to do the evolution of the quark mass numerically [104,50]. The method is starting to be used for unquenched QCD [103].

4.3 Comparison of the Results

I have outlined the basic ideas behind a lattice QCD calculation of the charm quark mass. In Table 5 I collect the state-of-the art results for the charm quark mass from (quenched) lattice QCD. These should be compared with the result quoted in the particle data table of $m_c^{\overline{MS}}(m_c)$ between 1.0 to 1.4 GeV.

5 The f_B Decay Constant

A crucial quantity for tests of the CKM matrix formalism is (44)

$$\frac{\Delta m_s}{\Delta m_d} = \mid \frac{V_{ts}}{V_{td}} \mid^2 \frac{m_{B_s}}{m_{B_d}} \xi^2 \tag{44}$$

The value of Δm_d has been measured experimentally, while Δm_s is expected to be measured at run II of the Tevatron [105]. There are already useful experimental limits on Δm_s. The hard part is extracting the ratios of QCD matrix elements in

$$\xi^2 = \frac{f_{Bs}^2 B_s}{f_B^2 B} \tag{45}$$

Table 6. Summary [5] of the results for heavy-light decay constants from quenched $N_f = 0$ and ($N_f = 2$) unquenched lattice QCD.

N_f	Decay constant
0	f_B = 173(23) MeV
2	f_B = 198(30) MeV
0	f_{Ds} = 230(14) MeV
2	f_{Ds} = 250(30) MeV

The quantity ξ can not be extracted from experiment and is non-perturbative.

The f_B decay is the QCD matrix element for the semi-leptonic decay of the B meson. It is analogous to the pion decay constant. It is claimed that f_B will never be measured experimentally, hence it must be computed from QCD. The B (bag) factors are also QCD matrix elements that have been computed from lattice QCD.

The computation of f_B shares many features to the calculation of the charm quark mass. The same data from supercomputers could be used for both the f_B and charm quark mass calculation. f_B is extracted from the matrix element

$$\langle 0 \mid A_0 \mid Qq, p = 0 \rangle = -i f_{Qq} M_{Qq} \tag{46}$$

This matrix element is simply related to the amplitudes (c_{ij}) in (7). The main additional complication over the charm mass calculation is the extrapolation of the decay constant to the bottom mass. As discussed by Blum, Bernard and Soni [106] it can be convenient to compute a matrix element for $f_B^2 B$, rather than to separately compute matrix elements for B and f_B.

Sinead Ryan reviewed the latest results for the f_B decay constant at the lattice 2001 conference [5]. Ryan's world average of the lattice data for heavy light decay constants is in Table 6.

The lattice methods can be checked by computing the f_{D_s} decay constant. The current experimental result for f_{Ds} = 280 (40)(19) MeV [85]. The CLEO-c experiment plans to reduce the experimental errors on f_{Ds} to the few percent level [107] to test lattice QCD. The actual comparison between theory and experiment will be ratios of matrix elements for leptonic and semi-leptonic decays, so that the test is independent of CKM matrix elements.

In her review article Ryan [5] quoted the errors on ξ from lattice QCD as

$$\xi = 1.15(6)^{+7}_{-0} \tag{47}$$

The first error in (47) is the statistical and systematic errors from quenched QCD. The asymmetric errors are from unquenching. It is instructive to compare the errors on ξ with the experimental errors on $\Delta m_d = 0.503 \pm 0.006 ps$. Although Δm_s has not yet been measured, it is expected to be measured to a few percent accuracy at the Tevatron. The final errors on $\mid \frac{V_{ts}}{V_{td}} \mid^2$ will be limited by the theoretical errors on ξ.

Unfortunately, during the last year the errors on ξ have gone up again, based on some observations by the JLQCD collaboration [108]. Kronfeld and Ryan [109]

Table 7. Summary of some results for g_π

Group	method	g_π
UKQCD [111]	Lattice QCD	0.42(4)(8)
Abada et al. [112]	Lattice QCD	0.69(18)
CLEO [113]	Experiment $D\star D\pi$	$0.59 \pm 0.01 \pm 0.07$

have suggested that a more realistic value of ξ is 1.32 ± 0.10 rather than the estimate in (47). Atwood and Soni [110] had previously suggested that larger errors on ξ than those in (47) were more realistic.

The key problem is that the light quarks in the current unquenched lattice QCD calculations are not so light. Lattice QCD calculations are typically done at unphysically large mass parameters. Physical results are obtained by extrapolating the results using effective field theories.

The effective field theory for heavy-light systems contains the light particles: π, K, and η, and a pseudoscalar and vector heavy-light state [2]. The Lagrangian is written so that it is invariant under heavy quark symmetry and $SU(3)_L \times SU(3)_R$ symmetry. This Lagrangian is for static quarks. The Lagrangian can be used to calculate masses of hadrons and decay constants in terms of the couplings in the Lagrangian.

The most important coupling at tree level is the g_π coupling that describes the $D^\star \to D + \pi$ decay (suitably extrapolated to the heavy quark limit). Table 7 contains some estimates of g_π from experiment and lattice QCD.

The first loop correction to the decay constant has the form

$$\sqrt{m_B} f_B = \Phi(1 + \Delta f_q) \tag{48}$$

where Φ is the quantity with zero light quark mass and Δf_q represents the deviation from the chiral limit due to the finite size of the light quark mass.

The problems with the chiral extrapolations of ξ are due to the ratio of the decay constants, so consider:

$$\xi_f - 1 \equiv \frac{f_{B_s}}{f_B} - 1 \tag{49}$$

$$= \delta f_s - \delta f_d \tag{50}$$

$$= (m_K^2 - m_\pi^2) f_2(\mu) + C \tag{51}$$

where $f_2(\mu)$ is a low energy constant of the effective field theory. The form of C is

$$C = \frac{1 + 3g_\pi^2}{(4\pi)^2} \left(\frac{1}{2} m_K^2 \ln(\frac{m_K^2}{\mu^2}) + \frac{1}{4} m_\eta^2 \ln(\frac{m_\eta^2}{\mu^2}) - \frac{3}{4} m_\pi^2 \ln(\frac{m_\pi^2}{\mu^2}\right) \tag{52}$$

The equivalent expression for the bag parameters B has the coefficient $1 - 3g_\pi^2$ in front of the chiral logs, from the values of g_π in Table 7, the $m_\pi^2 \ln(\frac{m_\pi^2}{\mu^2})$ term has a negligible effect.

Until recently most lattice QCD calculations extrapolated ξ with the C function set to zero. For example, the MILC collaboration used linear and quadratic chiral extrapolations into their fits for their original results [47]. The JLQCD collaboration tried to fit (51) to their unquenched data.

Kronfeld and Ryan [109] noted that once the g_π in known, then the chiral log term in C is known. Hence, they used the lattice data that is essentially consistent with linear quark mass dependence and add the log term by hand. The problem with this type of approach is that it assumes that the current lattice data is in the regime where there are no higher order corrections to (51). The definitive answer for the value of ξ will come from unquenched calculations with light quarks that explicitly see the chiral logs in f_B.

This "case study" demonstrates the importance of the parameters of the dynamical quarks to the computation of heavy-light matrix elements, particularly the masses of the sea quarks. This study also demonstrates that the use of quenched QCD to compute heavy-light matrix elements is coming to an end. The chiral structure of matrix elements in quenched QCD can be very different to that in unquenched QCD.

5.1 Computation of Form Factors from Lattice QCD

One of the best ways to extract the $\mid V_{cb} \mid$ CKM matrix element from experiment is to use the $B \to D^\star l \nu_l$ semi-leptonic decays [1,2]. The differential decay rate [114], based on HQET 53 is

$$\frac{d\Gamma}{dw}(B \to D^\star l \nu_l) = \frac{G_F^2 \mid V_{cb} \mid^2}{48\pi^2} \mathcal{K}(w) \mathcal{F}(w)^2 \tag{53}$$

where $K(w)$ is a known phase space factor and $\mathcal{F}(w)^2$ a form factor. The value of w is the dot product of the velocities of the two heavy-light mesons. The expression in equation is based on heavy quark effective field theory.

As the masses of the b and c quarks go to infinity the normalization point of the Isgur-Wise function is obtained $\mathcal{F}(1) = 1$. The form factor at zero recoil is broken into the following

$$\mathcal{F}(1) = \eta_{QED} \eta_A (1 + \delta_{1/m_Q^2} + ..) \tag{54}$$

where η_{QED} is a perturbative QED factor and η_A is the perturbative matching factor between QCD and HQET. The δ_{1/m_Q^2} term represents the breaking of heavy quark effective field theory. A term of the form $\frac{1}{m_Q}$ is forbidden by Luke'e theorem [115].

The value of $\mid V_{cb} \mid$ in the particle data table [114] from $B \to D^\star l \nu$ decays is:

$$\mid V_{cb} \mid_{exclusive} = (42.1 \pm 1.1_{expt} \pm 1.9_{theory}) x 10^{-3} \tag{55}$$

The theoretical error is dominated by the theoretical uncertainty in δ_{1/m_Q^2}. In the past δ_{1/m_Q^2} has been computed using sum rules and quark models. Without a systematic way of improving the results, 5% will be a lower limit on the accuracy of this CKM matrix element.

In Fig. 6 I show a space-time diagram that is used to calculate form factors for semi-leptonic using the path integral. The initial lattice studies [116–119]. mapped out the dependence of the $B \rightarrow D$ form factors on w. The experimental data is taken at nonzero recoil $w <> 1$. However, the extrapolation to $w = 1$ [120] is either done using a simple ansatz or using the results of a dispersion relation [121,122].

The Fermilab group [123] have concentrated on estimating δ_{1/m_Q^2} in (54). They compute the matrix element in Fig. 6 with all the mesons at rest. By taking clever combinations of matrix elements they can get a precise estimate of the form factor at zero recoil. The mass dependence of the form factor can be mapped out by varying the masses of the heavy quarks. The matching to the continuum is quite involved. The final results [123] have errors with the same order of magnitude as other approaches. Their error includes a 10% estimate for unquenching.

There are other ways that lattice QCD can contribute to the extraction of the $\mid V_{cb} \mid$ CKM matrix element from experimental data. There have been calculations of the semi-leptonic decays of the Λ_b baryon [124] from lattice QCD. There have been two calculation of the mass of the B_c meson [125,126]. Jones and Woloshyn [125] computed the decay constant for the leptonic decay of B_c to be 420(13) MeV (the error is statistical only), using NRQCD at a lattice spacing of 0.163(3) fm. Whether these additional channels can help reduce the theoretical uncertainty of $\mid V_{cb} \mid$ to below the 5% level is not clear.

6 Nonleptonic Decays

One of the main goals of the B physics experimental program is to check the CKM matrix formalism by measuring the CKM matrix elements many different

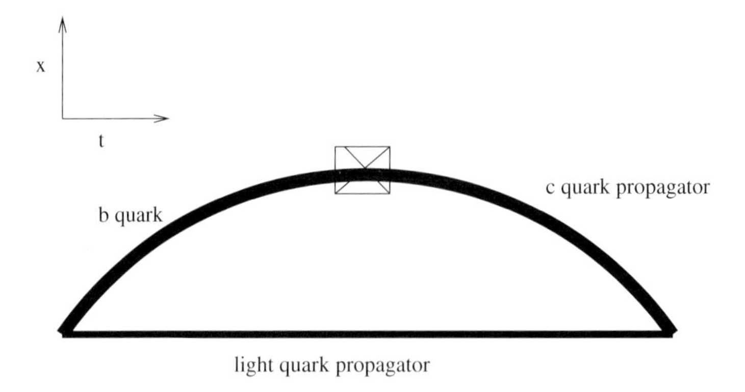

Fig. 6. Space time diagram of three point function

ways [127,128]. For example, the experimental measurements for $B^+ \to \pi^+\pi^0$ could be used to extract $\sin(2\alpha)$ if the hadronic uncertainties could be controlled.

The path integral in (1) is calculated in Euclidean space to regulate the oscillations in Minkowski space. This means that the amplitudes extracted from lattice QCD calculations are always real. Recently, there has been some theoretical work (see [129] for a review) on the non-leptonic decays of the kaons, motivated by the attempts to compute the hadronic matrix elements for ϵ'/ϵ.

There were some early attempts to study the decays $D \to K\pi$ on the lattice [130–132]. These type of lattice calculations stopped when the theoretical problems with making contact with experiment became apparent. In this section I briefly describe some of the old work on the $D \to K\pi$ decays and provide pointers to the new theoretical developments.

The correlator required is

$$G(t) = \langle 0 \mid (\pi K)(t) H_{eff}(0) D(t_K) \mid 0 \rangle \tag{56}$$

where $D(t_K)$ is the interpolating field for the D meson at time t_K and πK is the interpolating operator for the pion and kaon. The effective Hamiltonian is

$$H_{eff} = c_+(\mu) O_+^{cont} + c_-(\mu) O_-^{cont} \tag{57}$$

$$O_\pm^{cont} = (\overline{s_L}\gamma_\mu c_L)(\overline{u_l}\gamma_\mu d_L) \pm (\overline{s_L}\gamma_\mu d_L)(\overline{u_L}\gamma_\mu c_L) \tag{58}$$

where $c_\pm(\mu)$ are perturbative coefficients.

The diagrams for the Wick contraction of (56) are in Fig. 7. Although the diagrams are more complicated to compute than those for leptonic or semileptonic decays they can be calculated on a supercomputer. The problems occur trying to extract the pertinent amplitudes from $G(t)$ in (56).

As pointed out by Michael [133] and, Maiani and Testa [134] there is a complication with creating a pion and kaon state with definitive momentum. The operator

$$O_{\pi K}(t) = \pi(\underline{p}, t) K(-\underline{p}, t) \tag{59}$$

has the same quantum numbers as πK with all possible momentum values. The ground state of the operator in (59) will be the pion and kaon at rest. Hence, in the analogue of (7) for this graph, the required amplitude will not be the ground state. It is not easy to fit a multi-exponential model to data, although it is possible with a basis of interpolating operators [129].

Theoretical work, in the context of $K \to \pi\pi$ decays has shown how to get matrix elements in infinite volume from matrix elements computed in finite volumes [135,136]. There have also been proposals [133,137] to introduce some model independence to extract the complex phases of the matrix elements.

The methodology for non-leptonic decays will be further developed and tested on $K \to \pi\pi$ decays, before any attempts are made at the decays $D \to K\pi$,

7 Conclusions

The consumers of lattice QCD results need the error bars on current matrix elements to be reduced below 5%. The hardest error to reduce is from quenching.

Improved staggered quarks look like they will be the first to explore unquenched QCD with light sea quarks. This should motivate the champions of other light quark formalisms to speed up their unquenched calculations. The techniques that will reduce the error bars of heavy-light matrix elements will lie outside the domain of heavy quarks, in areas such as algorithms, improved computer hardware, and better grant writing.

My own, admittedly biased view, is that quenched QCD calculations are now of limited use for lattice QCD calculations with heavy quarks. As every experimentalist I have ever met has held this view, I am sure it will prevail.

The computation of matrix elements for two body hadronic decays still looks quite hard. Interesting things seem to be happening for kaon decays and in theory. It is not clear, whether these developments will be useful for non-leptonic decays of the B meson. It would obviously be a major breakthrough if this problem could be solved, however Mark Wise's wise [128] words, on the career ending nature of working on nonleptonic decays should be heeded.

The computation of the QCD matrix elements for the heavy flavour program is a well defined task. If we can't compute them reliably, then we will have failed. We will have to admit that we can't compute anything from QCD outside perturbation theory from first principles. I hope this doesn't happen.

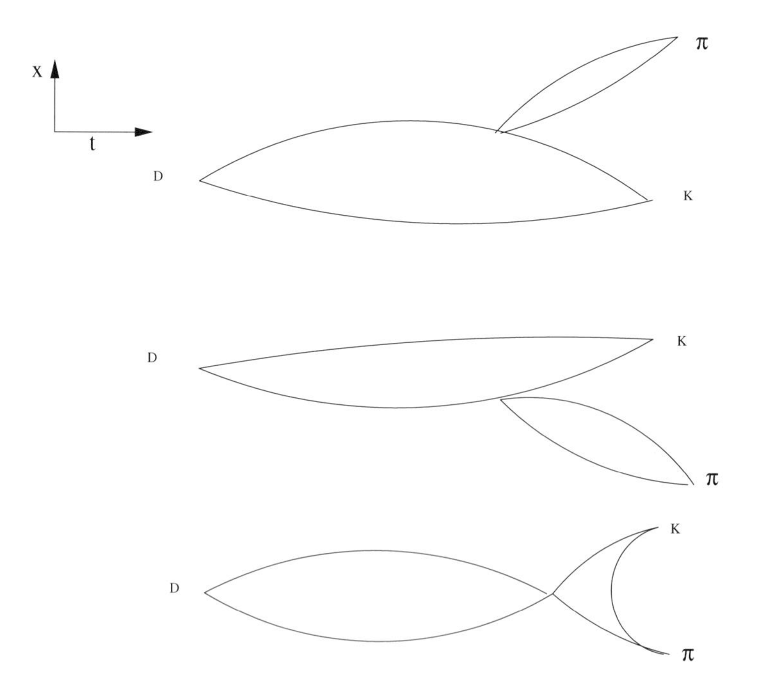

Fig. 7. Wick contractions for (56)

Acknowledgments

I thank Chris Michael and Alex Dougall for discussions.

References

1. M. Neubert, Phys. Rept. **245**, 259 (1994), hep-ph/9306320,
2. A. V. Manohar and M. B. Wise, Cambridge Monogr. Part. Phys. Nucl. Phys. Cosmol. **10**, 1 (2000),
3. T. Draper, Nucl. Phys. Proc. Suppl. **73**, 43 (1999), hep-lat/9810065,
4. C. W. Bernard, Nucl. Phys. Proc. Suppl. **94**, 159 (2001), hep-lat/0011064,
5. S. M. Ryan, Nucl. Phys. Proc. Suppl. **106**, 86 (2002), hep-lat/0111010,
6. C. T. H. Davies *et al.*, (1998), hep-lat/9801024,
7. M. Mueller-Preussker *et al.*, (2002), hep-lat/0203004,
8. A. S. Kronfeld, (2002), hep-lat/0205021,
9. C. Davies, (1997), hep-ph/9710394,
10. C. Davies, (2002), hep-ph/0205181,
11. J. M. Flynn and C. T. Sachrajda, Adv. Ser. Direct. High Energy Phys. **15**, 402 (1998), hep-lat/9710057,
12. R. Gupta, (1997), hep-lat/9807028,
13. J. Flynn, L. Lellouch, and G. Martinelli, (2002), hep-lat/0209167,
14. J. W. Negele, Nucl. Phys. Proc. Suppl. **73**, 92 (1999), hep-lat/9810053,
15. UKQCD, C. M. Maynard, Nucl. Phys. Proc. Suppl. **106**, 388 (2002), hep-lat/0109026,
16. TXL, T. Lippert, Nucl. Phys. Proc. Suppl. **106**, 193 (2002), hep-lat/0203009,
17. S. Gottlieb, Nucl. Phys. Proc. Suppl. **106**, 189 (2002), hep-lat/0112039,
18. M. Di Pierro, Nucl. Phys. Proc. Suppl. **106**, 1034 (2002), hep-lat/0110116,
19. UKQCD, C. McNeile, (2000), hep-lat/0003009,
20. UKQCD, C. T. H. Davies, A. C. Irving, R. D. Kenway, and C. M. Maynard, (2002), hep-lat/0209121,
21. MILC, C. W. Bernard *et al.*, Phys. Rev. **D61**, 111502 (2000), hep-lat/9912018,
22. CP-PACS, A. Ali Khan *et al.*, Phys. Rev. Lett. **85**, 4674 (2000), hep-lat/0004010,
23. UKQCD, C. R. Allton *et al.*, Phys. Rev. **D65**, 054502 (2002), hep-lat/0107021,
24. TXL, U. Glassner *et al.*, Phys. Lett. **B383**, 98 (1996), hep-lat/9604014,
25. S. R. Sharpe, (1998), hep-lat/9811006,
26. HPQCD, A. Gray *et al.*, (2002), hep-lat/0209022,
27. R. Sommer, Nucl. Phys. **B411**, 839 (1994), hep-lat/9310022,
28. P. H. Ginsparg and K. G. Wilson, Phys. Rev. **D25**, 2649 (1982),
29. M. Luscher, Phys. Lett. **B428**, 342 (1998), hep-lat/9802011,
30. H. Neuberger, Ann. Rev. Nucl. Part. Sci. **51**, 23 (2001), hep-lat/0101006,
31. P. Hasenfratz, Nucl. Phys. **B525**, 401 (1998), hep-lat/9802007,
32. P. Hernandez, Nucl. Phys. Proc. Suppl. **106**, 80 (2002), hep-lat/0110218,
33. K.-F. Liu, (2002), hep-lat/0206002,
34. CP-PACS, J. I. Noaki *et al.*, (2001), hep-lat/0108013,
35. RBC, T. Blum *et al.*, (2001), hep-lat/0110075,
36. K. Jansen, Nucl. Phys. Proc. Suppl. **106**, 191 (2002), hep-lat/0111062,
37. MILC, K. Orginos and D. Toussaint, Phys. Rev. **D59**, 014501 (1999), hep-lat/9805009,
38. CP-PACS, S. Aoki *et al.*, Phys. Rev. Lett. **84**, 238 (2000), hep-lat/9904012,

39. V. Lubicz, Nucl. Phys. Proc. Suppl. **94**, 116 (2001), hep-lat/0012003,
40. SPQ(CD)R, D. Becirevic, V. Lubicz, and C. Tarantino, (2002), hep-lat/0208003,
41. G. S. Bali, Phys. Rept. **343**, 1 (2001), hep-ph/0001312,
42. K. Symanzik, Nucl. Phys. **B226**, 187 (1983),
43. K. Symanzik, Nucl. Phys. **B226**, 205 (1983),
44. B. Sheikholeslami and R. Wohlert, Nucl. Phys. **B259**, 572 (1985),
45. M. Luscher, S. Sint, R. Sommer, P. Weisz, and U. Wolff, Nucl. Phys. **B491**, 323 (1997), hep-lat/9609035,
46. UKQCD, K. C. Bowler *et al.*, Nucl. Phys. **B619**, 507 (2001), hep-lat/0007020,
47. C. W. Bernard *et al.*, Phys. Rev. Lett. **81**, 4812 (1998), hep-ph/9806412,
48. E. Eichten and B. Hill, Phys. Lett. **B234**, 511 (1990),
49. UKQCD, C. Michael and J. Peisa, Phys. Rev. **D58**, 034506 (1998), hep-lat/9802015,
50. R. Sommer, (2002), hep-lat/0209162,
51. B. A. Thacker and G. P. Lepage, Phys. Rev. **D43**, 196 (1991),
52. G. P. Lepage, L. Magnea, C. Nakhleh, U. Magnea, and K. Hornbostel, Phys. Rev. **D46**, 4052 (1992), hep-lat/9205007,
53. G. P. Lepage and P. B. Mackenzie, Phys. Rev. **D48**, 2250 (1993), hep-lat/9209022,
54. C. J. Morningstar, Phys. Rev. **D50**, 5902 (1994), hep-lat/9406002,
55. H. D. Trottier and G. P. Lepage, Nucl. Phys. Proc. Suppl. **63**, 865 (1998), hep-lat/9710015,
56. H. D. Trottier, N. H. Shakespeare, G. P. Lepage, and P. B. Mackenzie, Phys. Rev. **D65**, 094502 (2002), hep-lat/0111028,
57. A. X. El-Khadra, A. S. Kronfeld, and P. B. Mackenzie, Phys. Rev. **D55**, 3933 (1997), hep-lat/9604004,
58. A. S. Kronfeld, Phys. Rev. **D62**, 014505 (2000), hep-lat/0002008,
59. T. R. Klassen, Nucl. Phys. Proc. Suppl. **73**, 918 (1999), hep-lat/9809174,
60. P. Chen, Phys. Rev. **D64**, 034509 (2001), hep-lat/0006019,
61. CP-PACS, M. Okamoto *et al.*, Phys. Rev. **D65**, 094508 (2002), hep-lat/0112020,
62. M. G. Alford, I. T. Drummond, R. R. Horgan, H. Shanahan, and M. J. Peardon, Phys. Rev. **D63**, 074501 (2001), hep-lat/0003019,
63. UKQCD, S. Collins *et al.*, Phys. Rev. **D64**, 055002 (2001), hep-lat/0101019,
64. J. Hein *et al.*, Phys. Rev. **D62**, 074503 (2000), hep-ph/0003130,
65. UKQCD, C. R. Allton *et al.*, Phys. Lett. **B292**, 408 (1992), hep-lat/9208018,
66. A. X. El-Khadra, Nucl. Phys. Proc. Suppl. **30**, 449 (1993), hep-lat/9211046,
67. QCD-TARO, S. Choe *et al.*, Nucl. Phys. Proc. Suppl. **106**, 361 (2002), hep-lat/0110104,
68. C. T. H. Davies *et al.*, Phys. Rev. **D52**, 6519 (1995), hep-lat/9506026,
69. H. D. Trottier, Phys. Rev. **D55**, 6844 (1997), hep-lat/9611026,
70. N. H. Shakespeare and H. D. Trottier, Phys. Rev. **D58**, 034502 (1998), hep-lat/9802038,
71. N. Mathur, R. Lewis, and R. M. Woloshyn, Phys. Rev. **D66**, 014502 (2002), hep-ph/0203253,
72. CP-PACS, T. Manke *et al.*, Phys. Rev. Lett. **82**, 4396 (1999), hep-lat/9812017,
73. K. Orginos, W. Bietenholz, R. Brower, S. Chandrasekharan, and U. J. Wiese, Nucl. Phys. Proc. Suppl. **63**, 904 (1998), hep-lat/9709100,
74. P. Hasenfratz, Nucl. Phys. Proc. Suppl. **63**, 53 (1998), hep-lat/9709110,
75. J. L. Richardson, Phys. Lett. **B82**, 272 (1979),
76. UKQCD, C. R. Allton *et al.*, Phys. Rev. **D60**, 034507 (1999), hep-lat/9808016,

77. C. W. Bernard *et al.*, Phys. Rev. **D62**, 034503 (2000), hep-lat/0002028,
78. A. X. El-Khadra, S. Gottlieb, A. S. Kronfeld, P. B. Mackenzie, and J. N. Simone, Nucl. Phys. Proc. Suppl. **83**, 283 (2000),
79. C. Stewart and R. Koniuk, Phys. Rev. **D63**, 054503 (2001), hep-lat/0005024,
80. B. Grinstein and I. Z. Rothstein, Phys. Lett. **B385**, 265 (1996), hep-ph/9605260,
81. UKQCD, C. McNeile, C. Michael, and K. J. Sharkey, Phys. Rev. **D65**, 014508 (2002), hep-lat/0107003,
82. C. J. Morningstar and M. J. Peardon, Phys. Rev. **D60**, 034509 (1999), hep-lat/9901004,
83. UKQCD, P. Boyle, Nucl. Phys. Proc. Suppl. **53**, 398 (1997),
84. UKQCD, P. Boyle, Nucl. Phys. Proc. Suppl. **63**, 314 (1998), hep-lat/9710036,
85. Particle Data Group, D. E. Groom *et al.*, Eur. Phys. J. **C15**, 1 (2000),
86. D. B. Leinweber, A. W. Thomas, K. Tsushima, and S. V. Wright, Phys. Rev. **D61**, 074502 (2000), hep-lat/9906027,
87. D. B. Leinweber, A. W. Thomas, K. Tsushima, and S. V. Wright, Phys. Rev. **D64**, 094502 (2001), hep-lat/0104013,
88. X. H. Guo and A. W. Thomas, Phys. Rev. **D65**, 074019 (2002), hep-ph/0112040,
89. D. Becirevic, V. Lubicz, and G. Martinelli, Phys. Lett. **B524**, 115 (2002), hep-ph/0107124,
90. ALPHA, J. Rolf and S. Sint, (2002), hep-ph/0209255,
91. A. S. Kronfeld, Nucl. Phys. Proc. Suppl. **63**, 311 (1998), hep-lat/9710007,
92. K. J. Juge, Nucl. Phys. Proc. Suppl. **106**, 847 (2002), hep-lat/0110131,
93. A. X. El-Khadra and M. Luke, (2002), hep-ph/0208114,
94. H. Fritzsch and Z.-z. Xing, Phys. Lett. **B506**, 109 (2001), hep-ph/0102295,
95. S. Sint and P. Weisz, Nucl. Phys. **B502**, 251 (1997), hep-lat/9704001,
96. R. Gupta and T. Bhattacharya, Phys. Rev. **D55**, 7203 (1997), hep-lat/9605039,
97. A. S. Kronfeld and J. N. Simone, Phys. Lett. **B490**, 228 (2000), hep-ph/0006345,
98. M. Gockeler *et al.*, Phys. Rev. **D57**, 5562 (1998), hep-lat/9707021,
99. I. T. Drummond, A. Hart, R. R. Horgan, and L. C. Storoni, (2002), hep-lat/0208010,
100. S. Sint, Nucl. Phys. Proc. Suppl. **94**, 79 (2001), hep-lat/0011081,
101. F. Di Renzo and L. Scorzato, JHEP **02**, 020 (2001), hep-lat/0012011,
102. G. Martinelli, C. Pittori, C. T. Sachrajda, M. Testa, and A. Vladikas, Nucl. Phys. **B445**, 81 (1995), hep-lat/9411010,
103. K. G. Chetyrkin, J. H. Kuhn, and M. Steinhauser, Comput. Phys. Commun. **133**, 43 (2000), hep-ph/0004189,
104. ALPHA, S. Capitani, M. Luscher, R. Sommer, and H. Wittig, Nucl. Phys. **B544**, 669 (1999), hep-lat/9810063,
105. K. Anikeev *et al.*, (2001), hep-ph/0201071,
106. C. W. Bernard, T. Blum, and A. Soni, Phys. Rev. **D58**, 014501 (1998), hep-lat/9801039,
107. I. Shipsey, (2002), hep-ex/0207091,
108. JLQCD, N. Yamada *et al.*, Nucl. Phys. Proc. Suppl. **106**, 397 (2002), hep-lat/0110087,
109. A. S. Kronfeld and S. M. Ryan, Phys. Lett. **B543**, 59 (2002), hep-ph/0206058,
110. D. Atwood and A. Soni, Phys. Lett. **B508**, 17 (2001), hep-ph/0103197,
111. UKQCD, G. M. de Divitiis *et al.*, JHEP **10**, 010 (1998), hep-lat/9807032,
112. A. Abada *et al.*, (2002), hep-ph/0206237,
113. CLEO, A. Anastassov *et al.*, Phys. Rev. **D65**, 032003 (2002), hep-ex/0108043,
114. M. Artuso and E. Barberio, (2002), hep-ph/0205163,

115. M. E. Luke, Phys. Lett. **B252**, 447 (1990),
116. C. W. Bernard, Y. Shen, and A. Soni, Phys. Lett. **B317**, 164 (1993), hep-lat/9307005,
117. UKQCD, S. P. Booth *et al.*, Phys. Rev. Lett. **72**, 462 (1994), hep-lat/9308019,
118. UKQCD, K. C. Bowler *et al.*, Phys. Rev. **D52**, 5067 (1995), hep-ph/9504231,
119. UKQCD, K. C. Bowler, G. Douglas, R. D. Kenway, G. N. Lacagnina, and C. M. Maynard, Nucl. Phys. **B637**, 293 (2002), hep-lat/0202029,
120. CLEO, R. A. Briere *et al.*, Phys. Rev. Lett. **89**, 081803 (2002), hep-ex/0203032,
121. I. Caprini, L. Lellouch, and M. Neubert, Nucl. Phys. **B530**, 153 (1998), hep-ph/9712417,
122. C. G. Boyd, B. Grinstein, and R. F. Lebed, Phys. Rev. **D56**, 6895 (1997), hep-ph/9705252,
123. S. Hashimoto, A. S. Kronfeld, P. B. Mackenzie, S. M. Ryan, and J. N. Simone, Phys. Rev. **D66**, 014503 (2002), hep-ph/0110253,
124. UKQCD, K. C. Bowler *et al.*, Phys. Rev. **D57**, 6948 (1998), hep-lat/9709028,
125. B. D. Jones and R. M. Woloshyn, Phys. Rev. **D60**, 014502 (1999), hep-lat/9812008,
126. UKQCD, H. P. Shanahan, P. Boyle, C. T. H. Davies, and H. Newton, Phys. Lett. **B453**, 289 (1999), hep-lat/9902025,
127. S. Stone, (2001), hep-ph/0112008,
128. M. B. Wise, (2001), hep-ph/0111167,
129. N. Ishizuka, (2002), hep-lat/0209108,
130. C. W. Bernard, J. Simone, and A. Soni, Nucl. Phys. Proc. Suppl. **17**, 504 (1990),
131. European Lattice, A. Abada *et al.*, Nucl. Phys. Proc. Suppl. **17**, 518 (1990),
132. C. W. Bernard, J. N. Simone, and A. Soni, Nucl. Phys. Proc. Suppl. **20**, 434 (1991),
133. C. Michael, Nucl. Phys. **B327**, 515 (1989),
134. L. Maiani and M. Testa, Phys. Lett. **B245**, 585 (1990),
135. L. Lellouch and M. Luscher, Commun. Math. Phys. **219**, 31 (2001), hep-lat/0003023,
136. C. J. D. Lin, G. Martinelli, C. T. Sachrajda, and M. Testa, Nucl. Phys. **B619**, 467 (2001), hep-lat/0104006,
137. M. Ciuchini, E. Franco, G. Martinelli, and L. Silvestrini, Phys. Lett. **B380**, 353 (1996), hep-ph/9604240,

BaBar Experiment Status and Recent Results

Guglielmo De Nardo, representing the BaBar Collaboration

Naples University and INFN, Naples, Dipartimento di Scienze Fisiche, Complesso Universitario di Monte Sant'Angelo, via Cintia, 80126 Napoli, Italy

Abstract. The BaBar detector at SLAC PEP-II asymmetric B-Factory has collected between 1999 and 2002 a data sample of 88 millions $\Upsilon(4S) \rightarrow B\bar{B}$ decays. We present here recent measurements of branching fractions and time-dependent CP-violating asymmetries of neutral B mesons decays to several CP eigenstates. We present the results on the decays to $(c\bar{c})$ K^0_S / K^0_L, which are related in the Standard Model to the angle β of the Unitarity Triangle of the Cabibbo-Kobayashi-Maskawa quark mixing matrix. Moreover we present the branching fractions and the CP-asymmetries of charmless two body decays related to the angle α.

1 Introduction

The source of CP symmetry violation within the Standard Model of electroweak interactions is provided in an elegant way by one non-negligible complex phase in the three generation Cabibbo-Kobayashi-Maskawa quark mixing matrix (CKM) [1]. In a convenient parameterization of the matrix due to Wolfenstein [2], the phase is placed in the V_{td} and V_{ub} elements. The unitarity of the CKM matrix, among the various relations between its rows and columns, implies in particular that $V_{td}V^*_{tb} + V_{cd}V^*_{cb} + V_{ud}V^*_{ub} = 0$, which can be visualized as the closure relation of a triangle in the complex plane. This triangle is called Unitarity Triangle(U.T.). The CP symmetry is violated if that triangle has non-zero area or, which is the same, its angles, called in the literature (α, β and γ), are different from zero or π.

The measurement of CP-violating time dependent asymmetries in neutral B meson decays to CP eigenstates with charm provides a theoretically clean determination of sin 2β [3], where $\beta = arg(-V_{cd}V^*_{cb}/V_{td}V^*_{tb})$, i.e. one of the angles of the Unitarity Triangle. Asymmetry in the $B^0 \rightarrow \pi\pi$ decay, instead, allows, although in a less direct way, the extraction of sin 2α, where $\alpha = arg(-V^*_{tb}V_{td}/V^*_{ub}V_{ud})$. The ratios of the branching fractions and the CP asymmetries for $\pi\pi$ and $K\pi$ modes are sensitive to $\gamma = \pi - \alpha - \beta$, if the CKM picture of CP violation is correct.

2 The BaBar Detector

The BaBar detector has been built and is operated by a large international team of scientists and engineers. It is taking data at the SLAC PEP-II B Factory, which is an asymmetric e^+e^- collider designed to operate at a luminosity of

G. De Nardo, BaBar Experiment Status and Recent Results, Lect. Notes Phys. **647**, 129–146 (2004)
http://www.springerlink.com/

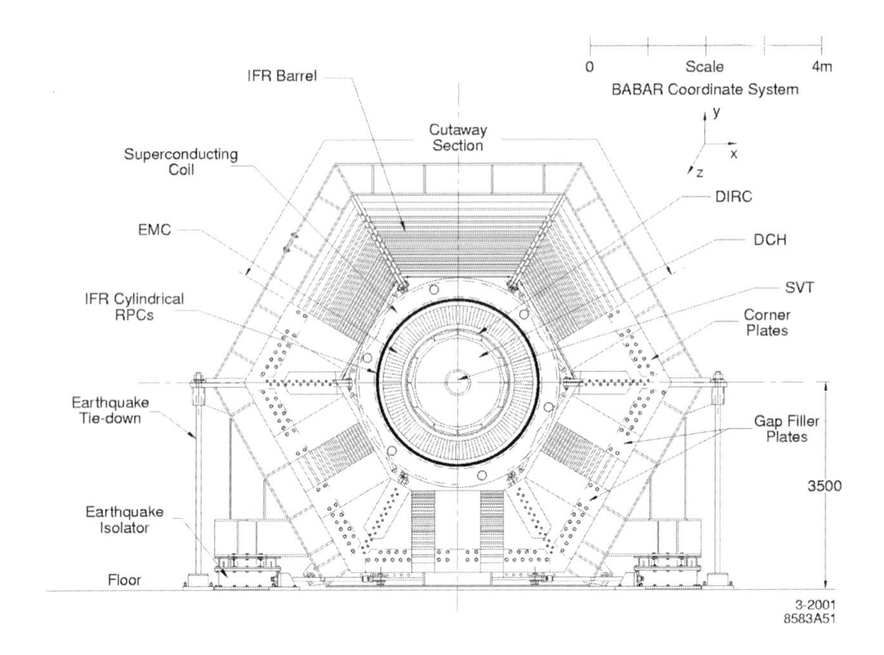

Fig. 1. BaBar detector longitudinal section.

$3 \times 10^{33}\mathrm{cm}^{-2}\mathrm{s}^{-1}$, at a center of mass energy of 10.58 GeV, the mass of the $\Upsilon(4S)$. The resonance decays exclusively in $B^0\bar{B}^0$ and B^+B^- pairs, providing a clean environment for B physics studies. The results reported in this paper are based on the data set collected between 1999 and June 2002, corresponding to 88 millions $\Upsilon(4S) \to B\bar{B}$ decays.

The electron beam of 9.0 GeV collides on with a positron beam of 3.1 GeV; because of the different beam energies, the center of mass frame moves in the laboratory frame with a Lorentz boost of $\beta\gamma = 0.56$. The boost makes it possible to reconstruct the decay vertices of the two B mesons, in order to measure the relative decay times and, consequently, the time dependent asymmetries.

The very small branching fractions of B mesons to CP eigenstates, the need of full reconstruction of final states, including charged and neutral particles, and the need of the measurement of the flavor of the companion B meson (B tagging), put stringent requirements in term of efficiencies and resolutions of the detector subsystems.

The detector, shown in Fig. 1, is divided in five subsytems: a vertex detector (SVT), a central drift chamber (DCH), a Cherenkov detector (DIRC), an electromagnetic calorimeter (EMC) and the muon and neutral hadron subsystem (IFR). A detailed description of the BaBar detector and PEP-II can be found in [4] and [5].

3 Time Dependent CP Asymmetries

The positrons and electrons beams of PEP-II collides on with a total energy equal to the mass of the $\Upsilon(4S)$. Therefore the B mesons at PEP-II are produced in pairs via the decay $\Upsilon(4S) \to B\bar{B}$. Since the $\Upsilon(4S)$ is a spin 1 particle and the B meson has spin 0, the $B\bar{B}$ pair is produced in a p-wave. Bose statistics, thus, enforces that they oscillate coherently: if the flavor of one B is measured at any time t, the flavor of the companion B meson must be the opposite, at the same instant of time.

This nice circumstance is exploited in the measurement of time dependent CP asymmetries in the B meson decays. More precisely the experimental technique proceeds as follows [6]:

- The final state of interest is fully reconstructed. All the kinematics constraints are taken into account to select an high purity sample of events containing this decay (i.e. a sample as much free of background as possible).
- In order to measure the flavor of the reconstructed B meson(B_{CP}), the flavor of the recoiling B meson (B_{tag}) is determined, examining signatures of flavor in the rest of the event. The Bose statistics argument given above guarantees that, at the same time t_{tag} at which the B_{tag} decayed, the B_{CP} meson had the opposite flavor.
- Being interested in the time evolution of the B_{CP} decay rate, the time difference $\Delta t = t_{CP} - t_{tag}$ has to be measured. This is accomplished by measuring the distance between the vertices of the two decays. The z_{CP} measurement is provided by means of a fit to a common vertex of the charged tracks in the CP final state, which takes into account all useful topological and kinematic constraints as well (intermediate state masses, decay vertices, non-negligible flight length of long-lived particles). The remaining charged particles in the event are used as input for a dedicated vertexing algorithm to provide the other needed position measurement z_{tag}. Of course, the latter measurement, being less constrained, dominates the uncertainty on the measurement of the distance between these two vertices. Relativistic kinematics connects this position measurements to a time measurement (to a very good approximation $\Delta t = \Delta z/\gamma\beta c$).

The decay rate distributions $f_+(\Delta t)$ ($f_-(\Delta t)$) of B decays to a CP eigenstate f, when the companion B is a B^0 ($\bar{B}^0$), are given by [7]

$$f_\pm(\Delta t) = \frac{e^{-|\Delta t|/\tau_{B^0}}}{4\tau_{B^0}} \left(1 \pm S_f \sin \Delta m_d \Delta t \mp C_f \cos \Delta m_d \Delta t\right), \tag{1}$$

where $\Delta t = t_{CP} - t_{tag}$ is the difference between the proper decay times of the $B \to f$ (B_{CP}) and the companion B meson (B_{tag}), τ_{B^0} is the B^0 lifetime and Δm_d is the B^0-$\bar{B}^0$ oscillation frequency.

The S_f and C_f coefficients in (1) are defined in terms of a complex parameter λ_f:

$$S_f = 2\frac{\text{Im}\lambda_f}{1 + |\lambda_f|^2}, \tag{2a}$$

$$C_f = \frac{1 - |\lambda_f|^2}{1 + |\lambda_f|^2}, \tag{2b}$$

and they vanish if CP is conserved.

If the imaginary part of λ_f, and thus S_f, is different from zero, there is CP violation in the interference between mixing and decay; if the absolute value of λ_f is different from unity, and thus C_f is non-zero, there is CP violation in decay, or direct CP violation. While the first type of CP violation has been already estabilished in the B meson system by the BABAR [8] and BELLE [9] collaborations, direct CP violation has not yet been observed only outside the neutral kaons system [10] [11].

The Standard Model predictions of CP violation can be tested determining the CP violation parameters C_f and S_f for various final states f, by means of the difference between B^0 and $\bar{B}^0$ tagged decay rates in the same CP final state f, as a function of the time difference Δt:

$$\mathcal{A}_{CP}(\Delta t) = \frac{f_+(\Delta t) - f_-(\Delta t)}{f_+(\Delta t) + f_-(\Delta t)}. \tag{3}$$

3.1 Selection of the CP Sample

The first step in the measurement of the time dependent asymmetry in (3) is the selection of the events containing the decay mode $B \to f$ regardless of the flavor of the parent B. This is accomplished by fully reconstructing all the particles present in the decay chain (after a preselection of multi-hadron events and basic cuts on event shape in order to suppress the background from non $B\bar{B}$ events). Since the final state f is completely reconstructed, all the available kinematic constraints (invariant masses of the intermediate non stable particles), topological constraints (vertexing of the charged tracks) and particle identification information (at various level of efficiency/purity optimized to the specific mode reconstructed) are used. This allows an accurate estimate of the B meson candidate four momentum. Moreover, since the B meson is produced in the process $e^+e^- \to \Upsilon(4S) \to B\bar{B}$, the selection can take advantage of the two body kinematics of the $\Upsilon(4S)$ decay and from the fact that the beam energy is well measured. Therefore, the analyses in BaBar which perform a full reconstruction of the B use the following two kinematic variables:

$$m_{ES} = \sqrt{\left(\frac{s/2 + \boldsymbol{p}_i \cdot \boldsymbol{p}_B}{E_i}\right)^2 - \boldsymbol{p}_B^2}, \tag{4a}$$

$$\Delta E = \sqrt{s}/2 - E_B, \tag{4b}$$

where $\sqrt{s}$ is the center of mass energy, $(E_B, \boldsymbol{p}_B)$ is the B reconstructed four momentum and $(E_i, \boldsymbol{p}_i)$ is the initial state four momentum in the laboratory frame. The beam energy substituted mass variable m_{ES} is the invariant mass of the B candidate evaluated from the known initial state total energy, to determine the energy of the B candidate, and from the total momentum of the B candidate decay products to determine the momentum. Since in this definition only the beam energy and the momenta of the particles appear, the m_{ES} variable does not depend on the mass hypotheses of the particles from which the B meson is reconstructed. Signal yields and sample purities are extracted from fits to the m_{ES} distributions of B candidates. Signal events follows a Gaussian distribution peaked at the B meson nominal mass, with a resolution dependent from the decay products. Combinatorial background, which arises from random combination of charged tracks and neutral showers from both the B mesons in $B\bar{B}$ events or from continuum events, follows an ARGUS distribution [12], whose shape has the following functional form:

$$\mathcal{A}(m_{ES}; m_0, \xi) = A_B m_{ES} \sqrt{1 - x_{ES}^2} e^{\xi(1-x_{ES}^2)}, \qquad (5)$$

where $x_{ES} = m_{ES}/m_0$, where m_0 represents a kinematic limit fixed by the beam energy at 5.291 GeV/c^2, and ξ and A_B are free parameters.

The ΔE distribution is the difference between the energy of the reconstructed B candidate and the energy of the B expected from the beam, and it is distributed following a Gaussian distribution for signal events, peaked at zero if all the mass hypotheses of the B meson candidate decay products are correct.

Signal region and background sidebands are delimited in the plane defined by this two uncorrelated variables, in order to select the CP sample and to study its backgrounds. Moreover, the same technique is used to select samples of events containing charged and neutral B decays, which have been completely reconstructed, in order to measure vertexing resolutions, tagging performances or to perform branching ratios measurements.

3.2 B Flavor Tagging and Δt Resolution

Several signatures of flavor can be found examining the decay products of the recoiling B meson. The most powerful is the charge of the primary leptons from the B semileptonic decays. Indeed, in the $b \to cl\nu$ transition of the b quark (charge -1/3) to the c quark (charge +2/3) proceeds by the emission of an intermediate virtual W^-, which decays to a negative lepton and an anti-neutrino. The charge of the lepton is the same of the charge of the b quark inside the B meson, tagging the B meson flavor. Similarly, charm decays determine flavor, as well: the charge of the best identified kaon, coming from a secondary decay $b \to c \to s$ is correlated to the B flavor; or, evidence of charm in the event can be found detecting the soft pion produced in the $D^{\star+}$ decay $D^{\star+} \to D^0\pi^+_{soft}$; in this case the pion charge is correlated to the B meson flavor.

In the BaBar experiment, multivariate tagging algorithms are defined to identify the flavor of the tagged B [6]. A neural network combines all the information from these physics based tagging algorithms in order to exploit all the correlations between the different sources of tagging information.

Table 1. Efficiencies ϵ_i, average mistag fraction w_i, mistag fraction differences $\Delta w_i = w_i(B^0) - w_i(\bar{B}^0)$, measured for each tagging category form the combined B_{flav} and B_{CP} samples, in $\sin 2\beta$ analysis.

Cetegory	ϵ (%)	w (%)	Δw(%)	Q (%)
Lepton	9.1 ± 0.2	3.3 ± 0.6	-1.5 ± 1.1	7.9 ± 0.3
Kaon-I	16.7 ± 0.2	10.0 ± 0.7	-1.3 ± 1.1	10.7 ± 0.4
Kaon-II	19.8 ± 0.3	20.9 ± 0.8	-4.4 ± 1.2	6.7 ± 0.4
Inclusive	20.0 ± 0.3	31.5 ± 0.9	-2.4 ± 1.3	2.7 ± 0.3
Total	65.6 ± 0.5			28.1 ± 0.7

If the tagging algorithm incorrectly determines the flavor with a probability w, the amplitudes of the observed $B^0\bar{B}^0$ oscillations and CP asymmetries are reduced by a dilution factor $\mathcal{D} = 1 - 2w$.

On the basis of the output of the physics-oriented algorithms and the estimated mistag probability, each event is assigned to one of four hierarchical and mutually exclusive categories. The Lepton category contains events with a well identified lepton and a supporting kaon tag, if present. Events with a kaon tag and a soft pion with opposite charge and consistent flight direction are assigned to the Kaon-I category. Events with a kaon without a soft pion are assigned to the Kaon-I or to the Kaon-II category on the basis of the estimated mistag probability. The rest of the events are excluded or assigned to the Inclusive category depending on the estimated mistag probability. The quality of the tag depends on both its efficiency ϵ (how many times the algorithm is able to give an answer) and its mistag probability (how many times the output is wrong). The quantitative figure of merit is the effective tagging efficiency $Q = \sum_i \epsilon_i(1 - 2w_i)^2$, since the contribution to the statistical uncertainty in the asymmetry measurement is $\sigma_{asym} = \sigma_0/\sqrt{NQ}$. The performances of the tagging algorithms in the $\sin 2\beta$ analysis are summarized in Table 1.

Another important effect that must be taken into account is the finite resolution of the detector in the measurement of the time difference Δt. The time evolution of the rates of tagged events $f_\pm(\Delta t)$ must be convolved with a resolution function $\mathcal{R}(\delta t = \Delta t - \Delta t_{true}; \boldsymbol{a})$, where Δt and Δt_{true} are the measured and true time difference between the tagging B decay and reconstructed B decay and $\boldsymbol{a}$ are the parameters of the resolution function.

In order to measure the w_i mistag rates and the a_i parameters of the Δt resolution functions, a data sample of events with a neutral B meson fully reconstructed in $B^0 \to D^{(*)-}\pi^+/\rho^+/a_1^+$ or $B^0 \to J/\psi K^{*0}(K^{*0} \to K^+\pi^-)$, and the corresponding flavor conjugates modes, has been used (B_{flav}). In these decays, the flavor of the reconstructed B meson is correlated to the sign of the D meson or the kaon in the final state. Therefore, the dilutions due to mistagging can be extracted studying the time dependent rate of the $B^0\bar{B}^0$ oscillations on these data sample.

The mistag rates w_i and the Δt resolution parameters $\boldsymbol{a_i}$, for each tagging category i, can be extracted performing an unbinned maximum likelihood fit to the time distribution of the fully reconstructed B_{flav} sample:

$$\ln \mathcal{L}_{mix} = \sum_i^{tagging} \left[\sum_{unmixed} \ln h_+(\Delta t; w_i, \boldsymbol{a}_i) + \sum_{mixed} \ln h_-(\Delta t; w_i, \boldsymbol{a}_i) \right] \otimes \mathcal{R}(\delta t, \boldsymbol{a}), \tag{6a}$$

$$h_\pm = \frac{e^{-|\Delta t|/\tau_B^0}}{4\tau_{B^0}} \left[1 \pm (1-2w) \cos \Delta m_d \Delta t\right], \tag{6b}$$

where the sum is over the taging category i, mixed (unmixed) is for events in which the B mesons pair is found to have opposite (same) flavor, and the $h_\mp$ are the probability to find the pair in opposite (same) flavor, as a function of the decay time difference, according to the known phenomenon of flavor oscillations.

In the limit of no dependence from the reconstructed side, the same mistag rate parameters w_i and resolution parameters a_i can be used for the CP asymmetry measurement, and the functions $f_\pm$ in (3), describing the decay rates have to be substituted by the following function which take into account the experimental effects ($f_\pm \to \mathcal{F}_\pm$):

$$\mathcal{A}_{CP}(\Delta t) = \frac{\mathcal{F}_+(\Delta t) - \mathcal{F}_-(\Delta t)}{\mathcal{F}_+(\Delta t) + \mathcal{F}_-(\Delta t)}, \tag{7a}$$

$$\mathcal{F}_\pm(\Delta t; w, \boldsymbol{a}) = \frac{e^{-|\Delta t|/\tau_{B^0}}}{4\tau_{B^0}} \left[1 \pm \mathcal{D}\left(\frac{2\mathrm{Im}\lambda}{1+|\lambda|^2} \sin \Delta m_d \Delta t - \frac{1-|\lambda|^2}{1+|\lambda|^2} \cos \Delta m_d \Delta t\right)\right] \tag{7b}$$

The value of the free parameter λ can be extracted using the B_{CP} sample with the tagging and vertexing requirements by maximizing the likelihood:

$$\ln \mathcal{L}_{CP} = \sum_i^{tagging} \left[\sum_{B^0 tagged} \ln \mathcal{F}_+(\Delta t; w_i, \boldsymbol{a}_i, \lambda) + \sum_{\bar{B}^0 tagged} \ln \mathcal{F}_-(\Delta t; w_i, \boldsymbol{a}_i, \lambda) \right] \otimes \mathcal{R}(\delta t, \boldsymbol{a}) \tag{8}$$

where B^0 tagged and $\bar{B}^0$ tagged is for events identified by the tagging algorithm as containing a recoiling B^0 or $\bar{B}^0$. In practice, the fit is performed simultaneously on the combined B_{flav} and B_{CP} sample with a likelihood constructed with the sum of the likelihoods in (6a) and (8), to determine the CP violating parameter, the mistag fractions, the vertex resolution parameters, including additional terms to account for backgrounds and their time dependence.

4 Measurement of the CP-Violating Asymmetry Amplitude $\sin 2\beta$

In the Standard Model the most abundant decays of neutral B meson, sensitive to the value of the β angle of the Unitarity Triangle, are due to the quark

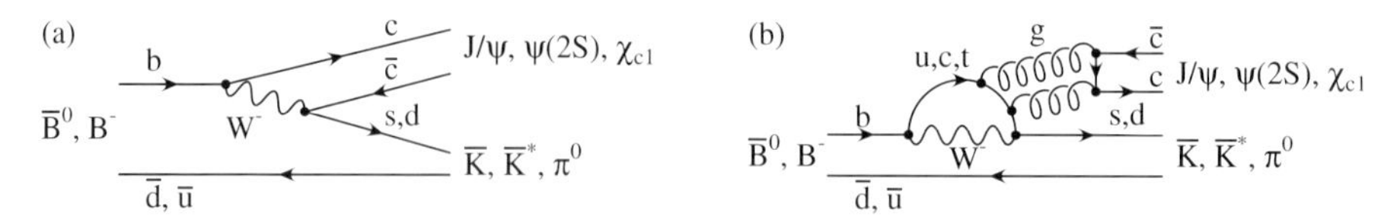

Fig. 2. (a) Tree level (a) and penguin (b) amplitudes for $b \to c\bar{c}s(d)$ transition and corresponding particles in the final state. (b) Penguin amplitude for

level process $b \to c\bar{c}s$, whose Feynman diagrams are shown in Fig. 2. The corresponding final states are a $c\bar{c}$ resonance $(J/\psi, \psi(2S), \chi_c, \eta_c, etc.)$ and a K^0 or a K^{*0}.

The penguin amplitude, which, in principle, contributes to these decays, actually does not modify the value of the parameter λ, since it shares the same weak phase with the leading tree amplitude. Therefore, λ can be written in terms of CKM matrix elements as:

$$\lambda_f = \eta_f^{CP} \times \left(\frac{V_{tb}^* V_{td}}{V_{tb} V_{td}^*} \right) \times \left(\frac{V_{cs}^* V_{cb}}{V_{cs} V_{cb}^*} \right) \times \left(\frac{V_{cd}^* V_{cs}}{V_{cd} V_{cs}^*} \right), \tag{9}$$

where the three CKM factors are due to CP violation in the $B^0 \bar{B}^0$ mixing, which is dominated by the top quark loop, to CP violation in the $b \to c\bar{c}s$ and $\bar{b} \to c\bar{c}\bar{s}$ amplitudes, and to the amplitude of $K^0 - \bar{K}^0$ mixing.

Using unitarity relations and the definition of the β angle, the expression simplifies to:

$$\lambda_f = \eta_f^{CP} e^{-2i\beta}, \tag{10}$$

where η_f^{CP} is the CP eigenvalue of the final state f, equal to -1 if the neutral kaon is a K_S^0, or equal to +1 if the neutral kaon is a K_L^0; the final state in K^{*0} is not a pure CP eigenstate, and requires an angular analysis, to separate the CP even and CP odd components and extract the CP violation parameter λ.

Since λ is a pure phase, the CP violating parameters of (2a) and (2b), and the time dependent CP asymmetry of (3) become:

$$\begin{aligned} C_f &= 0 \\ S_f &= -\eta_f^{CP} \sin 2\beta \\ \mathcal{A}_{CP}(\Delta t) &= -\eta_f^{CP} \sin 2\beta \sin(\Delta m_d \Delta t), \end{aligned} \tag{11}$$

In Fig. 3 are shown the m_{ES} distributions for the events containing a K_S^0 or a K^{*0} and the ΔE distribution for the $J/\psi K_L^0$ candidates. In the latter case, the background distribution is taken from Monte Carlo simulation for the $B^0 \to J/\psi X$ background, and from sidebands in data for the fake J/ψ background.

The measurement of $\sin 2\beta$ is performed following the method described in Sect. 3. The value of $\sin 2\beta$ is extracted from an unbinned maximum likelihood fit. The used Likelihood function is the sum of the likelihoods in (6a) and (8), assuming $|\lambda| = 1$, maximized simultaneously on the combined B_{CP} and B_{flav} samples:

$$\ln \mathcal{L} = \ln \mathcal{L}_{CP} + \ln \mathcal{L}_{mix}. \tag{12}$$

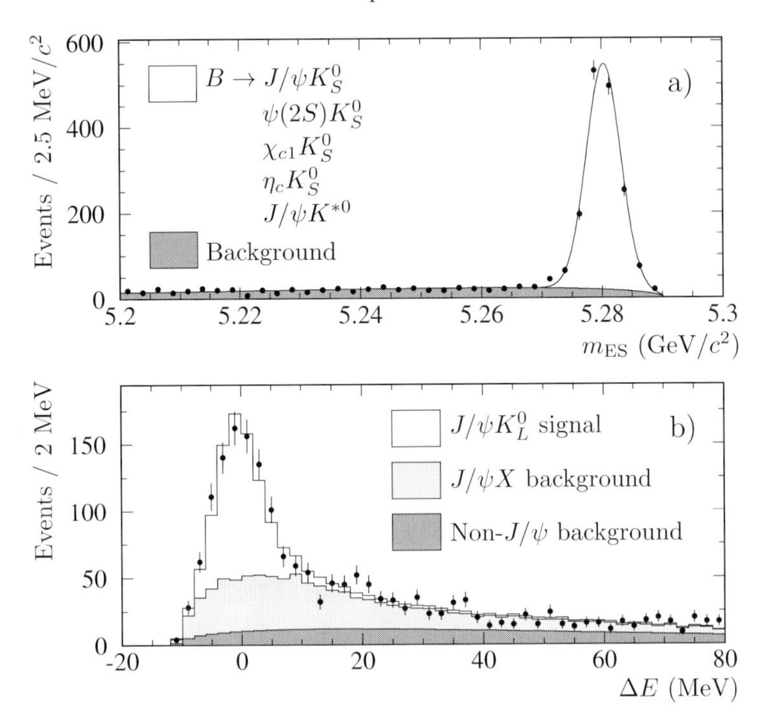

Fig. 3. Distributions for the CP sample after requiring the tagging and the vertexing: a) Energy Substituted mass m_{ES} for $J/\psi K_S^0$, $\psi(2S)K_S^0$, $\chi_{c1}K_S^0$, $\eta_c K_S^0$ and $J/\psi K^{*0}$ decay modes; b) ΔE for the $J/\psi K_L^0$ mode.

There are 34 free parameters in the fit: On the signal side the $\sin 2\beta$ parameter itself (1), the average mistag fraction w and the differences δw between mistag probability in B^0 and $\bar{B}^0$ for each tagging category (8), the parameters for the Δt resolution(8); on the background side parameters for the background time dependence (6), Δt resolution (3), and mistag fraction (8). The extracted value of $\sin 2\beta$ is [13]:

$$\sin 2\beta = 0.741 \pm 0.067(stat.) \pm 0.033(syst.). \tag{13}$$

The distributions of B^0 and $\bar{B}^0$ tagged decays as a function of Δt and the asymmetry, together with the fit result, are shown in Fig. 4. The dominant sources of systematic uncertainty are:

- uncertainties in the level, composition and CP asymmetry of the background (0.023);
- the assumed parameterization of the Δt resolution function (0.017) due to residual uncertainties in the vertex detector alignment;
- differences between the flavor and the CP sample mistag fraction (0.012).

The large size of the B_{CP} sample allows several consistency checks. This includes comparing the fit results by decay mode, tagging category, and B_{tag}

Table 2. Number of events N_{tag} after tagging and vertexing requirements, signal purity P and fitted value of $\sin 2\beta$ for various subsamples of the B_{CP}, B_{flav} and charged B control sample

Sample	N_{tag}	P (%)	$\sin 2\beta$
$(c\bar{c})K_S^0$	1506	94	0.76 ± 0.07
$J/\psi K_L^0$	988	55	0.72 ± 0.16
$J/\psi K^{*0}(K^* \to K_S^0\pi^0)$	147	81	0.22 ± 0.52
Full CP Sample	2641	78	0.74 ± 0.07
$(c\bar{c})K_S^0$ only			
$J/\psi K_S^0(K_S^0 \to \pi^+\pi^-)$	974	97	0.82 ± 0.08
$J/\psi K_S^0(K_S^0 \to \pi^0\pi^0)$	170	89	0.39 ± 0.24
$\psi(2S)K_S^0(K_S^0 \to \pi^+\pi^-)$	150	97	0.69 ± 0.24
$\chi_{c1}K_S^0$	80	95	1.01 ± 0.40
$\eta_c K_S^0$	132	73	0.59 ± 0.32
Lepton	220	98	0.79 ± 0.11
Kaon I	400	93	0.78 ± 0.12
Kaon II	444	93	0.73 ± 0.17
Inclusive	442	92	0.45 ± 0.28
B^0 tags	740	94	0.76 ± 0.10
$\bar{B}^0$ tags	766	93	0.75 ± 0.10
Fully Reconstructed Sample			
B_{flav}	25375	85	0.02 ± 0.02
B^+	22160	89	0.02 ± 0.02

flavor. The results of fits to these subsamples are found to be statistically consistent. Fits to the control samples of non-CP decay modes (the B_{flav} sample and fully reconstructed charged B decays sample) indicated no statistically significative asymmetry, as expected. This breakdown of the data sample and the results of checks are reported in Table 2.

The parameter $|\lambda|$ has been measured as well, repeating the fit procedure, with $|\lambda|$ allowed to float, on the $\eta_f = -1$ CP sample, for which the effect of the backgrounds is very limited. In this case, five additional parameters have been added, to account for differences in tagging and reconstruction efficiencies for B^0 and $\bar{B}^0$, which may simulate an artificial asymmetry. The result of the fit for $|\lambda|$ is:

$$|\lambda| = 0.948 \pm 0.051(stat.) \pm 0.017(syst.). \tag{14}$$

which is consistent with the hypothesis of pure phase and consequently no direct CP violation.

5 Measurement of the CP Violating Amplitude $\sin 2\alpha$

The time-dependent CP-violating asymmetries in the decay $B^0 \to \pi^+\pi^-$ are related to the angle α of the Unitarity Triangle. If the decay proceeds purely

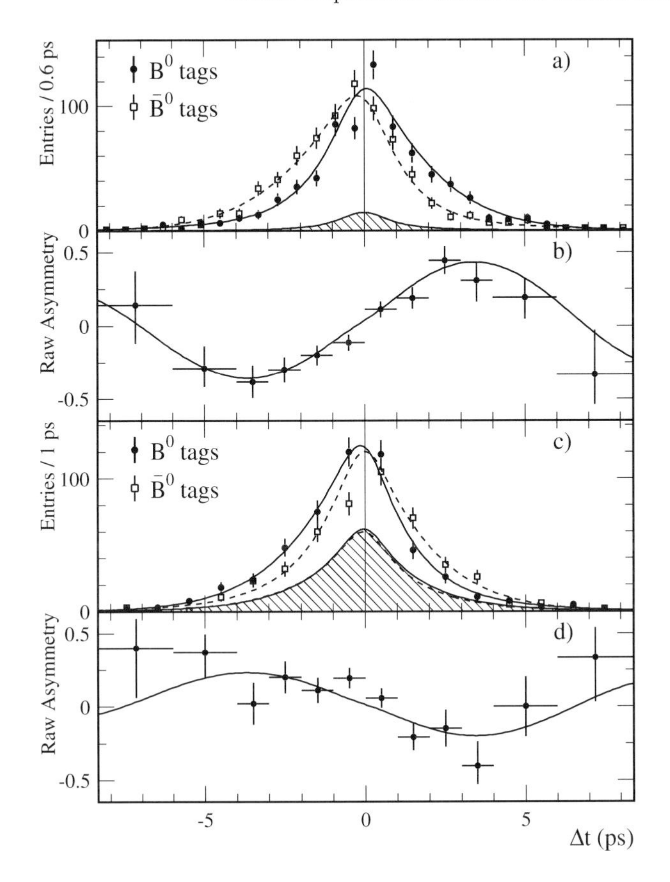

Fig. 4. a) Number of candidates for the $\eta_f = -1$ sample, in the signal region with a B^0 tag (full boxes data, solid line fit) and $\bar{B}^0$ tag (empty box data, dashed line fit). The shaded region represent the residual background. b) Raw asymmetry $(N_{B^0} - N_{\bar{B}^0})/(N_{B^0} + N_{\bar{B}^0})$, as a function of Δt. c) and d) Same content as a) and b) for the $J/\psi K_L^0$ mode ($\eta_f = +1$).

through the $b \to u$ tree amplitude, the complex parameter $\lambda_{\pi\pi}$, would be

$$\lambda(B^0 \to \pi^+\pi^-) = \left(\frac{V_{tb}^* V_{td}}{V_{tb} V_{td}^*}\right)\left(\frac{V_{ub}^* V_{ud}}{V_{ub} V_{ud}^*}\right). \tag{15}$$

Similarly to the $\sin 2\beta$ case, it would be $C_{\pi\pi} = 0$ and $S_{\pi\pi} = \sin 2\alpha$, where $\alpha = arg(-V_{tb}^* V_{td}/V_{ub}^* V_{ud})$.

Unfortunately, $b \to d$ penguins amplitudes can contribute in a significant way to the total amplitude of the process, so that they cannot be neglected. The tree and the penguin Feynman diagrams are showed in Fig. 5. Considering both the contributions, λ acquires a magnitude different from 1 and a shift in the phase:

$$\lambda(B^0 \to \pi^+\pi^-) = e^{-2i\alpha}\frac{1 + |P/T|\, e^{i\delta} e^{i\gamma}}{1 + |P/T|\, e^{i\delta} e^{-i\gamma}}$$

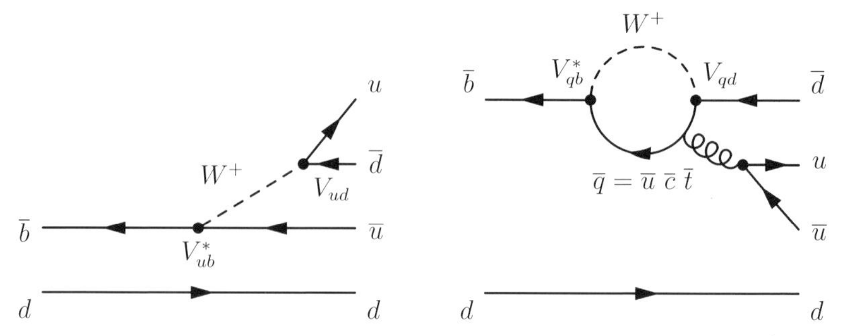

Fig. 5. Tree (left) and penguin (right) Feynman diagrams for the decay $B^0 \to \pi^+\pi^-$

$$\begin{aligned} C_{\pi\pi} &\propto \sin\delta \neq 0, \\ S_{\pi\pi} &= \sqrt{1 - C^2_{\pi\pi}} \sin 2\alpha_{eff}, \end{aligned}$$

where δ and γ are the the strong and weak phase differences between the tree amplitude T and the penguin amplitude P. Therefore, the time dependent CP-asymmetry permits the observation of α_{eff}, which depends on the magnitudes of relative strong phase δ and the relative weak phase γ between the tree and penguins amplitudes. The implication is that the relative magnitudes of the two contributions have to be determined in order to extract the value of $\sin 2\alpha$. Several approaches have been proposed to obtain information on α in the presence of penguins [14].

Moreover, from the experimental point of view, the smallness of $|V_{ub}|$ makes small the branching fractions of these decay modes ($10^{-5} \div 10^{-6}$); the necessity to suppress an high level of combinatoric background, makes the analysis a great experimental challenge.

5.1 Sample Selection of the Charmless Decays $B^0 \to h^+h^-$

The yields and the CP parameters are extracted from an unbinned maximum likelihood fit, which is described in the next section. The probability density functions used in the Likelihood function are chosen to discriminate between $B^0 \to h^+h^-$ signal and $q\overline{q}$ background, and among the various signals ($B^0 \to \pi\pi/K\pi/KK$).

For signal decays the ΔE and m_{ES} are Gaussian distributed with a resolution of 26 MeV and 2.6 MeV, respectively. Since the B candidate energy is determined in the pion hypothesis, the ΔE is shifted towards negative values, if a kaon is present in the final state. For example, the shift of mean of the Gaussian probability density function for $K\pi$ decays is $\delta\mu_{\Delta E} = -\gamma\left(\sqrt{m^2_K - p^2} - \sqrt{m^2_\pi - p^2}\right)$, where p is the kaon momentum. The average value of the shift is -45 MeV for the $K\pi$ case and -91 MeV for the KK case. The parameters of the m_{ES} and ΔE distributions are fitted from a sample of $B^- \to D^0\pi^-$ and the systematic uncertainty is determined by varying the m_{ES} peak position and resolution. The

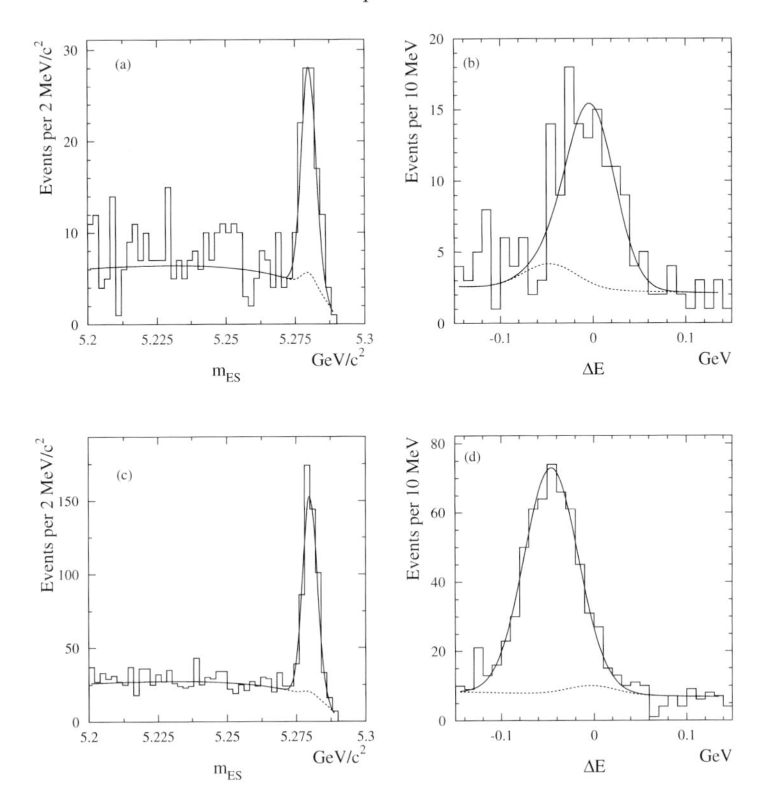

Fig. 6. Distributions of energy substituted mass m_{ES} (left) and energy difference ΔE (right) for events enhanced in signal. Top plots are $B^0 \to \pi\pi$ candidate events and bottom plots are $B^0 \to K\pi$ candidate events. Solid curve represent the projections of the maximum likelihood fit, while dashed curves represent the continuum and $\pi\pi \leftrightarrow K\pi$ cross-feed backgrounds.

m_{ES} and ΔE distributions for signal, enhanced in $B \to \pi\pi$ and $B \to K\pi$, are shown in Fig. 6.

Residual background from $e^+e^- \to q\bar{q}(q = u, d, s, c)$, is suppressed by event topology [15], since in $B\bar{B}$ events charged tracks and energy releases of neutral particles are more uniformly distributed over the solid angle with respect to the more jet-like continuum events.

Kinematics information from particles in the event, not used to build the B candidate, are combined in a Fisher discriminant [16]:

$$\mathcal{F} = 0.53 - 0.60 \times \sum_i p_i^* + 1.27 \times \sum_i p_i^* \left|\cos\theta_i^*\right|^2 , \tag{16}$$

where p_i^* is the momentum of the particle i in the center of mass frame and θ_i^* is the angle between the particle momentum and the B thrust axis in the center of mass frame. The value of the coefficients of the Fisher discriminant in

Table 3. Summary of the results for the branching fraction B and the asymmetry $\mathcal{A}$ in charmless B decays. The upper limits on B($B^0 \to K^+K^-$) and B($B^0 \to \pi^0\pi^0$) corresponds to a 90% C.L.

Mode	$B(10^{-6})$	$\mathcal{A}$	$\mathcal{A}$ 90% C.L.
$\pi^+\pi^-$	$4.6 \pm 0.6 \pm 0.2$		
$K^+\pi^-$	$17.9 \pm 0.9 \pm 0.7$	$-0.102 \pm 0.050 \pm 0.016$	[-0.188, -0.016]
K^+K^-	< 0.6		
$\pi^+\pi^0$	$5.5^{+1.0}_{-0.9} \pm 0.6$	$0.03^{+0.18}_{-0.17} \pm 0.02$	[-0.32,0.27]
$K^+\pi^0$	$12.8^{+1.2}_{-1.1} \pm 1.0$	$-0.09 \pm 0.09 \pm 0.01$	[-0.24,0.06]
$K^0\pi^0$	$10.4 \pm 1.5 \pm 0.8$	$0.03 \pm 0.36 \pm 0.09$	[-0.58,0.64]
$\pi^0\pi^0$	< 3.6		

(16) are, by definition, those which maximize the separation between signal and background. The shape of $\mathcal{F}$ is determined from Monte Carlo for the signal and m_{ES} sideband for background.

Particle identification is required to discriminate between the pion and kaon hypotheses. This is accomplished using the measurements from the Cherenkov detector.

The probability density function of the difference between the measured Cherenkov angle θ_c and the expected angle in the pion and in the kaon hypothesis, normalized by the error σ_{θ_c} is added to the Likelihood function for both the two charged particles. The parameters of the function are measured from a pure data sample of $D^* \to D^0\pi^+$, $D^0 \to K^-\pi^+$ decays.

5.2 CP Asymmetries and Branching Fractions Measurement of the Charmless Decays $B^0 \to h^+h^-$

The yields and the CP parameters are extracted from an unbinned maximum likelihood fit simultaneously on the B_{flav} and the h^+h^- sample, as usual. The sample is assumed to be composed in eight signal and background components: $\pi^+\pi^-$, $K^+\pi^-$, π^+K^-, K^+K^-. For both signal and backgrounds, $K\pi$ events are parameterized as the sum $N_{K\pi}$ and the asymmetry $\mathcal{A}_{K\pi} = (N_{K^-\pi^+} - N_{K^+\pi^-})/(N_{K^-\pi^+} + N_{K^+\pi^-})$. The probability for each event to be a given signal or background hypothesis is evaluated as the product of the probability density functions of the variables $(m_{ES}, \Delta E, \mathcal{F}, \theta_c^+, \theta_c^-, \Delta t)$.

The Likelihood for a candidate j in the tagging category k is defined to be the sum for every hypothesis i of the product of the yield N_i , the tagging efficiency $\epsilon_{i,k}$ and the probability $P_{i,k}$. The extended likelihood function for the category k is:

$$\mathcal{L}_k = exp\left(-\sum_i N_i\epsilon_{i,k}\right)\prod_j\left(\sum_i N_i\epsilon_{i,k}P_{i,k}(\boldsymbol{x}_j;\boldsymbol{\alpha}_i)\right). \quad (17)$$

The total likelihood is the product of the likelihood for each category, and the free parameter $\boldsymbol{\alpha}$ are determined by minimizing - ln $\mathcal{L}$.

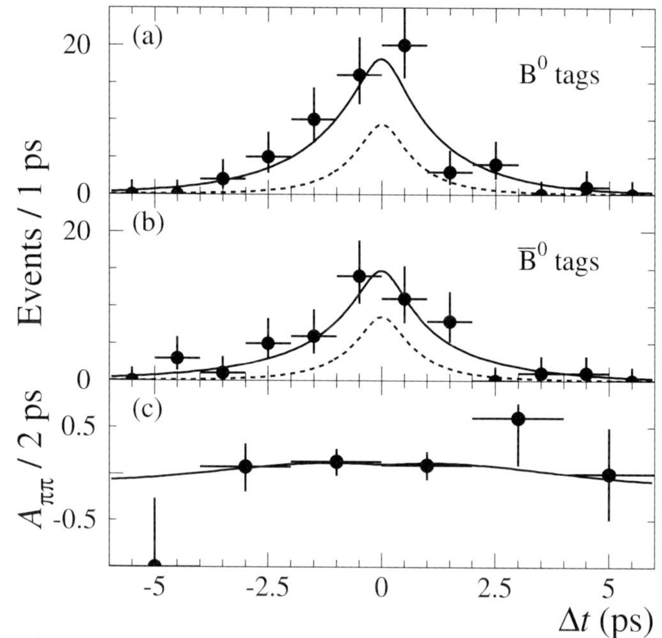

Fig. 7. Distributions of Δt for events enhanced in $\pi\pi$ decays tagged as (a) B^0, (b) $\bar{B}^0$.·(c) Distribution of the asymmetry $\mathcal{A}_{\pi\pi}$ as a function of Δt. Solid curve is the projection of the maximum likelihood fit, dashed curves represent the residual $q\bar{q}$ and $K\pi$ events.

In order to minimize the systematic uncertainty, the fit for the branching fraction measurement is performed not requiring tagging and vertexing of the recoiling B, and not requiring the Δt information. In this case the number of free parameters is 16 and includes the signal and background yields (6), the $K\pi$ asymmetries (2), and the background shape parameters for $m_{ES}, \Delta E, \mathcal{F}$ (8). Table 3 shows the summary of the results from the fit.

In order to extract the CP violating parameter $S_{\pi\pi}$ and $C_{\pi\pi}$, tagging and Δt informations are added. Since an asymmetry between tagging B^0 and $\bar{B}^0$ events can result in an artificial asymmetry in the $C_{\pi\pi}$ term, the mistag probabilities and the tagging efficiencies are included separately for B^0 and $\bar{B}^0$. The combined fit to signal and flavor sample is performed with 76 free parameters:

- The CP violating parameters $S_{\pi\pi}$ and $C_{\pi\pi}$ (2).
- Signal and background yields with N_{KK} fixed to zero (5).
- $K\pi$ asymmetries (2).
- signal and background tagging efficiencies (16).
- signal and background tagging efficiencies asymmetries (16).
- signal mistag and mistag asymmetries (8).
- signal resolution functions (9).
- the background shape parameters for m_{ES} (5), ΔE (2), $\mathcal{F}$ (5) and Δt (6).

Table 4. Expected statistical and systematic uncertainties on the CP violating parameters $\sin 2\beta$ and $sin2\alpha_{eff}$ at the present moment, and after 500 fb^{-1} and 2 ab^{-1} of integrated luminosity.

Parameter	Channel	σ (stat)/ σ (syst) at 81 fb^{-1}	σ (stat)/ σ (syst) at 0.5 ab^{-1}	σ (stat)/ σ (syst) at 2.0 ab^{-1}
$\sin 2\beta$	Golden	0.07 / 0.03	0.031/ 0.016	0.018 / 0.015
$\sin 2\alpha_{eff}$	$\pi^+\pi^-$	0.34 / 0.05	0.12 / 0.03	0.06 / 0.03
$C_{\pi\pi}$	$\pi^+\pi^-$	0.25 / 0.04	0.10 / 0.03	0.05 / 0.03

The fitted decay rates and asymmetry distributions for the $B^0 \to \pi^+\pi^-$ case are shown in Fig. 7. The fitted values for the CP violating parameters $C_{\pi\pi}$ and $S_{\pi\pi}$ are [17]:

$$S_{\pi\pi} = 0.02 \pm 0.34(stat) \pm 0.05(syst), \tag{18a}$$

$$C_{\pi\pi} = -0.30 \pm 0.25(stat) \pm 0.04(syst). \tag{18b}$$

The systematic uncertainty on $S_{\pi\pi}$ and $C_{\pi\pi}$ are dominated by the imperfect knowledge of the probability density functions shapes and fit bias.

Since the extraction of the α angle from the $B^0 \to \pi^+\pi^-$ CP asymmetry is complicated by the presence of the penguin amplitudes, additional measurements of isospin related decays $B^+ \to \pi^+\pi^0$ and $B^0 \to \pi^0\pi^0$ may help. Moreover, the measurements of $B \to K\pi$ decays branching fractions and asymmetries, can be related to α and γ angle, by means of various models [18] based on different theoretical assumptions. The detail of the analysis, which follows the general method described for h^+h^- modes, can be found in [19] and the results are reported in Table 3 for completeness.

6 Conclusions

The BaBar experiment has collected a data set of 88 millions $\Upsilon(4S)$ decays from 1999 to June 2002. The new measurement of $\sin 2\beta = 0.741 \pm 0.067 \pm 0.033$ shows that BaBar is starting to provide a precision measurement of this important parameter of the Standard Model. CP violation is now well established, and at the present moment it is fully consistent with the Standard Model expectation.

In Fig. 8 it is shown a comparison [20] between the BaBar $\sin 2\beta$ direct measurement and the indirect determination of an allowed region of the Unitarity Triangle apex position in the (ρ,η) plane from the measurements of $|\epsilon_K|$, $|V_{ub}/V_{cb}|$, Δm_d and Δm_s. The BaBar and Belle direct measurements differ qualitatively from the indirect constraint, because for the former the size of the region is determined by experimental uncertainties of statistical origin, while the latter is determined mainly by theoretical uncertainties of more difficult interpretation.

With the present data set, the experiment is already sensitive to more rare $\sin 2\beta$ modes, like $B^0 \to D^*D^*$ [21], $J/\psi\pi^0$ [22] or ΦK^0_S [23], which not only enrich the sample but may show up new physics, as well.

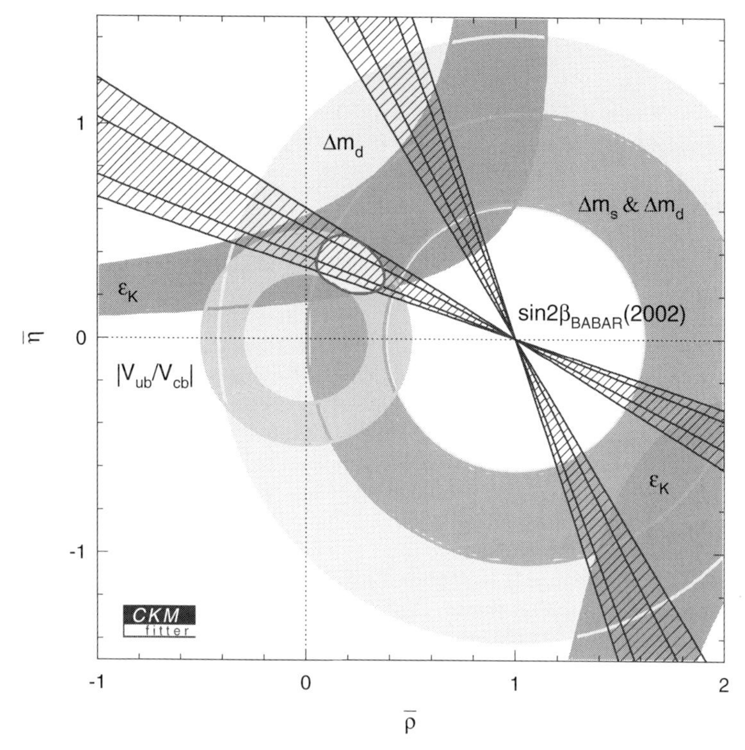

Fig. 8. Indirect constraints on the position of the Unitarity Triangle apex in the $\rho - \eta$ plane, not including the measurement of $\sin 2\beta$. The fit procedure is described in [20]. The BaBar measurement of $\sin 2\beta$ is represented by the two hatched regions corresponding to one and to two standard deviations.

Moreover, the $\sin 2\beta$ measurement on more abundant and clean sample of golden modes will continue to be not limited by the systematic uncertainties till 0.5 ab^{-1} of integrated luminosity has been collected. At the expected luminosity of $1.6 \times 10^{34} cm^{-2} s^{-1}$, this would happen in 2006. Before then, the measurement accuracy will continue to improve with the increasing statistics.

Measurements in the charmless B decays did not show any evidence of CP violation. Branching fraction of the various charged and neutral modes have been determined, allowing isospin analyses in order to extract, or put limits on, the parameter $\sin 2\alpha$.

Table 4 shows the uncertainties on $\sin 2\beta, \sin 2\alpha_{eff}$ and $C_{\pi\pi}$ at the present moment, and after an integrated luminosity of 0.5 ab^{-1} and 2.0 ab^{-1}, respectively.

References

1. N.Cabibbo, Phys. Rev. Lett. **10**, 531 (1963) ;
   M.Kobayashi and T.Maskawa Prog. Th. Phys. **49**, 652 (1973).
2. L. Wolfenstein, Phys. Rev. Lett. **51**, 1945 (1983).
3. A.B. Carter and A.I. Sanda, Phys. Rev. D **23**, 1567 (1981); I.I. Bigi and A.I. Sanda, Nucl. Phys. B **193**, 85 (1981).
4. BABAR Collaboration, B.Aubert et al., Nucl. Instr. and Meth. A **479** (2002) 1.
5. PEP-II: an Asymmetric B Factory, Conceptual Design Report, SLAC-418, LBL-5379 (1993)
6. BABAR Collaboration, B.Aubert et al., Phys. Rev. D **66**, 032003 (2002).
7. See for example the review by L.Wolfenstein in Phys. Rev. D **66** 010001 (2002).
8. BABAR collaboration, B.Aubert et al., Phys. Rev. Lett. **87**, 091801 (2001)
9. BELLE collaboration, K.Abe et al., Phys. Rev. Lett. **87**, 091802 (2001)
10. NA48 collaboration, Phys. Lett B **544** 97-112 (2002)
11. KTeV collaboration, hep-ex/0208007, submitted to Phys. Rev. D
12. ARGUS Collaboration, H. Albrecht et al., Phys. Lett. B **241**, 278 (1990).
13. BABAR Collaboration, B. Aubert et al., Phys. Rev. Lett. **89**, 201802 (2002).
14. M.Beneke, G.Buchalla, M.Neubert and C.T. Sachrajda, Nucl. Phys. B **606**, 245 (2001); Y. Grossman and H.R. Quinn, Phys. Rev. D **59**, 054007 (1999). M. Gronau, D.London, N.Sinha and R.Sinha, Phys. Lett. B **514**, 315 (2001). M. Gronau and J. Rosner, Phys. Rev. D **65**, 090012 (2002).
15. G.C. Fox and S. Wolfram, Phys. Rev. Lett. **41**, 1581 (1978).
16. R.A. Fisher, Annals of Eugenics **7** 179 (1936).
17. BABAR Collaboration, B. Aubert et al., SLAC-PUB-9317, hep-ex/0207055 submitted to Phys. Rev. Lett.
18. M.Ciuchini, E. Franco, G. Martinelli, M. Pierini and L.Silvestrini, Phys. Lett. B **515**, 33 (2001). Y.Y. Keum, H.N. Li and A.I. Sanda, Phys. Rev. D **63** 054008 (2001). C.Isola, M.Ladisa, G. Nardulli, T.N. Pham and P. Santorelli, Phys. Rev. D **65**, 094005 (2002).
19. BABAR Collaboration, B. Aubert et al., SLAC-PUB-9304 hep-ex/0207065 BABAR Collaboration, B. Aubert et al., SLAC-PUB-9310 hep-ex/0207063
20. A.Hocker et al., Eur. Phys. J. C **21**, 225 (2001)
21. BABAR Collaboration, B. Aubert et al., SLAC-PUB 9299, hep-ex/0207072
22. BABAR Collaboration, B. Aubert et al., SLAC-PUB 9298, hep-ex/0207058
23. BABAR Collaboration, B. Aubert et al., SLAC-PUB 9297, hep-ex/0207070

Part II

Modelling QCD

Unifying Aspects of Light- and Heavy-Systems

Craig D. Roberts[1,2]

[1] Physics Division, Bldg 203, Argonne National Laboratory, Argonne, Illinois 60439-4843, USA
[2] Fachbereich Physik, Universität Rostock, 18051 Rostock, Germany

Abstract. Dyson-Schwinger equations furnish a Poincaré covariant framework within which to study hadrons. A particular feature is the existence of a nonperturbative, symmetry preserving truncation that enables the proof of exact results. Key to the DSE's efficacious application is their expression of the materially important momentum-dependent dressing of parton propagators at infrared length-scales, which is responsible for the magnitude of constituent-quark masses and the length-scale characterising confinement in bound states. A unified quantitative description of light- and heavy-quark systems is achieved by capitalising on these features.

1 Introduction

This contribution provides an overview of one particular means by which a quantitative and intuitive understanding of strong interaction phenomena can be attained. The broad framework is that of continuum strong QCD, by which I mean the continuum nonperturbative methods and models that can address these phenomena, especially those where a direct connection with QCD can be established, in one true limit or another. Naturally, everyone has a favourite tool and, in this connection, the Dyson-Schwinger equations (DSEs) are mine [1]. The framework is appropriate here because the last decade has seen a renaissance in its phenomenological application, with studies of phenomena as apparently unconnected as low-energy $\pi\pi$ scattering, $B \to D^*$ decays and the equation of state for a quark gluon plasma [2–4]. Indeed, the DSEs promise a single structure applicable to the gamut of strong interaction observables.

Dyson-Schwinger equations provide a nonperturbative means of analysing a quantum field theory. Derived from a theory's Euclidean space generating functional, they are an enumerable infinity of coupled integral equations whose solutions are the n-point Schwinger functions (Euclidean Green functions), which are the same matrix elements estimated in numerical simulations of lattice-QCD. In theories with elementary fermions, the simplest of the DSEs is the *gap* equation, which is basic to studying dynamical symmetry breaking in systems as disparate as ferromagnets, superconductors and QCD. The gap equation is a good example because it is familiar and has all the properties that characterise each DSE. Its solution is a 2-point function (the fermion propagator) but its kernel involves higher n-point functions; e.g., in a gauge theory, the kernel is constructed from the gauge-boson 2-point function and fermion–gauge-boson vertex, a 3-point function. In addition, while a weak-coupling expansion yields all the diagrams of perturbation theory, a self-consistent solution of the gap equation

C.D. Roberts, Unifying Aspects of Light- and Heavy-Systems, Lect. Notes Phys. **647**, 149–188 (2004)
http://www.springerlink.com/

exhibits nonperturbative effects unobtainable at any finite order in perturbation theory; e.g, dynamical chiral symmetry breaking (DCSB).

The coupling between equations; namely, the fact that the equation for a given m-point function always involves at least one $n > m$-point function, necessitates a truncation of the tower of DSEs in order to define a tractable problem. It is unsurprising that the best known truncation scheme is just the weak coupling expansion which reproduces every diagram in perturbation theory. This scheme is systematic and valuable in the analysis of large momentum transfer processes because QCD is asymptotically free. However, it precludes the study of nonperturbative effects, and hence something else is needed for the investigation of strongly interacting systems and bound state phenomena.

In spite of the need for a truncation, gap equations have long been used effectively in obtaining nonperturbative information about many-body systems as, e.g., in the Nambu-Gorkov formalism for superconductivity. The positive outcomes have been achieved through the simple expedient of employing the most rudimentary truncation, e.g., Hartree or Hartree-Fock, and comparing the results with observations. Of course, agreement under these circumstances is not an unambiguous indication that the contributions omitted are small nor that the model expressed in the truncation is sound. However, it does justify further study, and an accumulation of good results is grounds for a concerted attempt to substantiate a reinterpretation of the truncation as the first term in a systematic and reliable approximation.

The modern application of DSEs, notably, comparisons with and predictions of experimental data, can properly be said to rest on model assumptions. However, those assumptions can be tested within the framework and also via comparison with lattice-QCD simulations, and the predictions are excellent. Furthermore, progress in understanding the intimate connection between symmetries and truncation schemes has enabled exact results to be proved. Herein I will briefly explain recent phenomenological applications and the foundation of their success, and focus especially on the links the approach provides between light- and heavy-quark phenomena. It will become apparent that the momentum-dependent *dressing* of the propagators of QCD's elementary excitations is a fundamental and observable feature of strong QCD.

The article is organised as follows: Sect. 2 [p. 151] – a review of DSE quiddities, especially in connection with the development of a nonperturbative, systematic and symmetry preserving truncation scheme, and the model-independent results whose proof its existence enables; Sect. 3 [p. 165] – an illustration of the efficacious application of DSE methods to light-meson systems and the connections that may be made with the results of lattice-QCD simulations; Sect. 4 [p. 170] – the natural extension of these methods to heavy-quark systems, with an explanation of the origin and derivation of heavy-quark symmetry limits and their confrontation with the real-world of finite quark masses; and Sect. 5 [p. 184] – an epilogue.

2 Dyson-Schwinger Equations

2.1 Gap Equation

The simplest DSE is the *gap* equation, which describes how the propagation of a fermion is modified by its interactions with the medium being traversed. In QCD that equation assumes the form:[1]

$$S^{-1}(p) = Z_2(\zeta,\Lambda)\, i\gamma\cdot p + Z_4(\zeta,\Lambda)\, m(\zeta) + \Sigma'(p,\Lambda), \tag{1}$$

wherein the dressed-quark self-energy is

$$\Sigma'(p,\Lambda) = Z_1(\zeta,\Lambda) \int_q^\Lambda g^2 D_{\mu\nu}(p-q)\, \frac{\lambda^i}{2}\gamma_\mu\, S(q)\, \Gamma^i_\nu(q,p). \tag{2}$$

Equations (1), (2) constitute the renormalised DSE for the dressed-quark propagator. In (2), $D_{\mu\nu}(k)$ is the renormalised dressed-gluon propagator, $\Gamma^a_\nu(q;p)$ is the renormalised dressed-quark-gluon vertex and $\int_q^\Lambda := \int^\Lambda d^4q/(2\pi)^4$ represents a *translationally-invariant* regularisation of the integral, with Λ the regularisation mass-scale.[2] In addition, $Z_1(\zeta,\Lambda)$, $Z_2(\zeta,\Lambda)$ and $Z_4(\zeta,\Lambda)$ are, respectively, Lagrangian renormalisation constants for the quark-gluon vertex, quark wave function and quark mass-term, which depend on the renormalisation point, ζ, and the regularisation mass-scale, as does the gauge-independent mass renormalisation constant,

$$Z_m(\zeta^2,\Lambda^2) = Z_4(\zeta^2,\Lambda^2)\, Z_2^{-1}(\zeta^2,\Lambda^2)\,, \tag{3}$$

whereby the renormalised running-mass is related to the bare mass:

$$m(\zeta) = Z_m^{-1}(\zeta^2,\Lambda^2)\, m_{\rm bm}(\Lambda)\,. \tag{4}$$

When ζ is very large the running-mass can be evaluated in perturbation theory, which gives

$$m(\zeta) = \frac{\hat{m}}{(\ln \zeta/\Lambda_{\rm QCD})^{\gamma_m}}\,,\; \gamma_m = 12/(33-2N_f)\,. \tag{5}$$

Here N_f is the number of current-quark flavours that contribute actively to the running coupling, and $\Lambda_{\rm QCD}$ and $\hat{m}$ are renormalisation group invariants.

The solution of (1) is the dressed-quark propagator and takes the form

$$S^{-1}(p) = i\gamma\cdot p\, A(p^2,\zeta^2) + B(p^2,\zeta^2) \equiv \frac{1}{Z(p^2,\zeta^2)}\left[i\gamma\cdot p + M(p^2)\right]. \tag{6}$$

[1] A Euclidean metric is employed throughout, wherewith the scalar product of two four vectors is $a\cdot b = \sum_{i=1}^4 a_i b_i$; and I employ Hermitian Dirac-γ matrices that obey $\{\gamma_\mu,\gamma_\nu\} = 2\delta_{\mu\nu}$ and ${\rm tr}\,\gamma_5\gamma_\mu\gamma_\nu\gamma_\rho\gamma_\sigma = -4\,\epsilon_{\mu\nu\rho\sigma}$, $\epsilon_{1234}=1$.

[2] Only with a translationally invariant regularisation scheme can Ward-Takahashi identities be preserved, something that is crucial to ensuring vector and axial-vector current conservation. The final stage of any calculation is to take the limit $\Lambda\to\infty$.

It is obtained by solving the gap equation subject to the renormalisation condition that at some large spacelike ζ^2

$$S^{-1}(p)\big|_{p^2=\zeta^2} = i\gamma \cdot p + m(\zeta) . \tag{7}$$

The observations made in in the Introduction are now manifest. The gap equation is a nonlinear integral equation for $S(p)$ and can therefore yield much-needed nonperturbative information. However, the kernel involves the two-point function $D_{\mu\nu}(k)$ and the three-point function $\Gamma^a_\nu(q;p)$. The equation is consequently coupled to the DSEs these functions satisfy and hence a manageable problem is obtained only once a truncation scheme is specified.

2.2 Nonperturbative Truncation

To understand why (1) is called a gap equation, consider the chiral limit, which is readily defined [5] because QCD exhibits asymptotic freedom and implemented in the gap equation by employing [6]

$$Z_2(\zeta^2, \Lambda^2)\, m_{\rm bm}(\Lambda) \equiv 0 \,, \quad \Lambda \gg \zeta \,. \tag{8}$$

It is noteworthy that for finite ζ and $\Lambda \to \infty$, the left hand side (l.h.s.) of (8) is identically zero, by definition, because the mass term in QCD's Lagrangian density is renormalisation-point-independent. The condition specified in (8), on the other hand, effects the result that at the (perturbative) renormalisation point there is no mass-scale associated with explicit chiral symmetry breaking, which is the essence of the chiral limit. An equivalent statement is that one obtains the chiral limit when the renormalisation-point-invariant current-quark mass vanishes; namely, $\hat{m} = 0$ in (5). In this case the theory is chirally symmetric, and a perturbative evaluation of the dressed-quark propagator from (1) gives

$$B^0_{\rm pert}(p^2) := \lim_{m\to 0} B_{\rm pert}(p^2) = \lim_{m\to 0} m \left(1 - \frac{\alpha}{\pi} \ln\left[\frac{p^2}{m^2}\right] + \ldots\right) \equiv 0 \,; \tag{9}$$

viz., the perturbative mass function is identically zero in the chiral limit. It follows that there is no gap between the top level in the quark's filled negative-energy Dirac sea and the lowest positive energy level.

However, suppose one has at hand a truncation scheme other than perturbation theory and that subject to this scheme (1) possessed a chiral limit solution $B^0(p^2) \not\equiv 0$. Then interactions between the quark and the virtual quanta populating the ground state would have nonperturbatively generated a mass gap. The appearance of such a gap breaks the theory's chiral symmetry. This shows that the gap equation can be an important tool for studying DCSB, and it has long been used to explore this phenomenon in both QED and QCD [1].

The gap equation's kernel is formed from a product of the dressed-gluon propagator and dressed-quark-gluon vertex but in proposing and developing a truncation scheme it is insufficient to focus only on this kernel [6,7]. The gap equation can only be a useful tool for studying DCSB if the truncation itself does not destroy chiral symmetry.

Chiral symmetry is expressed via the axial-vector Ward-Takahashi identity:

$$P_\mu \, \Gamma_{5\mu}(k;P) = S^{-1}(k_+)\, i\gamma_5 + i\gamma_5\, S^{-1}(k_-)\,,\; k_\pm = k \pm P/2, \tag{10}$$

wherein $\Gamma_{5\mu}(k;P)$ is the dressed axial-vector vertex. This three-point function satisfies an inhomogeneous Bethe-Salpeter equation (BSE):

$$[\Gamma_{5\mu}(k;P)]_{tu} = Z_2\, [\gamma_5\gamma_\mu]_{tu} + \int_q^\Lambda [S(q_+)\Gamma_{5\mu}(q;P)S(q_-)]_{sr} K_{tu}^{rs}(q,k;P)\,, \tag{11}$$

in which $K(q,k;P)$ is the fully-amputated quark-antiquark scattering kernel, and the colour-, Dirac- and flavour-matrix structure of the elements in the equation is denoted by the indices r,s,t,u. The Ward-Takahashi identity, (10), entails that an intimate relation exists between the kernel in the gap equation and that in the BSE. (This is another example of the coupling between DSEs.) Therefore an understanding of chiral symmetry and its dynamical breaking can only be obtained with a nonperturbative truncation scheme that preserves this relation, and hence guarantees (10) without a *fine-tuning* of model-dependent parameters.

Rainbow-Ladder Truncation. At least one such scheme exists [8]. Its leading-order term is the so-called renormalisation-group-improved rainbow-ladder truncation, whose analogue in the many body problem is an Hartree-Fock truncation of the one-body (Dyson) equation combined with a consistent ladder-truncation of the related two-body (Bethe-Salpeter) equation. To understand the origin of this leading-order term, observe that the dressed-ladder truncation of the quark-antiquark scattering kernel is expressed in (11) via

$$\begin{aligned}
[L(q,k;P)]_{tu}^{t'u'}\,[\Gamma_{5\mu}(q;P)]_{u't'} &:= [S(q_+)\Gamma_{5\mu}(q;P)S(q_-)]_{sr}\, K_{tu}^{rs}(q,k;P) \\
&= -g^2(\zeta^2)\, D_{\rho\sigma}(k-q) \\
&\quad \times \left[\Gamma_\rho^a(k_+,q_+)\, S(q_+)\right]_{tt'} \left[S(q_-)\, \Gamma_\sigma^a(q_-,k_-)\right]_{u'u} [\Gamma_{5\mu}(q;P)]_{t'u'}
\end{aligned} \tag{12}$$

wherein I have only made explicit the renormalisation point dependence of the coupling. One can exploit multiplicative renormalisability and asymptotic freedom to demonstrate that on the kinematic domain for which $Q^2 := (k-q)^2 \sim k^2 \sim q^2$ is large and spacelike

$$[L(q,k;P)]_{tu}^{t'u'} = -4\pi\alpha(Q^2)\, D_{\rho\sigma}^{\rm free}(Q) \left[\frac{\lambda^a}{2}\gamma_\rho\, S^{\rm free}(q_+)\right]_{tt'} \left[S^{\rm free}(q_-)\,\frac{\lambda^a}{2}\gamma_\sigma\right]_{u'u}, \tag{13}$$

where $\alpha(Q^2)$ is the strong running coupling and, e.g., $S^{\rm free}$ is the free quark propagator. It follows that on this domain the r.h.s. of (13) describes the leading contribution to the complete quark-antiquark scattering kernel, $K_{tu}^{rs}(q,k;P)$, with all other contributions suppressed by at least one additional power of $1/Q^2$.

The renormalisation-group-improved ladder-truncation supposes that

$$K_{tu}^{rs}(q,k;P) = -4\pi\,\alpha(Q^2)\, D_{\rho\sigma}^{\rm free}(Q) \left[\frac{\lambda^a}{2}\gamma_\rho\right]_{ts} \left[\frac{\lambda^a}{2}\gamma_\sigma\right]_{ru} \tag{14}$$

is also a good approximation on the infrared domain and is thus an assumption about the long-range ($Q^2 \lesssim 1$ GeV^2) behaviour of the interaction. Combining (14) with the requirement that (10) be automatically satisfied leads to the renormalisation-group-improved rainbow-truncation of the gap equation:

$$S^{-1}(p) = Z_2\,(i\gamma\cdot p + m_{\rm bm}) + \int_q^\Lambda 4\pi\,\alpha(Q^2)\,D_{\mu\nu}^{\rm free}(p-q)\frac{\lambda^a}{2}\gamma_\mu\,S(q)\,\frac{\lambda^a}{2}\gamma_\nu\,. \quad (15)$$

This rainbow-ladder truncation provides the foundation for an explanation of a wide range of hadronic phenomena [4].

2.3 Systematic Procedure

The truncation scheme of [8] is a dressed-loop expansion of the dressed-quark-gluon vertices that appear in the half-amputated dressed-quark-antiquark scattering matrix: S^2K, a renormalisation-group invariant [9]. All n-point functions involved thereafter in connecting two particular quark-gluon vertices are *fully dressed*. The effect of this truncation in the gap equation, (1), is realised through the following representation of the dressed-quark-gluon vertex, $i\Gamma^a_\mu = \frac{i}{2}\lambda^a\,\Gamma_\mu = l^a\Gamma_\mu$:

$$\begin{aligned} Z_1\Gamma_\mu(k,p) &= \gamma_\mu + \frac{1}{2N_c}\int_\ell^\Lambda g^2 D_{\rho\sigma}(p-\ell)\gamma_\rho S(\ell+k-p)\gamma_\mu S(\ell)\gamma_\sigma \\ &+ \frac{N_c}{2}\int_\ell^\Lambda g^2\,D_{\sigma'\sigma}(\ell)\,D_{\tau'\tau}(\ell+k-p)\,\gamma_{\tau'}\,S(p-\ell)\,\gamma_{\sigma'}\,\Gamma^{3g}_{\sigma\tau\mu}(\ell,-k,k-p) + [\ldots]\,. \end{aligned} \quad (16)$$

Here Γ^{3g} is the dressed-three-gluon vertex and it is readily apparent that the lowest order contribution to each term written explicitly is $\mathrm{O}(g^2)$. The ellipsis represents terms whose leading contribution is $\mathrm{O}(g^4)$; viz., the crossed-box and two-rung dressed-gluon ladder diagrams, and also terms of higher leading-order.

This expansion of S^2K, with its implications for other n-point functions, yields an ordered truncation of the DSEs that guarantees, term-by-term, the preservation of vector and axial-vector Ward-Takahashi identities, a feature that has been exploited [5,10,11] to establish exact results in QCD. It is readily seen that inserting (16) into (1) provides the rule by which the rainbow-ladder truncation can be systematically improved.

Planar Vertex. The effect of the complete vertex in (16) on the solutions of the gap equation is unknown. However, insights have been drawn from a study [9] of a more modest problem obtained by retaining only the sum of dressed-gluon ladders; i.e., the vertex depicted in Fig. 1. The elucidation is particularly transparent when one employs [12]

$$\mathcal{D}_{\mu\nu}(k) := g^2\,D_{\mu\nu}(k) = \left(\delta_{\mu\nu} - \frac{k_\mu k_\nu}{k^2}\right)(2\pi)^4\,\mathcal{G}^2\,\delta^4(k) \quad (17)$$

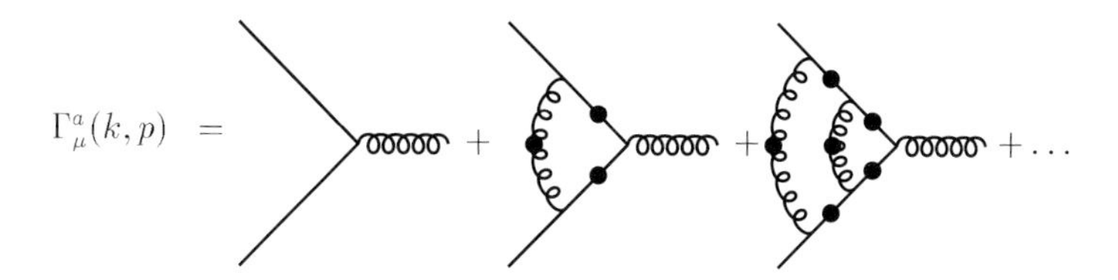

Fig. 1. Integral equation for a planar dressed-quark-gluon vertex obtained by neglecting contributions associated with explicit gluon self-interactions. Solid circles indicate fully dressed propagators. The vertices are not dressed. (Adapted from [9].)

for the dressed-gluon line, which defines an ultraviolet finite model so that the regularisation mass-scale can be removed to infinity and the renormalisation constants set equal to one.[3] This model has many positive features in common with the class of renormalisation-group-improved rainbow-ladder models and its particular momentum-dependence works to advantage in reducing integral equations to algebraic equations with similar qualitative features. There is naturally a drawback: the simple momentum dependence also leads to some model-dependent artefacts, but they are easily identified and hence not cause for concern.

The general form of the dressed-quark gluon vertex involves twelve distinct scalar form factors but using (17) only $\Gamma_\mu(p) := \Gamma_\mu(p,p)$ contributes to the gap equation, which considerably simplifies the analysis. The summation depicted in Fig. 1 is expressed via

$$\Gamma_\mu(p) = \gamma_\mu + \frac{1}{8}\,\gamma_\rho\,S(p)\,\Gamma_\mu(p)\,S(p)\,\gamma_\rho\,, \tag{18}$$

which supports a solution

$$\Gamma_\mu(p) = \alpha_1(p^2)\,\gamma_\mu + \alpha_2(p^2)\,\gamma\cdot p\,p_\mu - \alpha_3(p^2)\,i\,p_\mu\,. \tag{19}$$

One can re-express this vertex as

$$\Gamma_\mu(p) = \sum_{i=0}^{\infty} \Gamma^i_\mu(p) = \sum_{i=0}^{\infty} \left[\alpha^i_1(p^2)\,\gamma_\mu + \alpha^i_2(p^2)\,\gamma\cdot p\,p_\mu - \alpha^i_3(p^2)\,i\,p_\mu\right], \tag{20}$$

where the superscript enumerates the order of the iterate: $\Gamma^{i=0}_\mu$ is the bare vertex,

$$\alpha^0_1 = 1\,,\ \ \alpha^0_2 = 0 = \alpha^0_3\,; \tag{21}$$

$\Gamma^{i=1}_\mu$ is the result of inserting this into the r.h.s. of (18) to obtain the one-rung dressed-gluon correction; $\Gamma^{i=2}_\mu$ is the result of inserting $\Gamma^{i=1}_\mu$, and is therefore the two-rung dressed-gluon correction; etc. A key observation [9] is that each iterate

[3] The constant $\mathcal{G}$ sets the model's mass-scale and using $\mathcal{G} = 1$ simply means that all mass-dimensioned quantities are measured in units of $\mathcal{G}$.

is related to its precursor via a simple recursion relation and, substituting (20), that recursion yields ($s = p^2$)

$$\boldsymbol{\alpha}^{i+1}(s) := \left(\begin{array}{c} \alpha_1^{i+1}(s) \\ \alpha_2^{i+1}(s) \\ \alpha_3^{i+1}(s) \end{array} \right) = \mathcal{O}(s; A, B)\, \boldsymbol{\alpha}^i(s)\,, \tag{22}$$

$$\mathcal{O}(s; A, B) = \frac{1}{4}\frac{1}{\Delta^2} \left(\begin{array}{ccc} -\Delta & 0 & 0 \\ 2A^2 & sA^2 - B^2 & 2AB \\ 4AB & 4sAB & 2(B^2 - sA^2) \end{array} \right), \tag{23}$$

$\Delta = sA^2(s) + B^2(s)$. It follows that

$$\boldsymbol{\alpha} = \left(\sum_{i=1}^{\infty} \mathcal{O}^i \right) \boldsymbol{\alpha}^0 = \frac{1}{1 - \mathcal{O}}\, \boldsymbol{\alpha}^0 \tag{24}$$

and hence, using (21),

$$\begin{aligned} \alpha_1 &= \frac{4\,\Delta}{1 + 4\,\Delta}\,, \\ \alpha_2 &= \frac{-\,8\,A^2}{1 + 2\,(B^2 - s\,A^2) - 8\,\Delta^2}\, \frac{1 + 2\,\Delta}{1 + 4\,\Delta}\,, \\ \alpha_3 &= \frac{-8\,AB}{1 + 2(B^2 - s\,A^2) - 8\,\Delta^2}\,. \end{aligned} \tag{25}$$

The recursion relation thus leads to a closed form for the gluon-ladder-dressed quark-gluon vertex in Fig. 1; viz., (19), (25). Its momentum-dependence is determined by that of the dressed-quark propagator, which is obtained by solving the gap equation, itself constructed with this vertex. Using (17), that gap equation is

$$S^{-1}(p) = i\gamma \cdot p + m + \gamma_\mu\, S(p)\, \Gamma_\mu(p) \tag{26}$$

whereupon the substitution of (19) gives

$$A(s) = 1 + \frac{1}{sA^2 + B^2}\left[A\,(2\alpha_1 - s\alpha_2) - B\,\alpha_3\right]\,, \tag{27}$$

$$B(s) = m + \frac{1}{sA^2 + B^2}\left[B\,(4\alpha_1 + s\alpha_2) - sA\,\alpha_3\right]\,. \tag{28}$$

Equations (27), (28), completed using (25), form a closed algebraic system. It can easily be solved numerically, and that yields simultaneously the complete gluon-ladder-dressed vertex and the propagator for a quark fully dressed via gluons coupling through this nonperturbative vertex. Furthermore, it is apparent that in the chiral limit, $m = 0$, a realisation of chiral symmetry in the Wigner-Weyl mode, which is expressed via the $B \equiv 0$ solution of the gap equation, is always admissible. This is the solution anticipated in (9).

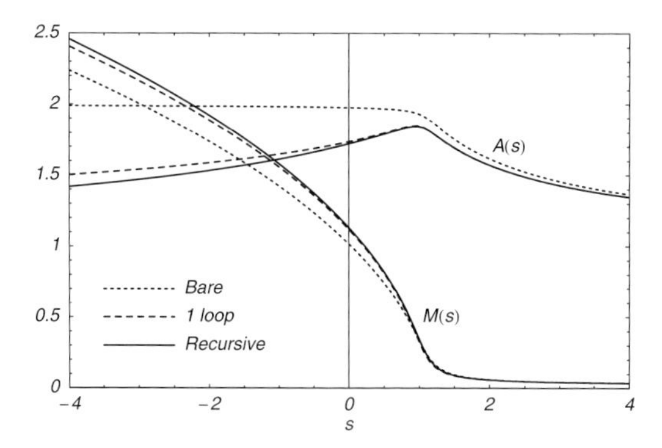

Fig. 2. $A(s)$, $M(s)$ obtained from (25), (27), (28) with $m = 0.023$ (*solid line*). All dimensioned quantities are expressed in units of $\mathcal{G}$ in (17). For comparison, the results obtained with the zeroth-order vertex (*dotted line*) and the one-loop vertex (*dashed line*) are also plotted. (Adapted from [9].)

The chiral limit gap equation also admits a Nambu-Goldstone mode solution whose $p^2 \simeq 0$ properties are unambiguously related to those of the $m \neq 0$ solution, a feature also evident in QCD [11]. A complete solution of (26) is available numerically, and results for the dressed-quark propagator are depicted in Fig. 2. It is readily seen that the complete resummation of dressed-gluon ladders gives a dressed-quark propagator that is little different from that obtained with the one-loop-corrected vertex; and there is no material difference from the result obtained using the zeroth-order vertex. Similar observations apply to the vertex itself. The scale of these modest effects can be quantified by a comparison between the values of $M(s=0) = B(0)/A(0)$ calculated using vertices dressed at different orders:

$\sum_{i=0,N} \Gamma^i_\mu$	$N=0$	$N=1$	$N=2$	$N=\infty$
$M(0)$	1	1.105	1.115	1.117

(29)

The rainbow truncation of the gap equation is accurate to within 12% and adding just one gluon ladder gives 1% accuracy. It is important to couple this with an understanding of how the vertex resummation affects the Bethe-Salpeter kernel.

Vertex-Consistent Bethe-Salpeter Kernel. The renormalised homogeneous BSE for the quark-antiquark channel denoted by M can be expressed

$$[\Gamma_M(k;P)]_{tu} = \int_q^\Lambda [\chi_M(q;P)]_{sr}\, [K(k,q;P)]^{rs}_{tu}\,, \tag{30}$$

where: $\Gamma_M(k;P)$ is the meson's Bethe-Salpeter amplitude, k is the relative momentum of the quark-antiquark pair, P is their total momentum; and

$$\chi_M(k;P) = S(k_+)\,\Gamma_M(k;P)\,S(k_-)\,. \tag{31}$$

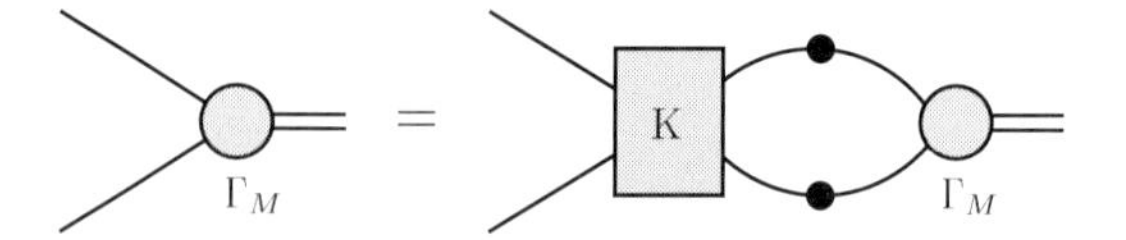

Fig. 3. Homogeneous BSE, (30). Filled circles: dressed propagators or vertices; K is the dressed-quark-antiquark scattering kernel. A systematic truncation of S^2K is the key to preserving Ward-Takahashi identities [8,14]. (Adapted from [9].)

Equation (30), depicted in Fig. 3, describes the residue at a pole in the solution of an inhomogeneous BSE; e.g., the lowest mass pole solution of (11) is identified with the pion.[4]

I noted on p. 153 that the automatic preservation of Ward-Takahashi identities in those channels related to strong interaction observables requires a conspiracy between the dressed-quark-gluon vertex and the Bethe-Salpeter kernel [8,14]. A systematic procedure for building that kernel follows [9] from the observation [14] that the gap equation can be expressed via

$$\frac{\delta \Gamma[S]}{\delta S} = 0 \,, \tag{32}$$

where $\Gamma[S]$ is a Cornwall-Jackiw-Tomboulis-like effective action [15]. The Bethe-Salpeter kernel is then obtained via an additional functional derivative:

$$K_{tu}^{rs} = -\frac{\delta \Sigma_{tu}}{\delta S_{rs}} \,. \tag{33}$$

With the vertex depicted in Fig. 1, the n-th order contribution to the kernel is obtained from the n-loop contribution to the self energy:

$$\Sigma^n(p) = -\int_q^\Lambda \mathcal{D}_{\mu\nu}(p-q)\, l^a \gamma_\mu\, S(q) l^a\, \Gamma_\nu^n(q,p). \tag{34}$$

Since $\Gamma_\mu(p,q)$ itself depends on S then (33) yields the Bethe-Salpeter kernel as a sum of two terms and hence (30) assumes the form

$$\Gamma_M(k;P) = \int_q^\Lambda \mathcal{D}_{\mu\nu}(k-q)\, l^a \gamma_\mu \left[\chi_M(q;P)\, l^a\, \Gamma_\nu(q_-,k_-) + S(q_+)\, \Lambda_{M\nu}^a(q,k;P)\right], \tag{35}$$

where I have used the mnemonic

$$\Lambda_{M\nu}^a(q,k;P) = \sum_{n=0}^{\infty} \Lambda_{M\nu}^{a;n}(q,k;P) \,. \tag{36}$$

[4] The canonical normalisation of a Bethe-Salpeter amplitude is fixed by requiring that the bound state contribute with unit residue to the fully-amputated quark-antiquark scattering amplitude: $M = K + K(SS)K + [\ldots]$. See, e.g., [13].

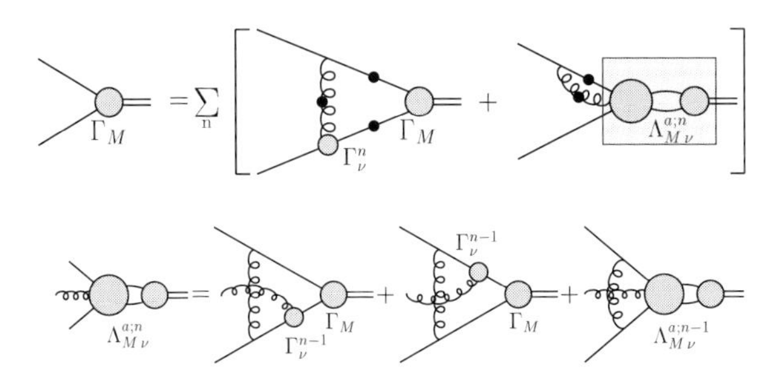

Fig. 4. Upper panel: BSE, (35), which is valid whenever Γ_μ can be obtained via a recursion relation. Lower panel: Recursion relation for $\Lambda^{a;n}_{M\nu}$. (Adapted from [9].)

Equation (35) is depicted in the upper panel of Fig. 4. The first term is instantly available once one has an explicit form for Γ^n_ν and the second term, identified by the shaded box in Fig. 4, can be obtained [9] via the inhomogeneous recursion relation depicted in the figure's lower panel. Combining these figures, it is apparent that to form the Bethe-Salpeter kernel the free gluon line is attached to the upper dressed-quark line. Consequently, the first term on the r.h.s. of the lower panel in Fig. 4 invariably generates crossed gluon lines; viz., nonplanar contributions to the kernel. The character of the vertex-consistent Bethe-Salpeter kernel is now clear: it consists of countably many contributions, a subclass of which are crossed-ladder diagrams and hence nonplanar. Only the rainbow gap equation, obtained with $i = 0$ in (20), yields a planar vertex-consistent Bethe-Salpeter kernel, namely the ladder kernel of (14). In this case alone is the number of diagrams in the dressed-vertex and kernel identical. Otherwise there are always more terms in the kernel.

Solutions for the π- and ρ-Mesons. I have recapitulated on a general procedure that provides the vertex-consistent channel-projected Bethe-Salpeter kernel once Γ^n_ν and the propagator functions; A, B, are known. That kernel must be constructed independently for each channel because, e.g., $\Lambda^a_{M\nu}$ depends on $\chi_M(q;P)$. As with the study of the vertex, an elucidation of the resulting BSEs' features is simplified by using the model of (17), for then the Bethe-Salpeter kernels are finite matrices [cf. $(1-\mathcal{O})^{-1}$ in (24)] and the homogeneous BSEs are merely linear, coupled algebraic equations.

Reference [9] describes in detail the solution of the coupled gap and Bethe-Salpeter equations for the π- and ρ-mesons. Herein I focus on the results, which are summarised in Table 1. It is evident that, irrespective of the order of the truncation; viz., the number of dressed gluon rungs in the quark-gluon vertex, the pion is massless in the chiral limit. This is in spite of the fact that it is composed of heavy dressed-quarks, as is clear in the calculated scale of the dynamically generated dressed-quark mass function: see Fig. 2, $M(0) \approx \mathcal{G} \approx 0.5\,\mathrm{GeV}$. These observations emphasise that the masslessness of the π is a model-independent consequence of consistency between the Bethe-Salpeter kernel and the kernel

Table 1. Calculated π and ρ meson masses, in GeV, quoted with $\mathcal{G} = 0.48\,\text{GeV}$, in which case $m = 0.023\,\mathcal{G} = 11\,\text{MeV}$. n is the number of dressed-gluon rungs retained in the planar vertex, see Fig. 1, and hence the order of the vertex-consistent Bethe-Salpeter kernel: the rapid convergence of the kernel is apparent from the tabulated results. (Adapted from [9].)

	$M_H^{n=0}$	$M_H^{n=1}$	$M_H^{n=2}$	$M_H^{n=\infty}$
π, $m = 0$	0	0	0	0
π, $m = 0.011$	0.152	0.152	0.152	0.152
ρ, $m = 0$	0.678	0.745	0.754	0.754
ρ, $m = 0.011$	0.695	0.762	0.770	0.770

in the gap equation. Furthermore, the bulk of the ρ-π mass splitting is present for $m = 0$ and with the simplest ($n = 0$; i.e., rainbow-ladder) kernel, which demonstrates that this mass difference is driven by the DCSB mechanism. It is not the result of a carefully contrived chromo-hyperfine interaction. Finally, the quantitative effect of improving on the rainbow-ladder truncation; namely, including more dressed-gluon rungs in the gap equation's kernel and consistently improving the kernel in the Bethe-Salpeter equation, is a 10% correction to the vector meson mass. Simply including the first correction (viz., retaining the first two diagrams in Fig. 1) yields a vector meson mass that differs from the fully resummed result by $\approx 1\%$. The rainbow-ladder truncation is clearly accurate in these channels.

Comments. While I have described results obtained with a rudimentary interaction model in order to make the construction transparent, the procedure is completely general. However, the algebraic simplicity of the analysis is naturally peculiar to the model. With a more realistic interaction, the gap and vertex equations yield a system of twelve coupled integral equations. The Bethe-Salpeter kernel for any given channel then follows as the solution of a determined integral equation.

The material reviewed covers those points in the construction of [8,9] that bear upon the fidelity of the rainbow-ladder truncation in pairing the gap equation and Bethe-Salpeter equations for the vector and flavour non-singlet pseudoscalar mesons. The error is small. In modelling it is therefore justified to fit one's parameters to physical observables at this level in these channels and then make predictions for other phenomena involving vector and pseudoscalar bound states in the expectation they will be reliable. That approach has been successful, as illustrated in [4].

Lastly, the placement of the rainbow-ladder truncation as the first term in a procedure that can methodically be improved explains why this truncation has been successful, the boundaries of its success, why it has failed outside these boundaries, and why sorting out the failures won't undermine the successes.

2.4 Selected Model-Independent Results

In the hadron spectrum the pion is identified as both a Goldstone mode, associated with DCSB, and a bound state composed of constituent u- and d-quarks, whose effective mass is $\sim 300 - 500\,\mathrm{MeV}$. Naturally, in quantum mechanics, one can fabricate a mass operator that yields a bound state whose mass is much less than the sum of the constituents' masses. However, that requires *fine tuning* and, without additional fine tuning, such models predict properties for spin- and/or isospin-flip relatives of the pion which conflict with experiment. A correct resolution of this apparent dichotomy is one of the fundamental challenges to establishing QCD as the theory underlying strong interaction physics, and the DSEs provide an ideal framework within which to achieve that end, as I now explain following the proof of [5]. It cannot be emphasised too strongly that the legitimate understanding of pion observables; including its mass, decay constant and form factors, requires an approach to contain a well-defined and valid chiral limit.

Proof of Goldstone's Theorem. Consider the BSE expressed for the isovector pseudoscalar channel:

$$\left[\Gamma_\pi^j(k;P)\right]_{tu} = \int_q^\Lambda [\chi_\pi^j(q;P)]_{sr}\, K_{tu}^{rs}(q,k;P)\,, \tag{37}$$

with $\chi_\pi^j(q;P) = S(q_+)\Gamma_\pi^j(q;P)S(q_-)$ obvious from (31) and j labelling isospin, of which the solution has the general form

$$\begin{aligned}\Gamma_\pi^j(k;P) = \tau^j\gamma_5 \Big[& iE_\pi(k;P) + \gamma\cdot P F_\pi(k;P) \\ & + \gamma\cdot k\, k\cdot P\, G_\pi(k;P) + \sigma_{\mu\nu}\, k_\mu P_\nu\, H_\pi(k;P)\Big]\,.\end{aligned} \tag{38}$$

It is apparent that the dressed-quark propagator, the solution of (1), is an important part of the BSE's kernel.

Chiral symmetry and its dynamical breaking are expressed in the axial-vector Ward-Takahashi identity, (10), which involves the axial-vector vertex:

$$\left[\Gamma_{5\mu}^j(k;P)\right]_{tu} = Z_2\left[\gamma_5\gamma_\mu\frac{\tau^j}{2}\right]_{tu} + \int_q^\Lambda [\chi_{5\mu}^j(q;P)]_{sr}\, K_{tu}^{rs}(q,k;P)\,, \tag{39}$$

that has the general form

$$\begin{aligned}\Gamma_{5\mu}^j(k;P) = & \frac{\tau^j}{2}\gamma_5\left[\gamma_\mu F_R(k;P) + \gamma\cdot k k_\mu G_R(k;P) - \sigma_{\mu\nu}\, k_\nu\, H_R(k;P)\right] \\ & + \tilde\Gamma_{5\mu}^j(k;P) + \frac{P_\mu}{P^2+m_\phi^2}\phi^j(k;P)\,,\end{aligned} \tag{40}$$

where F_R, G_R, H_R and $\tilde{\Gamma}^i_{5\mu}$ are regular as $P^2 \to -m_\phi^2$, $P_\mu \tilde{\Gamma}^i_{5\mu}(k;P) \sim \mathrm{O}(P^2)$ and $\phi^j(k;P)$ has the structure depicted in (38). Equation (40) admits the possibility of at least one pole term in the vertex but does not require it.

Substituting (40) into (39) and equating putative pole terms, it is clear that, if present, $\phi^j(k;P)$ satisfies (37). Since this is an eigenvalue problem that only admits a $\Gamma^j_\pi \neq 0$ solution for $P^2 = -m_\pi^2$, it follows that $\phi^j(k;P)$ is nonzero solely for $P^2 = -m_\pi^2$ and the pole mass is $m_\phi^2 = m_\pi^2$. Hence, if K supports such a bound state, the axial-vector vertex contains a pion-pole contribution. Its residue, r_A, however, is not fixed by these arguments. Thus (40) becomes

$$\begin{aligned} \Gamma^j_{5\mu}(k;P) = & \frac{\tau^j}{2}\gamma_5 \left[\gamma_\mu F_R(k;P) + \gamma\cdot k k_\mu G_R(k;P) - \sigma_{\mu\nu}\, k_\nu\, H_R(k;P)\right] \\ & + \tilde{\Gamma}^i_{5\mu}(k;P) + \frac{r_A P_\mu}{P^2 + m_\pi^2}\Gamma^j_\pi(k;P)\,. \end{aligned} \tag{41}$$

Consider now the chiral limit axial-vector Ward-Takahashi identity, (10). If one assumes $m_\pi^2 = 0$ in (41), substitutes it into the l.h.s. of (10) along with (6) on the right, and equates terms of order $(P_\nu)^0$ and P_ν, one obtains the chiral-limit relations [5]

$$\begin{aligned} r_A E_\pi(k;0) &= B(k^2)\,, & F_R(k;0) + 2\, r_A F_\pi(k;0) &= A(k^2)\,, \\ G_R(k;0) + 2\, r_A G_\pi(k;0) &= 2A'(k^2)\,, & H_R(k;0) + 2\, r_A H_\pi(k;0) &= 0\,. \end{aligned} \tag{42}$$

I have already explained that $B(k^2) \equiv 0$ in the chiral limit [remember (9)] and that a $B(k^2) \neq 0$ solution of (1) in the chiral limit signals DCSB. Indeed, in this case [16]

$$M(p^2) \stackrel{\text{large}-p^2}{=} \frac{2\pi^2\gamma_m}{3} \frac{\left(-\langle\bar{q}q\rangle^0\right)}{p^2\left(\frac{1}{2}\ln\left[p^2/\Lambda^2_{\mathrm{QCD}}\right]\right)^{1-\gamma_m}}\,, \tag{43}$$

where $\langle\bar{q}q\rangle^0$ is the renormalisation-point-independent vacuum quark condensate [17]. Furthermore, there is at least one nonperturbative DSE truncation scheme that preserves the axial-vector Ward-Takahashi identity, order by order. Hence (42) are exact quark-level Goldberger-Treiman relations, which state that when chiral symmetry is dynamically broken:

(i). the homogeneous isovector pseudoscalar BSE has a massless solution;
(ii). the Bethe-Salpeter amplitude for the massless bound state has a term proportional to γ_5 alone, with $E_\pi(k;0)$ completely determined by the scalar part of the quark self energy, in addition to other pseudoscalar Dirac structures, F_π, G_π and H_π, that are nonzero;
(iii). and the axial-vector vertex is dominated by the pion pole for $P^2 \simeq 0$.

The converse is also true. Hence DCSB is a sufficient and necessary condition for the appearance of a massless pseudoscalar bound state (of what can be very-massive constituents) that dominates the axial-vector vertex for $P^2 \approx 0$.

Mass Formula. When chiral symmetry is explicitly broken the axial-vector Ward-Takahashi identity becomes:

$$P_\mu \Gamma^j_{5\mu}(k;P) = S^{-1}(k_+) i\gamma_5 \frac{\tau^j}{2} + i\gamma_5 \frac{\tau^j}{2} S^{-1}(k_-) - 2i\, m(\zeta)\, \Gamma^j_5(k;P), \tag{44}$$

where the pseudoscalar vertex is obtained from

$$\left[\Gamma^j_5(k;P)\right]_{tu} = Z_4 \left[\gamma_5 \frac{\tau^j}{2}\right]_{tu} + \int_q^\Lambda \left[\chi^j_5(q;P)\right]_{sr} K^{rs}_{tu}(q,k;P)\,. \tag{45}$$

As argued in connection with (39), the solution of (45) has the form

$$\begin{aligned} i\Gamma^j_5(k;P) = {} & \frac{\tau^j}{2}\gamma_5 \left[iE^P_R(k;P) + \gamma \cdot P\, F^P_R + \gamma \cdot k\, k \cdot P\, G^P_R(k;P) \right. \\ & \left. + \sigma_{\mu\nu}\, k_\mu P_\nu\, H^P_R(k;P) \right] + \frac{r_P}{P^2 + m_\pi^2} \Gamma^j_\pi(k;P)\,, \end{aligned} \tag{46}$$

where E^P_R, F^P_R, G^P_R and H^P_R are regular as $P^2 \to -m_\pi^2$; i.e., the isovector pseudoscalar vertex also receives a contribution from the pion pole. In this case equating pole terms in the Ward-Takahashi identity, (44), entails [5]

$$r_A\, m_\pi^2 = 2\, m(\zeta)\, r_P(\zeta)\,. \tag{47}$$

This, too, is an exact relation in QCD. Now it is important to determine the residues r_A and r_P.

Study of the renormalised axial-vector vacuum polarisation shows [5]:

$$r_A\, \delta^{ij}\, P_\mu = f_\pi\, \delta^{ij}\, P_\mu = Z_2\, \mathrm{tr} \int_q^\Lambda \tfrac{1}{2}\tau^i \gamma_5 \gamma_\mu S(q_+) \Gamma^j_\pi(q;P) S(q_-)\,, \tag{48}$$

where the trace is over colour, Dirac and flavour indices; i.e., the residue of the pion pole in the axial-vector vertex is the pion decay constant. The factor of Z_2 on the r.h.s. in (48) is crucial: it ensures the result is gauge invariant, and cutoff and renormalisation-point independent. Equation (48) is the exact expression in quantum field theory for the pseudovector projection of the pion's wave function on the origin in configuration space.

A close inspection of (45), following its re-expression in terms of the renormalised, fully-amputated quark-antiquark scattering amplitude: $M = K + K(SS)K + \ldots$, yields [5]

$$i r_P\, \delta^{ij} = Z_4\, \mathrm{tr} \int_q^\Lambda \tfrac{1}{2}\tau^i \gamma_5 S(q_+) \Gamma^j_\pi(q;P) S(q_-)\,, \tag{49}$$

wherein the dependence of Z_4 on the gauge parameter, the regularisation mass-scale and the renormalisation point is exactly that required to ensure: 1) r_P is finite in the limit $\Lambda \to \infty$; 2) r_P is gauge-parameter independent; and 3) the renormalisation point dependence of r_P is just such as to guarantee the r.h.s. of (47) is renormalisation point *independent*. Equation (49) expresses the pseudoscalar projection of the pion's wave function on the origin in configuration space.

Focus for a moment on the chiral limit behaviour of (49) whereat, using (38), (42), one finds readily

$$-\langle \bar{q}q \rangle_\zeta^0 = f_\pi r_P^0(\zeta) = Z_4(\zeta,\Lambda)\, N_c\, \mathrm{tr_D} \int_q^\Lambda S_{\hat{m}=0}(q)\,. \tag{50}$$

Equation (50) is unique as the expression for the chiral limit *vacuum quark condensate*.[5] It is ζ-dependent but independent of the gauge parameter and the regularisation mass-scale, and (50) thus proves that the chiral-limit residue of the pion pole in the pseudoscalar vertex is $(-\langle \bar{q}q \rangle_\zeta^0)/f_\pi$. Now (47), (50) yield

$$(f_\pi^0)^2\, m_\pi^2 = -2\, m(\zeta)\, \langle \bar{q}q \rangle_\zeta^0 + \mathrm{O}(\hat{m}^2)\,, \tag{51}$$

where f_π^0 is the chiral limit value from (48). Hence what is commonly known as the Gell-Mann–Oakes–Renner relation is a *corollary* of (47).

One can now understand the results in Table 1: a massless bound state of massive constituents is a necessary consequence of DCSB and will emerge in any few-body approach to QCD that employs a systematic truncation scheme which preserves the Ward-Takahashi identities.

Upon review it will be apparent that (47)–(49) are valid for any values of the current-quark masses, and the generalisation to N_f quark flavours is [6,10,11]

$$f_H^2\, m_H^2 = -\, \langle \bar{q}q \rangle_\zeta^H \mathcal{M}_H^\zeta, \tag{52}$$

$\mathcal{M}_\zeta^H = m_{q_1}^\zeta + m_{q_2}^\zeta$ is the sum of the current-quark masses of the meson's constituents;

$$f_H\, P_\mu = Z_2\, \mathrm{tr} \int_q^\Lambda \tfrac{1}{2}(T^H)^T \gamma_5 \gamma_\mu \mathcal{S}(q_+)\, \Gamma^H(q;P)\, \mathcal{S}(q_-)\,, \tag{53}$$

with $\mathcal{S} = \mathrm{diag}(S_u, S_d, S_s, \ldots)$, T^H a flavour matrix specifying the meson's quark content, e.g., $T^{\pi^+} = \frac{1}{2}(\lambda^1 + i\lambda^2)$, $\{\lambda^i\}$ are N_f-flavour generalisations of the Gell-Mann matrices; and

$$\langle \bar{q}q \rangle_\zeta^H = -f_H\, r_H^\zeta = i f_H\, Z_4\, \mathrm{tr} \int_q^\Lambda \tfrac{1}{2}(T^H)^T \gamma_5 \mathcal{S}(q_+)\, \Gamma^H(q;P)\, \mathcal{S}(q_-)\,. \tag{54}$$

NB. Equation (50) means that in the chiral limit $\langle \bar{q}q \rangle_\zeta^H \to \langle \bar{q}q \rangle_\zeta^0$ and hence $\langle \bar{q}q \rangle_\zeta^H$ has been called an *in-hadron condensate*.

The formulae reviewed in this section also yield model-independent corollaries for systems involving heavy-quarks, as I relate in Sect. 4.

[5] The trace of the massive dressed-quark propagator is not renormalisable and hence there is no unique definition of a massive-quark condensate [17].

3 Basis for a Description of Mesons

The renormalisation-group-improved rainbow-ladder truncation has long been employed to study light mesons, and in Sects. 2.2, 2.3 it was shown to be a quantitatively reliable tool for vector and flavour nonsinglet pseudoscalar mesons. In connection with (14), (15), I argued that the truncation preserves the ultraviolet behaviour of the quark-antiquark scattering kernel in QCD but requires an assumption about that kernel in the infrared; viz., on the domain $Q^2 \lesssim 1\,\mathrm{GeV}^2$, which corresponds to length-scales $\gtrsim 0.2\,\mathrm{fm}$. The calculation of this behaviour is a primary challenge in contemporary hadron physics and there is progress [7,18–23]. However, at present the most efficacious approach is to model the kernel in the infrared, which enables quantitative comparisons with experiments that can be used to inform theoretical analyses. The most extensively applied model is specified by [24]

$$\frac{\alpha(Q^2)}{Q^2} = \frac{4\pi^2}{\omega^6} D\, Q^2 \mathrm{e}^{-Q^2/\omega^2} + \frac{8\pi^2\,\gamma_m}{\ln\left[\tau + \left(1 + Q^2/\Lambda^2_{\mathrm{QCD}}\right)^2\right]}\, \mathcal{F}(Q^2)\,, \qquad (55)$$

in (14), (15). Here, $\mathcal{F}(Q^2) = [1-\exp(-Q^2/[4m_t^2])]/Q^2$, $m_t = 0.5\,\mathrm{GeV}$; $\tau = \mathrm{e}^2 - 1$; $\gamma_m = 12/25$; and [25] $\Lambda_{\mathrm{QCD}} = \Lambda^{(4)}_{\overline{\mathrm{MS}}} = 0.234\,\mathrm{GeV}$. This simple form expresses the interaction strength as a sum of two terms: the second ensures that perturbative behaviour is preserved at short-range; and the first makes provision for the possibility of enhancement at long-range. The true parameters in (55) are D and ω, which together determine the integrated infrared strength of the rainbow-ladder kernel; i.e., the so-called interaction tension, σ^Δ [18]. However, I emphasise that they are not independent: in fitting to a selection of observables, a change in one is compensated by altering the other; e.g., on the domain $\omega \in [0.3, 0.5]\,\mathrm{GeV}$, the fitted observables are approximately constant along the trajectory [7]

$$\omega\, D = (0.72\,\mathrm{GeV})^3. \qquad (56)$$

Hence (55) is a one-parameter model. This correlation: a reduction in D compensating an increase in ω, ensures a fixed value of the interaction tension.

3.1 Rainbow Gap Equation

Equations (15) and (55) provide a model for QCD's gap equation and in hadron physics applications one is naturally interested in the nonperturbative DCSB solution. A familiar property of gap equations is that they only support such a solution if the interaction tension exceeds some critical value. In the present case that value is $\sigma^\Delta_c \sim 2.5\,\mathrm{GeV/fm}$ [18]. This amount of infrared strength is sufficient to generate a nonzero vacuum quark condensate *but only just.* An acceptable description of hadrons requires $\sigma^\Delta \sim 25\,\mathrm{GeV/fm}$ [6] and that is obtained with [24]

$$D = (0.96\,\mathrm{GeV})^2\,. \qquad (57)$$

Table 2. Comparison of experimental values with results for π and K observables calculated using the rainbow-ladder interaction specified by (55), quoted in MeV. The model's sole parameter and the current-quark masses were varied to obtain these results, and the best fit parameter values are given in (57), (58). Predictions for analogous vector meson observables are also tabulated. NB. A charged particle normalisation is used for f_H^V in (61), which differs from that in (53) by a multiplicative factor of $\sqrt{2}$. (Adapted from [24].)

	m_π	m_K	f_π	f_K	m_ρ	m_{K^*}	m_ϕ	f_ρ	f_{K^*}	f_ϕ
Calc. [24]	138	497	93	109	742	936	1072	207	241	259
Expt. [25]	138	496	92	113	771	892	1019	217	227	228
Rel. Error					0.04	-0.05	-0.05	0.05	-0.06	-0.14

This value of the model's infrared mass-scale parameter and the two current-quark masses

$$m_u(1\,\mathrm{GeV}) = 5.5\,\mathrm{MeV}\,, \quad m_s(1\,\mathrm{GeV}) = 125\,\mathrm{MeV}\,, \tag{58}$$

defined using the one-loop expression

$$\frac{m(\zeta)}{m(\zeta')} = Z_m(\zeta',\zeta) \stackrel{1-\mathrm{loop}}{=} \left(\frac{\ln[\zeta'/\Lambda_{\mathrm{QCD}}]}{\ln[\zeta/\Lambda_{\mathrm{QCD}}]}\right)^{\gamma_m} \tag{59}$$

to evolve $m_u(19\,\mathrm{GeV}) = 3.7\,\mathrm{MeV}$ and $m_s(19\,\mathrm{GeV}) = 85\,\mathrm{MeV}$, were obtained in [24] by requiring a least-squares fit to the π- and K-meson observables listed in Table 2. The procedure was straightforward: the rainbow gap equation [(7), (15), (55)] was solved with a given parameter set and the output used to complete the kernels in the homogeneous ladder BSEs for the π- and K-mesons [(14), (15), (37), (38) with τ^j for the π channel and $\tau^j \to T^{K^+} = \frac{1}{2}(\lambda^4 + i\lambda^5)$ for the K]. These BSEs were solved to obtain the π- and K-meson masses, and the Bethe-Salpeter amplitudes. Combining this information delivers the leptonic decay constants via (53). This was repeated as necessary to arrive at the results in Table 2. The model gives a vacuum quark condensate

$$-\langle \bar{q}q\rangle^0_{1\,\mathrm{GeV}} = (0.242\,\mathrm{GeV})^3\,, \tag{60}$$

calculated from (50) and evolved using the one-loop expression in (59).

With the model's single parameter fixed, and the dressed-quark propagator determined, it is straightforward to compose and solve the homogeneous BSE for vector mesons. This yields predictions, also listed in Table 2, for the vector meson masses and electroweak decay constants [11]

$$f_H^V\, M_H^V = \tfrac{1}{3} Z_2\, \mathrm{tr} \int_q^\Lambda (T^H)^T \gamma_\mu\, \mathcal{S}(q_+)\, \Gamma_\mu^H(q;P)\, \mathcal{S}(q_-)\,, \tag{61}$$

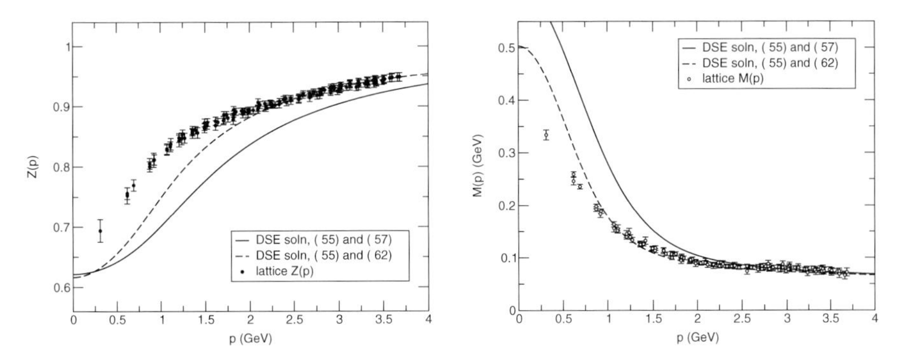

Fig. 5. Left panel – wave function renormalisation: solution of the gap equation using (55), (57) (*solid curve*); solution using (55), (62) (*dashed-curve*); quenched lattice-QCD simulations [26], obtained with $m = 0.036/a \sim 60\,\text{MeV}$ (*data*). The DSE study used a renormalisation point $\zeta = 19\,\text{GeV}$ and a current-quark mass $0.6\,m_s^{1\,\text{GeV}}$ [(58)], to enable a direct comparison with the lattice data. Right panel – mass function. (Adapted from [7].)

where M_H^V is the meson's mass and $P_\mu \Gamma_\mu^H(q;P) = 0$ for $P^2 = -(M_V^H)^2$; i.e., the Bethe-Salpeter amplitude is transverse. f_H^V characterises decays such as $\rho \to e^+e^-$, $\tau \to K^* \nu_\tau$.

Given the discussion in Sect. 2, the phenomenological success of the rainbow-ladder kernel, manifest in the results of Table 2, is unsurprising and, indeed, was to be expected.

3.2 Comparison with Lattice Simulations

The solution of the gap equation has long been of interest in grappling with DCSB in QCD and hence, in Figs. 5, I depict the scalar functions characterising the renormalised dressed-quark propagator: the wave function renormalisation, $Z(p^2)$, and mass function, $M(p^2)$, obtained by solving (15) using (55). The infrared suppression of $Z(p^2)$ and enhancement of $M(p^2)$ are longstanding predictions of DSE studies [1]. Indeed, this property of asymptotically free theories was elucidated in [16] and could be anticipated from studies of strong coupling QED [27]. The prediction has recently been confirmed in numerical simulations of quenched lattice-QCD, as is evident in the figures.

It is not yet possible to reliably determine the behaviour of lattice Schwinger functions for current-quark masses that are a realistic approximation to those of the u- and d-quarks. A veracious lattice estimate of m_π, f_π, $\langle \bar{q}q \rangle^0$ is therefore absent. To obtain such an estimate, [7] used the rainbow kernel described herein and varied (D,ω) in order to reproduce the quenched lattice-QCD data. A best fit was obtained with

$$D = (0.74\,\text{GeV})^2\,,\;\; \omega = 0.3\,\text{GeV}\,, \tag{62}$$

at a current-quark mass of $0.6\,m_s^{1\,\text{GeV}} \approx 14\,m_u$ [(58)] chosen to coincide with that employed in the lattice simulation. Constructing and solving the homogeneous

BSE for a pion-like bound state composed of quarks with this current-mass yields

$$m_\pi^{m_q \sim 14 m_u} = 0.48\,\text{GeV},\; f_\pi^{m_q \sim 14 m_u} = 0.094\,\text{GeV}\,. \tag{63}$$

The parameters in (62) give chiral limit results [7]:

$$f_\pi^0 = 0.068\,\text{GeV}\,,\; -\langle \bar{q}q\rangle^0_{1\,\text{GeV}} = (0.19\,\text{GeV})^3\,, \tag{64}$$

whereas (56), (57) give $f_\pi^0 = 0.088\,\text{GeV}$. These results have been confirmed in a more detailed analysis [28] and this correspondence suggests that chiral and physical pion observables are materially underestimated in the quenched theory: $|\langle \bar{q}q\rangle|$ by a factor of two and f_π by 30%.

The rainbow-ladder kernel has also been employed in an analysis of a trajectory of fictitious pseudoscalar mesons, all composed of equally massive constituents [29] (The only physical state on this trajectory is the pion.) The DSE study predicts [30]

$$\frac{m_{H_{m=2m_s}}}{m_{H_{m=m_s}}} = 2.2\,, \tag{65}$$

in agreement with a result of recent quenched lattice simulations [31]. The DSE study provides an intuitive understanding of this result, showing that it owes itself to a large value of the in-hadron condensates for light-quark mesons; e.g., $\langle \bar{q}q\rangle^{s\bar{s}}_{1\,\text{GeV}} = (-0.32\,\text{GeV})^3$ [6], and thereby confirms the large-magnitude condensate version of chiral perturbation theory, an observation also supported by (64). References [29,32] also provide vector meson trajectories.

3.3 *Ab Initio* Calculation of Meson Properties

The renormalisation-group-improved rainbow-ladder kernel defined with (55) has been employed to predict a wide range of meson observables, and this is reviewed in [4]. These results; e.g., those for vector mesons in Table 2, are true predictions, in the sense that the model's mass-scale was fixed, as described in connection with (57), and every element in each calculation was completely determined by, and calculated from, that kernel.

A particular success was the calculation of the electromagnetic pion form factor, which is described in [33,34]. The result is depicted in Fig. 6, wherein it is compared with the most recent experimental data [37]. It is noteworthy that all other pre-existing calculations are uniformly two – four standard deviations below that $Q^2 F_\pi(Q^2)$ data.[6]

In this connection one should also note that it is a model independent DSE prediction [40] that electromagnetic elastic meson form factors display

$$q^2 F(q^2) = \text{constant},\; q^2 \gg \Lambda^2_{\text{QCD}}, \tag{66}$$

[6] The nature and meaning of vector dominance is discussed in Sect. 2.3.1 of [39], Sect. 2.3 of [2] and Sect. 4.3 of [4]: the low-q^2 behaviour of the pion form factor is necessarily dominated by the lowest mass resonance in the $J^{PC} = 1^{--}$ channel. Any realistic calculation will predict that and also a deviation from dominance by the ρ-meson pole alone as spacelike-q^2 increases.

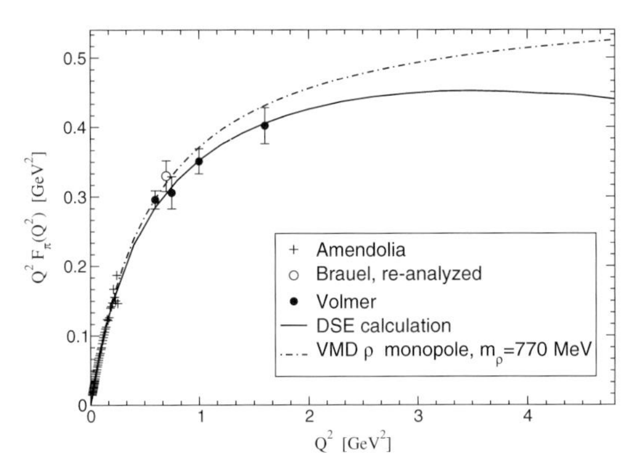

Fig. 6. Impulse approximation DSE prediction for $q^2 F_\pi(q^2)$ obtained in a parameter-free application of the renormalisation-group-improved rainbow-ladder truncation, (55), (57). The data are from [35–37]. (Adapted from [38].)

with calculable $(\ln q^2/\Lambda^2_{\rm QCD})^d$ corrections, where d is an anomalous dimension. This agrees with earlier perturbative QCD analyses [41,42]. However, to obtain this result in covariant gauges it is crucial to retain the pseudovector components of the Bethe-Salpeter amplitude in (38): F_π, G_π. (NB. The quark-level Goldberger-Treiman relations, (42), prove them to be nonzero.) Without these amplitudes [39], $q^2 F(q^2) \propto 1/q^2$. The calculation of [40] suggests that the perturbative behaviour of (66) is unambiguously evident for $q^2 \gtrsim 15\,\text{GeV}^2$. Owing to challenges in the numerical analysis, the *ab initio* calculations of [34] cannot yet make a prediction for the onset of the perturbative domain but progress in remedying that is being made [43].

Another very instructive success is the study of π-π scattering, wherein a range of new challenges arise whose quiddity and natural resolution via a symmetry-preserving truncation of the DSEs is explained in Sect. 4.6 of [4], which reviews the seminal work of [44]. It is worth remarking, too, that with a systematic and nonperturbative DSE truncation scheme, all consequences of the Abelian anomaly and Wess-Zumino term are obtained exactly, without fine tuning [39,40,45–48].

3.4 Heavier Mesons

The meson spectrum contains [25] four little-studied axial-vector mesons composed of u- and d-quarks. They appear as isospin $I = 0, 1$ partners (in the manner of the ω and ρ): $h_1(1170)$, $b_1(1235)$; and $f_1(1285)$, $a_1(1260)$, and differ in their charge-parity: $J^{PC} = 1^{+-}$ for h_1, b_1; and $J^{PC} = 1^{++}$ for f_1, a_1. In the $q\bar{q}$ constituent quark model the b_1 is represented as a constituent-quark and -antiquark with total spin $S = 0$ and angular momentum $L = 1$, while in the a_1 the quark and antiquark have $S = 1$ and $L = 1$. It is therefore apparent that in this model the b_1 is an orbital excitation of the π, and the a_1 is an orbital excitation and axial-vector partner of the ρ. In QCD the J^{PC} characteristics of a quark-antiquark bound state are manifest in the structure of its Bethe-Salpeter amplitude [13]. This amplitude is a valuable intuitive guide and, in cases where a $q\bar{q}$ constituent quark model analogue exists, it incorporates and extends the information present in that analogue's quantum mechanical wave function.

Three of the axial-vector mesons decay predominantly into two-body final states containing a vector meson and a pion: $h_1 \to \rho\pi$; $b_1 \to \omega\pi$; $a_1 \to \rho\pi$. With a $J = 1$ meson in both the initial and final state these three decays proceed via two partial waves (S, D), and therefore probe aspects of hadron structure inaccessible in simpler processes involving only spinless mesons in the final state, such as $\rho \to \pi\pi$. For example and of importance, in constituent-quark-like models the D/S amplitude ratio is very sensitive [49] to the nature of the phenomenological long-range confining interaction.

The additional insight and model constraints that such processes can provide is particularly important now as a systematic search and classification of "exotic" states in the light meson sector becomes feasible experimentally. I note that a meson is labelled "exotic" if it is characterised by a value of J^{PC} which is unobtainable in the $q\bar{q}$ constituent quark model; e.g., the experimentally observed [50] $\pi_1(1600)$, a 1.6 GeV $J^{PC} = 1^{-+}$ state. Such unusual charge parity states are a necessary feature of a field theoretical description of quark-antiquark bound states [13] with BSE studies typically yielding [51] masses approximately twice as large as that of the natural charge parity partner and, in particular, a $J^{PC} = 1^{-+}$ meson with a mass ~ 1.5 GeV [52].

In appreciation of these points, [53] used the simple DSE-based model of [51] in a simultaneous study of axial-vector meson decays, $\rho \to \pi\pi$ decay, and the electroweak decay constants of the mesons involved. The results are instructive. It was found that the rainbow-ladder truncation is capable of simultaneously providing a good description of these observables but that the D/S partial-wave ratio in the decays of axial-vector mesons is indeed very sensitive to details of the long-range part of a model interaction; i.e., to the expression of light-quark confinement. This is perhaps unsurprising given that the mass of each axial-vector meson mass is significantly greater than $2M(0)$; namely, twice the constituent-quark mass-scale. Unfortunately, more sophisticated calculations are lacking. This collection of experimentally well-understood mesons has many lessons to teach and should no longer be ignored.

4 Heavy Quarks

4.1 Features of the Mass Function

The DSE methods described hitherto have been applied to mesons involving heavy-quarks [11,54,55] and in this case there is a natural simplification. To begin, one focuses on the fact that mesons, whether heavy or light, are bound states of a dressed-quark and -antiquark, with the dressing described by the gap equation, (1), written explicitly again here with the addition of flavour label, $f\,(= u, d, s, c, b)$:

$$S_f(p)^{-1} = i\gamma \cdot p\, A_f(p^2) + B_f(p^2) = A_f(p^2)\left[i\gamma \cdot p + M_f(p^2)\right] \tag{67}$$

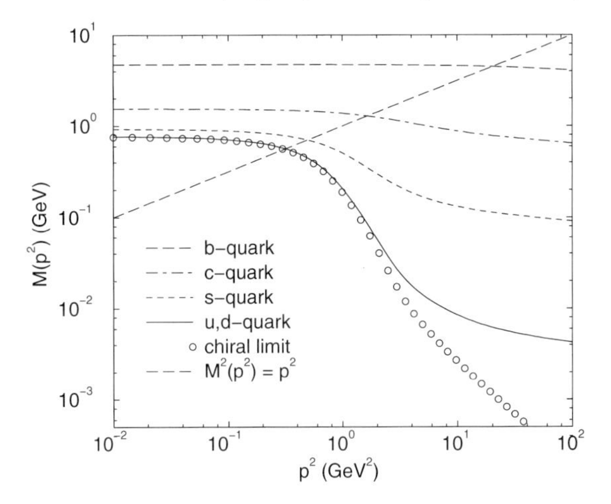

Fig. 7. Quark mass function obtained as a solution of (67) using the rainbow truncation, discussed in connection with (12) – (15), and the interaction of (55) with current-quark masses, fixed at $\zeta = 19\,\text{GeV}$: $m_{u,d}(\zeta) = 3.7\,\text{MeV}$, $m_s(\zeta) = 82\,\text{MeV}$, $m_c(\zeta) = 0.58\,\text{GeV}$ and $m_b(\zeta) = 3.8\,\text{GeV}$. The indicated solutions of $M^2(p^2) = p^2$ define a Euclidean constituent-quark mass, M_f^E, which takes the values: $M_u^E = 0.56\,\text{GeV}$, $M_s^E = 0.70\,\text{GeV}$, $M_c^E = 1.3\,\text{GeV}$, $M_b^E = 4.6\,\text{GeV}$.

$$= Z_2(i\gamma \cdot p + m_f^{\rm bm}) + Z_1 \int_q^\Lambda g^2 D_{\mu\nu}(p-q)\frac{\lambda^a}{2}\gamma_\mu S_f(q)\Gamma_\nu^{fa}(q,p)\,. \qquad (68)$$

The other elements of (68) will already be familiar.

The qualitative features of the gap equation's solution are known and typical mass functions, $M_f(p^2)$, are depicted in Fig. 7. There is some quantitative model-dependence in the momentum-evolution of the mass-function into the infrared. However, with any *Ansatz* for the effective interaction that provides an accurate description of $f_{\pi,K}$ and $m_{\pi,K}$, one obtains solutions with profiles like those illustrated in the figure. Owing to (13) the ultraviolet behaviour is naturally fixed, namely, it is given by (5) for massive quarks and by (43) in the chiral limit.

It is apparent in the figure that as p^2 decreases the chiral-limit and u,d-quark mass functions evolve to coincidence. This feature signals a transition from the perturbative to the nonperturbative domain. Furthermore, since the chiral limit mass-function is nonzero *only* because of the nonperturbative DCSB mechanism, whereas the u,d-quark mass function is purely perturbative at $p^2 > 20\,\text{GeV}^2$, it also indicates clearly that the DCSB mechanism has a significant impact on the propagation characteristics of u,d,s-quarks. However, it is conspicuous in Fig. 7 that this is not the case for the b-quark. Its large current-quark mass almost entirely suppresses momentum-dependent dressing so that $M_b(p^2)$ is nearly constant on a substantial domain. The same is true to a lesser extent for the c-quark.

The quantity $\mathcal{L}_f := M_f^E/m_f(\zeta)$ provides a single quantitative measure of the importance of the DCSB mechanism; i.e., nonperturbative effects, in modifying the propagation characteristics of a given quark flavour. In this particular illustration it takes the values

$$\begin{array}{c|cccc} f & u,d & s & c & b \\ \hline \mathcal{L}_f & 150 & 10 & 2.2 & 1.2 \end{array}\,, \tag{69}$$

which are representative: for light-quarks $\mathcal{L}_{q=u,d,s} \sim$ 10-100; while for heavy-quarks $\mathcal{L}_{Q=c,b} \sim 1$. They also highlight the existence of a mass-scale, M_χ, characteristic of DCSB: the propagation characteristics of a flavour with $m_f(\zeta) \leq M_\chi$ are significantly altered by the DCSB mechanism, while momentum-dependent dressing is almost irrelevant for flavours with $m_f(\zeta) \gg M_\chi$. It is evident and unsurprising that $M_\chi \sim 0.2\,\mathrm{GeV} \sim \Lambda_{\rm QCD}$. Consequently one anticipates that the propagation of c, b-quarks should be described well by replacing their mass-functions with a constant; viz., writing [11]

$$S_Q(p) = \frac{1}{i\gamma\cdot p + \hat{M}_Q}\,,\; Q = c, b\,, \tag{70}$$

where $\hat{M}_Q$ is a constituent-heavy-quark mass parameter.[7]

When considering a meson with an heavy-quark constituent one can proceed further, as in heavy-quark effective theory (HQET) [56], allow the heaviest quark to carry all the heavy-meson's momentum: $P_\mu =: m_H v_\mu =: (\hat{M}_Q + E_H)v_\mu$, and write

$$S_Q(k+P) = \frac{1}{2}\,\frac{1 - i\gamma\cdot v}{k\cdot v - E_H} + \mathrm{O}\left(\frac{|k|}{\hat{M}_Q}, \frac{E_H}{\hat{M}_Q}\right)\,, \tag{71}$$

where k is the momentum of the lighter constituent. It is apparent from the study of light-meson properties that in the calculation of observables the meson's Bethe-Salpeter amplitude will limit the range of $|k|$ so that (71) will only be a good approximation if *both* the momentum-space width of the amplitude, ω_H, and the binding energy, E_H, are significantly less than $\hat{M}_Q$.

In [55] the propagation of c- and b-quarks was described by (71), with a goal of exploring the fidelity of this idealisation, and it was found to allow for a uniformly good description of B_f-meson leptonic and semileptonic decays with heavy- and light-pseudoscalar final states. In that study, $\omega_{B_f} \approx 1.3\,\mathrm{GeV}$ and $E_{B_f} \approx 0.70\,\mathrm{GeV}$, both of which are small compared with $\hat{M}_b \approx 4.6\,\mathrm{GeV}$ in Fig. 7. Hence the accuracy of the approximation could be forseen. It is reasonable to suppose that $\omega_D \approx \omega_B$ and $E_D \approx E_B$, since they must be identical in the limit of exact heavy-quark symmetry. Thus in processes involving the weak decay of a c-quark ($\hat{M}_c \approx 1.3\,\mathrm{GeV}$) where a D_f-meson is the heaviest participant, (71) must be inadequate; an expectation verified in [55].

The failure of (71) for the c-quark complicates or precludes the development of a common understanding of D_f- and B_f-meson observables using such contemporary theoretical tools as HQET and light cone sum rules. However, as shown in [11] and I will illustrate, the constituent-like dressed-heavy-quark propagator of (70) can still be used to effect a unified, accurate simplification in the study of these observables.

[7] Although not illustrated explicitly, when $M_f(p^2) \approx \mathrm{const.}$, $A_f(p^2) \approx 1$ in (67).

4.2 Leptonic Decays

Pseudoscalar Mesons. The leptonic decay of a pseudoscalar meson, $P(p)$, is described by the matrix element (Sect. 2.4)

$$f_P\, p_\mu := \langle 0|\bar{\mathcal{Q}}\,(T^P)^{\rm T}\gamma_\mu\gamma_5\,\mathcal{Q}|P(p)\rangle = {\rm tr}\, Z_2 \int_k^\Lambda (T^P)^{\rm T}\gamma_5\gamma_\mu\,\chi_P(k;p), \qquad (72)$$

where $\mathcal{Q} = \text{column}(u,d,s,c,b)$ and here I have adopted a charged particle normalisation, which yields results for f_P a factor of $\sqrt{2}$ larger than (53) and is conventional in studying heavy-quark systems.

In (72), χ_P is the meson's Bethe-Salpeter wave function, related to its amplitude, Γ_P, via (31) and normalised canonically as described in connection with (30). Using (71), it follows from the canonical normalisation condition that

$$\mathcal{G}_P(k;p) := \frac{1}{\sqrt{m_P}}\,\Gamma_P(k;p) < \infty\,,\; m_P \to \infty\,; \qquad (73)$$

i.e., $\mathcal{G}_P(k;p)$ so-defined is mass-independent in the heavy-quark limit. Using this result plus (71) one finds from (72) [54]

$$f_P \propto \frac{1}{\sqrt{m_P}}\,,\; m_P \to \infty\,. \qquad (74)$$

Equation (74) is a model-independent result and a well-known general consequence of heavy-quark symmetry [56]. However, the value of the hadron mass at which this behaviour becomes evident is unknown. It is clear from Table 2 that, experimentally,

$$f_\pi = 131\,\text{MeV} \;<\; f_K = 160\,\text{MeV}. \qquad (75)$$

Furthermore, direct DSE studies following the method described in Sect. 3.3 show that for pseudoscalar mesons $u\bar{f}$, composed of a single u,d-quark and an antiquark of mass m_f, $f_P(m_P)$ is a monotonically increasing concave-down function on $m_P \in [0, 0.9]\,\text{GeV}$, where m_P is the calculated mass of this composite system, and likely on a larger domain [32]. On the other hand, numerical simulations of quenched lattice-QCD indicate [57]

$$f_D = 200 \pm 30\,\text{MeV} \;>\; f_B = 170 \pm 35\,\text{MeV}\,. \qquad (76)$$

In simulations of lattice-QCD with two flavours of sea quarks both of these decay constants increase in magnitude but there is no sign that the ordering is reversed [57,58]. The information in (75), (76) is depicted in Fig. 8. This and analysis to be reviewed subsequently suggest that D-mesons lie outside the domain on which (74) is a reliable tool.

Vector Mesons. The leptonic decay constant, f_V, for a vector meson with mass M_V is given in (61) and adapting the analysis that leads to (74) one finds readily

$$f_V \propto \frac{1}{\sqrt{M_V}}\,,\; M_V \to \infty\,, \qquad (77)$$

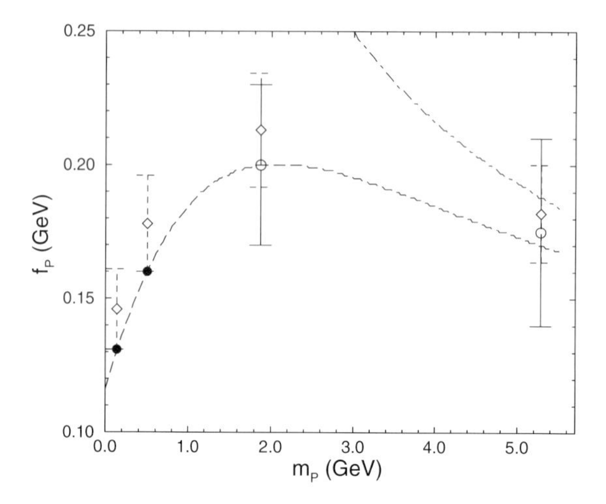

Fig. 8. Experimental values of $f_{\pi,K}$ ((75), *filled circles*); lattice estimates of $f_{D,B}$ ((76), *open circles*); values of $f_{\pi,K,D,B}$ calculated in [11] (*open diamonds*). Least-squares fit to the experimental values and lattice estimates (*dashed curve*): $f_P^2 = (0.013 + 0.028\, m_P)/(1 + 0.055\, m_P + 0.15\, m_P^2)$, which exhibits the large-m_P limit of (74); the large-m_P limit of this fit (*dot-dashed curve*). (Adapted from [11].)

which again is a model-independent result. Moreover, since the pseudoscalar and vector meson Bethe-Salpeter amplitudes become identical in the heavy-quark limit, it follows that [11]

$$f_V = f_P\,,\; M_V = m_P\,,\; \text{in the limit } m_P \to \infty\,. \tag{78}$$

4.3 Heavy-Meson Masses

More can be learnt from the pseudoscalar meson mass formula in (52). Using (74), and applying to (54) the analysis from which it follows, one obtains

$$-\langle \bar{q}q \rangle_\zeta^P = \text{constant}\,,\; \text{as } m_P \to \infty \tag{79}$$

and consequently [10,11]

$$m_P \propto \hat{m}_Q\,,\; \hat{m}_Q \to \infty\,, \tag{80}$$

where $\hat{m}_Q$ is the renormalisation-group-invariant current-quark mass of the flavour-nonsinglet pseudoscalar meson's heaviest constituent. This is the result one would have guessed from constituent-quark models but here I have outlined a direct proof in QCD.

Equation (47) is thus seen to be a single formula that unifies the masses of light- and heavy-quark mesons. This aspect has been quantitatively explored using the rainbow-ladder kernel described in Sect. 3, with the results illustrated in Fig. 9. Therein the calculated mass of a $u\bar{f}$ pseudoscalar meson is plotted as a

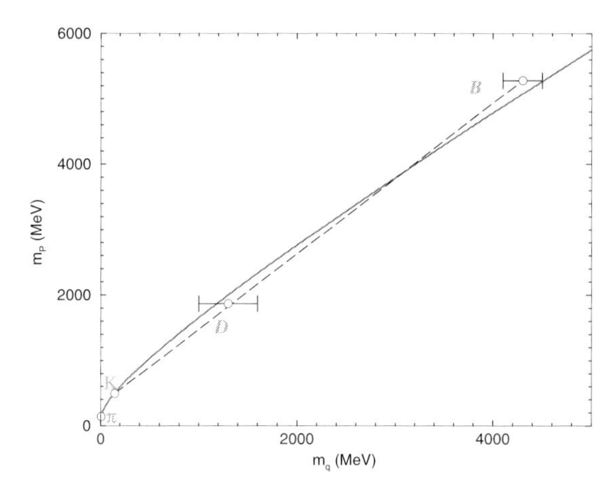

Fig. 9. Pseudoscalar $u\bar{q}$ meson's mass as a function of $m_q(\zeta)$, $\zeta = 19\,\text{GeV}$, with a fixed value of $m_u(\zeta)$ corresponding to $m_u(1\,\text{GeV}) = 5.5\,\text{MeV}$, (58) (*solid line*). The experimental data points are from [25] as are the errors assigned to the associated heavy-quark masses. A straight line is drawn through the K, D, B masses (*dashed curve*). (Adapted from [59]. See also [32].)

function of $m_f(\zeta)$, with $m_u(\zeta)$ fixed at the value in (58). The DSE calculations are depicted by the solid curve, which is [29] (in MeV)

$$m_P = 83 + 500\sqrt{\mathcal{X}} + 310\,\mathcal{X},\ \mathcal{X} = m_q^\zeta/\Lambda_{\rm QCD}. \tag{81}$$

The curvature appears slight in the figure but that is misleading: the nonlinear term in (81) accounts for almost all of m_π (the Gell-Mann–Oakes-Renner relation is nearly exact for the pion) and 80 % of m_K. NB. The dashed line in Fig. 9 fits the K, D, B subset of the data exactly. It is drawn to illustrate how easily one can be misled. Without careful calculation one might infer from this apparent agreement that the large-m_q limit of (47) is already manifest at the s-quark mass whereas, in reality, the linear term only becomes dominant for $m_q \gtrsim 1\,\text{GeV}$, providing 50 % of m_D and 67 % of m_B. The model predicts, via (59), $m_c^{1\,\text{GeV}} = 1.1\,\text{GeV}$ and $m_b^{1\,\text{GeV}} = 4.2\,\text{GeV}$, values that are typical of Poincaré covariant treatments.

4.4 Semileptonic Transition Form Factors

Pseudoscalar Meson in the Final State. The transition: $P_1(p_1) \to P_2(p_2)\,\ell\,\nu$, where P_1 represents either a B- or D-meson and P_2 can be a D, K or π, is described by the invariant amplitude

$$A(P_1 \to P_2\,\ell\,\nu) = \frac{G_F}{\sqrt{2}}\,V_{f'f}\;\bar{\ell}\,\gamma_\mu(1-\gamma_5)\nu\;M_\mu^{P_1P_2}(p_1,p_2)\,, \tag{82}$$

where $G_F = 1.166 \times 10^{-5}\,\text{GeV}^{-2}$, $V_{f'f}$ is the relevant element of the Cabibbo-Kobayashi-Maskawa (CKM) matrix, and the hadronic current is

$$M_\mu^{P_1P_2}(p_1,p_2) := \langle P_2(p_2)|\bar{f}'\gamma_\mu f|P_1(p_1)\rangle = f_+(t)\,(p_1+p_2)_\mu + f_-(t)\,q_\mu\,, \tag{83}$$

with $t := -q^2 = -(p_1 - p_2)^2$. The transition form factors, $f_\pm(t)$, contain all the information about strong-interaction effects in these processes, and their accurate calculation is essential for a reliable determination of the CKM matrix elements from a measurement of the decay width ($t_\pm := (m_{P_1} \pm m_{P_2})^2$):

$$\Gamma(P_1 \to P_2 \ell\nu) = \frac{G_F^2}{192\pi^3} |V_{f'f}|^2 \frac{1}{m_{P_1}^3} \int_0^{t_-} dt \, |f_+(t)|^2 \left[(t_+ - t)(t_- - t)\right]^{3/2}. \quad (84)$$

The related study of light-meson initial states is described in [60].

Vector Meson in the Final State. The transition: $P(p_1) \to V_\lambda(p_2)\,\ell\,\nu$, with P either a B or D and V_λ a D^*, K^* or ρ, is described by the invariant amplitude

$$A(P \to V_\lambda\,\ell\,\nu) = \frac{G_F}{\sqrt{2}} V_{f'f}\, \bar{\ell}\, i\gamma_\mu(1-\gamma_5)\nu\, \epsilon_\nu^\lambda(p_2)\, M_{\mu\nu}^{PV_\lambda}(p_1,p_2)\,, \quad (85)$$

in which the hadronic tensor involves four scalar functions

$$\begin{aligned} \epsilon_\nu^\lambda(p_2)\, M_{\mu\nu}^{PV_\lambda}(p_1,p_2) &= \epsilon_\mu^\lambda\, (m_P + M_V)\, A_1(t) + (p_1+p_2)_\mu\, \epsilon^\lambda \cdot q\, \frac{A_2(t)}{m_P + M_V} \\ &\quad + q_\mu\, \epsilon^\lambda \cdot q\, \frac{A_3(t)}{m_P + M_V} + \varepsilon_{\mu\nu\alpha\beta}\, \epsilon_\nu^\lambda\, p_{1\alpha}\, p_{2\beta}\, \frac{2V(t)}{m_P + M_V}\,. \end{aligned} \quad (86)$$

Introducing three helicity amplitudes

$$H_\pm = (m_P + M_V)\, A_1(t) \mp \frac{\lambda^{\frac{1}{2}}(m_P^2, M_V^2, t)}{m_P + M_V}\, V(t)\,, \quad (87)$$

$$H_0 = \frac{1}{2\, M_V \sqrt{t}} \left(\left[m_P^2 - M_V^2 - t\right] \left[m_P + M_V\right] A_1(t) - \frac{\lambda(m_P^2, M_V^2, t)}{m_P + M_V}\, A_2(t) \right), \quad (88)$$

where $\lambda(m_P^2, M_V^2, t) = [t_+ - t]\,[t_- - t]$, $t_\pm = (m_P \pm M_V)^2$, the transition rates

$$\frac{d\Gamma_{\pm,0}}{dt} = \frac{G_F^2}{192\pi^3 m_{P_1}^3} |V_{f'f}|^2\, t\, \lambda^{\frac{1}{2}}(m_P^2, M_V^2, t)\, |H_{\pm,0}(t)|^2\,, \quad (89)$$

from which one obtains the transverse and longitudinal rates

$$\frac{d\Gamma_T}{dt} = \frac{d\Gamma_+}{dt} + \frac{d\Gamma_-}{dt}\,,\; \Gamma_T = \int_0^{t_-} dt\, \frac{d\Gamma_T}{dt}\,, \quad (90)$$

$$\frac{d\Gamma_L}{dt} = \frac{d\Gamma_0}{dt}\,,\; \Gamma_L = \int_0^{t_-} dt\, \frac{d\Gamma_L}{dt}\,, \quad (91)$$

wherefrom the total width $\Gamma = \Gamma_T + \Gamma_L$. The polarisation ratio and forward-backward asymmetry are

$$\alpha = 2\,\frac{\Gamma_L}{\Gamma_T} - 1\,, \quad A_{\rm FB} = \frac{3}{4}\,\frac{\Gamma_- - \Gamma_+}{\Gamma}\,. \quad (92)$$

4.5 Impulse Approximation

As explained and illustrated in [4], the impulse approximation is accurate for three point functions and applied to these transition form factors it yields

$$\mathcal{H}_\mu^{PX}(p_1,p_2) = \\ 2N_c\,\mathrm{tr}_D \int_k^\Lambda \bar{\Gamma}_X(k;-p_2)\, S_q(k_2)\, i\mathcal{O}_\mu^{qQ}(k_2,k_1)\, S_Q(k_1)\, \Gamma_P(k;p_1)\, S_{q'}(k), \tag{93}$$

wherein the flavour structure is made explicit, $k_{1,2} = k + p_{1,2}$ and:

$$\mathcal{H}_\mu^{P_1X=P_2}(p_1,p_2) = M_\mu^{P_1P_2}(p_1,p_2)\,, \tag{94}$$

$$\mathcal{H}_\mu^{PX=V^\lambda}(p_1,p_2) = \epsilon_\nu^\lambda(p_2)\, M_{\mu\nu}^{PV_\lambda}(p_1,p_2)\,; \tag{95}$$

$\Gamma_{X=V^\lambda}(k;p) = \epsilon^\lambda(p)\cdot\Gamma^V(k;p)$; and $\mathcal{O}_\mu^{qQ}(k_2,k_1)$ is the dressed-quark-W-boson vertex, which in weak decays of heavy-quarks is well approximated by [54,55]

$$\mathcal{O}_\mu^{qQ}(k_2,k_1) = \gamma_\mu\,(1-\gamma_5) \tag{96}$$

because $A_Q(p^2)\approx$const. and $M_Q(p^2)\approx$const. for heavy-quarks (recall Fig. 7.)

Quark Propagators. It is plain that to evaluate $\mathcal{H}_\mu^{P_1X}(p_1,p_2)$ a specific form for the dressed-quark propagators is required. Equation (70) provides a good approximation for the heavier quarks, $Q = c, b$, as explained in Sect. 4.1, and this was used in [11] with $\hat{M}_Q$ treated as free parameters.

For the light-quark propagators:

$$S_f(p) = -i\gamma\cdot p\,\sigma_V^f(p^2) + \sigma_S^f(p^2) = \frac{1}{i\gamma\cdot p\,A_f(p^2) + B_f(p^2)}\,, \tag{97}$$

Reference [11] assumed isospin symmetry and employed the algebraic forms introduced in [39], which efficiently characterise the essential features of the gap equation's solutions:

$$\bar{\sigma}_S^f(x) = 2\bar{m}_f\mathcal{F}(2(x+\bar{m}_f^2)) + \mathcal{F}(b_1x)\mathcal{F}(b_3x)\left[b_0^f + b_2^f\mathcal{F}(\epsilon x)\right], \tag{98}$$

$$\bar{\sigma}_V^f(x) = \frac{2(x+\bar{m}_f^2) - 1 + e^{-2(x+\bar{m}_f^2)}}{2(x+\bar{m}_f^2)^2}\,, \tag{99}$$

$\mathcal{F}(y) = (1-\mathrm{e}^{-y})/y$, $x = p^2/\lambda^2$; $\bar{m}_f = m_f/\lambda$; and $\bar{\sigma}_S^f(x) = \lambda\,\sigma_S^f(p^2)$, $\bar{\sigma}_V^f(x) = \lambda^2\,\sigma_V^f(p^2)$, with λ a mass scale. The parameters are the current-quark mass, $\bar{m}$, and $b_{0,1,2,3}$, about which I shall subsequently explain more.

This algebraic form combines the effects of confinement[8] and DCSB with free-particle behaviour at large, spacelike p^2. One characteristic of DCSB is the

[8] The representation of $S(p)$ as an entire function is motivated by the algebraic solutions of (1) in [61]. The concomitant violation of the axiom of *reflection positivity* is a sufficient condition for confinement, as reviewed in Sect. 6.2 of [1], Sect. 2.2 of [2] and Sect. 2.4 of [3].

appearance of a nonzero vacuum quark condensate and using this parametrisation in (50) yields

$$-\langle \bar{u}u \rangle_\zeta = \lambda^3 \frac{3}{4\pi^2} \frac{b_0^u}{b_1^u\, b_3^u} \ln \frac{\zeta^2}{\Lambda_{\rm QCD}^2} . \tag{100}$$

The simplicity of this result emphasises the utility of an algebraic form for the dressed-quark propagator. That utility is amplified in the calculation of a form factor, which requires the repeated evaluation of a multidimensional integral whose integrand is a complex-valued function, and a functional of the propagator and the Bethe-Salpeter amplitudes.

Bethe-Salpeter Amplitudes. An algebraic parametrisation of the Bethe-Salpeter amplitudes also helps and the quark-level Goldberger-Treiman relation, (42), suggests a form for light pseudoscalar mesons:

$$\Gamma_P(k;p) = i\gamma_5\, \mathcal{E}_P(k^2) = i\gamma_5\, \frac{1}{\hat{f}_P}\, B_P(k^2)\,,\; P = \pi, K\,, \tag{101}$$

where $B_P := B_u|_{b_0^u \to b_0^P}$, obtained from (97), and $\hat{f}_P = f_P/\sqrt{2}$ because in this section I use the $f_\pi = 131\,{\rm MeV}$ normalisation. $b_0^{\pi,K}$ are two additional parameters. This *Ansatz* omits the pseudovector components of the amplitude but that is not a material defect in applications involving small to intermediate momentum transfers [40]. Equations (52), (54), (101) yield the following expression for the π- and K-meson masses:

$$\hat{f}_P^2\, m_P^2 = -(m_u + m_{f^P})\, \langle \bar{q}q \rangle^P_{1\,{\rm GeV}^2}\,, \tag{102}$$

where $m_{f^\pi} = m_d$, $m_{f^K} = m_s$, and

$$-\,\langle \bar{q}q \rangle^P_{1\,{\rm GeV}^2} = \lambda^3 \ln \frac{1}{\Lambda_{\rm QCD}^2}\, \frac{3}{4\pi^2}\, \frac{b_0^P}{b_1^u\, b_3^u}\,,\;\; P = \pi, K\,. \tag{103}$$

In studies of the type reviewed in Sect. 3, this in-hadron condensate takes values $\langle \bar{q}q \rangle^\pi_{1\,{\rm GeV}^2} \approx 1.05\, \langle \bar{u}u \rangle_{1\,{\rm GeV}^2}$ and $\langle \bar{q}q \rangle^K_{1\,{\rm GeV}^2} \approx 1.6\, \langle \bar{u}u \rangle_{1\,{\rm GeV}^2}$.

Employing algebraic parametrisations of the light vector meson Bethe-Salpeter amplitudes is also a useful expedient and that approach was adopted in [11,54,55]. Indeed, sophisticated calculations of light vector meson properties based on the rainbow-ladder truncation did not exist at the time of those studies, although it was clear that a given vector meson is narrower in momentum space than its pseudoscalar partner, and that for both vector and pseudoscalar mesons this width increases with the total current-mass of the constituents. These qualitative features were important in the explanation of meson electroproduction cross sections [62] and electromagnetic form factors [63], and can be realised in the simple expression

$$\Gamma^V_\mu(k;p) = \frac{1}{\mathcal{N}^V} \left(\gamma_\mu + p_\mu \frac{\gamma \cdot p}{M_V^2} \right) \varphi(k^2)\,, \tag{104}$$

where $\varphi(k^2) = 1/(1+k^4/\omega_V^4)$ with ω_V a parameter and $\mathcal{N}^V$ fixed by the canonical normalisation condition. One expects: $\omega_{K^*} \approx 1.6\,\omega_\rho$ [63].

In connection with the impulse approximation to semileptonic transition form factors it remains only to fix the heavy-meson Bethe-Salpeter amplitudes. In this case, too, algebraic parametrisations offer a simple, attractive and expeditious means of proceeding and that again was the approach adopted in [11]. Therein heavy vector mesons were described by (104), with $\varphi(k^2) \to \varphi_H(k^2)$, and heavy pseudoscalar mesons by its analogue:

$$\Gamma_P(k;p) = \frac{1}{\mathcal{N}^P}\, i\,\gamma_5\,\varphi_H(k^2)\,, \tag{105}$$

where $\varphi_H(k^2) = \exp\left(-k^2/\omega_H^2\right)$. The amplitudes are again normalised canonically. Such a parametrisation naturally introduces additional parameters; viz., the widths. The number is kept at two by acknowledging that Bethe-Salpeter amplitudes for truly heavy-mesons must be spin- and flavour-independent and assuming therefore that $\omega_B = \omega_{B^*} = \omega_{B_s}$ and $\omega_D = \omega_{D^*} = \omega_{D_s}$.

4.6 Additional Decay Processes

Many more decays were considered in [11], with the goal being to determine whether a unified description of light- and heavy-meson observables is possible based simply on the key DSE features of quark dressing and sensible bound state amplitudes. For example, there are experimental constraints on radiative decays $H^* \to H\,\gamma$, where $H = D_{(s)}$, $B_{(s)}$, and so these widths, $\Gamma_{H^*\to H\gamma}$, were calculated. The strong decays $H^* \to H\,\pi$ were also studied. They can be characterised by a coupling constant $g_{H^*H\pi}$, which is calculable even if the process is kinematically forbidden, as is $B^* \to B\pi$. Lastly, the width for the rare flavour-changing neutral current process $B \to K^*\gamma$, which proceeds predominantly via the local magnetic penguin operator [64] and can be characterised by a coupling $g_{BK^*\gamma}$, was calculated because data exists and this process might be expected to severely test the framework since it completely exceeds the scope of previous applications.

4.7 Heavy-Quark Symmetry Limits

Equation (71), and (73) and its natural analogues, can be used to elucidate the heavy-quark symmetry limit of the impulse approximation to any process and many were made explicit in [11,54,55]. I will only recapitulate on the most straightforward three-point case; namely, the semileptonic heavy $\to$ heavy transitions. From (83), (93) one obtains

$$\begin{aligned} f_\pm(t) &= \mathcal{T}_\pm\,\xi_f(w) := \frac{m_{P_2} \pm m_{P_1}}{2\sqrt{m_{P_2} m_{P_1}}}\,\xi_f(w)\,, \\ \xi_f(w) &= \kappa_f^2\,\frac{N_c}{4\pi^2}\int_0^1 d\tau\,\frac{1}{W}\int_0^\infty du\,\varphi_H(z_W)^2\left[\sigma_S^f(z_W) + \sqrt{\frac{u}{W}}\sigma_V^f(z_W)\right], \end{aligned} \tag{106}$$

where: $W = 1 + 2\tau(1-\tau)(w-1)$, $z_W = u - 2E_H\sqrt{u/W}$;

$$\frac{1}{m_H}\frac{1}{\kappa_f^2} := \mathcal{N}_P^2 = \mathcal{N}_V^2 = \frac{1}{m_H}\frac{N_c}{4\pi^2}\int_0^\infty du\,\varphi_H^2(z)\left\{\sigma_S^f(z) + \sqrt{u}\,\sigma_V^f(z)\right\}, \quad (107)$$

with $z = u - 2E_H\sqrt{u}$, f labelling the meson's lighter quark and all dimensioned quantities expressed in units of the mass-scale, λ; and

$$w = \frac{m_{P_1}^2 + m_{P_2}^2 - t}{2m_{P_1}m_{P_2}} = -v_{P_1}\cdot v_{P_2}\,. \quad (108)$$

The canonical normalisation of the Bethe-Salpeter amplitude automatically ensures

$$\xi_f(w=1) = 1 \quad (109)$$

and from (106) follows [54]

$$\rho^2 := -\left.\frac{d\xi_f}{dw}\right|_{w=1} \geq \frac{1}{3}\,. \quad (110)$$

Semileptonic transitions with heavy vector mesons in the final state, described by (86) and (93), can be analysed in the same way, and that yields

$$A_1(t) = \frac{1}{\mathcal{T}_+}\,\tfrac{1}{2}\,(1+w)\,\xi_f(w)\,,\; A_2(t) = -A_3(t) = V(t) = \mathcal{T}_+\,\xi_f(w)\,. \quad (111)$$

Equations (106), (111) are exemplars of a general result that in the heavy-quark symmetry limit the semileptonic $H_f \to H_f'$ transitions are described by a single, universal function: $\xi_f(w)$ [65]. In this limit the functions

$$R_1(w) := (1 - t/t_+)\,\frac{V(t)}{A_1(t)}\,,\; R_2(w) := (1 - t/t_+)\,\frac{A_2(t)}{A_1(t)} \quad (112)$$

are constant ($=1$), independent of w.

4.8 Survey of Results for Light- and Heavy-Meson Observables

With every necessary element defined, the calculation of observables is a straightforward numerical exercise. The algebraic *Ansätze* described above involve ten parameters plus four current-quark masses and in [11] they were fixed via a χ^2-fit to the $N_{\rm obs} = 42$ heavy- and light-meson observables in Table 3, a process which yielded [66]

	$\bar m_f$	b_1^f	b_2^f
u	0.00948	2.94	0.733
s	0.210	3.18	0.858

	b_0^P
π	0.204
K	0.319

	$\omega_V^{\rm GeV}$
ρ	0.515
K^*	0.817

	$\omega_H^{\rm GeV}$
D	1.81
B	1.81

	$\hat M_Q^{\rm GeV}$
c	1.32
b	4.65

(113)

with $\chi^2/{\rm d.o.f} = 1.75$ and $\chi^2/N_{\rm obs} = 1.17$. The dimensionless u, s current-quark masses correspond to $m_u = 5.4\,{\rm MeV}$, $m_s = 119\,{\rm MeV}$, and $M_u^E = 0.36\,{\rm GeV}$,

Table 3. The 16 dimension-GeV (*upper panel*) and 26 dimensionless (*lower panel*) quantities used in the χ^2-fit of [11]. The weighting error was the experimental error or 10% of the experimental value, if that is greater, which accounts for a realistic expectation of the model's accuracy. The light-meson electromagnetic form factors were calculated in impulse approximation [39,40,67] and $\xi(w)$ was obtained from $f_+^{B\to D}(t)$ via (106). The values in the "Obs." column were taken from [6,25,36,57,68–71]. (Adapted from [11].)

	Obs.	Calc.		Obs.	Calc.
f_π	0.131	0.146	m_π	0.138	0.130
f_K	0.160	0.178	m_K	0.496	0.449
$\langle\bar{u}u\rangle^{1/3}$	0.241	0.220	$\langle\bar{s}s\rangle^{1/3}$	0.227	0.199
$\langle\bar{q}q\rangle_\pi^{1/3}$	0.245	0.255	$\langle\bar{q}q\rangle_K^{1/3}$	0.287	0.296
f_ρ	0.216	0.163	f_{K^*}	0.244	0.253
$\Gamma_{\rho\pi\pi}$	0.151	0.118	$\Gamma_{K^*(K\pi)}$	0.051	0.052
f_D	0.200 ± 0.030	0.213	f_{D_s}	0.251 ± 0.030	0.234
f_B	0.170 ± 0.035	0.182	$g_{BK^*\gamma}\hat{M}_b$	2.03 ± 0.62	2.86
$f_+^{B\to D}(0)$	0.73	0.58	$f_\pi r_\pi$	0.44 ± 0.004	0.44
$F_{\pi\,(3.3\,\mathrm{GeV}^2)}$	0.097 ± 0.019	0.077	$\mathrm{B}(B \to D^*)$	0.0453 ± 0.0032	0.052
ρ^2	1.53 ± 0.36	1.84	$\alpha^{B\to D^*}$	1.25 ± 0.26	0.94
$\xi(1.1)$	0.86 ± 0.03	0.84	$A_{\mathrm{FB}}^{B\to D^*}$	0.19 ± 0.031	0.24
$\xi(1.2)$	0.75 ± 0.05	0.72	$\mathrm{B}(B \to \pi)$	$(1.8 \pm 0.6)_{\times 10^{-4}}$	2.2
$\xi(1.3)$	0.66 ± 0.06	0.63	$f_{+\,(14.9\,\mathrm{GeV}^2)}^{B\to\pi}$	0.82 ± 0.17	0.82
$\xi(1.4)$	0.59 ± 0.07	0.56	$f_{+\,(17.9\,\mathrm{GeV}^2)}^{B\to\pi}$	1.19 ± 0.28	1.00
$\xi(1.5)$	0.53 ± 0.08	0.50	$f_{+\,(20.9\,\mathrm{GeV}^2)}^{B\to\pi}$	1.89 ± 0.53	1.28
$\mathrm{B}(B \to D)$	0.020 ± 0.007	0.013	$\mathrm{B}(B \to \rho)$	$(2.5 \pm 0.9)_{\times 10^{-4}}$	4.8
$\mathrm{B}(D \to K^*)$	0.047 ± 0.004	0.049	$f_+^{D\to K}(0)$	0.73	0.61
$\dfrac{V(0)}{A_1(0)}^{D\to K^*}$	1.89 ± 0.25	1.74	$f_+^{D\to\pi}(0)$	0.73	0.67
$\dfrac{\Gamma_L}{\Gamma_T}^{D\to K^*}$	1.23 ± 0.13	1.17	$g_{B^*B\pi}$	23.0 ± 5.0	23.2
$\dfrac{A_2(0)}{A_1(0)}^{D\to K^*}$	0.73 ± 0.15	0.87	$g_{D^*D\pi}$	10.0 ± 1.3	11.0

$M_s^E = 0.49\,\mathrm{GeV}$. Furthermore, $\omega_{K^*}/\omega_\rho = 1.59$, which is identical to the value in [63].

It is evident that the fitted heavy-quark masses are consistent with the estimates in [25] and hence that the heavy-meson binding energy is large:

$$E_D := m_D - \hat{M}_c = 0.67\,\mathrm{GeV}\,,\; E_B := m_B - \hat{M}_b = 0.70\,\mathrm{GeV}\,. \tag{114}$$

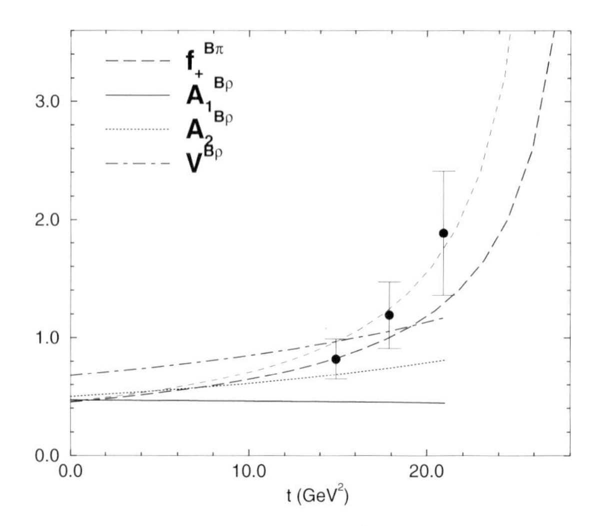

Fig. 10. Semileptonic $B \to \pi$ and $B \to \rho$ form factors with, for comparison, data from a lattice simulation [70] and a vector dominance, monopole model: $f_+^{B\to\pi}(t) = 0.46/(1 - t/m_{B^*}^2)$, $m_{B^*} = 5.325\,\mathrm{GeV}$ (*short-dashed line*). (Adapted from [11].)

These values yield $E_D/\hat{M}_c = 0.51$ and $E_B/\hat{M}_b = 0.15$, which furnishes another indication that while an heavy-quark expansion is accurate for the b-quark it will provide a poor approximation for the c-quark. This is emphasised by the value of $\omega_D = \omega_B$, which means that the Compton wavelength of the c-quark is greater than the length-scale characterising the bound state's extent.

With the parameters fixed, in [11] values for a wide range of other observables were calculated with the vast majority of the results being true predictions. The breadth of application is illustrated in Table 4, and in Fig. 10 which depicts the calculated t-dependence of $B \to \pi\,, \rho$ semileptonic transition form factors. I note that now there is a first experimental result for the D^{*+} width [72]: $\Gamma_{D^{*+}} = (96 \pm 4 \pm 22)\,\mathrm{keV}$, $g_{D^* D\pi} = 17.9 \pm 0.3 \pm 1.9$. Its confirmation and the gathering of additional information on c-quark mesons is crucial to improving our knowledge of the evolution from the light- to the heavy-quark domain, a transition whose true understanding will significantly enhance our grasp of nonperturbative dynamics.

Fidelity of Heavy-Quark Symmetry. The universal function characterising semileptonic transitions in the heavy-quark symmetry limit, $\xi(w)$ introduced in Sect. 4.7, can be estimated most reliably from $B \to D, D^*$ transitions. Using (106) to infer this function from $f_+^{B\to D}(t)$, one obtains

$$\xi^{f_+}(1) = 1.08\,, \tag{115}$$

which is a measurable deviation from (109). The ratio $\xi^{f_+}(w)/\xi^{f_+}(0)$ is depicted in Fig. 11, wherein it is compared with two experimental fits [69]:

$$\xi(w) = 1 - \rho^2\,(w-1), \qquad \rho^2 = 0.91 \pm 0.15 \pm 0.16\,, \tag{116}$$

Table 4. Predictions in [11] for a selection of observables. The "Obs." values are extracted from [25,57,68,72,73]. $t_{\max}$ is the maximum momentum transfer available in the process identified and $\omega_{\max} = \omega(t_{\max})$ calculated from (108). (Adapted from [11].)

	Obs.	Calc.		Obs.	Calc.
$f_K r_K$	0.472 ± 0.038	0.46	$-f_K^2 r_{K^0}^2$	$(0.19 \pm 0.05)^2$	$(0.10)^2$
$g_{\rho\pi\pi}$	6.05 ± 0.02	5.27	$\Gamma_{D^{*0}}$ (MeV)	< 2.1	0.020
$g_{K^* K \pi^0}$	6.41 ± 0.06	5.96	$\Gamma_{D^{*+}}$ (keV)	$96 \pm 4 \pm 22$	37.9
g_ρ	5.03 ± 0.012	5.27	$\Gamma_{D_s^* D_s \gamma}$ (MeV)	< 1.9	0.001
f_{D^*} (GeV)		0.290	$\Gamma_{B^{*+} B^+ \gamma}$ (keV)		0.030
$f_{D_s^*}$ (GeV)		0.298	$\Gamma_{B^{*0} B^0 \gamma}$ (keV)		0.015
f_{B_s} (GeV)	0.195 ± 0.035	0.194	$\Gamma_{B_s^* B_s \gamma}$ (keV)		0.011
f_{B^*} (GeV)		0.200	$\mathrm{B}(D^{*+} \to D^+ \pi^0)$	0.306 ± 0.025	0.316
$f_{B_s^*}$ (GeV)		0.209	$\mathrm{B}(D^{*+} \to D^0 \pi^+)$	0.683 ± 0.014	0.683
f_{D_s}/f_D	1.10 ± 0.06	1.10	$\mathrm{B}(D^{*+} \to D^+ \gamma)$	$0.011\ ^{+0.021}_{-0.007}$	0.001
f_{B_s}/f_B	1.14 ± 0.08	1.07	$\mathrm{B}(D^{*0} \to D^0 \pi^0)$	0.619 ± 0.029	0.826
f_{D^*}/f_D		1.36	$\mathrm{B}(D^{*0} \to D^0 \gamma)$	0.381 ± 0.029	0.174
f_{B^*}/f_B		1.10	$\mathrm{B}(B \to K^* \gamma)$	$(5.7 \pm 3.3)_{10^{-5}}$	11.4
$R_1^{B \to D^*}(1)$	1.30 ± 0.39	1.32	$R_2^{B \to D^*}(1)$	0.64 ± 0.29	1.04
$R_1^{B \to D^*}(w_{\max})$		1.23	$R_2^{B \to D^*}(w_{\max})$		0.98
$\mathrm{B}(D^+ \to \rho^0)$		0.032	$\alpha^{D \to \rho}$		1.03
$\mathrm{B}(D^0 \to K^-)$	0.037 ± 0.002	0.036	$\dfrac{\mathrm{B}(D \to \rho^0)}{\mathrm{B}(D \to K^*)}$	0.044 ± 0.034	0.065
$A_1^{D \to K^*}(0)$	0.56 ± 0.04	0.46	$A_1^{D \to K^*}(t_{\max}^{D \to K^*})$	0.66 ± 0.05	0.47
$A_2^{D \to K^*}(0)$	0.39 ± 0.08	0.40	$A_2^{D \to K^*}(t_{\max}^{D \to K^*})$	0.46 ± 0.09	0.44
$V^{D \to K^*}(0)$	1.1 ± 0.2	0.80	$V^{D \to K^*}(t_{\max}^{D \to K^*})$	1.4 ± 0.3	0.92
$\dfrac{\mathrm{B}(D^0 \to \pi)}{\mathrm{B}(D^0 \to K)}$	0.103 ± 0.039	0.098	$f_+^{D \to K}(t_{\max}^{D \to K})$	1.31 ± 0.04	1.11
$\dfrac{f_+^{D \to \pi}(0)}{f_+^{D \to K}(0)}$	1.2 ± 0.3	1.10	$f_+^{D \to \pi}(t_{\max}^{D \to \pi})$		2.18
$R_1^{D \to K^*}(1)$		1.72	$R_1^{D \to K^*}(w_{\max})$		1.74
$R_1^{D \to \rho}(1)$		2.08	$R_1^{D \to \rho}(w_{\max})$		2.03

$$\xi(w) = \frac{2}{w+1} \exp\left[(1-2\rho^2)\frac{w-1}{w+1}\right], \; \rho^2 = 1.53 \pm 0.36 \pm 0.14 . \quad (117)$$

The evident agreement was possible because [11] did not employ the heavy-quark expansion of (71), in particular and especially not for the c-quark. The calculated

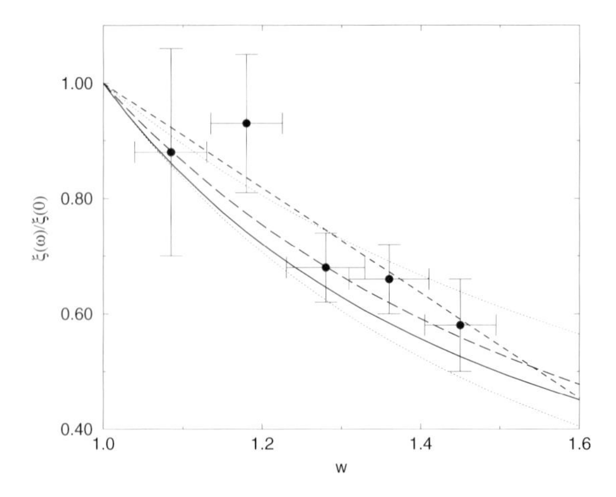

Fig. 11. Calculated [11] form of $\xi(w)$ (*solid line*) compared with experimental analyses: linear fit from [69], (116) (*short-dashed line*); nonlinear fit from [69], (117) (*long-dashed line*). The two lighter dotted lines are the nonlinear fit of [69] evaluated with the extreme values of ρ^2: upper line, $\rho^2 = 1.17$ and lower line, $\rho^2 = 1.89$. The data points are from [74]. (Adapted from [11].)

result (*solid curve*) in Fig. 11 is well approximated by

$$\xi^{f_+}(w) = \frac{1}{1 + \tilde{\rho}^2_{f_+}(w-1)}\,, \; \tilde{\rho}^2_{f_+} = 1.98\,. \tag{118}$$

Equations (111) were also used in [11] to extract $\xi(w)$ from $B \to D^*$. This gave $\xi^{A_1}(1) = 0.987$, $\xi^{A_2}(1) = 1.03$, $\xi^V(1) = 1.30$, an w-dependence well-described by (118) but with $\tilde{\rho}^2_{A_1} = 1.79$, $\tilde{\rho}^2_{A_2} = 1.99$, $\tilde{\rho}^2_V = 2.02$, and the ratios, (112), $R_1(1)/R_1(w_{\rm max}) = 1.08$, $R_2(1)/R_2(w_{\rm max}) = 1.06$.

This collection of results indicates the degree to which heavy-quark symmetry is respected in $b \to c$ processes. Combining them it is clear that even in this case, which is the nearest contemporary realisation of the heavy-quark symmetry limit, corrections of $\lesssim 30\%$ must be expected. In $c \to s\,, d$ transitions the corrections can be as large as a factor of two, as evident in Table 4.

5 Epilogue

This contribution provides a perspective on the modern application of Dyson-Schwinger equations (DSEs) to light- and heavy-meson properties. The keystone of this approach's success is an appreciation and expression of the momentum-dependence of dressed-parton propagators at infrared length-scales. That dependence is responsible for the magnitude of constituent-quark and -gluon masses, and the length-scale characterising confinement in bound states; and is now recognised as a fact.

It has recently become clear that the simple rainbow-ladder DSE truncation is the first term in a systematic and nonperturbative scheme that preserves the

Ward-Takahashi identities which express conservation laws at an hadronic level. This has enabled the proof of exact results in QCD, and explains why the truncation has been successful for light vector and flavour nonsinglet pseudoscalar mesons. Emulating more of these achievements with *ab initio* calculations of heavy-meson properties is a modern challenge.

However, at present, the study of heavy-meson systems using DSE methods stands approximately at the point occupied by those of light-meson properties seven – eight years ago. A Poincaré covariant treatment exploiting essential features, such as propagator dressing and sensible bound state Bethe-Salpeter amplitudes, has been shown capable of providing a unified and successful description of light- and heavy-meson observables. The goal now is to make the case compelling by tying the separate elements together; namely, relating the propagators and Bethe-Salpeter amplitudes via a single kernel. I am confident this will be accomplished, and the DSEs become a quantitatively reliable and intuition building tool as much in the heavy-quark sector as they are for light-quark systems.

While a more detailed understanding will be attained in pursuing this goal, certain qualitative results established already are unlikely to change. For example, it is plain that light- and heavy-mesons are essentially the same, they are simply bound states of dressed-quarks. Moreover, the magnitude of the b-quark's current-mass is large enough to sustain heavy-quark approximations for its propagator and the amplitudes for bound states of which it is a constituent. In addition, and unfortunately in so far as practical constraints on the Standard Model are concerned, the current-mass of the c-quark is too small to validate an heavy-quark approximation.

Acknowledgments

I am grateful for the hospitality and support of my colleagues and the staff in the Bogoliubov Laboratory of Theoretical Physics at the Joint Institute for Nuclear Research, Dubna, Russia. This work was supported by: the Department of Energy, Nuclear Physics Division, under contract no. W-31-109-ENG-38; Deutsche Forschungsgemeinschaft, under contract no. Ro 1146/3-1; and benefited from the resources of the National Energy Research Scientific Computing Center.

References

1. C.D. Roberts and A.G. Williams, Prog. Part. Nucl. Phys. **33**, 477 (1994).
2. C.D. Roberts and S.M. Schmidt, Prog. Part. Nucl. Phys. **45**, S1 (2000).
3. R. Alkofer and L.v. Smekal, Phys. Rept. **353**, 281 (2001).
4. P. Maris and C.D. Roberts, "Dyson-Schwinger equations: A tool for hadron physics," nucl-th/0301049.
5. P. Maris, C.D. Roberts and P.C. Tandy, Phys. Lett. **B 420**, 267 (1998).
6. P. Maris and C.D. Roberts, Phys. Rev. **C 56**, 3369 (1997).

7. P. Maris, A. Raya, C.D. Roberts and S.M. Schmidt, "Facets of confinement and dynamical chiral symmetry breaking," nucl-th/0208071.
8. A. Bender, C.D. Roberts and L.v. Smekal, Phys. Lett. **B 380**, 7 (1996).
9. A. Bender, W. Detmold, C.D. Roberts and A.W. Thomas, Phys. Rev. **C 65**, 065203 (2002).
10. P. Maris and C.D. Roberts, "QCD bound states and their response to extremes of temperature and density." In: *Proc. of the Wkshp. on Nonperturbative Methods in Quantum Field Theory, Adelaide, Australia, 2-13 Feb., 1998*, ed. by A.W. Schreiber, A.G. Williams and A.W. Thomas (World Scientific, Singapore 1998) pp. 132–151.
11. M.A. Ivanov, Yu.L. Kalinovsky and C.D. Roberts, Phys. Rev. **D 60**, 034018 (1999).
12. H.J. Munczek and A.M. Nemirovsky, Phys. Rev. **D 28**, 181 (1983).
13. C.H. Llewellyn-Smith, Annals Phys. (NY) **53**, 521 (1969).
14. H.J. Munczek, Phys. Rev. **D 52**, 4736 (1995).
15. R.W. Haymaker, Riv. Nuovo Cim. **14N8**, 1 (1991).
16. K.D. Lane, Phys. Rev. **D 10**, 2605 (1974); H.D. Politzer, Nucl. Phys. **B 117**, 397 (1976).
17. K. Langfeld, R. Pullirsch, H. Markum, C.D. Roberts and S.M. Schmidt, "Concerning the quark condensate," nucl-th/0301024.
18. C.D. Roberts, "Continuum strong QCD: Confinement and Dynamical Chiral Symmetry Breaking," nucl-th/0007054.
19. C. Alexandrou, P. De Forcrand and E. Follana, Phys. Rev. **D 65**, 117502 (2002); P.O. Bowman, U.M. Heller, D.B. Leinweber and A.G. Williams, Phys. Rev. **D 66**, 074505 (2002); and references therein.
20. J.C.R. Bloch, Phys. Rev. **D 64**, 116011 (2001).
21. R. Alkofer, C.S. Fischer and L.v. Smekal, Acta Phys. Slov. **52**, 191 (2002).
22. J. Skullerud and A. Kızılersü, JHEP **0209**, 013 (2002); J.I. Skullerud, P.O. Bowman, A. Kızılersü, D.B. Leinweber, A.G. Williams, "Nonperturbative structure of the quark-gluon vertex," hep-ph/0303176.
23. J.C.R. Bloch, A. Cucchieri, K. Langfeld and T. Mendes, "Running coupling constant and propagators in SU(2) Landau gauge," hep-lat/0209040.
24. P. Maris and P.C. Tandy, Phys. Rev. **C 60**, 055214 (1999).
25. L. Montanet, *et al.* [Part. Data Group Coll.], Phys. Rev. **D 50**, 1173 (1994); C. Caso *et al.* [Part. Data Group Coll.], Eur. Phys. J. **C 3**, 1 (1998).
26. P.O. Bowman, U.M. Heller and A.G. Williams, Phys. Rev. **D 66**, 014505 (2002).
27. K. Johnson, M. Baker and R. Willey, Phys. Rev. **136**, B1111 (1964).
28. M.S. Bhagwat, M.A. Pichowsky, C.D. Roberts and P.C. Tandy, "Analysis of a quenched lattice-QCD dressed-quark propagator," nucl-th/0304003.
29. P. Maris, "Continuum QCD and Light Mesons." In: *Wien 2000, Quark Confinement and the Hadron Spectrum — Proc. of the 4th Int. Conf., Vienna, Austria, 3-8 Jul 2000*, ed. by W. Lucha and K. Maung Maung (World Scientific, Singapore 2002) pp. 163-175.
30. M.B. Hecht, C.D. Roberts and S.M. Schmidt, "Contemporary Applications of Dyson-Schwinger Equations." In: *Wien 2000, Quark Confinement and the Hadron Spectrum – Proc. of the 4th Int. Conf., Vienna, Austria, 3-8 Jul 2000*, ed. by W. Lucha and K. Maung Maung (World Scientific, Singapore 2002) pp. 27-39.
31. K.C. Bowler, *et al.* [UKQCD Coll.], Phys. Rev. **D 62**, 054506 (2000).
32. P.C. Tandy, "Covariant QCD modeling of light meson physics," nucl-th/0301040.
33. P. Maris and P.C. Tandy, Phys. Rev. **C 61**, 045202 (2000).
34. P. Maris and P.C. Tandy, Phys. Rev. **C 62**, 055204 (2000).

35. P. Brauel, *et al.*, Z. Phys. **C 3**, 101 (1979).
36. S.R. Amendolia, *et al.* [NA7 Coll.], Nucl. Phys. **B 277**, 168 (1986).
37. J. Volmer, *et al.* [JLab F_π Coll.], Phys. Rev. Lett. **86**, 1713 (2001).
38. P. Maris, πN Newslett. **16**, 213 (2002).
39. C.D. Roberts, Nucl. Phys. **A 605**, 475 (1996).
40. P. Maris and C.D. Roberts, Phys. Rev. **C 58**, 3659 (1998).
41. G.R. Farrar and D.R. Jackson, Phys. Rev. Lett. **43**, 246 (1979).
42. G.P. Lepage and S.J. Brodsky, Phys. Rev. **D 22**, 2157 (1980).
43. M.S. Bhagwat, M.A. Pichowsky and P.C. Tandy, Phys. Rev. **D 67**, 054019 (2003).
44. S.R. Cotanch and P. Maris, Phys. Rev. **D 66**, 116010 (2002); P. Bicudo, Phys. Rev. **C 67**, 035201 (2003).
45. J. Praschifka, C.D. Roberts and R.T. Cahill, Phys. Rev. **D 36**, 209 (1987); C.D. Roberts, R.T. Cahill and J. Praschifka, Annals Phys. (NY) **188**, 20 (1988).
46. M. Bando, M. Harada and T. Kugo, Prog. Theor. Phys. **91**, 927 (1994).
47. R. Alkofer and C.D. Roberts, Phys. Lett. **B 369**, 101 (1996); B. Bistrović and D. Klabučar, Phys. Lett. **B 478**, 127 (2000).
48. P. Maris and P.C. Tandy, Phys. Rev. **C 65**, 045211 (2002).
49. E.S. Ackleh, T. Barnes and E.S. Swanson, Phys. Rev. **D 54**, 6811 (1996).
50. G.S. Adams *et al.* [E852 Coll.], Phys. Rev. Lett. **81**, 5760 (1998). S.U. Chung *et al.*, Phys. Rev. **D 65**, 072001 (2002).
51. C.J. Burden, Lu Qian, C.D. Roberts, P.C. Tandy and M.J. Thomson, Phys. Rev. **C 55**, 2649 (1997).
52. C.J. Burden and M.A. Pichowsky, Few Body Syst. **32**, 119 (2002).
53. J.C.R. Bloch, Yu.L. Kalinovsky, C.D. Roberts and S.M. Schmidt, Phys. Rev. **D 60**, 111502 (1999).
54. M.A. Ivanov, Yu.L. Kalinovsky, P. Maris and C.D. Roberts, Phys. Lett. **B 416**, 29 (1998).
55. M.A. Ivanov, Yu.L. Kalinovsky, P. Maris and C.D. Roberts, Phys. Rev. **C 57**, 1991 (1998).
56. M. Neubert, Phys. Rep. **245**, 259 (1994); M. Neubert, "Heavy quark masses, mixing angles, and spin flavor symmetry." In: *The Building Blocks of Creation: From Microfermis to Megaparsecs, Boulder, Colorado, 6/Jun - 2/Jul 1993*, ed. by S. Raby and T. Walker (World Scientific, Singapore 1994) pp. 125-206; and references therein.
57. J.M. Flynn and C.T. Sachrajda, Adv. Ser. Direct. High Energy Phys. **15**, 402 (1998).
58. C. McNeile, "Heavy quarks on the lattice," hep-lat/0210026.
59. C.D. Roberts, Nucl. Phys. Proc. Suppl. **108**, 227 (2002).
60. Yu.L. Kalinovsky, K.L. Mitchell and C.D. Roberts, Phys. Lett. **B 399**, 22 (1997); C.R. Ji and P. Maris, Phys. Rev. **D 64**, 014032 (2001).
61. H. Munczek, Phys. Lett. **B 175**, 215 (1986); C.J. Burden, C.D. Roberts and A.G. Williams, *ibid* **285**, 347 (1992).
62. M.A. Pichowsky and T.-S.H. Lee, Phys. Lett. B **379**, 1 (1996); M.A. Pichowsky and T.-S.H. Lee, Phys. Rev. **D 56**, 1644 (1997).
63. F.T. Hawes and M.A. Pichowsky, Phys. Rev. **C 59**, 1743 (1999).
64. G. Buchalla, A.J. Buras and M.E. Lautenbacher, Rev. Mod. Phys. **68**, 1125 (1996).
65. N. Isgur and M.B. Wise, Phys. Lett. **B 232**, 113 (1989); *ibid* **B 237**, 527 (1990).
66. The fitting used [25]: $V_{ub} = 0.0033$, $V_{cd} = 0.2205$, $V_{cs} = 0.9745$ and $V_{cb} = 0.039$; and, in GeV, $M_\rho = 0.77$, $M_{K^*} = 0.892$ and, except in the kinematic factor $\lambda(m_1^2, m_2^2, t)$ where the splittings are crucial, averaged D- and B-meson masses:

$m_D = 1.99$, $m_B = 5.35$ (from $m_D = 1.87$, $m_{D_s} = 1.97$, $M_{D^*} = 2.01$, $M_{D_s^*} = 2.11$, and $m_B = 5.28$, $m_{B_s} = 5.37$, $M_{B^*} = 5.32$, $M_{B_s^*} = 5.42$). Furthermore, $b_0^u = 0.131$, $b_0^s = 0.105$, $b_3^u = 0.185 = b_3^s$ and $\epsilon = 10^{-4}$ were not varied, being instead fixed at the values determined in [67].

67. C.J. Burden, C.D. Roberts and M.J. Thomson, Phys. Lett. **B 371**, 163 (1996).
68. J.D. Richman and P.R. Burchat, Rev. Mod. Phys. **67**, 893 (1995).
69. J.E. Duboscq *et al.* [CLEO Coll.], Phys. Rev. Lett. **76**, 3898 (1996).
70. D.R. Burford, *et al.*, [UKQCD Coll.], Nucl. Phys. **B 447**, 425 (1995).
71. V.M. Belyaev, V.M. Braun, A. Khodjamirian and R. Rückl, Phys. Rev. **D 51**, 6177 (1995).
72. A. Anastassov *et al.* [CLEO Coll.], Phys. Rev. **D 65**, 032003 (2002).
73. W.R. Molzon *et al.*, Phys. Rev. Lett. **41**, 1213 (1978) [Erratum-ibid. **41**, 1523 (1978)]; S.R. Amendolia *et al.*, Phys. Lett. **B 178**, 435 (1986).
74. H. Albrecht *et al.* [ARGUS Coll.], Z. Phys. **C 57**, 533 (1993).

An Introduction to 5-Dimensional Extensions of the Standard Model

Alexander Mück[1], Apostolos Pilaftsis[2], and Reinhold Rückl[1]

[1] Institut für Theoretische Physik und Astrophysik, Universität Würzburg, Am Hubland, 97074 Würzburg, Germany
[2] Department of Physics and Astronomy, University of Manchester, Manchester M13 9PL, United Kingdom

Abstract. We give a pedagogical introduction to the physics of large extra dimensions. We focus our discussion on minimal extensions of the Standard Model in which gauge fields may propagate in a single, compact extra dimension while the fermions are restricted to a 4-dimensional Minkowski subspace. First, the basic ideas, including an appropriate gauge-fixing procedure in the higher-dimensional context, are illustrated in simple toy models. Then, we outline how the presented techniques can be extended to more realistic theories. Finally, we investigate the phenomenology of different minimal Standard Model extensions, in which all or only some of the $SU(2)_L$ and $U(1)_Y$ gauge fields and Higgs bosons feel the presence of the fifth dimension. Bounds on the compactification scale between 4 and 6 TeV, depending on the model, are established by analyzing existing data.

1 Introduction

Why do we live in four dimensions? This fundamental question still cannot be answered. However, already at the beginning of the 20th century, Kaluza and Klein realized [1] that the question itself may be ill posed. It seems more appropriate to ask instead: In how many dimensions do we live?

In the modern physics point of view, a satisfactory answer to the above question may be found within the context of string theories or within a more unifiable framework, known as M theory. The reason is that string theories provide the only known theoretical framework within which gravity can be quantized and so undeniably plays a central rôle in our endeavours of unifying all fundamental forces of nature. A consistent quantum-mechanical formulation of a string theory, however, requires the existence of additional dimensions beyond the four ones we experience in our every-day life. These new dimensions must be sufficiently small, in some appropriate sense, so as to have escaped our detection. As we will see in detail, compactification, where additional dimensions are considered to be compact manifolds of a characteristic size R, provides a mechanism which can successfully hide them. In the original string-theoretic considerations [2], the inverse length $1/R$ of the extra compact dimensions and the string mass M_s turned out to be closely tied to the 4-dimensional Planck mass $M_{\mathrm{P}} = 1.9 \times 10^{16}$ TeV, with all involved mass scales being of the same order. More recent studies, however, have shown [3–5,53,7] that there could still be conceivable scenarios of stringy nature where $1/R$ and M_s may be lowered

A. Mück, A. Pilaftsis, and R. Rückl, An Introduction to 5-Dimensional Extensions of the Standard Model, Lect. Notes Phys. **647**, 189–211 (2004)
http://www.springerlink.com/

independently of $M_{\rm P}$ by several or many orders of magnitude. Taking such a realization to its natural extreme, [6] considers the radical scenario, in which M_s is of order TeV and represents the only fundamental scale in the universe at which unification of all forces of nature occurs. Thus, the so-called gauge hierarchy problem due to the high disparity between the electroweak and the 4-dimensional Planck scales can be avoided all together, as it does not appear right from the beginning.

Let us now try to understand why n extra dimensions with a large radius R can influence gravity. This question is tightly connected to the geometry of space-time. At distances small compared to R, the gravitational potential will simply change according to the Gauss law in $n+4$ dimensions, i.e.

$$V(r) \sim \frac{m_1\, m_2}{M_{\rm G}^{2+n}}\, \frac{1}{r^{n+1}}\,, \tag{1}$$

where $r \ll R$ and $M_{\rm G}$ is the true gravitational scale to be distinguished from the Planck scale $M_{\rm P}$. As the distance, gravity is probed at, becomes much larger than R, the potential will again look effectively four dimensional, i.e.

$$V(r) \stackrel{r \gg R}{\to} \frac{m_1\, m_2}{M_{\rm P}}\, \frac{1}{r}\,. \tag{2}$$

Matching the two potentials (1) and (2) to give the same answer at $r = R$, we derive an important relation among the parameters $M_{\rm P}$, $M_{\rm G}$ and R [6]:

$$M_{\rm P}^2 = M_{\rm G}^{2+n}\, R^n\,. \tag{3}$$

Hence, the weakness of gravity, observed by today's experiments, is not due to the enormity of the Planck scale $M_{\rm P}$, but thanks to the presence of a large radius R. As a result, the true fundamental gravity scale $M_{\rm G}$ is determined from (3) and is much smaller than $M_{\rm P}$. For example, extra dimensions of size

$$R \sim \left(\frac{M_{\rm P}}{M_{\rm G}}\right)^{2/n} \frac{1}{M_{\rm G}} \sim \begin{cases} \mathcal{O}(1\,{\rm mm}), & n = 2 \\ \mathcal{O}(10\,{\rm fm}), & n = 6 \end{cases} \tag{4}$$

are needed for a gravitational scale —typically of the order of a string scale M_s— in the TeV range. Therefore, even Cavendish-type experiments may potentially test the model by observing deviations from Newton's law [6] at distances smaller than a mm.

This low string-scale effective model could be embedded within e.g. type I string theories [5], where the Standard Model (SM) may be described as an intersection of Dp branes [53,7,8]. The Dp brane description implies that the SM fields do not necessarily feel the presence of all the extra dimension, but are restricted to some subspace of the full space-time. Especially mm-size dimensions, being clearly excluded for the SM by experimental evidence, are probed only by gravity. However, as such intersections may be higher-dimensional as well, in addition to gravitons the SM gauge fields could also propagate within at least a single extra dimension. Here, the bounds on the compactification radius from experimental data are much more severe and R has to be at least as small as an inverse TeV. In our introductory notes, we will abandon gravity and concentrate on the embedding of the Standard Model in a five dimensional space-time. Our main interest is to explain the basic ideas and techniques for constructing this kind of theories.

Note that this limited class of models with low string-scales may result in different higher-dimensional extensions of the SM [8,9], even if gravity is completely ignored. Hence, the actual experimental limits on the compactification radius are, to some extent, model dependent. In fact, most of the derived phenomenological limits in the literature were obtained by assuming that the SM gauge fields propagate all freely in a common higher-dimensional space [10–16]. Therefore, towards the end of our notes, we will also discuss the phenomenological consequences of models which minimally depart from the assumption of these higher-dimensional scenarios [17]. Specifically, we will consider 5-dimensional extensions of the SM compactified on an S^1/Z_2 orbifold, where the $\mathrm{SU}(2)_L$ and $\mathrm{U}(1)_Y$ gauge bosons may not both live in the same higher-dimensional space, the so-called bulk. In all our models, the SM fermions are localized on the 4-dimensional subspace, i.e. on a 3-brane or, as it is often simply called, brane.

The present introductory notes are organized as follows: in Sect. 2 we introduce the basic concepts of higher-dimensional theories in simple Abelian models. After compactifying the extra dimension on a particular orbifold, S^1/Z_2, we obtain an effective 4-dimensional theory, which in addition to the usual SM states contains infinite towers of massive Kaluza–Klein (KK) states of the higher-dimensional gauge fields. In particular, we consider the question how to consistently quantize the higher-dimensional models under study in the so-called R_ξ gauge. Such a quantization procedure can be successfully applied to theories that include both Higgs bosons living in the bulk and/or on the brane. After briefly discussing how these concepts can be applied to the SM in Sect. 3, we turn our attention to the phenomenological aspects of the models of our interest in Sect. 4. For each higher-dimensional model, we calculate the effects of the fifth dimension on electroweak observables and LEP2 cross sections and analyze their impact on constraining the compactification scale. Technical details are omitted here in favour of introducing the main concepts. A complete discussion, along with detailed analytic results and an extensive list of references, is given in our paper in [17]. Finally, we summarize in Sect. 5 our main results.

2 5-Dimensional Abelian Models

As a starting point, let us consider the Lagrangian of 5-dimensional Quantum Electrodynamics (5D-QED) given by

$$\mathcal{L}(x,y) = -\frac{1}{4}F_{MN}(x,y)F^{MN}(x,y) + \mathcal{L}_{\mathrm{GF}}(x,y)\,, \tag{5}$$

where

$$F_{MN}(x,y) = \partial_M A_N(x,y) - \partial_N A_M(x,y) \tag{6}$$

denotes the 5-dimensional field strength tensor, and $\mathcal{L}_{\mathrm{GF}}(x,y)$ is the gauge-fixing term. The Faddeev-Popov ghost terms have been neglected, because the ghosts are non-interacting in the Abelian case. Our notation for the Lorentz indices and space-time coordinates is: $M, N = 0,1,2,3,5$; $\mu,\nu = 0,1,2,3$; $x = (x^0, \boldsymbol{x})$; and $y = x^5$ denotes the coordinate of the additional dimension.

The structure of the conventional QED Lagrangian is simply carried over to the five-dimensional case. The field content of the theory is given by a single gauge-boson A_M transforming as a vector under the Lorentz group SO(1,4). In the absence of the gauge-fixing and ghost terms, the 5D-QED Lagrangian is invariant under a U(1) gauge transformation

$$A_M(x,y) \to A_M(x,y) + \partial_M \Theta(x,y)\,. \tag{7}$$

Hence, the defining features of conventional QED are present in 5D-QED as well. So far, we have treated all the spatial dimensions on the same footing. This is certainly an assumption in contradiction not only to experimental evidence but also to our daily experience. There has to be a mechanism in the theory which hides the additional dimension at low energies. As we will see in the following, the simplest approach accomplishing this goal is compactification, i.e., replace the infinitely extended extra dimension by a compact object.

A simple compact one dimensional manifold is a circle, denoted by S^1, with radius R. Asking for an additional reflection symmetry Z_2 with respect to the origin $y = 0$, one is led to the orbifold S^1/Z_2 which turns out to be especially well suited for higher dimensional physics. Thus, we consider the extra dimensional coordinate y to run only from 0 to $2\pi R$ where these two points are identified. Moreover, according to the Z_2 symmetry, y and $-y = 2\pi - y$ can be identified in a certain sense: knowing the field content for the segment $y \in [0,\pi]$ implies the knowledge of the whole system. For that reason, the fixed points $y = 0$ and $y = \pi$, which do not transform under Z_2, are also called boundaries of the orbifold.

The compactification on S^1/Z_2 reflects itself in certain restrictions for the fields. In order not to spoil the above property of gauge symmetry, we demand the fields to satisfy the following equalities:

$$\begin{aligned}
A_M(x,y) &= A_M(x,y+2\pi R)\,,\\
A_\mu(x,y) &= A_\mu(x,-y)\,,\\
A_5(x,y) &= -A_5(x,-y)\,,\\
\Theta(x,y) &= \Theta(x,y+2\pi R)\,,\\
\Theta(x,y) &= \Theta(x,-y)\,.
\end{aligned} \tag{8}$$

The field $A_\mu(x,y)$ is taken to be even under Z_2, so as to embed conventional QED with a massless photon into our 5D-QED, as we will see below. Notice that the reflection properties of the field $A_5(x,y)$ and the gauge parameter $\Theta(x,y)$ under Z_2 in (8) follow automatically if the theory is to remain gauge invariant after compactification.

Making the periodicity and reflection properties of A_μ and Θ in (8) explicit, we can expand these quantities in Fourier series

$$\begin{aligned} A^\mu(x,y) &= \frac{1}{\sqrt{2\pi R}} A^\mu_{(0)}(x) + \sum_{n=1}^{\infty} \frac{1}{\sqrt{\pi R}} A^\mu_{(n)}(x) \cos\left(\frac{ny}{R}\right), \\ \Theta(x,y) &= \frac{1}{\sqrt{2\pi R}} \Theta_{(0)}(x) + \sum_{n=1}^{\infty} \frac{1}{\sqrt{\pi R}} \Theta_{(n)}(x) \cos\left(\frac{ny}{R}\right). \end{aligned} \tag{9}$$

The Fourier coefficients $A^\mu_{(n)}(x)$ are the so-called Kaluza-Klein (KK) modes. The extra component of the gauge field is odd under the reflection symmetry and its expansion is given by

$$A^5(x,y) = \sum_{n=1}^{\infty} \frac{1}{\sqrt{\pi R}} A^5_{(n)}(x) \sin\left(\frac{ny}{R}\right). \tag{10}$$

Note that there is no zero mode, a phenomenologically important fact, as we will see below.

At this point, the theory is again formulated entirely in terms of four-dimensional fields, the KK modes. All the dependence of the Lagrangian density on the extra coordinate y is parameterized with simple Fourier functions. Finally, the physics is dictated by the Lagrangian anyway, not by its density, thus, one can go one step further and completely remove the explicit y dependence of the Lagrangian by integrating out the extra dimension. From now on, the quantity of interest will be

$$\mathcal{L}(x) = \int_0^{2\pi R} dy\ \mathcal{L}(x,y). \tag{11}$$

All the higher-dimensional physics is reflected by the infinite tower of KK modes for each field component. A simple calculation yields the 4-dimensional Lagrangian density

$$\begin{aligned} \mathcal{L}(x) &= -\frac{1}{4} F_{(0)\mu\nu} F^{\mu\nu}_{(0)} + \sum_{n=1}^{\infty} \left[-\frac{1}{4} F_{(n)\mu\nu} F^{\mu\nu}_{(n)} \right. \\ &\left. + \frac{1}{2} \left(\frac{n}{R} A_{(n)\mu} + \partial_\mu A_{(n)5} \right) \left(\frac{n}{R} A^\mu_{(n)} + \partial^\mu A_{(n)5} \right) \right] + \mathcal{L}_{\mathrm{GF}}(x), \end{aligned} \tag{12}$$

where $\mathcal{L}_{\mathrm{GF}}(x)$ is defined in analogy to (11). The first term in (12) represents conventional QED involving the massless field $A^\mu_{(0)}$. Note that all the other vector excitations $A^\mu_{(n)}$ from the infinite tower of KK modes come with mass terms, their mass being an integer multiple of the inverse compactification radius. Therefore, a small radius leads to a large mass or compactification scale $M = 1/R$ in the model. It is this large scale which is responsible for the fact that an extra dimension, as it may exist, has not yet been discovered. The extra dimension is, so to speak, hidden by its compactness.

Note that it is the absence of $A^5_{(0)}$ due to the odd Z_2 symmetry of $A^5(x,y)$ which allows us to recover conventional QED in the low energy limit of the model. For $n \geq 1$, the KK tower $A^5_{(n)}$ for the additional component of the five dimensional vector field mixes with the vector modes. The modes $A^5_{(n)}$, being scalars with respect to the four dimensional Lorentz group, play the rôle of the would-be Goldstone modes in a non-linear realization of an Abelian Higgs model, in which the corresponding Higgs fields are taken to be infinitely massive. Thus, one is tempted to view the mass generation for the heavy KK modes by compactification as a kind of geometric Higgs mechanism. Note, moreover, that the Lagrangian (12) is still manifestly gauge invariant under the transformation (7) which in terms of the KK modes reads

$$\begin{aligned} A_{(n)\mu}(x) &\rightarrow A_{(n)\mu}(x) + \partial_\mu \Theta_{(n)}(x)\,, \\ A_{(n)5}(x) &\rightarrow A_{(n)5}(x) - \frac{n}{R}\Theta_{(n)}(x)\,. \end{aligned} \tag{13}$$

The above observations motivate us to seek for a higher-dimensional generalization of 't-Hooft's gauge-fixing condition, for which the mixing terms bilinear in $A^\mu_{(n)}$ and $A^5_{(n)}$ are eliminated from the effective 4-dimensional Lagrangian (12). Taking advantage of the fact that orbifold compactification generally breaks SO(1,4) invariance [18], one can abandon the requirement of covariance of the gauge fixing condition with respect to the extra dimension and choose the following non-covariant generalized R_ξ gauge [17,19]:

$$\mathcal{L}_{\mathrm{GF}}(x,y) \;=\; -\frac{1}{2\xi}(\partial^\mu A_\mu \;-\; \xi\,\partial_5 A_5)^2\,. \tag{14}$$

Nevertheless, the gauge-fixing term in (14) is still invariant under ordinary 4-dimensional Lorentz transformations. Upon integration over the extra dimension, all mixing terms in (12) drop out up to irrelevant total derivatives. Thus, the gauge-fixed four dimensional Lagrangian of the 5-dimensional QED explicitly shows the different degrees of freedom in the model. It reads

$$\begin{aligned} \mathcal{L}(x) \;=\; & -\frac{1}{4}F_{(0)\mu\nu}\,F^{\mu\nu}_{(0)} \;-\; \frac{1}{2\xi}\,(\partial^\mu A_{(0)\mu})^2 \\ & + \sum_{n=1}^{\infty}\left[-\frac{1}{4}\,F_{(n)\mu\nu}\,F^{\mu\nu}_{(n)} + \frac{1}{2}\left(\frac{n}{R}\right)^2 A^\mu_{(n)}\,A_{(n)\mu} - \frac{1}{2\xi}\,(\partial^\mu A_{(n)\mu})^2\right] \\ & + \sum_{n=1}^{\infty}\left[\frac{1}{2}\,(\partial^\mu A_{(n)5})\,(\partial_\mu A_{(n)5}) \;-\; \frac{1}{2}\xi\left(\frac{n}{R}\right)^2 A_{(n)5}{}^2\right]. \end{aligned} \tag{15}$$

Gauge fixed QED is accompanied with a tower of its massive copies. The scalars $A_{(n)5}$ with gauge dependent masses resemble the would-be Goldstone bosons of an ordinary 4-dimensional Abelian-Higgs model in the R_ξ gauge. From this Lagrangian, it is obvious that the corresponding propagators take on their usual forms:

$$\mu \overset{(n)}{\sim\!\sim\!\sim\!\sim} \nu \qquad = \frac{i}{k^2 - \left(\frac{n}{R}\right)^2} \left[-g^{\mu\nu} + \frac{(1-\xi) k^\mu k^\nu}{k^2 - \xi \left(\frac{n}{R}\right)^2} \right]$$

$$\overset{(n)}{- - - - - - -} \qquad = \frac{i}{k^2 - \xi \left(\frac{n}{R}\right)^2} \tag{16}$$

Hereafter, we shall refer to the $A^5_{(n)}$ fields as Goldstone modes.

Having defined the appropriate R_ξ gauge through the gauge-fixing term in (14), we can recover the usual unitary gauge, in which the Goldstone modes decouple from the theory, in the limit $\xi \to \infty$ [20,21]. Thus, for the case at hand, we have seen how starting from a non-covariant higher-dimensional gauge-fixing condition, we can arrive at the known covariant 4-dimensional R_ξ gauge after compactification.

Having established a five dimensional gauge sector, we can now introduce fermions in the model. This is possible in the same spirit followed for the gauge field, leading to bulk fermions, i.e. fermions propagating into the extra dimension [22]. However, it turns out, that there is an even easier and phenomenologically challenging alternative to this approach. Moreover, problems with chiral fermions in five dimensions, to be included in more realistic theories, can be avoided. The S^1/Z_2 orbifold, as noted above, has the peculiar feature that there are fixed points $y = 0$ and $y = \pi$ not transforming under the Z_2 symmetry. These special points can be considered as so-called branes hosting localized fields which cannot penetrate the extra dimension. This concept can be easily formalized by introducing a δ-function in the Lagrangian density for the fermions, i.e.

$$\mathcal{L}_{\mathrm{F}}(x, y) = \delta(y) \overline{\Psi}(x) \left(i \gamma^\mu D_\mu - m_f \right) \Psi(x) , \tag{17}$$

where the covariant derivative

$$D_\mu = \partial_\mu + i\, e_5\, A_\mu(x, y) \tag{18}$$

contains the bulk gauge field and e_5 denotes the coupling constant of 5D-QED. The obvious generalization for the usual gauge-transformation properties of fermion fields reads

$$\Psi(x) \to \exp\left(-i\, e_5\, \Theta(x, 0)\right) \Psi(x) . \tag{19}$$

Again integrating out the fifth dimension, we are left with an effective four dimensional interaction Lagrangian

$$\mathcal{L}_{\mathrm{int}}(x) = -e \overline{\Psi} \gamma^\mu \Psi \left(A_{(0)\mu} + \sqrt{2} \sum_{n=1}^{\infty} A_{(n)\mu} \right) , \tag{20}$$

coupling all the KK modes to the fermion field on the brane. The coupling constant $e = e_5/\sqrt{2\pi R}$ is the QED coupling constant as measured by experiment. The factor $\sqrt{2}$ in (20) is a typical enhancement factor for the coupling of brane fields to heavy KK modes. Note that the scaler modes $A^5_{(n)}$ do not couple at all to

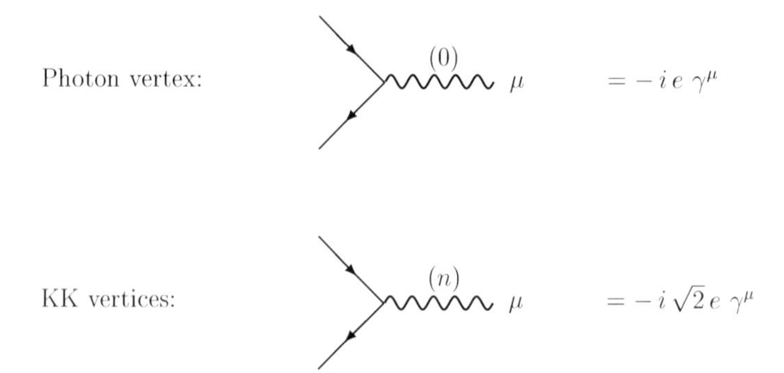

Fig. 1. Feynman rules for the vertices in 5D-QED

brane fermions because their wave functions vanish at $y = 0$ according to the odd Z_2-symmetry. These interaction terms together with completely standard kinetic terms for the fermion field complete 5D-QED. The corresponding Feynman rules for the electron-photon vertex and the analogous interaction of the KK modes are shown in Fig. 1.

If nature were described by QED up to energies probed so far by experiment an experimental signature of this five dimensional extension would be, e.g., a series of s-channel resonances in muon-pair production at an e^+e^--collider as shown in Fig. 2. Even though nature is not described by QED only, the generic signatures of extra dimensions are quite similar to those in more realistic theories.

The above quantization procedure can now be extended to more elaborate higher-dimensional models. If we want to extend the Standard Model by an extra dimension we have to understand spontaneous symmetry breaking in this context. Hence, adding a Higgs scalar in the bulk, the 5D Lagrangian of the theory reads

$$\mathcal{L}(x,y) \;=\; -\frac{1}{4}\, F^{MN}\, F_{MN} \;+\; (D_M \Phi)^* \,(D^M \Phi) \;-\; V(\Phi) \;+\; \mathcal{L}_{\rm GF}(x,y)\,, \quad (21)$$

where D_M again denotes the covariant derivative (18), e_5 the 5-dimensional gauge coupling,

$$\Phi(x,y) \;=\; \frac{1}{\sqrt{2}}\,(\, h(x,y) \,+\, i\,\chi(x,y)\,) \quad (22)$$

a 5-dimensional complex scalar field, and

$$V(\Phi) \;=\; \mu_5^2\, |\Phi|^2 \,+\, \lambda_5\, |\Phi|^4 \quad (23)$$

(with $\lambda_5 > 0$) the 5-dimensional Higgs potential. We consider $\Phi(x,y)$ to be even under Z_2, perform a corresponding Fourier decomposition, and integrate over y to obtain

$$\mathcal{L}_{\rm H}(x) = \frac{1}{2} \sum_{n=0}^{\infty} \Big[(\partial_\mu h_{(n)})\,(\partial^\mu h_{(n)}) \,-\, \frac{n^2}{R^2}\, h_{(n)}^2 \,-\, \mu^2\, h_{(n)}^2 \,+\, (\,h \leftrightarrow \chi\,) \Big] + \ldots\,, \quad (24)$$

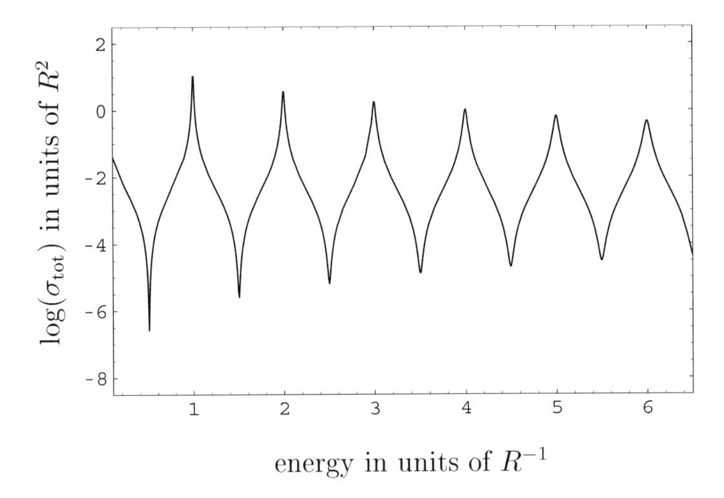

Fig. 2. Total cross section for $e^+e^- \to \mu^+\mu^-$ as a function of center-of-mass energy on a logarithmic scale. The width of the KK modes has been reasonably approximated

For $\mu^2 = \mu_5^2 < 0$, as in the usual 4-dimensional case, the zero KK Higgs mode acquires a non-vanishing vacuum expectation value (VEV) which breaks the U(1) symmetry. Moreover, it can be shown that as long as the phenomenologically relevant condition $v < 1/R$ is met, $h_{(0)}$ will be the only mode to receive a non-zero VEV

$$\langle h_{(0)} \rangle = v = \sqrt{2\pi R\, |\mu_5|^2/\lambda_5}\,. \tag{25}$$

The VEV introduces an additional mass term for each KK mode of the gauge fields. The zero mode turns from a massless to a massive degree of freedom, as usual for the Higgs mechanism. All the higher KK masses are slightly shifted. The gauge and self interactions of the Higgs fields, omitted in (24), only involve bulk fields, in contrast to the photon-fermion interaction introduced before. Although this leads to interesting effects we postpone their discussion until Sect. 3 where we investigate the phenomenologically more interesting gauge-boson self-couplings. After spontaneous symmetry breaking, it is instructive to introduce the fields

$$G_{(n)} = \left(\frac{n^2}{R^2} + e^2 v^2 \right)^{-1/2} \left(\frac{n}{R} A_{(n)5} + ev\, \chi_{(n)} \right), \tag{26}$$

where again $e = e_5/\sqrt{2\pi R}$, and the orthogonal linear combinations $a_{(n)}$. In the effective kinetic Lagrangian of the theory for the n-KK mode ($n > 0$)

$$\begin{aligned} \mathcal{L}^{(n)}_{\rm kin}(x) = & -\frac{1}{4} F^{\mu\nu}_{(n)} F_{(n)\mu\nu} \\ & + \frac{1}{2} \left(m_{A(n)} A_{(n)\mu} + \partial_\mu G_{(n)} \right) \left(m_{A(n)} A^\mu_{(n)} + \partial^\mu G_{(n)} \right) \\ & + \frac{1}{2} \left(\partial_\mu a_{(n)} \right) \left(\partial^\mu a_{(n)} \right) - \frac{1}{2} m^2_{a(n)} a^2_{(n)} + \dots , \end{aligned} \tag{27}$$

$G_{(n)}$ now plays the rôle of a Goldstone mode in an Abelian Higgs model. Both, $A_{(n)5}$ and $\chi_{(n)}$ take part in the mass generation for the heavy KK modes and, therefore, they are also mixed in the corresponding Goldstone mode. Because the mass contribution from spontaneous symmetry breaking is expected to be small compared to the KK masses, the Goldstone modes are dominated by the extra component of the gauge field. The pseudoscalar field $a_{(n)}$ describes an additional physical KK excitation degenerate in mass with the KK gauge mode $A_{(n)\mu}$, i.e.

$$m^2_{a(n)} = m^2_{A(n)} = (n^2/R^2) + e^2 v^2 \,. \tag{28}$$

The spectrum of the zero KK modes is simply identical to that of a conventional Abelian Higgs model, as it should be if we are to rediscover known physics in the low energy limit. It becomes clear that the appropriate gauge-fixing Lagrangian in (21) for a 5-dimensional generalized R_ξ-gauge should be

$$\mathcal{L}_{\rm GF}(x,y) \;=\; -\frac{1}{2\xi}\left[\partial_\mu A^\mu \;-\; \xi\left(\partial_5 A_5 \;+\; e_5\frac{v}{\sqrt{2\pi R}}\chi\right)\right]^2 . \tag{29}$$

All the mixing terms are removed and we again arrive at the standard kinetic Lagrangian for massive gauge bosons and the corresponding would-be Goldstone modes

$$\begin{aligned}
\mathcal{L}^{(n)}_{\rm kin}(x) \;=\; & -\frac{1}{4}F^{\mu\nu}_{(n)}F_{(n)\mu\nu} + \frac{1}{2}m^2_{A(n)}A_{(n)\mu}A^\mu_{(n)} \;-\; \frac{1}{2\xi}(\partial_\mu A^\mu_{(n)})^2 \\
& +\frac{1}{2}(\partial_\mu G_{(n)})(\partial^\mu G_{(n)}) \;-\; \frac{\xi}{2}m^2_{A(n)}G^2_{(n)} \\
& +\frac{1}{2}(\partial_\mu a_{(n)})(\partial^\mu a_{(n)}) \;-\; \frac{1}{2}m^2_{a(n)}a^2_{(n)} \\
& +\frac{1}{2}(\partial_\mu h_{(n)})(\partial^\mu h_{(n)}) \;-\; \frac{1}{2}m^2_{h(n)}h^2_{(n)} \;.
\end{aligned} \tag{30}$$

The CP-odd scalar modes $a_{(n)}$ and the Higgs KK-modes $h_{(n)}$ with mass

$$m_{h(n)} = \sqrt{(n^2/R^2) + \lambda_5 v^2/\pi R} \tag{31}$$

are not affected by the gauge fixing procedure. Observe finally that the limit $\xi \to \infty$ consistently corresponds to the unitary gauge.

As a qualitatively different way of implementing the Higgs sector in a higher-dimensional Abelian model, we can localize the Higgs field at the $y = 0$ boundary of the S^1/Z_2 orbifold, following the example of the fermions in 5D-QED. Introducing the appropriate δ-function in the 5-dimensional Lagrangian, this amounts to

$$\mathcal{L}(x,y) \;=\; -\frac{1}{4}F^{MN}F_{MN} + \delta(y)\,\left[(D_\mu\Phi)^*(D^\mu\Phi) \;-\; V(\Phi)\right] + \mathcal{L}_{\rm GF}(x,y)\,, \tag{32}$$

where the covariant derivative is given by (18) and the Higgs potential has its familiar 4-dimensional form. Because the Higgs potential is effectively four dimensional the Higgs field, not having KK excitations as a brane field, acquires the usual VEV. Notice that the bulk scalar field $A_5(x,y)$, as a result of its odd Z_2-parity, does not couple to the Higgs sector on a brane.

After compactification and integration over the y-dimension, spontaneous symmetry breaking again generates masses for all the KK gauge modes $A^\mu_{(n)}$. However, the mass matrix for the simple Fourier modes in (9) is no longer diagonal because of the δ-function in (32). Instead, it is given by

$$M_A^2 = \begin{pmatrix} m^2 & \sqrt{2}\,m^2 & \sqrt{2}\,m^2 & \cdots \\ \sqrt{2}\,m^2 & 2\,m^2 + (1/R)^2 & 2\,m^2 & \cdots \\ \sqrt{2}\,m^2 & 2\,m^2 & 2\,m^2 + (2/R)^2 & \cdots \\ \vdots & \vdots & \vdots & \ddots \end{pmatrix}, \tag{33}$$

where $m = ev$. Therefore, the Fourier modes are no longer mass eigenstates. By diagonalization of the mass matrix the mass eigenvalues $m_{(n)}$ of the KK mass eigenstates are found to obey the transcendental equation

$$m_{(n)} = \pi\, m^2\, R \cot\left(\pi\, m_{(n)}\, R\right). \tag{34}$$

Hence, the zero-mode mass eigenvalues are slightly shifted from what we expect in a 4D model. An approximate calculation, to first order in m^2/M^2, yields

$$m_{(0)} \approx \left(1 - \frac{\pi^2}{6}\frac{m^2}{M^2}\right) m. \tag{35}$$

The respective KK mass eigenstates can also be calculated analytically. They are given by

$$\hat{A}^\mu_{(n)} = \left(1 + \pi^2\, m^2\, R^2 + \frac{m^2_{(n)}}{m^2}\right)^{-1/2} \sum_{j=0}^{\infty} \frac{2\, m_{(n)}\, m}{m^2_{(n)} - (j/R)^2} \left(\frac{1}{\sqrt{2}}\right)^{\delta_{j,0}} A^\mu_{(j)}. \tag{36}$$

The couplings of these mass eigenstates to fermions will be slightly shifted with respect to the couplings of the Fourier modes in (20). To be specific, the interaction Lagrangian can be parameterized by

$$\mathcal{L}_{\rm int} = -\overline{\Psi}\,\gamma_\mu \Psi \sum_{n=0}^{\infty} e_{(n)}\, \hat{A}^\mu_{(n)}, \tag{37}$$

where the couplings $e_{(n)}$ of the different mass eigenstates are given by

$$e_{(n)} = \sqrt{2}\, e \left(1 + \frac{m^2}{m^2_{(n)}} + \pi^2 \frac{m^2}{M^2}\frac{m^2}{m^2_{(n)}}\right)^{-\frac{1}{2}}. \tag{38}$$

For example, the shift in the zero mode coupling is approximately given by

$$e_{(0)} \approx \left(1 - \frac{\pi^2}{3}\frac{m^2}{M^2}\right) e. \tag{39}$$

Here, in the Abelian model, the shifts in masses and couplings may seem to be a mere matter of redefinition of the measured masses and coupling in terms of the fundamental constants of the 5D-theory. However, they lead to important phenomenological implications in the context of the higher-dimensional Standard Model, where the various couplings are affected differently, as we will see below.

To find the appropriate form of the gauge-fixing term $\mathcal{L}_{\rm GF}(x,y)$ in (32), we follow (29), but restrict the scalar field χ to the brane $y=0$, viz.

$$\mathcal{L}_{\rm GF}(x,y) = -\frac{1}{2\xi}\Big[\partial_\mu A^\mu - \xi\left(\partial_5 A_5 + e_5 v\,\chi\,\delta(y)\right)\Big]^2 . \tag{40}$$

As is expected from a generalized R_ξ gauge, all mixing terms of the gauge modes $A^\mu_{(n)}$ with $A_{(n)5}$ and χ disappear up to total derivatives if $\delta(0)$ is appropriately interpreted on S^1/Z_2. Determining the unphysical mass spectrum of the Goldstone modes, we find a one-to-one correspondence of each physical vector mode of mass $m_{(n)}$ to an unphysical Goldstone mode with gauge-dependent mass $\sqrt{\xi}\,m_{(n)}$. In the unitary gauge $\xi\to\infty$, the would-be Goldstone modes are absent from the theory. The present brane-Higgs model does not predict other KK massive scalars apart from the physical Higgs boson h.

At this point, we cannot decide by any means which of the two possibilities for the Higgs sector, brane or bulk Higgs fields, could be realized in nature. Thus, we have to be ready to analyze both of them phenomenologically when we move on to SM extensions.

3 5-Dimensional Extensions of the Standard Model

It is a straightforward exercise to generalize the ideas introduced in Sect. 2 for non-Abelian theories

$$\begin{aligned}\mathcal{L}(x,y) = & -\frac{1}{4}F^a_{MN}F^{aMN} + \delta(y)\,\overline{\Psi}(x)\,\left(i\gamma^\mu D_\mu - m_f\right)\Psi(x)\\ & + \mathcal{L}_{\rm H} + \mathcal{L}_{\rm GF} + \mathcal{L}_{\rm FP}\,,\end{aligned} \tag{41}$$

where the field strength for the non-Abelian gauge field of a group with structure constants f^{abc} and coupling constant g_5 is given by

$$F^a_{MN} = \partial_M A^a_N - \partial_N A^a_M + g_5 f^{abc}A^b_M A^c_N\,. \tag{42}$$

Compactification, spontaneous symmetry breaking and gauge fixing [17,23] are very analogous to the Abelian case and the non-decoupling ghost sector can be easily included [17]. Hence, in the effective 4D theory, we arrive at a particle spectrum being similar to the Abelian case.

In addition, the self-interactions of the gauge bosons, induced by the bilinear terms in the non-Abelian field strength (42), lead to self-interactions of the KK modes which are restricted by selection rules, i.e., there are certain conditions for the KK numbers to be obeyed at each triple and quartic gauge-boson vertex.

3-boson vertex:

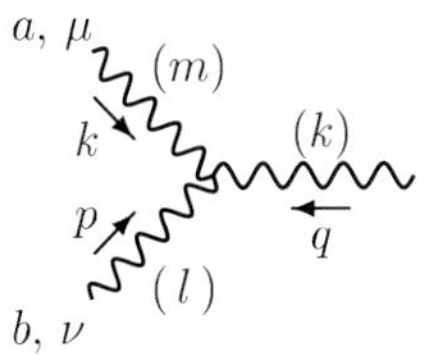

$$= g \left(\frac{1}{\sqrt{2}}\right)^{(\delta_{k,0}+\delta_{l,0}+\delta_{m,0}+1)} \delta_{k,l,m} \; f^{abc} \left[g^{\mu\nu} (k-p)^\rho + g^{\nu\rho} (p-q)^\mu + g^{\rho\mu} (q-k)^\nu \right]$$

vertex with 1 scalar:

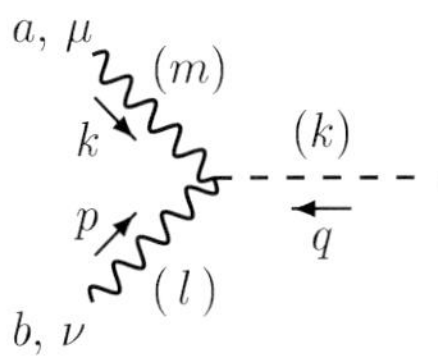

$$= -i\, g\, f^{abc}\, g^{\mu\nu} \left[\left(\frac{m}{R}\right)\left(\frac{1}{\sqrt{2}}\right)^{(\delta_{l,0}+1)} \tilde{\delta}_{k,l,m} - \left(\frac{l}{R}\right)\left(\frac{1}{\sqrt{2}}\right)^{(\delta_{m,0}+1)} \tilde{\delta}_{k,m,l} \right]$$

vertex with 2 scalars:

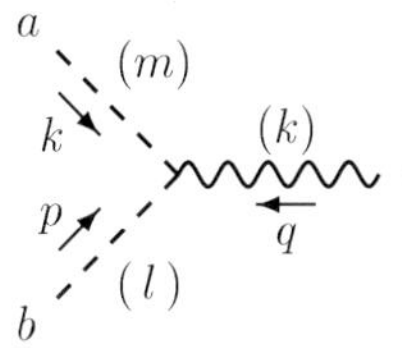

$$= g \left(\frac{1}{\sqrt{2}}\right)^{(\delta_{k,0}+1)} \tilde{\delta}_{l,k,m} \; f^{abc} (p-k)^\mu$$

4-boson vertex:

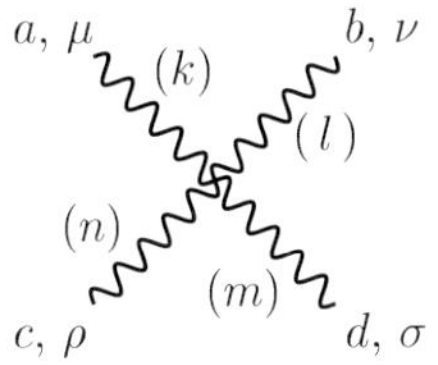

$$= -ig^2 \delta_{k,l,m,n} \left(\frac{1}{\sqrt{2}}\right)^{(\delta_{k,0}+\delta_{l,0}+\delta_{m,0}+\delta_{n,0}+2)} \left[f^{abe} f^{cde} \left(g^{\mu\rho} g^{\nu\sigma} - g^{\mu\sigma} g^{\nu\rho} \right) \; f^{ace} f^{bde} \left(g^{\mu\nu} g^{\rho\sigma} - g^{\mu\sigma} g^{\nu\rho} \right) \; f^{ade} f^{bce} \left(g^{\mu\nu} g^{\rho\sigma} - g^{\mu\rho} g^{\nu\sigma} \right) \right]$$

vertex with 2 scalars:

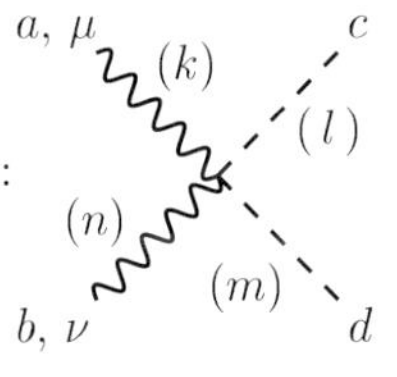

$$= i\, g^2 \left(\frac{1}{\sqrt{2}}\right)^{(\delta_{k,0}+\delta_{n,0})} \tilde{\delta}_{k,n,l,m} \; g^{\mu\nu}\, 2 \left[f^{ace} f^{bde} + f^{ade} f^{bce} \right]$$

Fig. 3. Feynman rules for the triple and quartic gauge boson couplings. The scalar modes correspond to the extra component of the higher-dimensional gauge fields. As in the Abelian model, they are would-be Goldstone modes and can be gauged away in unitary gauge. In parenthesis, the KK number of the interacting modes is shown. The selection rules are enforced by the prefactors defined in (43)

This is a general feature for interactions in which only bulk modes take part. A brane completely breaks the translational invariance of the orbifold and, thus, a brane field can couple to any bulk mode. In contrast, interactions between bulk fields obey a kind of quasi-momentum conservation with respect to the extra dimension, reflecting the special structure of the S^1/Z_2 orbifold and leading to the selection rules. The corresponding Feynman rules are displayed in Fig. 3. The selection rules are enforced by the prefactors

$$\begin{aligned} \delta_{k,l,m} &= \delta_{k+l+m,0} + \delta_{k+l-m,0} + \delta_{k-l+m,0} + \delta_{k-l-m,0} \,, \\ \tilde{\delta}_{k,l,m} &= -\delta_{k+l+m,0} + \delta_{k+l-m,0} - \delta_{k-l+m,0} + \delta_{k-l-m,0} \,, \end{aligned} \tag{43}$$

where $\delta_{i,j}$ denotes a standard Kronecker symbol. $\delta_{k,l,m,n}$ and $\tilde{\delta}_{k,l,m,n}$ are defined analogously.

At this point, we have considered all the important generic aspects of higher-dimensional theories. Therefore, we can now turn our attention to the theory we are really interested in, the electroweak sector of the Standard Model. Its gauge structure $\mathrm{SU}(2)_L \otimes \mathrm{U}(1)_Y$ opens up several possibilities for 5-dimensional extensions, because the $\mathrm{SU}(2)_L$ and $\mathrm{U}(1)_Y$ gauge fields do not necessarily both propagate in the extra dimension. As the fermion or Higgs fields we encountered before, one of the gauge groups can be confined to a brane at $y = 0$. Such a realization of a higher-dimensional model may be encountered within specific stringy frameworks [8,9].

However, in the most frequently investigated scenario, $\mathrm{SU}(2)_L$ and $\mathrm{U}(1)_Y$ gauge fields live in the bulk of the extra dimension (bulk-bulk model). The Lagrangian density is simply an application of (41) to the Standard Model gauge groups. Here, it is possible, even before integrating out the extra dimension, to choose a basis for the fields, where the photon and Z-boson field become explicit. The photon sector resembles exactly the 5D-QED discussed in Sect. 2, while for the Z boson and its KK modes spontaneous symmetry breaking leads to the effects also presented in Sect. 2. In the bulk-bulk model, both a localized (brane) and a 5-dimensional (bulk) Higgs doublet can be included in the theory. For generality, we will consider a 2-doublet Higgs model, where the one Higgs field Φ_1 propagates in the fifth dimension, while the other one Φ_2 is localized. The phenomenology presented in this note is not sensitive to details of the Higgs potential but only to their vacuum expectation values v_1 and v_2, or equivalently to $\tan\beta = v_2/v_1$ and $v^2 = v_1^2 + v_2^2$. Hence, β is the only additional free parameter introduced in the model.

The chiral structure of the Standard Model can be easily incorporated as long as one only considers fermions restricted to a brane. A simple extension of (20) leads to

$$\mathcal{L}_{\mathrm{int}}(x) \;=\; g\overline{\Psi}\gamma^\mu \left(g_V + g_A\gamma^5\right)\Psi\left(A_{(0)\mu} + \sqrt{2}\sum_{n=1}^{\infty} A_{(n)\mu}\right), \tag{44}$$

where A_μ generically denotes some gauge boson and g the respective coupling constant. The coupling parameters g_V and g_A are set by the electroweak quantum numbers of the fermions and receive their SM values. Because the KK mass eigenmodes $\hat{A}^{\mu}_{(n)}$ generally differ from the Fourier modes $A^{\mu}_{(n)}$ in (44), as we have seen before, their couplings to fermions $g_{V(n)}$ and $g_{A(n)}$ have to be individually calculated for each model. The photon and its possible KK modes are not affected by spontaneous symmetry breaking and keep their simple couplings, already presented in Fig. 1. For the bulk-bulk model, the shifts in the vector and axial-vector couplings of the Z boson are actually the same, such that it is sufficient to replace the SU(2) coupling constant g by $g_{Z(n)}$ for each KK mode, in analogy to the Abelian Higgs model. The mass generation in the Yukawa sector, involving brane fermions, hardly changes at all.

An even more minimal 5-dimensional extension of electroweak physics constitutes a model in which only the $\mathrm{U}(1)_Y$ -sector feels the extra dimension while the $\mathrm{SU}(2)_L$ gauge fields are localized at $y = 0$ (brane-bulk model). It is described by the Lagrangian density

$$\begin{aligned}\mathcal{L}(x,y) = -\frac{1}{4}\ B_{MN}\, B^{MN} + \delta(y)\Big[-\frac{1}{4}\, F^{a}_{\mu\nu} F^{a\mu\nu} + (D_\mu \Phi)^\dagger (D^\mu \Phi) - V(\Phi)\Big] \\ + \mathcal{L}_{\mathrm{GF}}(x,y) \ + \ \mathcal{L}_{\mathrm{FP}}(x,y)\,,\end{aligned} \tag{45}$$

where Φ denotes the Standard Model Higgs doublet on the brane and the covariant derivative

$$D_\mu \ = \ \partial_\mu \ - \ i\, g\, A^{a}_{\mu}(x)\, \tau^{a} \ - i\, \frac{g'_5}{2}\, B_\mu(x,y) \tag{46}$$

involves a brane as well as a bulk field. The W bosons are brane fields and their physics is completely SM-like. In this case, the Higgs field being charged with respect to both gauge groups has to be localized at $y = 0$ in order to preserve gauge invariance of the (classical) Lagrangian. A gauge field on the brane cannot compensate the variation of a Higgs field under gauge transformations in the whole bulk. For the same reason, a bulk Higgs is forbidden in the third possible model in which $\mathrm{U}(1)_Y$ is localized while $\mathrm{SU}(2)_L$ propagates in the fifth dimension (bulk-brane model), i.e., the gauge groups interchange their rôle. Consequently, the W bosons are bulk fields and are described in analogy to the models discussed in Sect. 2. In both models, after spontaneous symmetry breaking, there is a single massless gauge field protected by the residual unbroken gauge symmetry, the photon. A second light neutral mode can be identified with the Z boson. However, in contrast to the bulk-bulk model, there is only a single neutral tower of heavy KK modes. Up to small admixtures due to the brane VEV (see Sect. 2), it mainly contains the $\mathrm{U(1)_Y}$ or the neutral $\mathrm{SU(2)_L}$ gauge field, respectively. Nevertheless, for simplicity, we will refer to it as Z-boson KK tower. Note, however, that $g_{V(n)}$ and $g_{A(n)}$ in (44) are affected differently for the Z-boson and its KK modes. Most easily they are parameterized by introducing effective quantum numbers $T_{3(n)}$ and $Q_{(n)}$ to absorb all the higher-dimensional effects.

After the setup of all the models, we can finally turn our attention to the actual predictions of higher-dimensional theories for experiment.

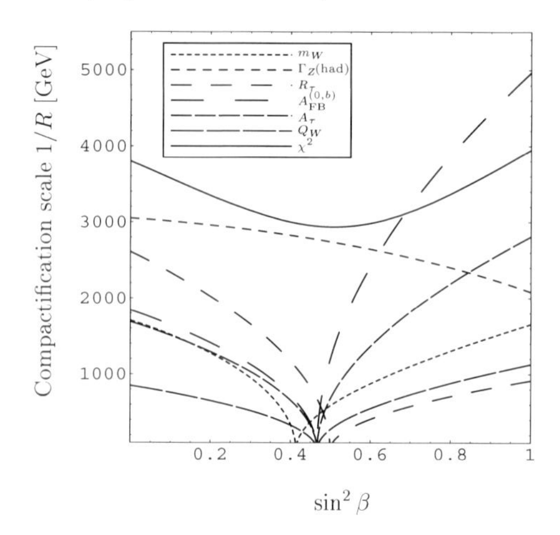

Fig. 4. Lower bounds on $M = 1/R$ (in TeV) from different observables at the 3σ confidence level for the bulk-bulk model

4 Effects on Electroweak Observables

In this section, we will concentrate on the phenomenology and present bounds on the compactification scale $M = 1/R$ of minimal higher-dimensional extensions of the SM, calculated by analyzing a large number of observables. To be specific, we proceed as follows. We relate the SM prediction $\mathcal{O}^{\text{SM}}$ [24,25] for an observable to the prediction $\mathcal{O}^{\text{HDSM}}$ for the same observable obtained in the higher-dimensional SM under investigation through

$$\mathcal{O}^{\text{HDSM}} = \mathcal{O}^{\text{SM}}\left(1 + \Delta_{\mathcal{O}}^{\text{HDSM}}\right). \tag{47}$$

Here, $\Delta_{\mathcal{O}}^{\text{HDSM}}$ is the tree-level modification of a given observable $\mathcal{O}$ from its SM value due to the presence of one extra dimension. The tree-level modifications can be expanded in powers of the typical scale factor

$$X = \frac{\pi^2}{3}\frac{m_Z^2}{M^2}\,. \tag{48}$$

We work to first order in X being a very good approximation for phenomenologically viable compactification scales in the TeV region. On the other hand, to enable a direct comparison of our predictions with precise data [24,25], we include SM radiative corrections to $\mathcal{O}^{\text{SM}}$. However, we neglect SM- as well as KK-loop contributions to $\Delta_{\mathcal{O}}^{\text{HDSM}}$ as higher order effects.

As input SM parameters for our numerical predictions, we choose the most accurately measured ones, namely the Z-boson mass m_Z, the electromagnetic fine structure constant α, and the Fermi constant G_F. While α is not affected in the models under study, $m_Z = m_{Z(0)}$, the mass of the lightest mode in the Z boson KK tower, generally deviates from its SM form, where we have

Table 1. Lower bounds on $M = 1/R$ (in TeV) at the 3σ confidence level for the brane-bulk and bulk-brane models

Observable	$\mathrm{U}(1)_Y$ in bulk	$\mathrm{SU}(2)_L$ in bulk
m_W	1.2	1.2
$\Gamma_Z(\mathrm{had})$	0.8	2.3
$Q_W(\mathrm{Cs})$	0.4	0.8
$A_{\mathrm{FB}}^{(0,b)}$	4.4	2.4
A_τ	2.5	1.4
R_τ	1.0	0.5
global analysis	3.5	2.6

$m_Z^{\mathrm{SM}} = \sqrt{g^2 + g'^2}\, v/2$ at tree level. To first order in X, m_Z may be parameterized by

$$m_Z \;=\; m_Z^{\mathrm{SM}}\, (\,1\,+\,\Delta_Z\, X\,)\;, \tag{49}$$

where Δ_Z is a model-dependent parameter. For the bulk-bulk, brane-bulk and bulk-brane models with respect to the $\mathrm{SU}(2)_L$ and $\mathrm{U}(1)_Y$ gauge groups, Δ_Z is given by

$$\Delta_Z = \left\{ -\frac{1}{2}\, \sin^4\beta\,,\; -\frac{1}{2}\, \sin^2\hat{\theta}_W\,,\; -\frac{1}{2}\, \cos^2\hat{\theta}_W \right\}, \tag{50}$$

where $\hat{\theta}_W$ is an effective weak mixing angle to be introduced below in (52). These shifts in the Z-boson mass are induced by the VEV of a brane Higgs.

The Fermi constant G_F, as determined by the muon lifetime, may receive additional direct contributions due to KK states mediating the muon decay. We may account for this modification of G_F by writing

$$G_F \;=\; G_F^{\mathrm{SM}}\, (\,1\,+\,\Delta_G\, X\,)\;, \tag{51}$$

where Δ_G is again model-dependent and has to be calculated consistently to first order in X.

The relation between the weak mixing angle θ_W and the input variables is also affected by the fifth dimension. Hence, it is useful to define an effective mixing angle $\hat{\theta}_W$ by

$$\sin^2\hat{\theta}_W \;=\; \sin^2\theta_W\, (\,1\,+\,\Delta_\theta\, X\,)\;, \tag{52}$$

such that the effective angle still fulfills the tree-level relation

$$G_F \;=\; \frac{\pi\alpha}{\sqrt{2}\sin^2\hat{\theta}_W\, \cos^2\hat{\theta}_W\, m_Z^2}\;, \tag{53}$$

of the Standard Model.

Table 2. Lower bounds (in TeV) on the compactification scale $M = 1/R$ at 2σ, 3σ and 5σ confidence levels from combined precision observables

model	2σ	3σ	5σ
$\mathrm{SU}(2)_L$-brane, $\mathrm{U}(1)_Y$-bulk	4.3	3.5	2.7
$\mathrm{SU}(2)_L$-bulk, $\mathrm{U}(1)_Y$-brane	3.0	2.6	2.1
$\mathrm{SU}(2)_L$-bulk, $\mathrm{U}(1)_Y$-bulk (brane Higgs)	4.7	4.0	3.1
$\mathrm{SU}(2)_L$-bulk, $\mathrm{U}(1)_Y$-bulk (bulk Higgs)	4.6	3.8	3.0

For the tree-level calculation of $\Delta_{\mathcal{O}}^{\mathrm{HDSM}}$, we have to carefully consider the effects from mixing of the Fourier modes on the masses of the Standard-Model gauge bosons as well as on their couplings to fermions. In addition, we have to keep in mind that the mass spectrum of the KK gauge bosons also depends on the model under consideration.

Within the framework outlined above, we first compute $\Delta_{\mathcal{O}}^{\mathrm{HDSM}}$ for the following high precision observables to first order in X: the W-boson mass m_W, the Z-boson invisible width $\Gamma_Z(\nu\overline{\nu})$, Z-boson leptonic widths $\Gamma_Z(l^+l^-)$, the Z-boson hadronic width $\Gamma_Z(\mathrm{had})$, the weak charge of cesium Q_W measuring atomic parity violation, various ratios R_l and R_q involving partial Z-boson widths, fermionic asymmetries A_f at the Z pole, and various fermionic forward-backward asymmetries $A_{\mathrm{FB}}^{(0,f)}$. For example, for the invisible Z width $\Gamma_Z(\nu\overline{\nu})$ we obtain

$$\Delta_{\Gamma_Z(\nu\overline{\nu})}^{\mathrm{HDSM}} = \begin{cases} \sin^2\hat{\theta}_W \left(\sin^2\beta - 1\right)^2 - 1 & \text{for the bulk-bulk model,} \\ -\sin^2\hat{\theta}_W & \text{for the brane-bulk model,} \\ -\cos^2\hat{\theta}_W & \text{for the bulk-brane model.} \end{cases} \tag{54}$$

Employing the results for $\Delta_{\mathcal{O}}^{\mathrm{HDSM}}$ and calculating all the electroweak observables considered in our analysis by virtue of (47), we confront these predictions with the respective experimental results. We can either test each variable individually or perform a χ^2 test to obtain combined bounds on the compactification scale $M = 1/R$, where

$$\chi^2(R) = \sum_i \frac{\left(\mathcal{O}_i^{\mathrm{exp}} - \mathcal{O}_i^{\mathrm{HDSM}}\right)^2}{\left(\Delta\mathcal{O}_i\right)^2}, \tag{55}$$

i runs over all the observables, and $\Delta\mathcal{O}_i$ is the combined experimental and theoretical error. A compactification radius is considered to be compatible at the $n\sigma$ confidence level (CL) if $\chi^2(R) - \chi^2_{\mathrm{min}} < n^2$, where χ^2_{min} is the minimum of χ^2 for a compactification radius in the physical region, i.e. for $R^2 > 0$.

Table 3. 2σ bounds in TeV inferred from fermion-pair production

model	$\mu^+\mu^-$	$\tau^+\tau^-$	hadrons	e^+e^-
$SU(2)_L$-brane, $U(1)_Y$-bulk	2.0	2.0	2.6	3.0
$SU(2)_L$-bulk, $U(1)_Y$-brane	1.5	1.5	4.7	2.0
$SU(2)_L$-bulk, $U(1)_Y$-bulk (brane Higgs)	2.5	2.5	5.4	3.6
$SU(2)_L$-bulk, $U(1)_Y$-bulk (bulk Higgs)	2.5	2.5	5.8	3.5

Figure 4 summarizes the lower bounds on the compactification scale M inferred from different types of observables for the bulk-bulk model. In this model, we present the bounds as a function of $\sin^2\beta$ parameterizing the Higgs sector. In Table 1, we summarize the bounds obtained by our calculations for the two bulk-brane models. The bounds from the global analysis at different confidence levels are shown in Table 2. The global 2σ bounds lie in the $4 \sim 5$ TeV region, only the bulk-brane model is less restricted. Here, a compactification scale of 3 TeV cannot be excluded by electroweak precision observables.

While the precision observables, analyzed so far, are measured at the Z pole or even at low energies, LEP2 provides us with data on cross sections at higher energies, up to more than 200 GeV. At these energies, the interference terms between Standard Model and KK contributions to a process like fermion-pair production dominate the higher-dimensional effects. In a first approximation, they are only suppressed by a factor of order s/M^2, where $\sqrt{s}$ is the center-of-mass energy, compared to the typical scale factor X of mass mixings and coupling shifts. Higher energies naturally lead to more sensitivity with respect to a possible fifth dimension [16,26,27]. As the simplest example for LEP2 processes, let us have a closer look at fermion-pair production. The relevant differential cross section for these observables is given at tree level by

$$\frac{\sigma(e^+e^- \to f\overline{f})}{d\cos\theta} = \frac{s}{128\,\pi}\left[(1+\cos\theta)^2\left(|M^{ef}_{LL}(s)|^2 + |M^{ef}_{RR}(s)|^2\right) + (1-\cos\theta)^2\left(|M^{ef}_{LR}(s)|^2 + |M^{ef}_{RL}(s)|^2\right)\right], \tag{56}$$

where θ is the scattering angle between the incoming electron and the negatively charged outgoing fermion, and

$$M^{ef}_{\alpha\beta}(s) = \sum_{n=0}^{\infty}\left(e^2_{(n)}\,\frac{Q_e Q_f}{s - m^2_{\gamma(n)}} + \frac{g^e_{\alpha(n)}g^f_{\beta(n)}}{\cos^2\theta_W}\,\frac{1}{s - m^2_{Z(n)}}\right). \tag{57}$$

Table 4. 2σ bounds in TeV inferred from precision observables and LEP2 cross sections

model	LEP1	LEP2	combined
SU(2)$_L$-brane, U(1)$_Y$-bulk	4.3	3.5	4.7
SU(2)$_L$-bulk, U(1)$_Y$-brane	3.0	4.4	4.3
SU(2)$_L$-bulk, U(1)$_Y$-bulk (brane Higgs)	4.7	5.4	6.1
SU(2)$_L$-bulk, U(1)$_Y$-bulk (bulk Higgs)	4.6	5.7	6.4

The couplings $g^f_{L(n)}$ and $g^f_{R(n)}$ in turn are given by

$$\begin{aligned} g^f_{L(n)} &= g_{Z(n)} \left(T_{3f(n)} - Q_{f(n)} \sin^2\theta_W \right) , \\ g^f_{R(n)} &= g_{Z(n)} \left(-Q_{f(n)} \sin^2\theta_W \right) . \end{aligned} \tag{58}$$

$T_{3f(n)}$, $Q_{f(n)}$, $g_{Z(n)}$, $e_{(n)}$, $m_{Z(n)}$ and $m_{\gamma(n)}$, as introduced in Sect. 3, can be calculated to first order in X, e.g. from the exact analytic expressions in [17]. For Bhabha scattering the t-channel exchange also has to be taken into account, however, there are no fundamental differences. The above parameterization for the cross section is particularly convenient because it clearly separates the higher-dimensional effects from well known physics. All the higher-dimensional physics manifests itself in the effective sum of s-channel propagators (57) differing with respect to the Standard Model.

From (56), we calculate $\Delta^{\mathrm{HDSM}}_{\mathcal{O}}$ for the different fermion-pair production channels at LEP2 energies and finally the bounds shown in Table 3. In Table 4, all the bounds are combined for final lower limits on the compactification scale. The bounds for the brane-bulk and the bulk-brane models from LEP1 and LEP2 observables are kind of complementary, such that the combined limit is larger than 4 TeV in both models. The compactification scale for the bulk-bulk model, no matter where the Higgs lives, is still more restricted to lie above 6 TeV at the 2σ confidence level.

5 Conclusions

The aim of the present notes has been to give an introduction to the model-building of low-energy 5-dimensional electroweak models. We have derived step by step the corresponding four dimensional theory by compactifying on the orbifold S^1/Z_2 and by integrating out the extra dimension. We have paid special attention to consistently quantize the higher-dimensional models in the generalized R_ξ gauges. The 5-dimensional R_ξ gauge fixing conditions introduced here lead, after compactification, to a 4-dimensional Lagrangian in the standard R_ξ gauge for each KK mode. The latter also clarifies the rôle of the different degrees of freedom. One of the main advantages of our gauge-fixing procedure is that one can now derive manifest gauge-independent analytic expressions for the KK-mass spectrum of the gauge bosons and for their interactions to the fermionic matter. Most importantly, one may even apply an analogous gauge-fixing approach to spontaneous symmetry breaking theories.

To render the topics under discussion more intuitive, we have analyzed all the main ideas in simple Abelian toy models. However, we have pointed out how to generalize these ideas to new possible 5-dimensional extensions of the SM in which the $\mathrm{SU}(2)_L$ and $\mathrm{U}(1)_Y$ gauge fields and Higgs bosons may or may not all experience the presence of the fifth dimension. The fermions in all the models are considered to be confined to one of the two boundaries of the S^1/Z_2 orbifold.

After introducing a framework for deriving predictions of possible observables, we have given a glimpse of higher-dimensional phenomenology. Electroweak precision observables are considered as well as cross sections for fermion-pair production at LEP2. In particular, we have presented bounds on the compactification scale $M = 1/R$ in three different 5-dimensional extensions of the SM: (i) the $\mathrm{SU}(2)_L \otimes \mathrm{U}(1)_Y$-bulk model, where all SM gauge bosons are bulk fields; (ii) the $\mathrm{SU}(2)_L$-brane, $\mathrm{U}(1)_Y$-bulk model, where only the W bosons are restricted to the brane, and (iii) the $\mathrm{SU}(2)_L$-bulk, $\mathrm{U}(1)_Y$-brane model, where only the $\mathrm{U}(1)_Y$ gauge field is confined to the brane. For the often-discussed first model, we find the 2σ lower bounds on M: $M \gtrsim 6.4$ and 6.1 TeV, for a Higgs boson living in the bulk and on the brane, respectively. For the second and third models, the corresponding 2σ lower limits are 4.7 and 4.3 TeV. Hence, the bounds for different models can differ significantly.

Any non-stringy field-theoretic treatment of higher-dimensional theories, as the one presented here, involves a number of assumptions. Although the results obtained in the higher-dimensional models with one compact dimension are convergent at the tree level, they become divergent if more than one extra dimensions are considered. Also, the analytic results are ultra-violet (UV) divergent at the quantum level, since the higher-dimensional theories are not renormalizable. Within a string-theoretic framework, the above UV divergences are expected to be regularized by the string mass scale M_s. Therefore, from an effective field-theory point of view, the phenomenological predictions will depend to some extend on the UV cut-off procedure [28] related to the string scale M_s. Nevertheless, assuming validity of perturbation theory, we expect that quantum corrections due to extra dimensions will not exceed the 10% level of the tree-level effects we have been studying here. Finally, we have ignored possible model-dependent winding-number contributions [29] and radiative brane effects [30] that might also affect to some degree our phenomenological predictions.

The lower limits on the compactification scale derived by the present global analysis indicate that resonant production of the first KK state may be at the edge of accessibility at the LHC, at which heavy KK masses up to 6–7 TeV [8,14] might be explored. Hence, the phenomenological analysis has to be carried further in order to be able to discriminate possible higher-dimensional signals from other Standard Model extensions.

Acknowledgements

This work was supported by the Bundesministerium für Bildung and Forschung (BMBF, Bonn, Germany) under the contract number 05HT1WWA2.

References

1. T. Kaluza: Sitzungsber. d. Preuss. Akad. d. Wiss. Berlin, 966 (1921) O. Klein: Zeitschrift f. Physik **37** 895 (1926)
2. For a review, see e.g., M.B. Green, J.H. Schwarz, E. Witten: *Superstring Theory.* (Cambridge University Press, Cambridge 1987).
3. I. Antoniadis: Phys. Lett. B **246**, 377 (1990)
4. J.D. Lykken: Phys. Rev. D **54** 3693 (1996)
5. E. Witten: Nucl. Phys. B **471** 135 (1996) P. Hořava, E. Witten: Nucl. Phys. B **460** 506 (1996); Nucl. Phys. B **475** 94 (1996)
6. N. Arkani-Hamed, S. Dimopoulos, G. Dvali: Phys. Lett. B **429** 263 (1998) I. Antoniadis, N. Arkani-Hamed, S. Dimopoulos, G. Dvali: Phys. Lett. B **436** 257 (1998); N. Arkani-Hamed, S. Dimopoulos, G. Dvali: Phys. Rev. D **59** 086004 (1999)
7. K.R. Dienes, E. Dudas, T. Gherghetta: Phys. Lett. B **436** 55 (1998); Nucl. Phys. B **537** 47 (1999)
8. I. Antoniadis, K. Benakli: Int. J. Mod. Phys. A **15** 4237 (2000)
9. I. Antoniadis, E. Kiritsis, T.N. Tomaras: Phys. Lett. B **486** 186 (2000)
10. P. Nath, M. Yamaguchi: Phys. Rev. D **60** 116006 (1999); Phys. Lett. B **466** 100 (1999)
11. W.J. Marciano: Phys. Rev. D **60** 093006 (1999); M. Masip, A. Pomarol: Phys. Rev. D **60** 096005 (1999)
12. R. Casalbuoni, S. De Curtis, D. Dominici, R. Gatto: Phys. Lett. B **462** 48 (1999); C. Carone: Phys. Rev. D **61** 015008 (2000)
13. A. Delgado, A. Pomarol, M. Quiros: JHEP **0001** 030 (2000)
14. T. Rizzo, J. Wells: Phys. Rev. D **61** 016007 (2000); A. Strumia: Phys. Lett. B **466** 107 (1999)
15. A. Delgado, A. Pomarol, M. Quiros: Phys. Rev. D **60** 095008 (1999)
16. K. Cheung, G. Landsberg: Phys. Rev. D **65** 076003 (2002)
17. A. Mück, A. Pilaftsis, R. Rückl: Phys. Rev. D **65** 085037 (2002)
18. For example, see H. Georgi, A.K. Grant, G. Hailu: Phys. Lett. B **506** 207 (2001)
19. D.M. Ghilencea, S. Groot Nibbelink, H.P. Nilles: Nucl. Phys. B **619** 385 (2001)
20. J. Papavassiliou, A. Santamaria: Phys. Rev. D **63** 125014 (2001)
21. D. Dicus, C. McMullen, S. Nandi: Phys. Rev. D **65** 076007 (2002)
22. T. Appelquist, H.C. Cheng and B.A. Dobrescu: Phys. Rev. D **64** 035002 (2001)
23. R.S. Chivukula, D.A. Dicus, H.-J. He: Phys. Lett. B **525** 175 (2002)
24. Particle Data Group (D.E. Groom et al.): European Physical Journal C **15** 1 (2000)
25. The LEP Collaborations ALEPH, DELPHI, L3, OPAL, the LEP Electroweak Working Group and the SLD Heavy Flavor and Electroweak Groups: hep-ex/0112021
26. A. Mück, A. Pilaftsis, R. Rückl: work in preparation
27. C.D. McMullen, S. Nandi: hep-ph/0110275

28. T. Kobayashi, J. Kubo, M. Mondragon and G. Zoupanos, Nucl. Phys. B **550** 99 (1999)
29. I. Antoniadis, K. Benakli and A. Laugier, JHEP **0105** 044 (2001)
30. H. Georgi, A.K. Grant, G. Hailu, Phys. Lett. B **506** 207 (2001) M. Carena, T. Tait, C.E.M. Wagner: hep-ph/0207056

One-Loop Corrections to Polarization Observables

J.G. Körner and M.C. Mauser

Institut für Physik, Johannes Gutenberg–Universität, Staudinger Weg 7, 55099 Mainz, Germany

Abstract. We review the physics of polarization observables in high energy reactions in general and discuss the status of *NLO* one–loop corrections to these observables in specific. Many high order radiative corrections exists for rates but not many *NLO* radiative corrections exist for polarization observables. The radiative correction calculations for polarization observables are somewhat more complicated than those for rates. They tend to be smaller than those for the rates. In most of the examples we discuss we include mass effects which significantly complicate the radiative correction calculations. We elaborate a general scheme which allows one to enumerate the number of independent density matrix elements in a reaction and provide explicit examples of angular decay distributions in self–analyzing decays that allow one to experimentally measure the density matrix elements. We provide examples of reactions where certain density matrix elements are only populated at *NLO* or by mass effects. In our discussion we concentrate on semileptonic bottom and top quark decays which are linked to leptonic μ and τ decays through a Fierz transformation.

1 Introduction

Most of the *NLO* radiative corrections to rates have been done quite some years ago. In fact, two–loop *NNLO* calculations for rates are now becoming quite standard. In e^+e^-–annihilation one is even pushing hard to determine the *NNNNLO* corrections to the R ratio [1,2]. Contrary to this, many *NLO* radiative correction calculations to polarization observables involving also massive quarks have only been done in the last few years.

It is clear that higher order radiative corrections to unpolarized observables will always be at the center of attention because data on unpolarized observables are as a rule much more accurate than data for polarized observables. Nevertheless, as experimental data on polarization observables has been accumulating over the years there is an evolving need for radiative corrections to polarization observables. One of the reasons that polarized radiative correction were lagging behind unpolarized radiative corrections is that the computational effort in the calculation of radiative corrections to polarization observables is larger than that for rates. For once, one cannot sum over the spins of intermediate states whose polarization one wants to calculate. One therefore cannot make use of the powerful unitarity method to calculate radiative corrections from the absorptive parts of higher order loop graphs. Further, the definition of polarization observables brings in extra momentum factors in the integrands of the requisite phase space

J.G. Körner and M.C. Mauser, One-Loop Corrections to Polarization Observables, Lect. Notes Phys. **647**, 212–244 (2004)
http://www.springerlink.com/

integrals which makes life more difficult. This is particularly true if the masses of particles in the process cannot be neglected. For example, in the process $t \to b + W^+$ the longitudinal component of the polarization vector of the top quark along the W–direction is given by

$$s_t^{l,\mu} = \frac{1}{|\boldsymbol{q}|}\left(q^\mu - \frac{p_t \cdot q}{m_t^2} p_t^\mu\right), \tag{1}$$

where the denominator factor $|\boldsymbol{q}| = \sqrt{q_0^2 - m_W^2}$ comes in for normalization reasons (q is the momentum of the W^+). It is quite clear that such square root factors lead to nontrivial complications in the phase space integrations. Similar square root factors appear when projecting onto the polarization states of the W^+.

Inclusion of mass effects as e.g. in the semileptonic decay $b \to c + l^- + \bar{\nu}_l$ ($m_c/m_b \approx 0.30$) or in the leptonic decay $\tau^- \to \mu^- + \nu_\tau + \bar{\nu}_\mu$ ($m_\mu/m_\tau \approx 0.07$) render the analytical calculations considerably more complicated. The villain is the Kállen function $(m_1^4 + m_2^4 + m_3^4 - 2(m_1^2 m_2^2 + m_1^2 m_3^2 + m_2^2 m_3^2))^{1/2}$ which is brought in by extra three–momentum factors. In the case that e.g. $m_3 \to 0$ the Kállen function simplifies to $(m_1 - m_2)^2$ which leads to an enormous simplification in the phase space integrals.

Physically speaking mass effects become large in regions of phase space where the massive particles become nonrelativistic. For example, in the leptonic decay of the muon $\mu^- \to e^- + \nu_\mu + \bar{\nu}_e$, the mass of the electron cannot be neglected in the threshold region where the energy of the electron is small. This is illustrated in Fig. 1 where the longitudinal polarization of the electron is plotted against the scaled energy $x = 2E_e/m_\mu$ of the electron [3]. We have chosen a logarithmic scale for x in order to enhance the threshold region. In the threshold region the longitudinal polarization deviates considerably from the naive value $P_e^l = -1$. The radiative corrections in the threshold region can be seen to be quite large.

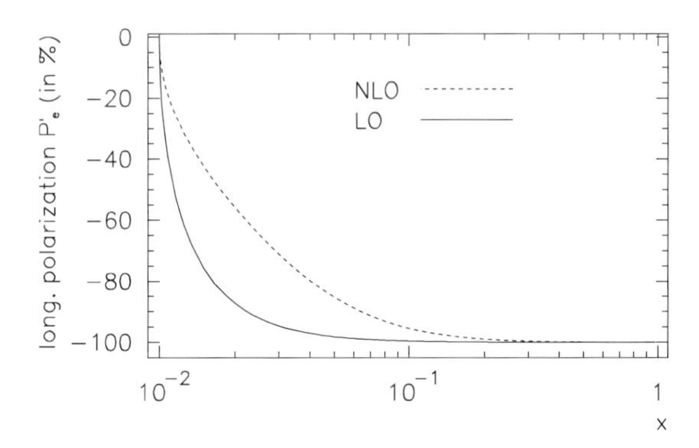

Fig. 1. Longitudinal polarization of the electron in leptonic muon decays at *LO* and *NLO* as a function of the scaled energy $x = 2E_e/m_\mu$ [3]. The *LO* curve is very well described by the functional behaviour $P_e^l = -\beta$.

All our results are given in closed analytic form. In the days of fast numerical computers one may rightfully ask what the advantage of having closed form expressions is. One of the advantages is that all requisite mass and momentum limits can be taken in analytic form. Thus the analytic results can be subjected to tests against known limiting cases providing for the necessary checks of the full results. A second issue is user–friendliness. The analytic expressions are simple enough to be incorporated in numerical programs by the prospective user. Parameter values such as masses and coupling constants can be varied at will by the user in his own program without having to refer to numerical programs written by others. We have checked that our analytical formulas are numerically stable even in the small and large mass limits.

NLO corrections to rates can be quite large. *NLO* corrections to polarization observables have a tendency to be somewhat smaller. The reason is that polarization observables are normalized quantities. They are normalized with regard to the total rate. The numerators and denominators in the relevant polarization expressions tend to go in the same direction. The reason is that the *NLO* numerators and denominators are dominated by the soft gluon or soft photon contributions which are universal in the sense that they multiply the relevant Born term expressions and thus cancel out in the ratio. For example, for $t \to b + W^+$ the $O(\alpha_s)$ correction to the rate is -8.5% and $(-2.5\% \div +3\%)$ for the polarization observables [4–6].

At *NLO* some density matrix elements become populated which vanish at leading order. An example is again the decay $t \to b + W^+$ where the rate into the right–handed W^+ vanishes ($\Gamma(W_R) = 0$) for $m_b = 0$ at *LO* but where $\Gamma(W_R) \neq 0$ at *NLO*. Generally speaking, Standard Model radiative corrections can populate density matrix elements elements that vanish at leading order. Consequently radiative corrections change angular decay distributions as do new physics effects. The lesson to be learned is clear. Before ascribing a given polarization effect to new physics one has to make sure that it does not result from radiative corrections of old physics. In the above example a non–zero $\Gamma(W_R) \neq 0$ could result from an admixture of a right–handed charged current (new physics) or from radiative corrections. Only the precise knowledge of the magnitude of the radiative corrections allows one to exclude or conclude for new physics effects.

In this review we will mostly be concerned with radiative corrections to the current–induced transitions $t \to b$, $b \to c$ and $l \to l'$. It should be clear that the additional gluons (or photons) in radiative correction calculations only couple to the $q_1 \to q_2$ (or $l \to l'$) side of the relevant semileptonic (or leptonic) transitions. This implies that the structure of the *NLO* radiative corrections is the same in the two classes of processes. The current–induced transition $l \to l'$ is not in the Standard Model form but in the charge retention form. However, the charge retention form can be linked to the Standard Model form by the remarkable property that the $(V-A)^\mu(V-A)_\mu$ interaction is an eigenvector under Fierz crossing (see Sect. 3). One therefore has

$$\mathcal{L} = \frac{G_F}{\sqrt{2}}[\bar{\nu}_\mu \gamma^\alpha (\mathbb{1} - \gamma_5)\mu][\bar{e}\gamma_\alpha(\mathbb{1} - \gamma_5)\nu_e] + \text{h.c.} \tag{2}$$

$$= \frac{G_F}{\sqrt{2}}[\bar{\nu}_\mu \gamma^\alpha (\mathbb{1} - \gamma_5)\nu_e][\bar{e}\gamma_\alpha(\mathbb{1} - \gamma_5)\mu] + \text{h.c.} \tag{3}$$

All of the above three transitions are therefore governed by the same matrix elements. In particular the *NLO* QED and QCD radiative corrections have the same structure for all three transitions. In Table 1 we provide a list of the processes for which the radiative corrections to the rates and the polarization observables can all be calculated from essentially the same set of *NLO* matrix elements. In this review we shall only present a few sample results from the processes in Table 1 because the results are to numerous to fit into a review of the present size. Instead we attempt to share our insights into the general features of spin physics and illustrate these with sample results taken from the processes in Table 1.

Table 1. List of processes based on the same current–induced matrix elements

parton level	particle level	references
$t(\uparrow) \longrightarrow b + W^+(\uparrow)$	$t(\uparrow) \longrightarrow X_b + W^+(\uparrow)(\longrightarrow l^+ + \nu_l)$	[4–7]
$t(\uparrow) \longrightarrow b + H^+$	$t(\uparrow) \longrightarrow X_b + H^+$	[8]
$b(\uparrow) \longrightarrow c(\uparrow) + l^- + \bar{\nu}_l$	$B \longrightarrow X_c + D_s^{(*)}(\uparrow)$	[9]
	$\Lambda_b(\uparrow) \longrightarrow X_c + D_s^{(*)}(\uparrow)$	[10]
$b(\uparrow) \longrightarrow u(\uparrow) + l^- + \bar{\nu}_l$	$B \longrightarrow X_u + \pi(\rho(\uparrow))$	[9]
	$\Lambda_b(\uparrow) \longrightarrow X_u + \pi(\rho(\uparrow))$	[10]
$l^-(\uparrow) \longrightarrow l'^-(\uparrow) + \nu_l + \bar{\nu}_{l'}$		[3]
$(l, l') = (\mu, e), (\tau, \mu), (\tau, e)$		

2 Miscellaneous Remarks on Polarization Effects

2.1 Examples of 100% Polarization

In this subsection we discuss examples of 100% polarization. Cases of 100% polarization are usually associated with limits when masses or energies of particles become small or large compared to a given scale in a process. For example, an on–shell or off–shell gauge bosons radiated off massless fermions (leptons or quarks) is purely transverse since fermion helicity is conserved by the vector and axial vector Standard Model couplings of the gauge bosons. Contrary to this, massive gauge bosons become purely longitudinal in the high energy limit (see Sect. 11). We discuss two examples of 100% polarization in more detail. Namely, the case of a left–chiral fermion which becomes purely left–handed in the chiral limit when $m_f \to 0$ and the case of soft gluon radiation from heavy quarks.

Relativistic Left–Chiral Fermions

Consider a fermion moving along the z–axis. The positive and negative helicity polarization four–vectors of the fermion are given by $s^{\mu}_{\pm} = \pm(p; 0, 0, E)/m$. At first sight it appears to be problematic to take the limit $m \to 0$ in expressions involving the polarization four–vectors because of the denominator mass factor. However, the saving feature is that $s^{\mu}_{\pm}$ becomes increasingly parallel to the momentum $\pm p^{\mu}$ when $m \to 0$. In fact, one has

$$s^{\mu}_{\pm} = \pm p^{\mu} + O(m/E), \tag{4}$$

since $E = p + (m^2)/(E+p)$. Therefore the projectors onto the two helicity states simplify to

$$u(\pm)\bar{u}(\pm) = \frac{1}{2}(\not p + m)(\mathbb{1} + \gamma_5 \not s_{\pm}) \xrightarrow{m \to 0} \frac{1}{2}\not p(\mathbb{1} \mp \gamma_5). \tag{5}$$

This shows that a right/left–chiral fermion is purely right/left–handed in the chiral limit. Thus the final state electron emerging from leptonic μ–decay is purely left–handed in the chiral limit except for the anomalous spin–flip contribution appearing first at *NLO* which populate also the right–handed state. The anomalous spin–flip contribution will be discussed in Sect. 12.

The approach to the chiral limit is well described by an approximate formula frequently discussed in text books. The argument goes as follows.

Introduce left–chiral fermion spinors according to $u_L(\pm) = 1/2(\mathbb{1} - \gamma_5)u(\pm)$, where $u(\pm)$ are helicity $\lambda = +1/2, -1/2$ spinors. The longitudinal polarization of a left–chiral electron is then calculated by taking the ratio of the difference and the sum of the $\lambda = +1/2$ and $\lambda = -1/2$ scalar densities according to

$$P^l_e = \frac{u^{\dagger}_L(+)u_L(+) - u^{\dagger}_L(-)u_L(-)}{u^{\dagger}_L(+)u_L(+) + u^{\dagger}_L(-)u_L(-)} \tag{6}$$

$$= \frac{(E+m-p)^2 - (E+m+p)^2}{(E+m-p)^2 + (E+m+p)^2} = -\frac{p}{E} = -\beta. \tag{7}$$

A corresponding result holds for left–chiral positrons where $P^l_e = \beta$.

The accuracy of the above result can be checked with the corresponding expression for the Born term polarization of the electron in leptonic μ–decay. One finds (see e.g. [3])

$$P^l_e = -\beta \frac{x(3 - 2x + y^2)}{x(3-2x) - (4-3x)y^2}, \tag{8}$$

where $x = 2E_e/m_{\mu}$ and $y = m_e/m_{\mu}$. For the Born term contribution the correction to the approximate result (6) $P^l_e = -\beta$ is of $O(1\%)$ or less such that the correct and approximate curves are not discernible at the scale of Fig. 1. Contrary to this the *NLO* polarization deviates substantially from a simple $P^l_e = -\beta$

behaviour (see Fig. 1) even though one has again $P_e^l \propto -\beta$ at *NLO* [3]. This may have to do with the fact that the radiative corrections involve an extra photon emission from the electron leg. The electron is therefore no longer left–chiral at *NLO* as in the *LO* Born term case.

Soft Gluon or Soft Photon Radiation

When a soft gluon (or a soft photon) is radiated off a fermion line it is 100% polarized in the plane spanned by the fermion and the gluon (or photon). To see this in an exemplary way consider the tree graph matrix element of a soft gluon radiated off a top or antitop in the process $e^+e^- \to t\bar{t}g$. The soft gluon matrix element reads

$$T_\mu^{(\text{s.g.})}(\pm) = T_\mu^{(\text{Born})}\left(\frac{p_t^\alpha}{p_t \cdot k} - \frac{p_{\bar{t}}^\alpha}{p_{\bar{t}} \cdot k}\right)\epsilon_\alpha^*(\pm) \tag{9}$$

with $\epsilon_\alpha^*(\pm) = (0; \mp 1, i, 0)/\sqrt{2}$.

The 2×2 density matrix of the gluon can be expanded in terms of the unit matrix $\mathbb{1}$ and the Pauli matrices σ_i according to

$$\begin{pmatrix} h_+h_+^* & h_+h_-^* \\ h_-h_+^* & h_-h_-^* \end{pmatrix} = \frac{1}{2}(\sigma \cdot \mathbb{1} + \boldsymbol{\xi} \cdot \boldsymbol{\sigma}), \tag{10}$$

where $\boldsymbol{\xi}$ denotes the Stokes "vector". We have set "vector" in quotation marks because the Stokes "vector" does not transform as a vector but rather as a spin 2 object under three–dimensional rotations. From (9) one sees that the helicity amplitudes are relatively real and that $h_+ = -h_-$. The normalized Stokes "vector" is then given by

$$\boldsymbol{\xi}/\sigma = (-1, 0, 0), \tag{11}$$

i.e., in the terminology of classifying the polarization states of a gluon (or a photon), the polarization of the gluon is 100% linearly polarized in the production plane.

In Fig. 2 we show a plot of the linear polarization of the gluon as a function of the scaled energy of the gluon for $e^+e^- \to t\,\bar{t}\,g$ and for $e^+e^- \to c\,\bar{c}\,g$ [11]. At the soft end of the spectrum the gluon is 100% polarized and then slowly drops to zero polarization at the hard end of the spectrum. It is quite remarkable that a high degree of polarization is maintained over a large part of the spectrum.

2.2 Examples of Zero Polarization

The mean of a single spin polarization observable such as $\langle \boldsymbol{\sigma} \cdot \boldsymbol{p} \rangle$ is a parity–odd measure. In strong interaction processes, which are parity conserving, there are no single spin polarization effects, i.e. $\langle \boldsymbol{\sigma} \cdot \boldsymbol{p} \rangle = 0$ (except for the T–odd single spin effects $\propto \langle \boldsymbol{\sigma} \cdot (\boldsymbol{p_i} \times \boldsymbol{p_j}) \rangle$ to be discussed later on). This is different in weak and electroweak interactions where the presence of parity violations induces a wide variety of single spin polarization phenomena. As an example we take

e^+e^-–annihilation into heavy quarks where single spin polarization phenomena occur due to (γ, Z)–exchange. *NLO* corrections to these single spin polarization phenomena have been discussed in [12–17]. Another example is the semileptonic decay of a polarized heavy quark where the *NLO* corrections to the lepton spectrum have been presented in [18,19].

However, there are double spin polarization effects in strong interactions since $\langle \boldsymbol{\sigma} \cdot \boldsymbol{\sigma} \rangle$ is a parity even measure. An example is the correlation between the spins of the top and the antitop produced in hadronic collisions [20–23]. Naturally, double spin polarization effects also occur in weak interaction processes such as in high energy e^+e^-–annihilation (see [16,24–26]). However, even in weak processes single spin polarization effects can vanish when one approaches phase space boundaries. Consider, for example, the process $e^+e^- \to t\,\bar{t}$ at threshold. In general, the vector current induced amplitude is s– and d–wave, while the axial vector current amplitude is p–wave. As one approaches threshold the s–wave amplitude will dominate and thus there will be no vector–axial vector interference. Therefore single top polarization goes to zero in $e^+e^- \to t\,\bar{t}$ as one approaches the threshold region.

Another example of zero polarization are *NLO* T–odd measures in the process $e^+e^- \to q\,\bar{q}\,(g)$ for massless quarks. T–odd effects arise from the imaginary parts of one–loop contributions. However, in the case $e^+e^- \to q\,\bar{q}\,(g)$ the imaginary parts can be shown to be proportional to the Born term contribution in the mass zero limit. Hence, the T–odd measures are zero in this case [27]. When $m_q \neq 0$ the imaginary parts are no longer proportional to the Born term and, consequently, one obtains nonzero values for the T–odd measures [28,29]. In the crossed channels (deep inelastic scattering, Drell–Yan process) the imaginary parts no longer have Born term structure and one obtains non–vanishing T–odd effects even for zero mass quarks [30–33].

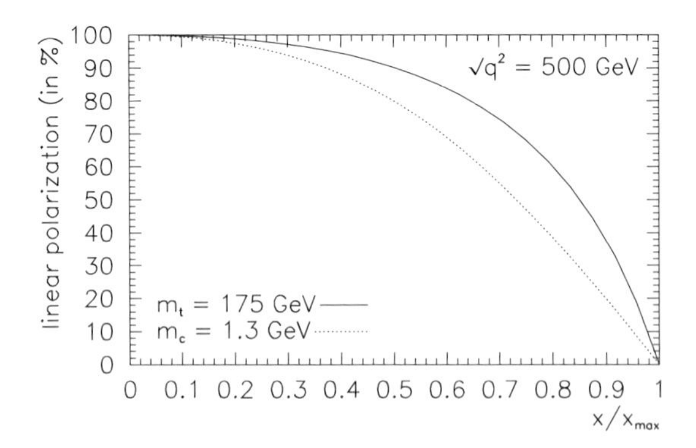

Fig. 2. Linear polarization of the gluon as a function of the scaled energy of the gluon in $e^+e^- \to Q\,\bar{Q}\,g$ [11].

2.3 Mass Effects

Mass effects make the radiative correction more complicated. We attempt to illustrate the complication brought about by an additional mass scale by listing the $(LO + NLO)$ results for the total rate of leptonic μ decays $\mu^- \to e^- + \bar{\nu}_e + \nu_\mu$ for (i) $y = m_e/m_\mu = 0$ and for (ii) $y = m_e/m_\mu \neq 0$. In case (i) one has

$$\Gamma = \frac{G_F^2 m_\mu^5}{192\pi^3}\left(1 + \frac{\alpha}{\pi}\left\{\frac{25 - 4\pi^2}{8}\right\}\right). \tag{12}$$

For case (ii) one obtains

$$\begin{aligned}
\Gamma = \; & \frac{G_F^2 m_\mu^5}{192\pi^3}\Bigg((1-y^4)(1-8y^2+y^4) - 12y^4\ln(y) + \frac{\alpha}{\pi}\Bigg\{ \\
& + \frac{1}{24}(1-y^4)(75-956y^2+75y^4) - 2y^4(36+y^4)\ln^2(y) \\
& - \frac{\pi^2}{2}(1-32y^3+16y^4-32y^5+y^8) \\
& - \frac{1}{3}(60+270y^2-4y^4+17y^6)y^2\ln(y) \\
& - \frac{1}{6}(1-y^4)(17-64y^2+17y^4)\ln(1-y^2) \\
& + 4(1-y)^4(1+4y+10y^2+4y^3+y^4)\ln(1-y)\ln(y) \\
& + 4(1+y)^4(1-4y+10y^2-4y^3+y^4)\ln(1+y)\ln(y) \\
& + 2(3+32y^3+48y^4+32y^5+3y^8)\mathrm{Li}_2(-y) \\
& + 2(3-32y^3+48y^4-32y^5+3y^8)\mathrm{Li}_2(y)\Bigg\}\Bigg).
\end{aligned} \tag{13}$$

The length of the *NLO* term in (13) illustrates but does not describe the added complication when introducing a new mass scale. The inclusion of full mass effects is not of much relevance for leptonic $\mu^- \to e^-$ decays (except in the threshold region) where $(m_e/m_\mu)^2 = 2.34 \cdot 10^{-5}$ but may be relevant for the leptonic $\tau \to \mu$ decays, where e.g. $(m_\mu/m_\tau)^2 = 3.54 \cdot 10^{-3}$. In fact, experimental data on leptonic τ decays do show a mass dependence of the partial rates on the daughter lepton's mass. For the decays $\tau^- \to \mu^- + \bar{\nu}_\mu + \nu_\tau$ and $\tau^- \to e^- + \bar{\nu}_e + \nu_\tau$ one finds branching ratios of $17.37 \pm 0.06\%$ and $17.84 \pm 0.06\%$ [34]. In agreement with experiment the *LO* result predicts a reduction of the partial rate by 2.82% for the $\tau^- \to \mu^-$ mode relative to the $\tau^- \to e^-$ mode. As concerns *NLO* effects, the experimental errors on the two partial rates are still too large to allow for checks on the mass dependence of the *NLO* result. Quite naturally, for semileptonic $b \to c$ decays, where $y = m_c/m_b \approx 0.3$, the mass dependence of the daughter quark must be kept.

Mass effects also populate density matrix elements which are zero when a given mass is set to zero. A prominent example is again $e^+e^- \to q\bar{q}$ which is purely transverse for mass zero quarks but acquires a longitudinal component when the quark becomes massive. In the same vein one has the Callan–Gross relation $F_2 = xF_1$ in deep inelastic scattering for mass zero quarks which is spoiled when the quarks become massive. Another example is the density matrix of the electron in leptonic muon decay. Its off–diagonal elements contribute to the transverse component of the electron's polarization (see (20)). The off–diagonal elements are proportional to the positive helicity amplitude $h_{1/2}$ which in turn is proportional to the mass of the electron ($h_{1/2} \sim m_e$). One therefore has $P_e^{\perp} \sim m_e$ [3]. A third example is the transition of the top quark into a right–handed W^+ (W_R) and a bottom quark. For $m_b = 0$ the left–chiral bottom quark becomes purely left–handed and, from angular momentum conservation, one therefore finds $\Gamma(W_R) = 0$ (see Fig. 3) (see e.g. [5]).

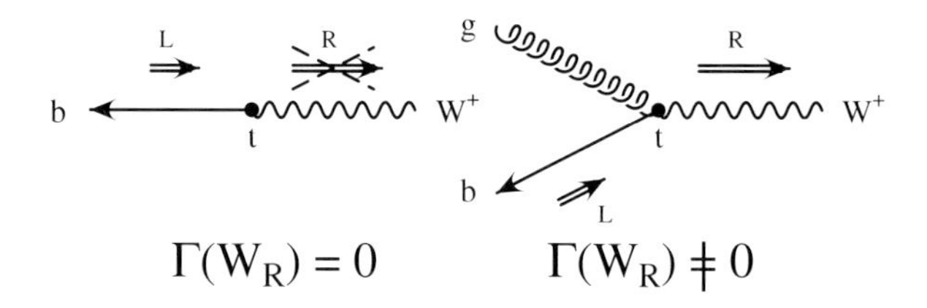

Fig. 3. The decay $t \to b + W_R^+$ at *LO* and *NLO*

2.4 Some Polarization Measures Are *NLO* Effects

Whenever a given polarization component is zero at the *LO* Born term level, and the vanishing of this polarization component is not due to general symmetry principles, it is very likely that this polarization component will be populated by gluon or photon emission at *NLO*. We shall list a few examples where this is the case.

A prominent example is the Callan–Gross relation which is violated by gluon emission at *NLO*. Similarly, a longitudinal contribution is generated in e^+e^-–annihilation into massless quarks by gluon emission. T–odd measures derive from imaginary parts of one–loop contributions, i.e. they start only at *NLO*. As mentioned before, a transverse–plus polarization of the W^+ in the decay $t \to b + W^+$ is generated by gluon emission as illustrated in Fig. 3. At *LO* the longitudinal polarization of massive quarks in $e^+e^- \to Q\bar{Q}$ is purely transverse, i.e. the polar distribution of this polarization component is proportional to $(1 + \cos^2\theta)$. The appearance of a longitudinal dependence proportional to $\sin^2\theta$ is a *NLO* effect [13]. Anomalous helicity flip contributions involving massless quarks or leptons occur only at *NLO* as discussed in Sect. 12.

We have already mentioned the T–odd measures which, in a CP invariant theory, obtain contributions only from the imaginary parts of *NLO* one–loop contributions. Apart from the examples listed in Sect. 2.2 we mention the case of the transverse polarization of a top quark produced in hadronic collisions [35–37].

2.5 Presentation of *NLO* Results of Polarization Observables

We have always chosen to present our analytical and numerical *NLO* results on polarization observables in the form

$$\langle P\rangle = \frac{N}{D} = \frac{N^{\mathrm{LO}}\big(1+\alpha_s\hat{N}^{\mathrm{NLO}}\big)}{D^{\mathrm{LO}}\big(1+\alpha_s\hat{D}^{\mathrm{NLO}}\big)}, \tag{14}$$

and not as

$$\langle P\rangle = \frac{N^{\mathrm{LO}}}{D^{\mathrm{LO}}}\big(1+\alpha_s(\hat{N}^{\mathrm{NLO}}-\hat{D}^{\mathrm{NLO}})\big). \tag{15}$$

An argument for our preference of the form (14) over (15) is the following. It is common usage to try and extend a perturbation series beyond a given known order by Padé improving the perturbation series. For example, for a second order perturbation series the Padé $P(1,1)$ improvement reads

$$1+\alpha_s a_1+\alpha_s^2 a_2 \approx \frac{1+\alpha_s\frac{a_1^2-a_2}{a_1}}{1-\alpha_s\frac{a_2}{a_1}}. \tag{16}$$

What has been done in the Padé $P(1,1)$ improvement is to fix the first order coefficients in the numerator and denominator of the r.h.s. of (16) by expanding the r.h.s. in powers of α_s and then equating coefficients in (16). Note that at the *NLO* level Padé's method does not yet allow one to reconstruct (14) from (15). Considering the successes of the Padé improvement program in other applications we nevertheless believe that the form (14) is a better approximation to the whole perturbation series than (15).

As an example where the use of the form (15) can lead to a misunderstanding is the *NLO* result of the mean forward–backward asymmetry $\langle A_{\mathrm{FB}}\rangle$ in e^+e^-–annihilation into a pair of massless quarks which, when using the form (15), is sometimes stated as

$$\langle A_{\mathrm{FB}}\rangle = \frac{\sigma_{\mathrm{FB}}^{\mathrm{LO}}}{\sigma^{\mathrm{LO}}}(1-\alpha_s/\pi). \tag{17}$$

This result is very suggestive of a de facto non–vanishing radiative correction to σ_{FB}. However, the true result is $\sigma_{\mathrm{FB}}^{\mathrm{NLO}}=0$ [38]. The non–vanishing radiative correction to $\langle A_{\mathrm{FB}}\rangle$ in (17) is just a reflection of the well–known *NLO* result for the rate $\sigma=\sigma^{\mathrm{LO}}(1+\alpha_s/\pi)$ in the denominator of (14). To our knowledge the vanishing of $\sigma_{\mathrm{FB}}^{\mathrm{NLO}}$ evades a simple explanation and must be termed to be a dynamical accident.

3 Fierz Transformation

Fierz crossing exchanges two Dirac indices in the contracted product of two strings of Γ–matrices according to (i not summed)

$$\big[\Gamma_i^{\{\mu\}}\big]_{\alpha\beta}\big[\Gamma_{i\{\mu\}}\big]_{\gamma\delta} \to \big[\Gamma_i^{\{\mu\}}\big]_{\alpha\delta}\big[\Gamma_{i\{\mu\}}\big]_{\gamma\beta}, \tag{18}$$

where the five currents in the set are conventionally labelled by $i = S, V, T, A, P$ with $\Gamma_i^{\{\mu\}} = \{1, \gamma^\mu, \sigma^{\mu\nu}, \gamma^\mu\gamma_5, \gamma_5\}$. The set of crossed configurations can then be expressed in terms of the set of uncrossed configurations. The five–by–five matrix relating the two sets is called the Fierz crossing matrix C_{Fierz}. It is clear that one gets back to the original configuration when crossing twice, i.e. $C_{Fierz}^2 = 1$. The eigenvalues λ of the Fierz crossing matrix are thus $\lambda = \pm 1$ as can easily be seen by going to the diagonal representation. In explicit form the Fierz crossing matrix is given by

$$C_{Fierz} = \frac{1}{4}\begin{pmatrix} 1 & 1 & 1/2 & -1 & 1 \\ 4 & -2 & 0 & -2 & -4 \\ 12 & 0 & -2 & 0 & 12 \\ -4 & -2 & 0 & -2 & 4 \\ 1 & -1 & 1/2 & 1 & 1 \end{pmatrix}, \tag{19}$$

where the rows and columns are labeled in the order $i = S, V, T, A, P$.

As (19) shows the trace of the Fierz crossing matrix is $Tr(C_{Fierz}) = -1$. This implies that the Fierz crossing matrix has three eigenvalues $\lambda = -1$ and two eigenvalues $\lambda = +1$. When discussing physics applications one also needs the corresponding Fierz crossing matrix for parity odd products of currents. This can easily be derived from multiplying the parity even case by γ_5.

Amazingly, the charged current interaction of the Standard Model $(V - A)_\mu(V - A)^\mu$ is an eigenvector of Fierz crossing, with eigenvalue $\lambda = -1$. When relating the Standard Model and charge retention forms of the Lagrangian as in (2) the minus sign from Fierz crossing is cancelled from having to commute the Fermion fields an odd number of times in order to relate the two forms.

In pre–QCD days it was considered unfortunate that the eigenvalue of the $(V - A)_\mu(V - A)^\mu$ current–current product is negative. If it were positive one would have a very natural explanation of the so called $\Delta I = 1/2$– or octet rule in weak nonleptonic decays. To see this consider the direct products $\mathbf{3}\otimes\mathbf{3} = \bar{\mathbf{3}}_\mathbf{a} \oplus \mathbf{6}_\mathbf{s}$ and $\bar{\mathbf{3}} \otimes \bar{\mathbf{3}} = \mathbf{3}_\mathbf{a} \oplus \bar{\mathbf{6}}_\mathbf{s}$. With the wrong eigenvalue $\lambda = +1$ (and the minus sign from Fermi statistics) one would then remain with $\bar{\mathbf{3}}_\mathbf{a} \oplus \mathbf{3}_\mathbf{a} = \mathbf{1} \oplus \mathbf{8}$ and one thus would have explained the famous octet rule for nonleptonic weak interactions. Nevertheless, with the advent of QCD and colour, it was realized that an important class of diagrams in nonleptonic baryon transitions between ground state baryons was subject to the octet rule due to the simple Fierz property of the nonleptonic current–current product. This discovery is sometimes referred to as the Pati–Woo theorem [39] even though the discovery of Pati and Woo was predated by the paper [40].

4 Counting Spin Observables

When setting up a problem involving the spin of particles it is always instructive to first denumerate the complexity of the problem and count the number of independent structures of the problem. A particularly efficient and physical way to do so is to count the number of independent elements of the spin density matrices involved in the process.

The single spin density matrix $\rho_{\lambda\lambda'}$ is a $(2J+1)(2J+1)$ hermitian matrix ($\rho_{\lambda\lambda'} = \rho^*_{\lambda'\lambda}$). It thus has $(2J+1)^2$ independent components of which $(J+1)(2J+1)$ are real and $J(2J+1)$ are imaginary. The rate is represented by the trace of the density matrix. In Table 2 we list the corresponding degrees of freedom for the first few spin cases $J = 1/2,\ 1,\ 3/2$. The density matrix is also frequently represented in terms of its angular momentum content. For the three cases listed in Table 2 the angular momentum content is given by $1/2 \otimes 1/2 = 0 \oplus 1$, $1 \otimes 1 = 0 \oplus 1 \oplus 2$ and $3/2 \otimes 3/2 = 0 \oplus 1 \oplus 2 \oplus 3$ where the one–dimensional representation "0" is equivalent to the trace or rate. It is clear that one obtains the correct respective number of spin degrees of freedom $(2J+1)^2$ in Table 2 when adding up the dimensions of the angular momentum spaces appearing in the decomposition $J \otimes J = 0 \oplus \ldots \oplus 2J$.

Table 2. Independent components of the single spin density matrix

spin J	rate trace	real $J(2J+3)$	imaginary $J(2J+1)$	sum $(2J+1)^2$
$J=1/2$	1	2	1	4
$J=1$	1	5	3	9
$J=3/2$	1	9	6	16

In the spin $1/2$ case the four spin degrees of freedom are the rate, the two real components and the imaginary component of the spin $1/2$ density matrix. The unnormalized density matrix ρ is usually written as

$$\begin{aligned} \rho_{\lambda\lambda'} &= \begin{pmatrix} h_{+1/2}h^*_{1/2} & h_{+1/2}h^*_{-1/2} \\ h_{-1/2}h^*_{1/2} & h_{-1/2}h^*_{-1/2} \end{pmatrix} \\ &= \frac{1}{2}(\sigma \cdot \mathbb{1} + \boldsymbol{\xi} \cdot \boldsymbol{\sigma}). \end{aligned} \tag{20}$$

In the helicity basis the four spin degrees of freedom are labelled by

$$\begin{aligned} \sigma \quad &: \ \text{rate}\ , \\ \xi_x/\sigma &: \ \text{"perpendicular" polarization } P^{\perp}, \\ \xi_y/\sigma &: \ \text{"normal" polarization } P^{N}, \\ \xi_z/\sigma &: \ \text{"longitudinal" polarization } P^{l}. \end{aligned} \tag{21}$$

The y–component P^N can be seen to be a T–odd measure since one needs a vector product to construct the normal to a given plane. For example, in μ–decay P^N is determined by the expectation value $\langle \boldsymbol{\sigma}_e \cdot (\boldsymbol{\sigma}_\mu \times \boldsymbol{p}_e)\rangle$ which is a

T–odd measure. In the context of this review P^N obtains contributions only from the imaginary parts of one–loop diagrams. For all processes listed in Table 1 in Sect. 1 the one–loop diagrams are real and thus $P^N = 0$ in these processes. This is different in $e^+e^- \to q\,\bar{q}\,g$ where the one–loop diagrams possess imaginary parts and where one therefore has $P^N \neq 0$ (see e.g. [14,41]).

Returning to Table 2 we shall see that not all $(2J+1)^2$ spin degrees of freedom are accessible (angular momentum conservation) or measurable (parity) in general.

In the following we will be concerned with double spin density matrices. Their spin degrees of freedom are given by the products of the single spin degrees of freedom. For the cases discussed in this review the double density matrix is a sparsely populated matrix, due to angular momentum conservation and the absence of imaginary parts from loop contributions. Take, for example, the double spin density matrix of the decay $t(\uparrow) \to X_b + W^+(\uparrow)$. On naive counting one would expect $N = (1(\text{trace}) + 20(\text{real}) + 15(\text{imaginary}))$ spin degrees of freedom. This number is considerably reduced by considering the following two facts.

1. The one–loop amplitude does not possess an imaginary (absorptive) part. This can be seen by taking a look at the one–loop vertex corrections for the transitions discussed in Table 1. The vertex correction does not admit of real on–shell intermediate states, i.e. it does not have an absorptive part. The 15 imaginary components of the double density matrix are zero.
2. Angular momentum conservation. In the rest frame of the top the decay into X_b and W^+ is back–to–back and thus anti–collinear. Since one is summing over the helicities of the X_b one has $\lambda_{X_b} = \lambda'_{X_b}$ and thus $\lambda_t - \lambda_W = \lambda'_t - \lambda'_W$ (see Fig. 3). Taking this constraint into consideration one remains with six diagonal and two non–diagonal double spin density matrix elements (see Table 3) where one refers to diagonal and non–diagonal elements when $(\lambda_t, \lambda_W) = (\lambda'_t, \lambda'_W)$ and $(\lambda_t, \lambda_W) \neq (\lambda'_t, \lambda'_W)$, respectively. One has non–diagonal transitions for the two cases $\lambda_t - \lambda_W = \pm 1/2$. Concerning the non–diagonal density matrix elements it is easy to see from the master formula (26) to be discussed in Sect. 5 that they generate azimuthal dependences in angular decay distributions.

We shall now go through the exercise and count the spin degrees of freedom by a different method, namely by counting the number of covariants of the process. We emphasize that counting spin degrees of freedom via density matrix elements is generally safer than counting the number of independent covariants because there exist nontrivial identities between covariants in $D = 4$ dimensions. The decay $t(\uparrow) \to X_b + W^+(\uparrow)$ will illustrate this point.

Consider the expansion of the spin dependent hadron tensor of the process into covariants constructed from the metric tensor, the Levi–Civita tensor, the independent momenta of the process and the spin four–vector of the top. One has

$$H^{\mu\nu} = \big(- g^{\mu\nu}\, H_1 + p_t^\mu p_t^\nu\, H_2 - i\epsilon^{\mu\nu\rho\sigma} p_{t,\rho} q_\sigma\, H_3 \big)$$

$$- \ (q\cdot s_t)\big(-g^{\mu\nu}\,G_1 + p_t^\mu p_t^\nu\,G_2 - i\epsilon^{\mu\nu\rho\sigma}p_{t,\rho}q_\sigma\,G_3\big)$$
$$+ \ \big(s_t^\mu p_t^\nu + s_t^\nu p_t^\mu\big)\,G_6 + i\epsilon^{\mu\nu\rho\sigma}p_{t\rho}s_{t\sigma}\,G_8 + i\epsilon^{\mu\nu\rho\sigma}q_\rho s_{t\sigma}\,G_9. \tag{22}$$

There are nine covariants and associated with it nine invariants which obviously does not agree with the number eight counted before using helicity counting. The discrepancy arises because there exist a nontrivial four–dimensional identity between three of the nine covariants which reads

$$q\cdot s_t\,\epsilon^{\mu\nu\rho\sigma}p_{t,\rho}q_\sigma - q^2\epsilon^{\mu\nu\rho\sigma}p_{t,\rho}s_{t\sigma} + q\cdot p_t\,\epsilon^{\mu\nu\rho\sigma}q_\rho s_{t,\sigma} = 0. \tag{23}$$

This identity can be derived using the Schouten identity which reads

Table 3. Non–vanishing double density matrix elements in $t(\uparrow) \longrightarrow X_b + W^+(\uparrow)$

λ_t	λ_W	$\lambda_t - \lambda_W$
1/2	1	−1/2
1/2	0	1/2
1/2	−1	3/2
−1/2	1	−3/2
−1/2	0	−1/2
−1/2	−1	1/2

$$T_{\mu[\mu_1\mu_2\mu_3\mu_4\mu_5]} = 0, \tag{24}$$

where the symbol "[...]" denotes antisymmetrization and where

$$T_{\mu[\mu_1\mu_2\mu_3\mu_4\mu_5]} = g_{\mu\mu_1}\epsilon_{\mu_2\mu_3\mu_4\mu_5} + \text{cycl.}(\mu_1,\mu_2,\mu_3,\mu_4,\mu_5). \tag{25}$$

The Schouten identity is just the statement that it is impossible to place four index values in an antisymmetric fifth rank tensor, or in the language of Young Tableaux, that a Young Tableau with five vertical boxes is identically zero in four dimensions.

This illustration is not only of academic interest but has numerical implications since one can get into a terrible mess numerically if one works with a redundant set of covariants and tries to do matrix inversions involving the over-counted set of degrees of freedom. Although the Schouten identity seems rather obvious nowadays there have been examples in the literature where Schouten–type of identities have been overlooked.

We conclude this section by enumerating the number of spin degrees of freedom in the various decay processes discussed in this review. The processes are listed in Table 4 together with the number of spin degrees of freedom. The corresponding references to the papers in which these decays were treated can be found in Table 1. In this list we have taken into account that the decay $D^* \to \pi(\gamma)$ is parity conserving. In parantheses we list the number of T–odd observables for each process which are zero due to the absence of imaginary contributions in the one–loop vertex correction.

Table 4. Number of measurable double density matrix elements for the processes listed in Table 1

process	spin degrees of freedom
$t(\uparrow) \longrightarrow b + W^+(\uparrow)$	$N = 1 + 7 \ \ (+2)$
$t(\uparrow) \longrightarrow b + H^+$	$N = 1 + 1$
$B \longrightarrow X_c + D_s$	$N = 1$
$B \longrightarrow X_c + D_s^*(\uparrow)$	$N = 1 + 1$
$\Lambda_b(\uparrow) \longrightarrow X_c + D_s$	$N = 1 + 1$
$\Lambda_b(\uparrow) \longrightarrow X_c + D_s^*(\uparrow)$	$N = 1 + 4 \ \ (+1)$
$l(\uparrow) \longrightarrow l'(\uparrow) + \nu_l + \nu_{l'}$	$N = 1 + 4 \ \ (+1)$

5 Angular Decay Distributions

It should be clear that one needs to do polarization measurements in order to disentangle the full structure of particle interactions. Polarization measurements are particularly simple when the particle whose polarization one wants to measure decays. The angular decay distribution of the decay products reveals information on the state of polarization of the decaying particle. The information contained in the angular decay distribution is maximal when the particle decay is weak. The fact that the angular decay distribution reveals information on the polarization of the decaying particle is sometimes referred to as that the particle decay is self–analyzing.

There are principally two ways to obtain angular decay distributions which we will refer to as the non–covariant and the covariant methods. In the non–covariant method one makes use of rotation matrices whereas in the covariant method one evaluates scalar products of four–vectors involving momenta and spin four–vectors in given reference frames. We shall discuss one example each of the two methods.

As an example for the non–covariant method we write down the angular decay distribution of polarized top decay $t(\uparrow) \to b + W^+$ followed by $W^+ \to l^+ + \nu_l$.

The angular decay distribution can be obtained from the master formula

$$W(\theta_P,\theta,\phi) \propto \sum_{\lambda_W-\lambda'_W=\lambda_t-\lambda'_t} e^{i(\lambda_W-\lambda'_W)\phi}\, d^1_{\lambda_W 1}(\theta)\, d^1_{\lambda'_W 1}(\theta)\, H^{\lambda_t\,\lambda'_t}_{\lambda_W\lambda'_W}\, \rho_{\lambda_t\,\lambda'_t}(\theta_P), \tag{26}$$

where $\rho_{\lambda_t\,\lambda'_t}(\theta_P)$ is the density matrix of the top quark which reads

$$\rho_{\lambda_t\,\lambda'_t}(\theta_P) = \frac{1}{2}\begin{pmatrix} 1+P\cos\theta_P & P\sin\theta_P \\ P\sin\theta_P & 1-P\cos\theta_P \end{pmatrix}. \tag{27}$$

P is the magnitude of the polarization of the top quark. The $H^{\lambda_t\,\lambda'_t}_{\lambda_W\lambda'_W}$ are helicity matrix elements of the hadronic structure function $H_{\mu\nu}$. The sum in (26) extends over all values of $\lambda_W, \lambda'_W, \lambda_t$ and λ'_t compatible with the constraint $\lambda_W-\lambda'_W = \lambda_t-\lambda'_t$. The second lower index in the small Wigner $d(\theta)$–function $d^1_{\lambda_W 1}$ is fixed at $m=1$ for zero mass leptons because the total m–quantum number of the lepton pair along the l^+ direction is $m=1$. Because there exist different conventions for Wigner's d–functions we explicate the requisite components that enter (26): $d^1_{11}=(1+\cos\theta)/2$, $d^1_{01}=\sin\theta/\sqrt{2}$ and $d^1_{-11}=(1-\cos\theta)/2$. The fact that one has to specify the phase convention of the Wigner's d–functions points to one of the weaknesses of the non–covariant method to obtain the correct angular decay distributions: one has to use one set of consistent phase conventions to obtain the correct signs for the angular factors. For someone not not so familiar with the angular momentum apparatus this is not always simple. For the polar angle dependencies the correctness of a sign can always be checked by using physics arguments. This is more difficult for the azimuthal signs.

Including the appropriate normalization factor the four–fold decay distribution is given by [4,6]

$$\begin{aligned}\frac{d\Gamma}{dq_0 d\cos\theta_P d\cos\theta d\phi} &= \frac{1}{4\pi}\frac{G_F|V_{tb}|^2 m_W^2}{\sqrt{2}\,\pi}|\boldsymbol{q}|\Big\{ \\ &+\frac{3}{8}(H_U+P\cos\theta_p H_{U^P})(1+\cos^2\theta) \\ &+\frac{3}{4}(H_L+P\cos\theta_p H_{L^P})\sin^2\theta+\frac{3}{4}(H_F+P\cos\theta_p H_{F^P})\cos\theta \\ &+\frac{3}{2\sqrt{2}}P\sin\theta_p H_{I^P}\sin 2\theta\cos\phi+\frac{3}{\sqrt{2}}P\sin\theta_p H_{A^P}\sin\theta\cos\phi\Big\},\end{aligned} \tag{28}$$

where the helicity structure functions H_U, H_L etc. are linear combinations of the helicity matrix elements $H^{\lambda_t\,\lambda'_t}_{\lambda_W\lambda'_W}$ We have taken the freedom to normalize the differential rate such that one obtains the total $t\to b+W^+$ rate upon integration *and not* the total rate multiplied by the branching ratio of the respective W^+ decay channel.

The polar angles θ_P and θ, and the azimuthal angle ϕ that arise in the full cascade–type description of the two–stage decay process $t(\uparrow)\to W^+(\to$

$l^+ + \nu_l) + X_b$ are defined in Fig. 4. For better visibility we have oriented the lepton plane with a negative azimuthal angle relative to the hadron plane. For the hadronic decays of the W into a pair of light quarks one has to replace (l^+, ν_l) by $(\bar{q}, q)$ in Fig. 4. We mention that we have checked the signs of the angular decay distribution (28) using covariant techniques.

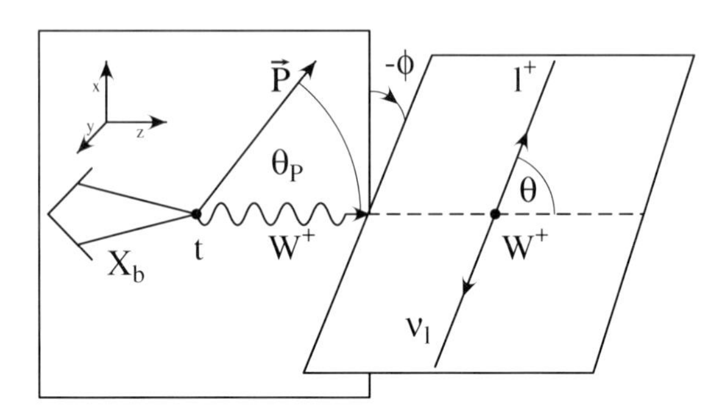

Fig. 4. Definition of the angles θ_P, θ and ϕ in the cascade decay $t(\uparrow) \to X_b + W^+$ and $W^+ \to l^+ + \nu_l$

At first sight it seems rather strange that the angular analysis is done in two different coordinate systems, namely the top quark rest system and the W rest system. This runs counter to common wisdom that invariants should always be evaluated in one reference system. Some insight into the problem may be gained by using the orthonormality and completeness relation of polarization vectors to rewrite the contraction of the hadron and lepton tensors in a form which exhibits the correctness of using two different reference systems for the cascade decay. The orthonormality and completeness relation read

$$\text{Orthonormality:} \qquad g_{\mu\nu}\epsilon^{*\mu}(m)\epsilon^{\nu}(m') = g_{mm'} \quad m, m' = S, \pm, 0, \tag{29}$$

$$\text{Completeness:} \qquad \sum_{m,m'=S,\pm,0} \epsilon^{\mu}(m)\epsilon^{*\nu}(m')g_{mm'} = g^{\mu\nu} \quad \mu, \nu = 0, 1, 2, 3, \tag{30}$$

where $g_{mm'} = \text{diag}(+,-,-,-)$, $(m, m' = S, \pm, 0)$ and $g^{\mu\nu} = \text{diag}(+,-,-,-)$, $(\mu, \nu = 0, 1, 2, 3)$. The scalar or time component of the polarization four–vector is denoted by S. On using the completeness relation one then has

$$\begin{aligned} L^{\mu\nu}H_{\mu\nu} &= L_{\mu'\nu'}g^{\mu'\mu}g^{\nu'\nu}H_{\mu\nu} = L_{\mu'\nu'}\epsilon^{\mu'}(m)\epsilon^{*\mu}(m')g_{mm'}\epsilon^{*\nu'}(n)\epsilon^{\nu}(n')g_{nn'}H_{\mu\nu} \\ &= \left(L_{\mu'\nu'}\epsilon^{\mu'}(m)\epsilon^{*\nu'}(n)\right)\left(H_{\mu\nu}\epsilon^{*\mu}(m')\epsilon^{\nu}(n')\right) g_{mm'}g_{nn'} . \end{aligned} \tag{31}$$

The point is that the two Lorentz contractions appearing on the second line of (31) can be evaluated in two different Lorentz frames. The leptonic invariant $L_{\mu'\nu'}\epsilon^{\mu'}(m)\epsilon^{*\nu'}(n)$ can be evaluated in the $(l\nu)$ CM frame (or in the W^+ rest frame) while the hadronic invariant $H_{\mu\nu}\epsilon^{*\mu}(m')\epsilon^{\nu}(n')$ can be evaluated in the rest frame of the top quark. Another advantage of this method is that one can easily incorporate lepton mass effects as is mandatory for $b \to c$ or the rare $b \to s$ transitions if the charged lepton or leptons in the final state are τ leptons. This technique was used to derive angular decay distributions including leptonic polarization effects for semileptonic and rare bottom meson decays in [42] and [43]. Note also that the correct phase choice for the polarization vectors is no longer crucial since the polarization vectors always appear as squares in the orthonormality and completeness relations.

As an example of the covariant method we discuss the angular decay distribution of polarized μ decay into a polarized e. From helicity counting as described in Sect. 4 one knows that there are altogether five spin–dependent structure functions and one spin–independent structure function describing the leptonic decay of a polarized muon into a polarized electron. We thus define a spin–dependent differential rate in terms of six invariant structure functions A_i. Accordingly one has [3]

$$\frac{d\Gamma}{dx\, d\cos\theta_P} = A_1 + \frac{1}{m_\mu} A_2 (p_e \cdot s_\mu) + \frac{1}{m_\mu} A_3 (p_\mu \cdot s_e) + \frac{1}{m_\mu^2} A_4 (p_e \cdot s_\mu)\,(p_\mu \cdot s_e)$$
$$+ A_5 (s_\mu \cdot s_e) + \frac{1}{m_\mu^2} A_6\, \epsilon_{\alpha\beta\gamma\delta}\, p_\mu^\alpha\, p_e^\beta\, s_\mu^\gamma\, s_e^\delta. \tag{32}$$

Equation (32) will be evaluated in the rest system of the muon where $p_\mu = (m_\mu; 0,0,0)$ and $p_e = (E_e; 0, 0, |\boldsymbol{p}_e|) = (m_\mu/2)(x; 0, 0, x\beta)$. The velocity of the electron is denoted by $\beta = \sqrt{1 - 4y^2/x^2}$ where $y = m_e/m_\mu$ and $x = 2E_e/m_\mu$ denotes the scaled energy of the electron. In the rest frame of the μ^- the polarization four–vectors of the μ^- and e^- are given by

$$s_\mu^\alpha = (0; \boldsymbol{\zeta}_\mu), \tag{33}$$

$$s_e^\alpha = \Big(\frac{\boldsymbol{n}_e \cdot \boldsymbol{p}_e}{m_e}; \boldsymbol{n}_e + \frac{\boldsymbol{n}_e \cdot \boldsymbol{p}_e}{m_e (E_e + m_e)} \boldsymbol{p}_e\Big), \tag{34}$$

where the polarization three–vector $\boldsymbol{\zeta}_\mu$ of the μ^- and the quantization axis $\boldsymbol{n}_e$ of the spin of the e^- in their respective rest frames read (see Fig. 5)

$$\boldsymbol{\zeta}_\mu = (\sin\theta_P, 0, \cos\theta_P) \tag{35}$$

and

$$\boldsymbol{n}_e = (\sin\theta\cos\chi, \sin\theta\sin\chi, \cos\theta). \tag{36}$$

Equation (35) holds for 100% polarized muons. For partially polarized muons with magnitude of polarization P the representation (35) has to be multiplied by P such that $\boldsymbol{P}_\mu = P\boldsymbol{\zeta}_\mu$.

The scalar products in (32) can then be evaluated with the result

$$p_e \cdot s_\mu = -\frac{m_\mu}{2} x\beta P \cos\theta_P,$$
$$p_\mu \cdot s_e = \frac{m_\mu}{2y} x\beta \cos\theta,$$

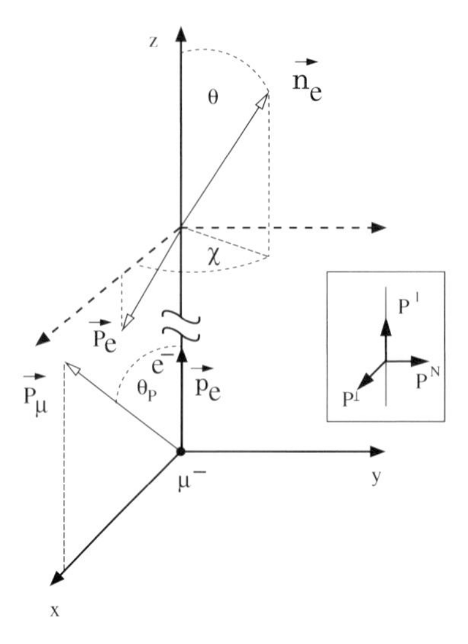

Fig. 5. Definition of the angles θ_P, θ and χ in polarized muon decay

$$\begin{aligned} s_\mu \cdot s_e &= -P \sin\theta_P \sin\theta \cos\chi, \\ \epsilon_{\alpha\beta\gamma\delta}\, p_\mu^\alpha\, p_e^\beta\, s_\mu^\gamma\, s_e^\delta &= \frac{m_\mu^2}{2} x\beta P \sin\theta_P \sin\theta \sin\chi. \end{aligned} \tag{37}$$

Including the correct normalization one finally arrives at the angular decay distribution [3]

$$\begin{aligned} \frac{d\Gamma}{dx\, d\cos\theta_P} &= \beta x\, \Gamma_0 (G_1 + G_2 P \cos\theta_P + G_3 \cos\theta + G_4 P \cos\theta_P \cos\theta \\ &\quad + G_5 P \sin\theta_P \sin\theta \cos\chi + G_6 P \sin\theta_P \sin\theta \sin\chi), \end{aligned} \tag{38}$$

where the relation between the invariant structure functions A_i and the frame–dependent spectrum functions G_i can be found in [3]. G_1 is the unpolarized spectrum function, G_2 and G_3 are single spin polarized spectrum functions referring to the spins of the μ^- and e^-, resp., and G_4, G_5 and G_6 describe spin–spin correlations between the spin vectors of the muon and electron. G_6 represents a so–called T–odd observable. This is evident when rewriting the angular factor multiplying G_6 in (38) in triple–product form, i.e. $\sin\theta_P \sin\theta \sin\chi = |\boldsymbol{p}_e|^{-1} \boldsymbol{p}_e \cdot (\boldsymbol{\zeta}_\mu \times \boldsymbol{n}_e)$. As stated before G_6 is identically zero in the Standard Model since, on the one hand, the weak coupling constant G_F is real and, on the other hand, the loop contributions do not generate imaginary parts. In the terminology of Sect. 4 $G_1 - G_4$ are diagonal structure functions, while the azimuthally dependent structure functions G_5 and G_6 are non–diagonal. As mentioned before, the structure function G_5 is proportional to the electron mass and therefore vanishes for $m_e \to 0$. Naturally, the angular decay distribution (38) can also be derived in the helicity formalism.

In the two examples discussed in this section we chose to orient the z–axis along the three–momentum of one of the particles. In this case one says that the angular analysis is done in the helicity system. Of course, other choices of the orientation of the z–axis are possible. A popular choice is the transversity system where the z–axis is normal to a given plane spanned by the momenta of the decay products. A more detailed discussion of the choice of frames can be found in [42] where the angular decay distribution in the semileptonic decays $B \to (D, D^*) + l + \nu$ including lepton mass effects was investigated. We mention that the transversity system is the preferred choice when discussing the *CP* properties of the final state [44].

6 One-Loop Amplitude

We present our results in terms of the three vector current amplitudes F_i^V $(i = 1, 2, 3)$ and the three axial vector current amplitudes F_i^A $(i = 1, 2, 3)$ defined by $(\, J_\mu^V = \bar{q}_b \gamma_\mu q_t,\ J_\mu^A = \bar{q}_b \gamma_\mu \gamma_5 q_t)$

$$\langle b(p_b)|J_\mu^V|t(p_t)\rangle = \bar{u}_b(p_b)\Big\{\gamma_\mu F_1^V + p_{t,\mu} F_2^V + p_{b,\mu} F_3^V\Big\} u_t(p_t), \tag{39a}$$

$$\langle b(p_b)|J_\mu^A|t(p_t)\rangle = \bar{u}_b(p_b)\Big\{\gamma_\mu F_1^A + p_{t,\mu} F_2^A + p_{b,\mu} F_3^A\Big\} \gamma_5 u_t(p_t). \tag{39b}$$

In this section we choose to label the current transition according to the transition $t \to b$ as in [6]. As emphasized before, up to the colour factor identical expressions are obtained for the $\mu \to e$ transition if the leptonic four–fermion interaction is written in the charge retention form. For the form factors one obtains [6]

$$\begin{aligned} F_1^V &= 1 + \frac{\alpha_s}{4\pi} C_F \Bigg\{ -\frac{m_t^2 + m_b^2 - q^2}{m_t^2 \sqrt{\lambda}} \Bigg[2\,\mathrm{Li}_2(1 - w_1^2) - 2\,\mathrm{Li}_2\left(1 - \frac{w_1}{w_\mu}\right) \\ &+ \frac{1}{2} \ln\left(\frac{\Lambda^4}{m_b^2 m_t^2}\right) \ln(w_1 w_\mu) + \ln\left(\frac{w_1^3}{w_\mu}\right) \ln\left(\frac{w_\mu (1 - w_1^2)}{w_\mu - w_1}\right)\Bigg] - \ln\left(\frac{\Lambda^4}{m_b^2 m_t^2}\right) \\ &- \frac{m_t^2 - m_b^2}{2q^2} \ln\left(\frac{m_b^2}{m_t^2}\right) - 4 + \ln(w_1 w_\mu)\left(\frac{m_t^2\sqrt{\lambda}}{2q^2} - \frac{(m_t + m_b)^2 - q^2}{m_t^2 \sqrt{\lambda}}\right)\Bigg\}, \end{aligned} \tag{40a}$$

$$\begin{aligned} F_2^V &= \frac{\alpha_s}{4\pi} C_F \frac{m_t - m_b}{q^2} \Bigg\{ 2 - \left(\frac{m_t + 2m_b}{m_t - m_b} - \frac{m_t^2 - m_b^2}{q^2}\right) \ln\left(\frac{m_b^2}{m_t^2}\right) \\ &- \left(\frac{m_t^2\sqrt{\lambda}}{q^2} - \frac{m_b}{m_t - m_b} \frac{q^2 + (m_t - m_b)(3m_t + m_b)}{m_t^2 \sqrt{\lambda}}\right) \ln(w_1 w_\mu)\Bigg\}, \end{aligned} \tag{40b}$$

$$F_3^V = F_3^V(m_t, m_b) = F_2^V(m_b, m_t), \tag{40c}$$

where we have denoted the scaled (small) gluon mass by $\Lambda = m_g/m_t$. We define $\lambda = 1 + x^4 + y^4 - 2x^2y^2 - 2x^2 - 2y^2$ ($x = q^2/m_t^2$, $y = m_b/m_t$) and use the abbreviations

$$w_1 = \frac{x}{y} \cdot \frac{1 - x^2 + y^2 - \sqrt{\lambda}}{1 + x^2 - y^2 + \sqrt{\lambda}}, \qquad w_\mu = \frac{x}{y} \cdot \frac{1 - x^2 + y^2 - \sqrt{\lambda}}{1 + x^2 - y^2 - \sqrt{\lambda}}. \tag{41}$$

The axial vector amplitudes F_i^A can be obtained from the vector amplitudes by the replacement $m_t \to -m_t$, i.e. one has $F_i^A(m_t) = F_i^V(-m_t)$ $(i = 1, 2, 3)$. Our one–loop amplitudes are linearly related to the one–loop amplitudes given in [45] after correcting for a typo in [45] (see also [46]).

Note that the infrared singularities proportional to $\ln \Lambda$ and the would–be mass singularities (also called collinear singularities) proportional to $\ln m_b$ all reside in the Born term form factors F_1^V and F_1^A. They are eventually cancelled by the corresponding singularities in the tree graph contribution. A look at the arguments of the log and dilog functions in 40a shows that the one–loop contribution is purely real as remarked on earlier.

In the case of the electroweak radiative corrections to $t \to b + W^+$ there are altogether 18 different vertex correction diagrams in the Feynman–'t Hooft gauge [7] as compared to the one vertex correction diagram in the QCD and the μ–decay cases discussed in this section. In addition to the massive one–loop three–point functions one has to calculate the many massive one–loop two–point functions needed in the renormalization program. We have recalculated all one–loop contributions analytically and have checked them analytically and numerically with the help of a XLOOPS/GiNaC package that automatically calculates one–loop two–point and three–point functions [47]. Our one–loop results agree with the results of [48]. The results are too lengthy to be reproduced here in analytical form. The full analytical results will be given in a forthcoming publication [49].

7 Tree-Graph Contribution

As emphasized earlier on, the *NLO* tree graph contributions for the QED and QCD cases are identical up to *NLO* except for a trivial colour factor in the case of QCD. Differences set in only at *NNLO* where three–gluon coupling contributions come in. We choose to present and discuss the *NLO* tree graph contributions in the QED radiative corrections to leptonic μ–decay as written down in [3].

We begin with the Born term contribution. This is an exercise that everyone has probably gone through before in the unpolarized case. The point that not everyone is familiar with is that it is very simple to include the spin of fermions at the Born term level. All that is needed is the substitution $p \to \bar{p}$ for the fermion's momenta where the notation $\bar{p}$ is explained in the following. The $p \to \bar{p}$ rule is best explained by looking at the trace expression of the charge–side tensor which reads

$$C^{\alpha\beta}_{Born} = \frac{1}{4}\mathrm{Tr}\Big\{(\not{p}_e + m_e)(\mathbb{1} + \gamma_5 \not{s}_e)\gamma^\alpha(\mathbb{1} - \gamma_5)(\not{p}_\mu + m_\mu)(\mathbb{1} + \gamma_5 \not{s}_\mu)\gamma^\beta(\mathbb{1} - \gamma_5)\Big\}. \tag{42}$$

The dependence on the polarization four–vectors of the μ^- and e^- has been retained in (42).

Since only even–numbered γ–matrix strings survive between the two $(\mathbb{1}-\gamma_5)$–factors in (42) one can compactly write the result of the trace evaluation as

$$C^{\alpha\beta}_{Born} = 2(\bar{p}^{\beta}_{\mu}\bar{p}^{\alpha}_{e} + \bar{p}^{\alpha}_{\mu}\bar{p}^{\beta}_{e} - g^{\alpha\beta}\,\bar{p}_{\mu}\!\cdot\!\bar{p}_{e} + i\epsilon^{\alpha\beta\gamma\delta}\bar{p}_{e,\gamma}\bar{p}_{\mu,\delta}), \tag{43}$$

where

$$\bar{p}^{\alpha}_{\mu} = p^{\alpha}_{\mu} - m_{\mu}s^{\alpha}_{\mu}, \tag{44a}$$

$$\bar{p}^{\alpha}_{e} = p^{\alpha}_{e} - m_{e}s^{\alpha}_{e}, \tag{44b}$$

and where s^{α}_{μ} and s^{α}_{e} are the polarization four–vectors of the μ^- and e^-.

In the QED case one has the simplifying feature that the contribution of the last term antisymmetric in $(\alpha\beta)$ can be dropped. The reason is that the dependence on the momentum directions of the $\bar{\nu}_e$– and ν_μ–neutrinos is completely integrated out in the differential rate. Thus the neutrino–side of the interaction can only depend on the spatial piece of the second rank tensor build from the momentum transfer to the neutrinos (the neutrinos are treated as massless) which is symmetric in $(\alpha\beta)$. Upon contraction with the charge–side tensor the antisymmetric piece drops out.

Unfortunately the $p \to \bar{p}$ trick no longer works at *NLO* order. But still, by using the $\bar{p}$–notation, the result can be presented in a very compact form. The result for the *NLO* $\mu \to e$ tree graph contribution reads

$$\begin{aligned}
C^{\alpha\beta} &= \sum_{\gamma-\text{spin}} \mathcal{M}^{\alpha}\mathcal{M}^{\beta\dagger} = \frac{e^2}{2}\Bigg\{\left(\frac{k\!\cdot\!\bar{p}_e - m_e^2}{k\!\cdot\!p_e} + \frac{p_\mu\!\cdot\!\bar{p}_e}{k\!\cdot\!p_\mu}\right)\frac{k^\alpha\bar{p}^\beta_\mu + k^\beta\bar{p}^\alpha_\mu - k\!\cdot\!\bar{p}_\mu g^{\alpha\beta}}{k\!\cdot\!p_e} \\
&+ \left(\frac{k\!\cdot\!\bar{p}_\mu + m_\mu^2}{k\!\cdot\!p_\mu} - \frac{p_e\!\cdot\!\bar{p}_\mu}{k\!\cdot\!p_e}\right)\frac{k^\alpha\bar{p}^\beta_e + k^\beta\bar{p}^\alpha_e - k\!\cdot\!\bar{p}_e g^{\alpha\beta}}{k\!\cdot\!p_\mu} \\
&+ (k\!\cdot\!\bar{p}_\mu)\frac{p^\alpha_e\bar{p}^\beta_e + p^\beta_e\bar{p}^\alpha_e - m_e^2 g^{\alpha\beta}}{(k\!\cdot\!p_e)(k\!\cdot\!p_\mu)} - (k\!\cdot\!\bar{p}_e)\frac{p^\alpha_\mu\bar{p}^\beta_\mu + p^\beta_\mu\bar{p}^\alpha_\mu - m_\mu^2 g^{\alpha\beta}}{(k\!\cdot\!p_e)(k\!\cdot\!p_\mu)} \\
&+ (k\!\cdot\!\bar{p}_e)\frac{p^\alpha_e\bar{p}^\beta_\mu + p^\beta_e\bar{p}^\alpha_\mu - p_e\!\cdot\!\bar{p}_\mu g^{\alpha\beta}}{(k\!\cdot\!p_e)^2} - (k\!\cdot\!\bar{p}_\mu)\frac{p^\alpha_\mu\bar{p}^\beta_e + p^\beta_\mu\bar{p}^\alpha_e - p_\mu\!\cdot\!\bar{p}_e g^{\alpha\beta}}{(k\!\cdot\!p_\mu)^2}\Bigg\} \\
&- \frac{e^2}{2}\left(\frac{m_\mu^2}{(k\!\cdot\!p_\mu)^2} + \frac{m_e^2}{(k\!\cdot\!p_e)^2} - \frac{2p_e\!\cdot\!p_\mu}{(k\!\cdot\!p_e)(k\!\cdot\!p_\mu)}\right)(\bar{p}^\alpha_e\bar{p}^\beta_\mu + \bar{p}^\beta_e\bar{p}^\alpha_\mu - \bar{p}_e\!\cdot\!\bar{p}_\mu g^{\alpha\beta}).
\end{aligned} \tag{45}$$

The momentum of the radiated photon is denoted by k. For the aforementioned reason we have dropped the contribution of antisymmetric terms. When the antisymmetric ϵ–tensor pieces are kept, and when one replaces $e \to g_s$ and $1 \to N_c C_F = 4$ to account for colour, one recovers the *NLO* QCD corrected hadronic tensor for the $t \to b$ transition listed in [6].

In the last line of (45) we have isolated the infrared singular piece of the charge–side tensor which is given by the usual soft photon factor multiplying the Born term contribution. Technically this is done by writing

$$C^{(\alpha)\alpha\beta} = \Big(C^{(\alpha)\alpha\beta} - C^{(\alpha)\alpha\beta}(\text{softphoton}) \Big) + C^{(\alpha)\alpha\beta}(\text{softphoton}). \tag{46}$$

The remaining part of the charge–side tensor in (45) is referred to as the hard photon contribution. It is infrared finite and can thus be integrated without a regulator photon mass.

Analytic phase space integrations without a regulator photon mass are much simpler. The regulator photon mass would introduce a new mass scale into the problem which, as emphasized before, would complicate the phase space integrations.

In the phase space integration over the photon momentum the infrared singular piece is regularized by introducing a (small) photon mass resulting in a logarithmic mass divergence in the photon mass. Since the integrand of the soft photon piece is much simpler the phase space integration can be done analytically by carefully taking the appropriate mass zero limits in the integrations. The infrared divergence shows up as a logarithmic mass divergence in the photon mass. In addition, since the soft photon piece factors the Born term tensor, the Born term tensor can be pulled out of the phase space integrations. The resulting singular soft photon piece is therefore universal in the sense that it is the same for all spin structure functions, i.e. once the soft photon integration has been done for the rate the work is done. Eventually the infrared singular piece is cancelled by the corresponding singular piece in the one–loop contributions.

The approximation where only the soft photon (or gluon) piece is retained in the *NLO* tree–graph hadron tensor is called the soft photon (or gluon) approximation. From what has been said before it is clear that a *NLO* radiative correction calculation is much simplified in the soft photon (or gluon) approximation. In the next section we shall take a specific example, namely the process $e^+e^- \to t\,\bar{t}\,(g)$, to investigate the quality of the soft gluon approximation.

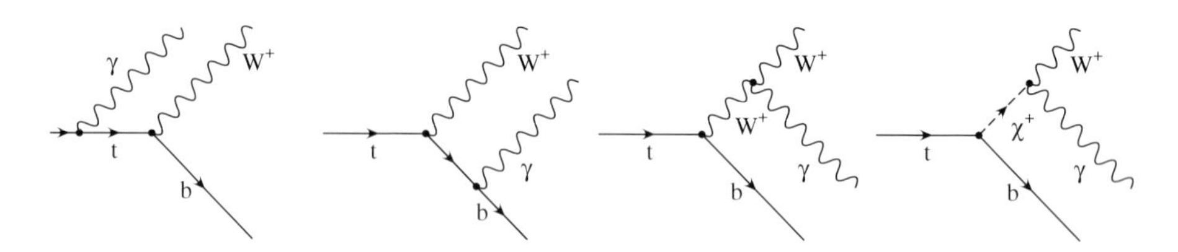

Fig. 6. Tree–level Feynman graphs for $t \to b + W^+ + \gamma$ in the Feynman–'t Hooft gauge

It is interesting to note that the *NLO* hadronic tensor in the elektroweak corrections to the process $t \to b + W^+$ is only marginally more complicated than the corresponding tree graph contribution (45) although the number of contributing diagrams has doubled to four [7] (see Fig. 6). We do not list the finite piece of the hadronic tensor in this review but only write down the soft photon factor which multiplies the Born term tensor just as in (45). When discussing the decay $t \to b + W^+$ one has to of course include the antisymmetric ϵ–tensor piece in the Born term tensor. The soft photon contribution reads

$$-g^{\mu\nu} A_\mu^{(\text{s.ph.})} A_\nu^{(\text{s.ph.})} = -e^2 \left(\frac{Q_t^2 \, m_t^2}{(p_t \cdot k)^2} + \frac{Q_b^2 \, m_b^2}{(p_b \cdot k)^2} + \frac{Q_W^2 \, m_W^2}{(q \cdot k)^2} \right.$$
$$\left. - \frac{2Q_t Q_b \, p_t \cdot p_b}{(p_t \cdot k)(p_b \cdot k)} - \frac{2Q_t Q_W \, p_t \cdot q}{(p_t \cdot k)(q \cdot k)} + \frac{2Q_b Q_W \, p_b \cdot q}{(p_b \cdot k)(q \cdot k)} \right), \quad (47)$$

where the soft photon amplitude is given by (see e.g. [51])

$$A^{(\text{s.ph.})\mu} \epsilon_\mu^* = e \left(\frac{Q_t \, p_t^\mu}{p_t \cdot k} - \frac{Q_b \, p_b^\mu}{p_b \cdot k} - \frac{Q_W \, q^\mu}{q \cdot k} \right) \epsilon_\mu^* . \quad (48)$$

$Q_t = 2/3$, $Q_b = -1/3$ and $Q_W = 1$ are the electric charges of the top quark, the bottom quark and the W–boson, resp., in units of the elementary charge e. Gauge invariance of the soft–photon amplitude is easily verified when replacing $\epsilon_\mu^* \to k_\mu$ in (48). It is then just a statement about charge conservation $Q_t - Q_b - Q_W = 0$. It is noteworthy that by setting $Q_t = Q_b = 1$ and $Q_W = 0$ in (47), one recovers the soft photon contribution in (45). In fact, the whole *NLO* electroweak tree–graph tensor in [7] reduces to the corresponding *NLO* QED tensor with the above charge replacements. In the same vein, the replacements $e \to g_s$ and $1 \to N_c C_F = 4$, to account for colour, and the replacements of the charge factors by $Q_t = Q_b = 1$ and $Q_W = 0$ will bring one from the electroweak case to the QCD case.

8 *NLO* Radiative Corrections to $e^+e^- \to t\,\bar{t}\,(g)$ in the Soft Gluon Approximation

In the soft gluon (or photon) approximation one keeps only the soft gluon (or photon) piece in the tree graph contribution but includes the full one–loop contribution. New structure is thus only generated by the one–loop contribution since the soft gluon piece has Born term structure. The structure contained in the hard part of the tree graph contribution is lost in the soft gluon approximation. A brief glance at the corresponding hard photon piece in (45) shows that it is considerably more difficult to do the analytic phase space integration for the hard part than for the soft photon piece. In particular, the integration of the hard photon piece has to be done separately for each density matrix element. Contrary to this, the integration of the soft photon piece has to be done only once since the Born term factor can be factored out of the integral. Technically, the soft gluon (or photon) approximation is much simpler than the full calculation if done analytically. Of course, if the calculation is done numerically, the integration of the hard gluon (or photon) part causes no additional problems. When the soft gluon (or photon) approximation is used this is done at the cost of loosing interesting structure contained in the hard gluon (or photon) part.

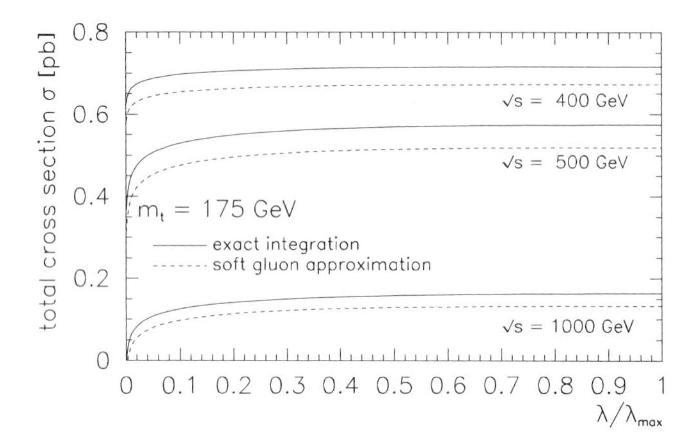

Fig. 7. Total cross section for $e^+e^- \to t\bar{t}(g)$ as a function of the scaled gluon energy cut λ/λ_{max} for different c.m. energies. Full line: full calculation; dashed line: soft gluon approximation

In order to be able to judge the quality of the soft gluon approximation we have calculated the rate for $e^+e^- \to t\,\bar{t}\,(g)$ and have compared the results of the full calculation with the results of the soft gluon approximation using different cut–off values for the gluon energy [52]. Fig. 7 shows that the soft gluon result is below the full result for the whole range of cut–off values $\lambda = E_g/\sqrt{q^2}$ ($\lambda_{max} = 1 - 2m_t/\sqrt{q^2}$). For maximal cut–off values $\lambda = \lambda_{max}$ the soft gluon approximation is 3.4%, 5% and 11.5% below the full result at 400 GeV, 500 GeV and 1000 GeV, respectively. As emphasized above, much of the rate that is being missed by the soft gluon approximation has interesting structure.

9 Unpolarized Top Decay $t \to b + W^+$

We have already discussed various aspects of the decay $t \to b + W^+$ in previous sections. In this section we concentrate on unpolarized top decay. In particular, we want to discuss the mass dependence of the longitudinal piece of the W^+ boson the measurement of which could lead to an independent determination of the mass of the top quark.

The angular decay distribution for unpolarized top decay can be obtained by setting $P = 0$ in (28). The contribution of the three remaining structure functions H_L, H_U and H_F can be disentangled by a measurement of the shape of the lepton energy spectrum from top decay. Given enough data, one can hope to determine the longitudinal contribution with 1% accuracy [50].

At *LO* the mass dependence of the longitudinal contribution is given by $\Gamma_L/\Gamma = 1/(1 + 2(m_W/m_t)^2)$ which gives $\Gamma_L/\Gamma = 0.703$ using $m_t = 175\,\mathrm{GeV}$ and $m_W = 80.419\,\mathrm{GeV}$. *NLO* corrections to the longitudinal contribution were calculated in [5,6] (QCD) and in [7] (electroweak and finite width). Curiously enough the electroweak and finite width corrections tend to cancel each other in the structure functions. In Fig. 8 we show the top mass dependence of the ratio Γ_L/Γ. The Born term and the corrected curves are practically straight line curves. The horizontal displacement of the two curves is $\approx 3.5\,\mathrm{GeV}$. One would thus make the corresponding mistake in the top mass determination from a measurement of Γ_L/Γ if the Born term curve were used instead of the corrected curve. If we take $m_t = 175\,\mathrm{GeV}$ as central value, a 1% relative error on the measurement of Γ_L/Γ would allow one to determine the top quark mass with $\approx 3\,\mathrm{GeV}$ accuracy. Such a top mass measurement would be a welcome alternative to the usual invariant mass determination of the top quark mass since the Γ_L/Γ measurement is a completely independent measurement of the top quark mass.

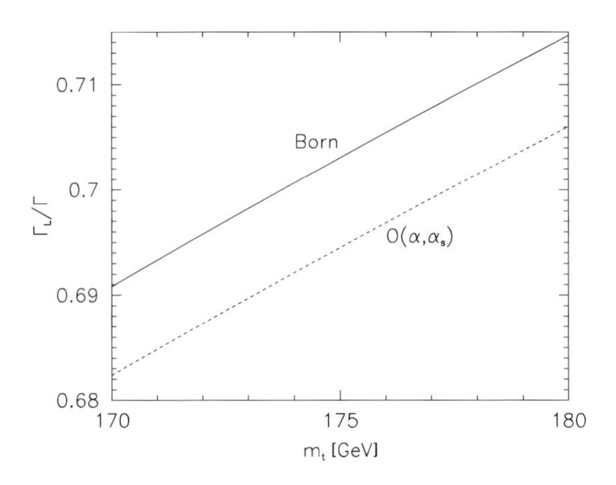

Fig. 8. Top mass dependence of the ratio Γ_L/Γ (full line: LO, dashed line: corrections including QCD, electroweak (G_F–scheme), finite–width and ($m_b \neq 0$) Born term corrections)

10 The Decay $t(\uparrow) \rightarrow b + H^+$

Certain extensions of the minimal Standard Model with two Higgs–doublets (as e.g. the Minimal Supersymmetric Standard Model) contain charged Higgs bosons such that the decay $t(\uparrow) \rightarrow b + H^+$ would be possible if kinematically allowed. The Lagrangian for the decay reads

$$\mathcal{L} = \bar{\Psi}_b(x)(a + b\gamma_5)\Psi_t(x)\phi_{H^+}(x) + \text{h.c.}\,. \tag{49}$$

For the differential decay rate one obtains

$$\frac{d\Gamma}{d\cos\theta_P} = \frac{1}{2}\Big(\Gamma + P\,\Gamma^P \cos\theta_P\Big) = \frac{1}{2}\Gamma\Big(1 + P\alpha_H \cos\theta_P\Big), \tag{50}$$

where θ_P is the angle between the polarization vector of the top quark and the Higgs. The $\cos\theta_P$–dependence in (50) is determined by the asymmetry parameter $\alpha_H = \Gamma/\Gamma_P$. At *LO* one obtains ($m_b = 0$)

$$\alpha_H = \frac{2ab}{a^2 + b^2}. \tag{51}$$

At this point we want to emphasize that one needs the polarization information contained in (50) to be able to determine the relative sign of the two coupling constants a and b. The rate is proportional to $(a^2 + b^2)$ and is therefore insensitive to the relative sign of the coupling constants.

We now specify to the so called model 2 where the coupling structure is expressed in terms of $\tan\beta = v_2/v_1$, and where v_1 and v_2 are the vacuum expectation values of the two neutral components of the two Higgs doublets. In model 2 the coupling constants are given by

$$a = \frac{g_w}{2\sqrt{2}m_W} V_{tb}(m_t \cot\beta + m_b \tan\beta), \tag{52}$$

$$b = \frac{g_w}{2\sqrt{2}m_W} V_{tb}(m_t \cot\beta - m_b \tan\beta). \tag{53}$$

The weak coupling factor g_w is related to the usual Fermi coupling constant G_F

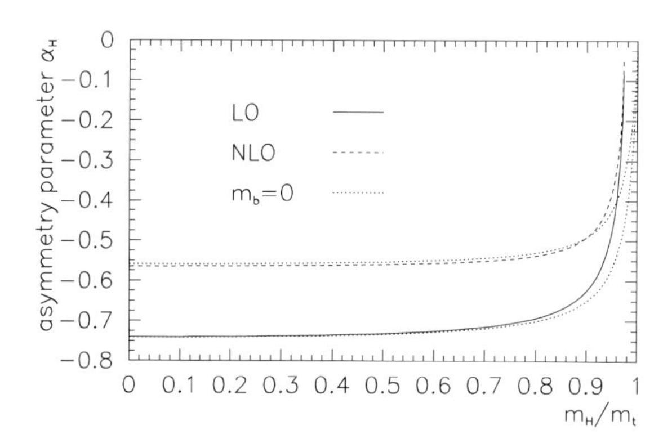

Fig. 9. Asymmetry parameter α_H for model 2 with $m_b = 4.8\,\text{GeV}$ and $m_t = 175\,\text{GeV}$ (LO: full line, NLO: dashed line) as function of m_{H^+}/m_t and $\tan\beta = 10$. The barely visible dotted lines show the corresponding $m_b \to 0$ curves.

by $g_w = 2m_W\sqrt[4]{2}\sqrt{G_F}$. In Fig. 9 we show a plot of the asymmetry parameter α_H as a function of the mass ratio m_H^+/m_t for a fixed value of $\tan\beta = 10$. The asymmetry parameter is large and negative over most of the range of mass ratios. The radiative corrections can be seen to be quite substantial.

11 Goldstone Equivalence Theorem

Consider the decay $t \to b + J_\mu^+(q^2)$ where the current $J_\mu^+(q^2)$ can be thought of as representing an off–shell W^+ containing spin 1 and spin 0 pieces. Taking $q^\mu = (q_0; 0, 0, |\boldsymbol{q}|)$ the longitudinal polarization vector of the spin 1 piece is given by $\epsilon^\mu(L) = (|\boldsymbol{q}|; 0, 0, q_0)/\sqrt{q^2}$ while the scalar polarization vector reads $\epsilon^\mu(S) = q^\mu/\sqrt{q^2}$. In the limit $\sqrt{q^2}/m_t \to 0$ the longitudinal polarization vector becomes increasingly parallel to its momentum. In fact, one finds $\epsilon^\mu(L) = q^\mu/\sqrt{q^2} + O(\sqrt{q^2}/q_0)$. In this limit the amplitude $A(L)$ dominates over the transverse amplitudes $A(T)$ since $A(T)/A(L) \sim \sqrt{q^2}/q_0$. At the same time one finds $A(S) \sim A(L)$ in this limit. This provides a useful check on the high energy limits of the relevant unpolarized and polarized rate functions.

Next consider the scalar projection of the $t \to b$ current transition at lowest order. Using the Dirac equation one obtains ($q = p_t - p_b$)

$$\begin{aligned} q_\mu \bar{u}_b \gamma^\mu (\mathbb{1} - \gamma_5) u_t &= (p_t - p_b)_\mu \bar{u}_b \gamma^\mu (\mathbb{1} - \gamma_5) u_t \\ &= \bar{u}_b (m_t(\mathbb{1} + \gamma_5) + m_b u_t (\mathbb{1} - \gamma_5)) u_t . \end{aligned} \tag{54}$$

The second line of (54) has the coupling structure of a Goldstone boson in the Standard Model. It is a very instructive exercise to follow this argument through to one–loop order including renormalization [53]. In fact, this generalizes to any order in perturbation theory: the scalar piece of the charged transition current has the coupling structure of a Goldstone boson. Together with the above statement that the longitudinal rate dominates at high energies and that it is equal to the scalar rate in this very limit one arrives at the Goldstone Boson Equivalence Theorem. In the high energy limit the coupling of a gauge vector boson to a fermion current is equivalent to that of a Goldstone boson. This simplifying feature has frequently been used in the literature to dramatically reduce the effort needed in the computation of high energy processes involving vector gauge bosons. For example, the Goldstone equivalence theorem was used in [54] to calculate the bosonic two–loop electroweak radiative corrections to the decay $H \to \gamma + \gamma$ in the limit of a large Higgs mass. The same approximation was used to calculate the dominant bosonic two–loop electroweak radiative corrections to the decay of a heavy Higgs into pairs of W and Z gauge bosons [55].

As concerns the coupling of charged Higgs bosons in $t \to b + H^+$ discussed in Sect. 10 the coupling structure of the charged Higgs is equal to that of a Goldstone boson in the Standard Model if one specifies to the so called model 1 with $\cot\beta = 1$ (see e.g. [8]). In this case one has

$$\Gamma_{t \to b + H^+} \sim \Gamma_{t \to b + W^+} \qquad \text{as} \qquad m_{W^+}, m_{H^+}/m_t \to 0 . \tag{55}$$

The same relation holds true for the corresponding polarized top rates. All of the statements made in this section have been explicitly verified using our *NLO* results on $t \to b + W^+$ [6] and $t \to b + H^+$ [8].

12 Leptonic Decays of the μ and the τ and Anomalous Helicity Flip Contributions

We have already discussed various aspects of the leptonic decays of the μ and the τ lepton in previous sections. A complete solution to the problem has been given in [3] where, for the first time, all mass and polarization effects have been included at *NLO*. Partial *NLO* results on polarization can be found in [57,58]. In this section we concentrate on one aspect of the problem, namely on the so–called anomalous contribution that, in the chiral limit, flips the helicity of the final–state lepton at *NLO*.

Collinear photon emission from a massless fermion line can flip the helicity of the massless fermion contrary to naive expectation. This has been discussed in a variety of physical contexts. This is a "m_e/m_e" effect where the m_e in the numerator is a spin flip factor and the m_e in the denominator arises from the collinear configuration. In the limit $m_e \to 0$ the helicity flip contribution survives whereas it is not seen in massless QED.

We shall discuss this phenomenon in the context of the left–chiral $\mu \to e$ transition. At the Born term level an electron emerging from a weak $(V - A)$ vertex is purely left–handed in the limit $m_e = 0$. Naively, one would expect this to be true also at $O(\alpha)$ or at any order in α because in massless QED photon emission from the electron is helicity conserving. Let us make this statement more precise by looking at the string of γ–matrices between the initial state μ spinor and final state e antispinor of the left–chiral $\mu \to e$ transition. Using $(1-\gamma_5) = (1-\gamma_5)(1-\gamma_5)/2$, and the fact that in massless QED every photon emission brings in two γ–factors (one from the vertex and one from the fermion propagator) one finds

$$\bar{u}_e \Gamma^{(n)} \gamma^\mu (\mathbb{1} - \gamma_5) u_\mu = \frac{1}{2} \bar{u}_e (\mathbb{1} + \gamma_5) \Gamma^{(n)} \gamma^\mu (\mathbb{1} - \gamma_5) u_\mu, \tag{56}$$

where $\Gamma^{(n)}$ stands for the (even) γ matrix string brought in by the emission of n photons. We have commuted the left–chiral factor $(1-\gamma_5)$ to the left end where it then projects out the helicity state $\lambda_e = -1/2$ from the electron antispinor thus proving the above assertion.

Let us take a closer look at the anomalous helicity flip contribution in leptonic $\mu \to e$ decays by considering the unnormalized density matrix element ρ_{++} of the final state electron which is obtained by setting $\cos\theta = 1$ in (38) (remember that G_5 vanishes for $m_e \to 0$ and $G_6 = 0$ in the Standard Model). One has

$$\frac{d\Gamma^{(+)}}{dx\, d\cos\theta_P} = \frac{1}{2} \beta x\, \Gamma_0 \Big((G_1 + G_3) + (G_2 + G_4) P \cos\theta_P \Big). \tag{57}$$

Contrary to naive expectations one finds non–vanishing right–handed (+) contributions which survive the $m_e \to 0$ limit when one takes the $m_e \to 0$ limit of the *NLO* contributions to (57) [3]. In fact, one finds

$$\frac{d\Gamma^{(+)}}{dx\, d\cos\theta_P} = \frac{\alpha}{12\pi} \Gamma_0 \Big(\Big[(1-x)^2(5-2x)\Big] - \Big[(1-x)^2(1+2x)\Big] P \cos\theta_P \Big). \tag{58}$$

The result is rather simple. In particular, it does not contain any logarithms or dilogarithms. The simplicity of the right–handed contribution becomes manifest in the equivalent particle description of μ–decay where, in the peaking approximation, μ–decay is described by the two–stage process $\mu^- \to e^-$ followed by the branching process $e^- \to e^- + \gamma$ characterized by universal splitting functions $D_{nf/hf}(z)$ [56]. The symbols nf and hf stand for a helicity non–flip and helicity flip of the helicity of the electron. In the splitting process z is the fractional energy of the emitted photon. The off–shell electron in the propagator is replaced by an equivalent on–shell electron in the intermediate state. Since the helicity flip contribution arises entirely from the collinear configuration it can be calculated in its entirety using the equivalent particle description.

The helicity flip splitting function is given by $D_{hf}(z) = \alpha z/(2\pi)$, where $z = k_0/E' = (E' - E)/E' = 1 - x/x'$, and where k_0 is the energy of the emitted photon. E' and E denote the energies of the initial and final electron in the splitting process. The helicity flip splitting function has to be folded with the appropriate $m_e = 0$ Born term contribution. The lower limit of the folding integration is determined by the soft photon point where $E' = E$. The upper limit is determined by the maximal energy of the initial electron $E' = m_\mu/2$. One obtains

$$\begin{aligned}\frac{d\Gamma^{(+)}}{dx\, d\cos\theta_P} &= \frac{\alpha}{2\pi}\int_x^1 dx' \frac{1}{x'}\frac{d\Gamma^{\text{Born};(-)}(x')}{dx'\, d\cos\theta_P}(1 - \frac{x}{x'}) \\ &= \frac{\alpha}{2\pi}\Gamma_0 \int_x^1 dx'(x' - x)\Big((3 - 2x') + (1 - 2x')P\cos\theta_P\Big) \\ &= \frac{\alpha}{12\pi}\Gamma_0\Big((1 - x)^2(5 - 2x) - (1 - x)^2(1 + 2x)P\cos\theta_P\Big), \quad (59)\end{aligned}$$

which exactly reproduces the result (58).

Numerically, the flip spectrum function is rather small compared to the $O(\alpha_s)$ no–flip spectrum function. However, when averaging over the spectrum the ratio of the $O(\alpha_s)$ flip and no–flip contributions amounts to a non–negligible (−12%), due to cancellation effects in the $O(\alpha_s)$ no–flip contribution.

13 Summary and Concluding Remarks

We have discussed *NLO* corrections to a multitude of polarization observables in different processes including nonzero mass effects. The results are available in compact analytical form. They are ready for use in physics simulation programs which are reliable even at corners of phase space where mass effects become important. Present and planned experiments (TEVATRON Run2, LHC, BELLE, BABAR, τ–charm factories) will be sensitive to these *NLO* Standard Model effects. Standard Model *NLO* corrections to polarization observables are needed as background for possible new physics contributions. Last but not least the *NLO* results are needed for *NLO* sum rule analysis' involving polarization observables. All calculations are based on the same one–loop and tree–graph matrix elements. Results on rates agree with previous calculations. Results on polarization observables agree with previous calculations where available. We have checked various limits and found agreement with previous mass zero calculations, the Goldstone boson equivalence theorem and the equivalent particle description of the anomalous helicity flip contribution. Because of the various checks and the fact that we have essentially used one set of matrix elements as input to the calculations we feel quite confident that our results are correct.

Acknowledgements

We would like to thank the organizers of this meeting, D. Blaschke, M.A. Ivanov and S. Nedelko for providing a most graceful conference setting. We acknowledge informative discussions and e–mail exchanges with A.B. Arbuzov, F. Berends, W. van Neerven, F. Scheck, K. Schilcher, H. Spiesberger and O.V. Teryaev. We would like to thank B. Lampe, M. Fischer, S. Groote and H.S. Do for their collaborative effort. M.C. Mauser is supported by the DFG (Germany) through the Graduiertenkolleg "Eichtheorien" at the University of Mainz.

References

1. P.A. Baikov, K.G. Chetyrkin and J.H. Kühn: Phys. Rev. Lett. **88** 012001 (2002)
2. P.A. Baikov, K.G. Chetyrkin and J.H. Kühn: hep-ph/0212299
3. M. Fischer, S. Groote, J.G. Körner and M.C. Mauser: hep-ph/0203048 (2003), to be published in Phys. Rev. D
4. M. Fischer, S. Groote, J. G. Körner, B. Lampe and M. C. Mauser: Phys. Lett. B **451** 406 (1999)
5. M. Fischer, S. Groote, J.G. Körner and M.C. Mauser: Phys. Rev. D **63** 031501 (2001)
6. M. Fischer, S. Groote, J.G. Körner and M.C. Mauser: Phys. Rev. D **65** 013205 (2002)
7. H.S. Do, S. Groote, J.G. Körner and M.C. Mauser: Phys. Rev. D **67** 091501 (2003)
8. J.G. Körner and M.C. Mauser: hep-ph/0211098
9. M. Fischer, S. Groote, J.G. Körner and M.C. Mauser: Phys. Lett. B **480** 265 (2000)
10. M. Fischer, S. Groote, J.G. Körner and M.C. Mauser: "Inclusive decays $\Lambda_b \to X_c + D_s^{(*)}$ at $O(\alpha_s)$ including Λ_b and D_s^* polarization effects", to be published
11. S. Groote, J.G. Körner and J.A. Leyva: Phys. Rev. D **56** 6031 (1997)
12. J.G. Körner, A. Pilaftsis and M.M. Tung: Z. Phys. C **63** 575 (1994)
13. S. Groote, J.G. Körner and M.M. Tung: Z. Phys. C **70** 281 (1996)
14. S. Groote and J.G. Körner: Z. Phys. C **72** 255 (1996)
15. S. Groote, J.G. Korner and M.M. Tung: Z. Phys. C **74** 615 (1997)
16. A. Brandenburg, M. Flesch and P. Uwer: Phys. Rev. D **59** 014001 (1999)
17. V. Ravindran and W.L. van Neerven: Phys. Lett. B **445** 214 (1998); Nucl. Phys. B **589** 507 (2000)
18. A. Czarnecki, M. Jezabek and J.H. Kühn: Nucl. Phys. B **351** 70 (1991)
19. A. Czarnecki, M. Jezabek, J.G. Körner and J.H. Kühn: Phys. Rev. Lett. **73** 384 (1994)

20. A. Brandenburg: Phys. Lett. B **388** 626 (1996)
21. G. Mahlon and S. Parke: Phys. Lett. B **411** 173 (1997)
22. W. Bernreuther, A. Brandenburg, Z.G. Si and P. Uwer: Phys. Rev. Lett. **87** 242002 (2001)
23. W. Bernreuther, A. Brandenburg, Z.G. Si and P. Uwer: Phys. Lett. B **509** 53 (2001)
24. S. Parke and Y. Shadmi: Phys. Lett. B **387** 199 (1996)
25. S. Groote, J.G. Körner and J.A. Leyva: Phys. Lett. B **418** 192 (1998)
26. M.M. Tung, J. Bernabeu and J. Penarrocha: Phys. Lett. B **418** 181 (1998)
27. J.G. Körner, G. Kramer, G. Schierholz, K. Fabricius and I. Schmitt: Phys. Lett. B **94** 207 (1980)
28. K. Fabricius, I. Schmitt, G. Kramer and G. Schierholz: Phys. Rev. Lett. **45** 867 (1980)
29. A. Brandenburg, L.J. Dixon and Y. Shadmi: Phys. Rev. D **53** 1264 (1996)
30. K. Hagiwara, K.i. Hikasa and N. Kai: Phys. Rev. Lett. **47** 983 (1981)
31. K. Hagiwara, K.i. Hikasa and N. Kai: Phys. Rev. D **27** 84 (1983)
32. K. Hagiwara, K. i. Hikasa and N. Kai: Phys. Rev. Lett. **52** 1076 (1984)
33. J.G. Körner, B. Melic and Z. Merebashvili: Phys. Rev. D **62** 096011 (2000)
34. K. Hagiwara *et al.* [Particle Data Group Collaboration]: Phys. Rev. D **66** 010001 (2002)
35. W.G. Dharmaratna and G.R. Goldstein: Phys. Rev. D **41** 1731 (1990);
36. W. Bernreuther, A. Brandenburg and P. Uwer: Phys. Lett. B **368** 153 (1996)
37. W.G. Dharmaratna and G.R. Goldstein: Phys. Rev. D **53** 1073 (1996)
38. J.G. Körner, G. Schuler, G. Kramer and B. Lampe: Z. Phys. C **32** 181 (1986) 181
39. J. C. Pati and C. H. Woo: Phys. Rev. D **3** 2920 (1971)
40. J.G. Körner: Nucl.Phys.**32** 282 (1970)
41. J.H. Kühn, A. Reiter and P.M. Zerwas: Nucl. Phys. B **272** 560 (1986)
42. J.G. Körner and G.A. Schuler: Z. Phys. C **46** 93 (1990)
43. A. Faessler, T. Gutsche, M.A. Ivanov, J.G. Körner and V.E. Lyubovitskij: Eur. Phys. J. directC **4**18 (2002)
44. I. Dunietz, H. R. Quinn, A. Snyder, W. Toki and H. J. Lipkin: Phys. Rev. D **43** 2193 (1991)
45. G.J. Gounaris and J.E. Paschalis: Nucl. Phys. B **222** 473 (1983)
46. K. Schilcher, M.D. Tran and N.F. Nasrallah, Nucl. Phys. B **181** 91 (1981), Erratum *ibid.* B **187** 594 (1981)
47. C. Bauer and H. S. Do: Comput. Phys. Commun. **144** 154 (2002)
48. A. Denner and T. Sack: Nucl. Phys. B **358** 46 (1991)
49. H.S. Do, S. Groote, J.G. Körner and M.C. Mauser: to be published
50. Future Electroweak Physics at the Fermilab Tevatron: Report of the TeV2000 Group, edited by D. Amidei and R. Brock, Fermilab-Pub-96/046, http://www-theory.fnal.gov/TeV2000.html
51. S. Weinberg: "The Quantum Theory Of Fields. Vol. 1: Foundations," (Cambridge University Press, Cambridge 1995)
52. S. Groote, J.G. Körner: "Analytical results for $O(\alpha_s)$ radiative corrections to $e^+e^- \to \bar{t}t^\uparrow$ up to a given gluon energy cut", to be published
53. A. Czarnecki and S. Davidson: Proceedings of the Lake Louise Winter Institute, 330 (1993)
54. J.G. Körner, K. Melnikov and O.I. Yakovlev: Phys. Rev. D **53** 3737 (1996)
55. A. Frink, B.A. Kniehl, D. Kreimer and K. Riesselmann: Phys. Rev. D **54** 4548 (1996)

56. B. Falk and L.M. Sehgal: Phys. Lett. B **325** 509 (1994)
57. W.E. Fischer and F. Scheck: Nucl. Phys. B **83** 25 (1974)
58. A.B. Arbuzov: Phys. Lett. B **524** 99 (2002)

Exclusive Rare Decays of B and B_c Mesons in a Relativistic Quark Model

M.A. Ivanov[1] and V.E. Lyubovitskij[2]

[1] Bogoliubov Laboratory of Theoretical Physics, Joint Institute for Nuclear Research, 141980 Dubna, Russia
[2] Institut für Theoretische Physik, Universität Tübingen, Auf der Morgenstelle 14, 72076 Tübingen, Germany

Abstract. In these lectures we give, first, the model-independent analysis of the exclusive rare decays $B \to K\bar{l}l$ and $B_c \to D(D^*)\bar{l}l$ with special emphasis on the cascade decay $B_c \to D^*(\to D\pi)\bar{l}l$. We derive a four-fold angular decay distribution for this process in terms of helicity amplitudes including lepton mass effects. The four-fold angular decay distribution allows to define a number of physical observables which are amenable to measurement. Second, we calculate the relevant form factors within a relativistic constituent quark model, for the first time without employing the impulse approximation. The calculated form factors are used to evaluate differential decay rates and polarization observables. We present results on a set of observables with and without long-distance contributions. and compare them with the results of other studies.

1 Introduction

The flavor-changing neutral current transitions $B \to K+X$ and $B_c \to D(D^*)+X$ with $X = \gamma, l^+l^-, \bar{\nu}\nu$ are of special interest because they proceed at the loop level in the Standard Model (SM) involving also the top quark. They may therefore be used for a determination of the Cabibbo-Kobayashi-Maskawa (CKM) matrix elements V_{tq} ($q = d, s, b$). The available experimental measurements of the branching ratio of the inclusive radiative B-meson decay

$$\mathrm{Br}\,(B \to X_s\gamma) = \begin{cases} (3.11 \pm 0.80(\mathrm{stat}) \pm 0.72(\mathrm{syst})) \times 10^{-4} & \text{ALEPH [1]} \\ \big(3.36 \pm 0.53(\mathrm{stat}) \pm 0.42(\mathrm{syst})^{+0.50}_{-0.54}(\mathrm{th})\big) \times 10^{-4} & \text{BELLE [2]} \\ \big(3.21 \pm 0.43(\mathrm{stat}) \pm 0.27(\mathrm{syst})^{+0.18}_{-0.10}(\mathrm{th})\big) \times 10^{-4} & \text{CLEO [3]} \end{cases}$$

are consistent with the next-to-leading order prediction of the standard model (see, e.g. [4] and references therein):

$$\mathrm{Br}(B \to X_s\gamma)_{\mathrm{SM}} = (3.35 \pm 0.30) \times 10^{-4}\,. \tag{1}$$

The decay $B \to K\, l^+l^-$ ($l = e, \mu$) has been observed by the BELLE Collaboration [5] with a branching ratio of

$$\mathrm{Br}\left(B \to K\, l^+l^-\right) = (0.75^{+0.25}_{-0.21} \pm 0.09) \times 10^{-6}\,. \tag{2}$$

M.A. Ivanov and V.E. Lyubovitskij, Exclusive Rare Decays of B and B_c Mesons in a Relativistic Quark Model, Lect. Notes Phys. **647**, 245–263 (2004)
http://www.springerlink.com/

The recent observation of the B_c meson by the CDF Collaboration at Tevatron in Fermilab [6] raises hopes that one may also explore the rare decays of the bottom-charm meson in the future.

The theoretical study of the exclusive rare decays proceeds in two steps. First, the effective Hamiltonian for such transitions is derived by calculating the leading and next-to-leading loop diagrams in the SM and by using the operator product expansion and renormalization group techniques. The modern status of this part of the calculation is described in the review [7] (and references therein). Second, one needs to evaluate the matrix elements of the effective Hamiltonian between hadronic states. This part of the calculation is model dependent since it involves nonperturbative QCD. There are many papers on this subject. The decay rates, dilepton invariant mass spectra and the forward-backforward asymmetry in the decays $B \to K\, l^+ l^-$ ($l = e, \mu, \tau$) have been investigated in the SM and its supersymmetric extensions by using improved form factors from light-cone QCD sum rules [8]. An updated analysis of these decays has been done in [4] by including explicit $O(\alpha_s)$ and $\Lambda_{\rm QCD}/m_b$ corrections. The invariant dilepton mass spectrum and the Dalitz plot for the decay $B \to K\, l^+ l^-$ have been studied in [9] by using quark model form factors. The $B \to K\, l^+ l^-$ decay form factors were studied via QCD sum rules in [10] and within the lattice-constrained dispersion quark model in [11]. Various aspects of these decays were discussed in numerous papers by Aliev *et al.* [12]. The exclusive semileptonic rare decays $B \to K\, l^+ l^-$ were analyzed in supersymmetric theories in [13]. The angular distribution and CP asymmetries in the decays $B \to K\pi e^+ e^-$ were investigated in [14]. The lepton polarization for the inclusive decay $B \to X_s l^+ l^-$ was discussed in [15] and [16]. The rare decays of $B_c \to D(D^*)\, l^+ l^-$ were studied in [17] by using the form factors evaluated in the light front and constituent quark models.

We employ a relativistic quark model [18,19] to calculate the decay form factors. This model is based on an effective Lagrangian which describes the coupling of hadrons H to their constituent quarks. The coupling strength is determined by the compositeness condition $Z_H = 0$ [20,21] where Z_H is the wave function renormalization constant of the hadron H. One starts with an effective Lagrangian written down in terms of quark and hadron fields. Then, by using Feynman rules, the S-matrix elements describing the hadronic interactions are given in terms of a set of quark diagrams. In particular, the compositeness condition enables one to avoid a double counting of hadronic degrees of freedom. The approach is self-consistent and universally applicable. All calculations of physical observables are straightforward. The model has only a small set of adjustable parameters given by the values of the constituent quark masses and the scale parameters that define the size of the distribution of the constituent quarks inside a given hadron. The values of the fit parameters are within the window of expectations.

The shape of the vertex functions and the quark propagators can in principle be found from an analysis of the Bethe-Salpeter and Dyson-Schwinger equations as was done e.g. in [22]. In this paper, however, we choose a phenomenological approach where the vertex functions are modelled by a Gaussian form, the size parameter of which is determined by a fit to the leptonic and radiative decays of the lowest lying charm and bottom mesons. For the quark propagators we use the local representation. In the present calculations we do not employ the so-called impulse approximation used previously [19]. The numerical results obtained with and without the impulse approximation are close to each other for light-to-light and heavy-to-heavy transitions but differ considerably from one another for heavy-to-light transitions as e.g. in the $B \to \pi$ transitions.

We calculate the form factors of the transition $B \to K$ and use them to evaluate differential decay rates and polarization observables with and without long-distance contributions which include the lower-lying charmonium states according to [23]. We extend our analysis to the exclusive rare decay $B_c \to D(D^*)\bar{l}l$. We derive a four-fold angular decay distribution for the cascade $B_c \to D^*(\to D\pi)\bar{l}l$ process in the helicity frame including lepton mass effects following the method outlined in [24]. The four-fold angular decay distribution allows one to define a number of physical observables which are amenable to measurement. We compare our results with the ones of other studies.

2 Effective Hamiltonian

The starting point of the description of the rare exclusive decays is the effective Hamiltonian obtained from the SM-diagrams by using the operator product expansion and renormalization group techniques. It allows one to separate the short-distance contributions and isolate them in the Wilson coefficients which can be studied systematically within perturbative QCD. The long-distance contributions are contained in the matrix elements of local operators. Contrary to the short-distance contributions the calculation of such matrix elements requires nonperturbative methods and is therefore model dependent.

We will follow [7] in writing down the analytical expressions for the effective Hamiltonian and paper [8] in using the numerical values of the input parameters characterizing the short-distance contributions. At the quark level, the rare semileptonic decay $b \to s(d)l^+l^-$ can be described in terms of the effective Hamiltonian:

$$H_{\text{eff}} = -\frac{G_F}{\sqrt{2}}\lambda_t \sum_{i=1}^{10} C_i(\mu)Q_i(\mu)\,. \tag{3}$$

where $\lambda_t \equiv V^\dagger_{ts(d)}V_{tb}$ is the product of CKM elements. For example, the standard set [7] of local operators for $b \to sl^+l^-$ transition is written as

$$\begin{array}{ll} Q_1 = (\bar{s}_i c_j)_{V-A}\,(\bar{c}_j b_i)_{V-A}\,, & Q_2 = (\bar{s}c)_{V-A}(\bar{c}b)_{V-A}\,, \\ Q_3 = (\bar{s}b)_{V-A}\sum_q(\bar{q}q)_{V-A}\,, & Q_4 = (\bar{s}_i b_j)_{V-A}\sum_q(\bar{q}_j q_i)_{V-A}\,, \\ Q_5 = (\bar{s}b)_{V-A}\sum_q(\bar{q}q)_{V+A}\,, & Q_6 = (\bar{s}_i b_j)_{V-A}\sum_q(\bar{q}_j q_i)_{V+A}\,, \\ Q_7 = \frac{e}{8\pi^2}m_b\,\bar{s}\sigma^{\mu\nu}(1+\gamma_5)b\,F_{\mu\nu}\,, & Q_8 = \frac{g}{8\pi^2}m_b\,\bar{s}_i\sigma^{\mu\nu}(1+\gamma_5)\mathbf{T}_{ij}b_j\,\mathbf{G}_{\mu\nu}\,, \\ Q_9 = \frac{e}{8\pi^2}(\bar{s}b)_{V-A}(\bar{l}l)_V & Q_{10} = \frac{e}{8\pi^2}(\bar{s}b)_{V-A}(\bar{l}l)_A \end{array} \tag{4}$$

where $\mathbf{G}_{\mu\nu}$ and $F_{\mu\nu}$ are the gluon and photon field strengths, respectively; $\mathbf{T}_{ij}$ are the generators of the $SU(3)$ color group; i and j denote color indices (they are omitted in the color-singlet currents). Labels $(V \pm A)$ stand for $\gamma^\mu(1 \pm \gamma^5)$. $Q_{1,2}$ are current-current operators, Q_{3-6} are QCD penguin operators, $Q_{7,8}$ are "magnetic penguin" operators, and $Q_{9,10}$ are semileptonic electroweak penguin operators.

The effective Hamiltonian leads to the free quark $b \to s l^+ l^-$-decay amplitude:

$$M(b \to s\ell^+\ell^-) = \frac{G_F \alpha}{2\sqrt{2}\pi} \lambda_t \left\{ C_9^{\rm eff} \, (\bar{s}b)_{V-A} \left(\bar{l}l\right)_V + C_{10} \, (\bar{s}b)_{V-A} \left(\bar{l}l\right)_A \right. \tag{5}$$
$$\left. - \frac{2m_b}{q^2} \, C_7^{\rm eff} \, \left(\bar{s} \, i\sigma^{\mu\nu} \, (1+\gamma^5) \, q^\nu \, b\right) \left(\bar{l}l\right)_V \right\}.$$

where $C_7^{\rm eff} = C_7 - C_5/3 - C_6$. The Wilson coefficient $C_9^{\rm eff}$ effectively takes into account, first, the contributions from the four-quark operators Q_i (i=1,...,6) and, second, the nonperturbative effects coming from the $c\bar{c}$-resonance contributions which are as usual parametrized by a Breit-Wigner ansatz [23]:

$$C_9^{\rm eff} = C_9 + C_0 \left\{ h(\hat{m}_c, s) + \frac{3\pi}{\alpha^2} \kappa \sum_{V_i=\psi(1s),\psi(2s)} \frac{\Gamma(V_i \to l^+l^-) \, m_{V_i}}{m_{V_i}{}^2 - q^2 - i m_{V_i} \Gamma_{V_i}} \right\}$$
$$- \frac{1}{2} h(1,s) \, (4C_3 + 4C_4 + 3C_5 + C_6) \tag{6}$$
$$- \frac{1}{2} h(0,s) \, (C_3 + 3C_4) + \frac{2}{9} \, (3C_3 + C_4 + 3C_5 + C_6) \, .$$

where $C_0 \equiv 3C_1 + C_2 + 3C_3 + C_4 + 3C_5 + C_6$, $\hat{m}_c = m_c/m_B$, $s = q^2/m_B^2$ and $\kappa = 1/C_0$. Explicit expressions for the function $h(\hat{m}_c, s)$, $m_b = m_b(\mu)$ and $\alpha_s(\mu)$ can be found in [7]. The numerical values of the input parameters are taken from [8] and the corresponding values of the Wilson coefficients used in the numerical calculations are listed in Table 1.

Table 1. Central values of the input parameters and the corresponding values of the Wilson coefficients used in the numerical calculations.

m_W	80.41 GeV	C_1	-0.248
m_Z	91.1867 GeV	C_2	1.107
$\sin^2\theta_W$	0.2233	C_3	0.011
m_c	1.4 GeV	C_4	-0.026
m_t	173.8 GeV	C_5	0.007
$m_{b,\rm pole}$	4.8 GeV	C_6	-0.031
μ	$m_{b,\rm pole}$	$C_7^{\rm eff}$	-0.313
Λ_{QCD}	0.220 GeV	C_9	4.344
α^{-1}	129	C_{10}	-4.669
$\alpha_s(m_Z)$	0.119	C_0	0.362
$\lvert V_{ts}^\dagger V_{tb}\rvert$	0.0385		
$\lvert V_{td}^\dagger V_{tb}\rvert$	0.008		
$\lvert V_{ts}^\dagger V_{tb}\rvert/\lvert V_{cb}\rvert$	1		

3 Form Factors and Differential Decay Distributions

We specify our choice of the momenta as $p_1 = p_2 + k_1 + k_2$ with $p_1^2 = m_1^2$, $p_2^2 = m_2^2$ and $k_1^2 = k_2^2 = \mu^2$ where k_1 and k_2 are the l^+ and l^- momenta, and m_1, m_2, μ are the masses of initial and final mesons and lepton, respectively.

We define dimensionless form factors by

$$< K(D)(p_2)\,|\,\bar{s}(d)\,\gamma_\mu\,b\,|B(B_c)(p_1)\,>= F_+(q^2)P_\mu + F_-(q^2)q_\mu\,, \tag{7}$$

$$< K(D)(p_2)\,|\,\bar{s}(d)\,i\sigma_{\mu\nu}q^\nu\,b\,|\,B(B_c)(p_1)\,>= -\frac{1}{m_1+m_2}\,P_\mu^\perp\,q^2\,F_T(q^2)\,,$$

$$i < D^*(p_2,\epsilon_2)\,|\,\bar{d}\,O_\mu\,b\,|\,B_c(p_1)\,>= \frac{1}{m_1+m_2}\,\epsilon_2^{\dagger\nu}$$
$$\times\{-g_{\mu\nu}\,Pq\,A_0(q^2) + P_\mu P_\nu\,A_+(q^2) + q_\mu P_\nu\,A_-(q^2) + i\varepsilon_{\mu\nu\alpha\beta}P^\alpha q^\beta\,V(q^2)\}\,,$$

$$i < D^*(p_2,\epsilon_2)\,|\,\bar{d}\,i\sigma_{\mu\nu}q^\nu(1+\gamma_5)\,b\,|\,B_c(p_1)\,>=$$
$$= \epsilon_2^{\dagger\nu}\{\,g_{\mu\nu}^\perp\,Pq\,a_0(q^2) - P_\mu^\perp\,P_\nu\,a_+(q^2) - i\varepsilon_{\mu\nu\alpha\beta}P^\alpha q^\beta\,g(q^2)\}$$

where $P = p_1 + p_2$, $q = p_1 - p_2$, $P_\mu^\perp \doteq P_\mu - q_\mu Pq/q^2$, $g_{\mu\nu}^\perp \doteq g_{\mu\nu} - q_\mu q_\nu/q^2$, and $\epsilon_2^\dagger$ is the polarization four-vector of the D^*.

The matrix elements of the exclusive transitions $B \to K\bar{l}l$ and $B_c \to D(D^*)\bar{l}l$ are written as

$$M\left(B(B_c) \to K(D^*)\bar{l}l\right) = \frac{G_F}{\sqrt{2}}\cdot\frac{\alpha\lambda_t}{2\pi}\,\left\{T_1^\mu\,(\bar{l}\gamma_\mu l) + T_2^\mu\,(\bar{l}\gamma_\mu\gamma_5 l)\right\} \tag{8}$$

where the quantities T_i^μ are expressed through the form factors and the Wilson coefficients in the following manner:

(a) $B(B_c) \to K(D)\bar{l}l$-decay:

$$T_i^\mu = \mathcal{F}_+^{(i)}\,P^\mu + \mathcal{F}_-^{(i)}\,q^\mu \qquad (i=1,2)\,, \tag{9}$$

$$\mathcal{F}_+^{(1)} = C_9^{\mathrm{eff}}\,F_+ + C_7^{\mathrm{eff}}\,F_T\,\frac{2m_b}{m_1+m_2}\,,$$
$$\mathcal{F}_-^{(1)} = C_9^{\mathrm{eff}}\,F_- - C_7^{\mathrm{eff}}\,F_T\,\frac{2m_b}{m_1+m_2}\,\frac{Pq}{q^2}\,,$$

$$\mathcal{F}_\pm^{(2)} = C_{10}\,F_\pm\,.$$

(b) $B_c \to D^*\bar{l}l$-decay:

$$T_i^\mu = T_i^{\mu\nu}\,\epsilon_{2\nu}^\dagger\,, \qquad (i=1,2)\,, \tag{10}$$

$$T_i^{\mu\nu} = \frac{1}{m_1+m_2}\,\left\{-Pq\,g^{\mu\nu}\,A_0^{(i)} + P^\mu P^\nu\,A_+^{(i)} + q^\mu P^\nu\,A_-^{(i)} + i\varepsilon^{\mu\nu\alpha\beta}P_\alpha q_\beta\,V^{(i)}\right\}$$

$$
\begin{aligned}
V^{(1)} &= C_9^{\rm eff}\, V + C_7^{\rm eff}\, g\, \frac{2m_b(m_1+m_2)}{q^2}\,,\\
A_0^{(1)} &= C_9^{\rm eff}\, A_0 + C_7^{\rm eff}\, a_0\, \frac{2m_b(m_1+m_2)}{q^2}\,,\\
A_+^{(1)} &= C_9^{\rm eff}\, A_+ + C_7^{\rm eff}\, a_+\, \frac{2m_b(m_1+m_2)}{q^2}\,,\\
A_-^{(1)} &= C_9^{\rm eff}\, A_- + C_7^{\rm eff}\, (a_0 - a_+)\, \frac{2m_b(m_1+m_2)}{q^2}\,\frac{Pq}{q^2}\,,\\
V^{(2)} &= C_{10}\, V\,, \qquad A_0^{(2)} = C_{10}\, A_0\,, \qquad A_\pm^{(2)} = C_{10}\, A_\pm\,.
\end{aligned}
$$

Let us first consider the polar angle decay distribution differential in the momentum transfer squared q^2. The polar angle is defined by the angle between $\mathbf{q} = \mathbf{p}_1 - \mathbf{p}_2$ and $\mathbf{k}_1$ (l^+l^- rest frame) as shown in Fig. 1. One has

$$
\begin{aligned}
\frac{d^2\Gamma}{dq^2 d\cos\theta} &= \frac{|\mathbf{p_2}|\, v}{(2\pi)^3\, 4\, m_1^3}\cdot\frac{1}{8}\sum_{\rm pol}|M|^2 \qquad (11)\\
&= \frac{G_F^2}{(2\pi)^3}\,(\frac{\alpha|\lambda_t|}{2\,\pi})^2\,\frac{|\mathbf{p_2}|\, v}{8m_1^2}\cdot\frac{1}{2}\Big\{L_{\mu\nu}^{(1)}\cdot(H_{11}^{\mu\nu} + H_{22}^{\mu\nu})\\
&- \frac{1}{2}\, L_{\mu\nu}^{(2)}\cdot(q^2\, H_{11}^{\mu\nu} + (q^2 - 4\mu^2)\, H_{22}^{\mu\nu}) + L_{\mu\nu}^{(3)}\cdot(H_{12}^{\mu\nu} + H_{21}^{\mu\nu})\Big\}
\end{aligned}
$$

where $|\mathbf{p_2}| = \lambda^{1/2}(m_1^2, m_2^2, q^2)/2m_1$ is the momentum of the final meson and $v = \sqrt{1 - 4\mu^2/q^2}$ is the lepton velocity both given in the $B(B_c)$-rest frame. We have introduced lepton and hadron tensors as

$$
\begin{aligned}
L_{\mu\nu}^{(1)} &= k_{1\mu}k_{2\nu} + k_{2\mu}k_{1\nu}\,, \qquad L_{\mu\nu}^{(2)} = g_{\mu\nu}\,, \qquad L_{\mu\nu}^{(3)} = i\varepsilon_{\mu\nu\alpha\beta}k_1^\alpha k_2^\beta\,, \qquad (12)\\
H_{ij}^{\mu\nu} &= T_i^\mu\, T_j^{\dagger\nu}\,.
\end{aligned}
$$

4 Helicity Amplitudes and Two-Fold Distributions

The Lorentz contractions in (11) can be evaluated in terms of helicity amplitudes as described in [24]. First, we define an orthonormal and complete helicity basis $\epsilon^\mu(m)$ with the three spin 1 components orthogonal to the momentum transfer q^μ, i.e. $\epsilon^\mu(m)q_\mu = 0$ for $m = \pm, 0$, and the spin 0 (time)-component $m = t$ with $\epsilon^\mu(t) = q^\mu/\sqrt{q^2}$.

The orthonormality and completeness properties read

$$
\epsilon_\mu^\dagger(m)\epsilon^\mu(n) = g_{mn}\,, \qquad \epsilon_\mu(m)\epsilon_\nu^\dagger(n)g_{mn} = g_{\mu\nu} \qquad (13)
$$

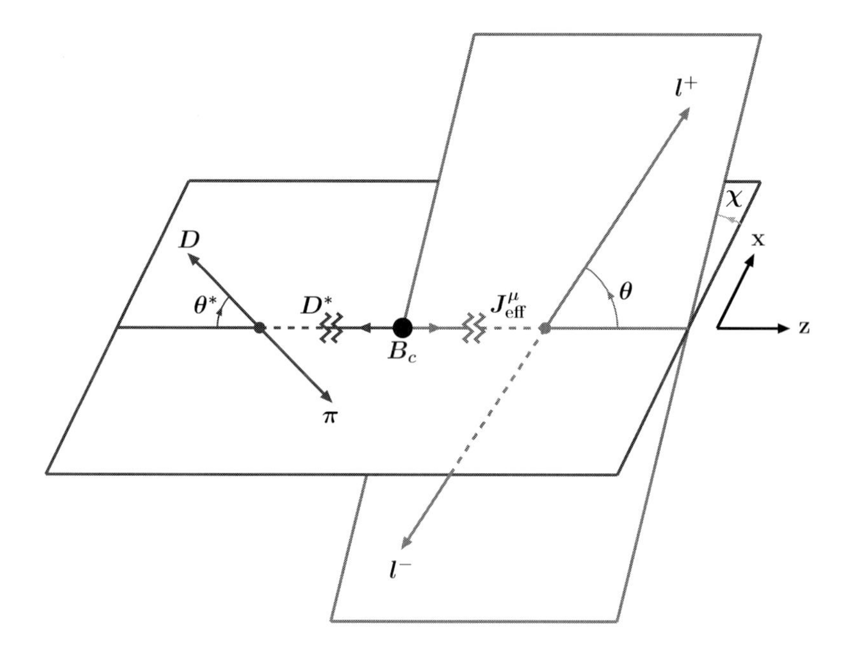

Fig. 1. Definition of angles θ, θ^* and χ in the cascade decay $B_c \to D^*(\to D\pi)\bar{l}l$.

with $(m,n = t,\pm,0)$ and $g_{mn} = \text{diag}\,(+\,,\,-\,,\,-\,,\,-\,)$. We include the time component polarization vector $\epsilon^\mu(t)$ in the set because we want to discuss lepton mass effects in the following.

Using the completeness property we rewrite the contraction of the lepton and hadron tensors in (11) according to

$$\begin{aligned} L^{(k)\mu\nu} H^{ij}_{\mu\nu} &= L^{(k)}_{\mu'\nu'} g^{\mu'\mu} g^{\nu'\nu} H^{ij}_{\mu\nu} = L^{(k)}_{\mu'\nu'} \epsilon^{\mu'}(m)\epsilon^{\dagger\mu}(m') g_{mm'} \epsilon^{\dagger\nu'}(n)\epsilon^{\nu}(n') g_{nn'} H^{ij}_{\mu\nu} \\ &= L^{(k)}(m,n) g_{mm'} g_{nn'} H^{ij}(m',n') \end{aligned} \tag{14}$$

where we have introduced the lepton and hadron tensors in the space of the helicity components

$$L^{(k)}(m,n) = \epsilon^\mu(m)\epsilon^{\dagger\nu}(n) L^{(k)}_{\mu\nu}, \qquad H^{ij}(m,n) = \epsilon^{\dagger\mu}(m)\epsilon^{\nu}(n) H^{ij}_{\mu\nu}. \tag{15}$$

The point is that the two tensors can be evaluated in two different Lorentz systems. The lepton tensors $L^{(k)}(m,n)$ will be evaluated in the $\bar{l}l$-CM system whereas the hadron tensors $H^{ij}(m,n)$ will be evaluated in the $B(B_c)$ rest system.

In the $B(B_c)$ rest frame one has

$$\begin{aligned} p_1^\mu &= (\,m_1\,,\;0,\;0,\;0\,)\,, \\ p_2^\mu &= (\,E_2\,,\;0,\;0,\;-|\mathbf{p_2}|\,)\,, \\ q^\mu &= (\,q_0\,,\;0,\;0,\;|\mathbf{p_2}|\,)\,, \end{aligned} \tag{16}$$

where $E_2 = (m_1^2+m_2^2-q^2)/2m_1$ and $q_0 = (m_1^2-m_2^2+q^2)/2m_1$. In the $B(B_c)$-rest frame the polarization vectors of the effective current read

$$\begin{aligned}
\epsilon^\mu(t) &= \frac{1}{\sqrt{q^2}}(\,q_0\,,\;0,\;0,\;|\mathbf{p_2}|\,)\,,\\
\epsilon^\mu(\pm) &= \frac{1}{\sqrt{2}}(\,0,\;\mp 1\,,\;-i\,,\;0\,)\,,\\
\epsilon^\mu(0) &= \frac{1}{\sqrt{q^2}}(\,|\mathbf{p_2}|\,,\;0,\;0,\;q_0\,)\,.
\end{aligned} \tag{17}$$

Using this basis one can express the components of the hadronic tensors through the invariant form factors defined in (7).

(a) $B(B_c) \to K(D)$ transition:

$$H^{ij}(m,n) = \left(\epsilon^{\dagger\mu}(m)T^i_\mu\right)\cdot\left(\epsilon^{\dagger\nu}(n)T^j_\nu\right)^\dagger \equiv H^i(m)H^{\dagger j}(n) \tag{18}$$

The helicity form factors $H^i(m)$ are given in terms of the invariant form factors. One has

$$\begin{aligned}
H^i(t) &= \frac{1}{\sqrt{q^2}}(Pq\,\mathcal{F}^i_+ + q^2\,\mathcal{F}^i_-)\,,\\
H^i(\pm) &= 0\,,\\
H^i(0) &= \frac{2\,m_1\,|\mathbf{p_2}|}{\sqrt{q^2}}\,\mathcal{F}^i_+\,.
\end{aligned} \tag{19}$$

(b) $B_c \to D^*$ transition:

$$\begin{aligned}
H^{ij}(m,n) &= \epsilon^{\dagger\mu}(m)\epsilon^\nu(n)H^{ij}_{\mu\nu} = \epsilon^{\dagger\mu}(m)\epsilon^\nu(n)T^i_{\mu\alpha}\left(-g^{\alpha\beta}+\frac{p_2^\alpha p_2^\beta}{m_2^2}\right)T^{\dagger j}_{\beta\nu}\\
&= \epsilon^{\dagger\mu}(m)\epsilon^\nu(n)T^i_{\mu\alpha}\epsilon_2^{\dagger\alpha}(r)\epsilon_2^\beta(s)\delta_{rs}T^{\dagger j}_{\beta\nu}\\
&= \epsilon^{\dagger\mu}(m)\epsilon_2^{\dagger\alpha}(r)T^i_{\mu\alpha}\cdot\left(\epsilon^{\dagger\nu}(n)\epsilon_2^{\dagger\beta}(s)T^j_{\nu\beta}\right)^\dagger\delta_{rs} = H^i(m)H^{\dagger j}(n).
\end{aligned} \tag{20}$$

From angular momentum conservation one has $r = m$ and $s = n$ for $m, n = \pm, 0$ and $r, s = 0$ for $m, n = t$. For further evaluation one needs to specify the helicity components $\epsilon_2(m)$ $(m = \pm, 0)$ of the polarization vector of the D^*. They read

$$\epsilon_2^\mu(\pm) = \frac{1}{\sqrt{2}}(0,\;\pm 1\,,\;-i\,,\;0\,)\,, \qquad \epsilon_2^\mu(0) = \frac{1}{m_2}(|\mathbf{p_2}|\,,\;0\,,\;0\,,\;-E_2\,)\,. \tag{21}$$

They satisfy the orthonormality and completeness properties:

$$\epsilon_2^{\dagger\mu}(r)\epsilon_{2\mu}(s) = -\delta_{rs}, \qquad \epsilon_{2\mu}(r)\epsilon^\dagger_{2\nu}(s)\delta_{rs} = -g_{\mu\nu} + \frac{p_{2\mu}p_{2\nu}}{m_2^2}. \tag{22}$$

Finally one obtains the non-zero components of the hadron tensors

$$\begin{aligned}
H^i(t) &= \epsilon^{\dagger\mu}(t)\epsilon_2^{\dagger\alpha}(0)T^i_{\mu\alpha} = \frac{1}{m_1+m_2}\frac{m_1\,|\mathbf{p_2}|}{m_2\sqrt{q^2}}\left(Pq\,(-A^i_0+A^i_+)+q^2A^i_-\right),\\
H^i(\pm) &= \epsilon^{\dagger\mu}(\pm)\epsilon_2^{\dagger\alpha}(\pm)T^i_{\mu\alpha} = \frac{1}{m_1+m_2}\left(-Pq\,A^i_0 \pm 2\,m_1\,|\mathbf{p_2}|\,V^i\right), \qquad (23)\\
H^i(0) &= \epsilon^{\dagger\mu}(0)\epsilon_2^{\dagger\alpha}(0)T^i_{\mu\alpha}\\
&= \frac{1}{m_1+m_2}\frac{1}{2\,m_2\sqrt{q^2}}\left(-Pq\,(m_1^2+m_2^2-q^2)\,A^i_0+4\,m_1^2\,|\mathbf{p_2}|^2\,A^i_+\right).
\end{aligned}$$

The lepton tensors $L^{(k)}(m,n)$ are evaluated in the $\bar{l}l$-CM system $\mathbf{k}_1+\mathbf{k}_2=0$. One has (see Fig. 1)

$$\begin{aligned}
q^\mu &= (\,\sqrt{q^2}\,,\;0\,,\;0\,,\;0\,)\,,\\
k_1^\mu &= (\,E_1\,,\;|\mathbf{k_1}|\sin\theta\cos\chi\,,\;|\mathbf{k_1}|\sin\theta\sin\chi\,,\;|\mathbf{k_1}|\cos\theta\,)\,, \qquad (24)\\
k_2^\mu &= (\,E_1\,,\;-|\mathbf{k_1}|\sin\theta\cos\chi\,,\;-|\mathbf{k_1}|\sin\theta\sin\chi\,,\;-|\mathbf{k_1}|\cos\theta\,)\,,
\end{aligned}$$

with $E_1=\sqrt{q^2}/2$ and $|\mathbf{k_1}|=\sqrt{q^2-4\mu^2}/2$. The longitudinal and time component polarization vectors in the $\bar{l}l$ rest frame can be read off from (17) and are given by $\epsilon^\mu(0)=(0,0,0,1)$ and $\epsilon(t)=(1,0,0,0)$ whereas the transverse parts remain unchanged from (17).

The differential $(q^2,\cos\theta)$ distribution finally reads

$$\begin{aligned}
\frac{d\Gamma(H_{in}\to H_f\bar{l}l)}{dq^2d(\cos\theta)} &= \frac{3}{8}\,(1+\cos^2\theta)\cdot\frac{1}{2}\left(\frac{d\Gamma^{11}_U}{dq^2}+\frac{d\Gamma^{22}_U}{dq^2}\right) \qquad (25)\\
&+\frac{3}{4}\sin^2\theta\cdot\frac{1}{2}\left(\frac{d\Gamma^{11}_L}{dq^2}+\frac{d\Gamma^{22}_L}{dq^2}\right)\\
&-\;v\cdot\frac{3}{4}\cos\theta\cdot\frac{d\Gamma^{12}_P}{dq^2}\\
&+\frac{3}{4}\sin^2\theta\cdot\frac{1}{2}\frac{d\tilde{\Gamma}^{11}_U}{dq^2}-\frac{3}{8}\,(1+\cos^2\theta)\cdot\frac{d\tilde{\Gamma}^{22}_U}{dq^2}\\
&+\frac{3}{2}\cos^2\theta\cdot\frac{1}{2}\frac{d\tilde{\Gamma}^{11}_L}{dq^2}-\frac{3}{4}\sin^2\theta\cdot\frac{d\tilde{\Gamma}^{22}_L}{dq^2}+\frac{1}{4}\frac{d\tilde{\Gamma}^{22}_S}{dq^2}\,.
\end{aligned}$$

Integrating over $\cos\theta$ one obtains

$$\begin{aligned}
\frac{d\Gamma(H_{in}\to H_f\bar{l}l)}{dq^2} &= \frac{1}{2}\left(\frac{d\Gamma^{11}_U}{dq^2}+\frac{d\Gamma^{22}_U}{dq^2}+\frac{d\Gamma^{11}_L}{dq^2}+\frac{d\Gamma^{22}_L}{dq^2}\right) \qquad (26)\\
&+\frac{1}{2}\frac{d\tilde{\Gamma}^{11}_U}{dq^2}-\frac{d\tilde{\Gamma}^{22}_U}{dq^2}+\frac{1}{2}\frac{d\tilde{\Gamma}^{11}_L}{dq^2}-\frac{d\tilde{\Gamma}^{22}_L}{dq^2}+\frac{1}{2}\frac{d\tilde{\Gamma}^{22}_S}{dq^2}\,,
\end{aligned}$$

where the partial helicity rates $d\Gamma^{ij}_X/dq^2$ and $d\tilde{\Gamma}^{ij}_X/dq^2$ ($X = U, L, P, S$; $i, j = 1, 2$) are defined as

$$\frac{d\Gamma^{ij}_X}{dq^2} = \frac{G_F^2}{(2\pi)^3}\left(\frac{\alpha|\lambda_t|}{2\pi}\right)^2 \frac{|\mathbf{p_2}|\, q^2\, v}{12\, m_1^2}\, H^{ij}_X\,, \qquad \frac{d\tilde{\Gamma}^{ij}_X}{dq^2} = \frac{2\,\mu^2}{q^2}\frac{d\Gamma^{ij}_X}{dq^2}\,. \tag{27}$$

The relevant bilinear combinations of the helicity amplitudes are defined in Table 2.

5 The Four-Fold Angle Distribution in the Cascade Decay $B_c \to D^*(\to D\pi)\bar{l}l$

The lepton-hadron correlation function $L_{\mu\nu}H^{\mu\nu}$ reveals even more structure when one uses the cascade decay $B_c \to D^*(\to D\pi)\bar{l}l$ to analyze the polarization of the D^*. The hadron tensor now reads

$$H^{ij}_{\mu\nu} = T^i_{\mu\alpha}(T^j_{\nu\beta})^\dagger \frac{3}{2\,|\mathbf{p_3}|}\mathrm{Br}(K^* \to K\pi)p_{3\alpha'}p_{3\beta'}S^{\alpha\alpha'}(p_2)S^{\beta\beta'}(p_2) \tag{28}$$

where $S^{\alpha\alpha'}(p_2) = -g^{\alpha\alpha'} + p_2^\alpha p_2^{\alpha'}/m_2^2$ is the standard spin 1 tensor, $p_2 = p_3 + p_4$, $p_3^2 = m_D^2$, $p_4^2 = m_\pi^2$, and p_3 and p_4 are the momenta of the D and the π, respectively. The relative configuration of the (D,π)- and $(\bar{l}l)$-planes is shown in Fig. 1.

In the rest frame of the D^* one has

$$\begin{aligned} p_2^\mu &= (m_{D^*}, \mathbf{0}), \\ p_3^\mu &= (\,E_D\,,\ |\mathbf{p_3}|\,\sin\theta^*\,,\ 0,\ -|\mathbf{p_3}|\,\cos\theta^*\,)\,, \\ p_4^\mu &= (\,E_\pi\,,\ -|\mathbf{p_3}|\,\sin\theta^*\,,\ 0,\ |\mathbf{p_3}|\,\cos\theta^*\,)\,, \\ |\mathbf{p_3}| &= \lambda^{1/2}(m_{D^*}^2, m_D^2, m_\pi^2)/(2\,m_{D^*})\,. \end{aligned} \tag{29}$$

Without loss of generality we set the azimuthal angle χ^* of the (D,π)-plane to zero. According to (21) the rest frame polarization vectors of the D^* are given by

$$\epsilon_2^\mu(\pm) = \frac{1}{\sqrt{2}}(\,0\,,\ \pm,\ -i\,,\ 0\,)\,, \qquad \epsilon_2^\mu(0) = (\,0\,,\ 0\,,\ 0\,,\ -1\,)\,. \tag{30}$$

The spin 1 tensor $S^{\alpha\alpha'}(p_2)$ is then written as

$$S^{\alpha\alpha'}(p_2) = -g^{\alpha\alpha'} + \frac{p_2^\alpha p_2^{\alpha'}}{m_2^2} = \sum_{m=\pm,0}\epsilon_2^\alpha(m)\epsilon_2^{\dagger\alpha'}(m) \tag{31}$$

Following basically the same trick as in (14) the contraction of the lepton and hadron tensors may be written through helicity components as

$$\begin{aligned} L^{(k)\mu\nu}H^{ij}_{\mu\nu} &= \epsilon^{\mu'}(m)\epsilon^{\dagger\nu'}(n)L^k_{\mu'\nu'}g_{mn'}g_{nn'}\epsilon^{\dagger\mu}(m')\epsilon^\nu(n')H^{ij}_{\mu\nu} \\ &= L^k(m,n)g_{mm'}g_{nn'}\left(\epsilon^{\dagger\mu}(m')\epsilon_2^{\dagger\alpha}(r)T^i_{\mu\alpha}\right)\left(\epsilon^{\dagger\nu}(n')\epsilon_2^{\dagger\alpha}(s)T^j_{\nu\beta}\right)^\dagger \end{aligned} \tag{32}$$

Table 2. Bilinear combinations of the helicity amplitudes that enter in the four-fold decay distribution (27).

Definition	Property	Title
$H_U^{ij} = \mathrm{Re}\left(H_+^i H_+^{\dagger\, j}\right) + \mathrm{Re}\left(H_-^i H_-^{\dagger\, j}\right)$ $H_{IU}^{ij} = \mathrm{Im}\left(H_+^i H_+^{\dagger\, j}\right) + \mathrm{Im}\left(H_-^i H_-^{\dagger\, j}\right)$	$H_U^{ij} = H_U^{ji}$ $H_{IU}^{ij} = -H_{IU}^{ji}$	**U**npolarized-transverse
$H_P^{ij} = \mathrm{Re}\left(H_+^i H_+^{\dagger\, j}\right) - \mathrm{Re}\left(H_-^i H_-^{\dagger\, j}\right)$ $H_{IP}^{ij} = \mathrm{Im}\left(H_+^i H_+^{\dagger\, j}\right) - \mathrm{Im}\left(H_-^i H_-^{\dagger\, j}\right)$	$H_P^{ij} = H_P^{ji}$ $H_{IP}^{ij} = -H_{IP}^{ji}$	**P**arity-odd
$H_T^{ij} = \mathrm{Re}\left(H_+^i H_-^{\dagger\, j}\right)$ $H_{IT}^{ij} = \mathrm{Im}\left(H_+^i H_-^{\dagger\, j}\right)$		**T**ransverse-interference
$H_L^{ij} = \mathrm{Re}\left(H_0^i H_0^{\dagger\, j}\right)$ $H_{IL}^{ij} = \mathrm{Im}\left(H_0^i H_0^{\dagger\, j}\right)$	$H_L^{ij} = H_L^{ji}$ $H_{IL}^{ij} = -H_{IL}^{ji}$	**L**ongitudinal
$H_S^{ij} = 3\,\mathrm{Re}\left(H_t^i H_t^{\dagger\, j}\right)$ $H_{IS}^{ij} = 3\,\mathrm{Im}\left(H_t^i H_t^{\dagger\, j}\right)$	$H_S^{ij} = H_S^{ji}$ $H_{IS}^{ij} = -H_{IS}^{ji}$	**S**calar
$H_{SL}^{ij} = \mathrm{Re}\left(H_t^i H_0^{\dagger\, j}\right)$ $H_{ISL}^{ij} = \mathrm{Im}\left(H_t^i H_0^{\dagger\, j}\right)$		**S**calar-**L**ongitudinal-interference
$H_I^{ij} = \frac{1}{2}\left[\mathrm{Re}\left(H_+^i H_0^{\dagger\, j}\right) + \mathrm{Re}\left(H_-^i H_0^{\dagger\, j}\right)\right]$ $H_{II}^{ij} = \frac{1}{2}\left[\mathrm{Im}\left(H_+^i H_0^{\dagger\, j}\right) + \mathrm{Im}\left(H_-^i H_0^{\dagger\, j}\right)\right]$		transverse-longitudinal-**I**nterference
$H_A^{ij} = \frac{1}{2}\left[\mathrm{Re}\left(H_+^i H_0^{\dagger\, j}\right) - \mathrm{Re}\left(H_-^i H_0^{\dagger\, j}\right)\right]$ $H_{IA}^{ij} = \frac{1}{2}\left[\mathrm{Im}\left(H_+^i H_0^{\dagger\, j}\right) - \mathrm{Im}\left(H_-^i H_0^{\dagger\, j}\right)\right]$		parity-**A**symmetric
$H_{ST}^{ij} = \frac{1}{2}\left[\mathrm{Re}\left(H_+^i H_t^{\dagger\, j}\right) + \mathrm{Re}\left(H_-^i H_t^{\dagger\, j}\right)\right]$ $H_{IST}^{ij} = \frac{1}{2}\left[\mathrm{Im}\left(H_+^i H_t^{\dagger\, j}\right) + \mathrm{Im}\left(H_-^i H_t^{\dagger\, j}\right)\right]$		**S**calar-**T**ransverse-interference
$H_{SA}^{ij} = \frac{1}{2}\left[\mathrm{Re}\left(H_+^i H_t^{\dagger\, j}\right) - \mathrm{Re}\left(H_-^i H_t^{\dagger\, j}\right)\right]$ $H_{ISA}^{ij} = \frac{1}{2}\left[\mathrm{Im}\left(H_+^i H_t^{\dagger\, j}\right) - \mathrm{Im}\left(H_-^i H_t^{\dagger\, j}\right)\right]$		**S**calar-**A**symmetric-interference

$$\times \; p_3\epsilon_2(r)\cdot p_3\epsilon_2^\dagger(s)\frac{3\,\mathrm{Br}(D^*\to D\pi)}{2\,|\mathbf{p_3}|}$$
$$= \frac{3\,\mathrm{Br}(D^*\to D\pi)}{2\,|\mathbf{p_3}|}\biggl(L^k(t,t)|H^{ij}(t)|^2\cdot(p_3\epsilon_2^\dagger(0))^2$$
$$+\sum_{m,n=\pm,0} L^k(m,n)H^i(m)H^{\dagger j}(n)\cdot p_3\epsilon_2(m)\cdot p_3\epsilon_2^\dagger(n)$$
$$-\sum_{n=\pm,0} L^k(t,n)H^i(t)H^{\dagger j}(n)\cdot p_3\epsilon_2(0)\cdot p_3\epsilon_2^\dagger(n)$$
$$-\sum_{m=\pm,0} L^k(m,t)H^i(m)H^{\dagger j}(t)\cdot p_3\epsilon_2(m)\cdot p_3\epsilon_2^\dagger(0)\biggr)$$

Using these results one obtains the full four-fold angular decay distribution

$$\frac{d\Gamma(B_c\to D^*(\to D\pi)\bar{l}l)}{dq^2\,d\cos\theta\,d(\chi/2\pi)\,d\cos\theta^*} = \mathrm{Br}(D^*\to D\pi) \tag{33}$$
$$\times\biggl\{\frac{3}{8}(1+\cos^2\theta)\cdot\frac{3}{4}\sin^2\theta^*\cdot\frac{1}{2}\left(\frac{d\Gamma_U^{11}}{dq^2}+\frac{d\Gamma_U^{22}}{dq^2}\right)$$
$$+\frac{3}{4}\sin^2\theta\cdot\frac{3}{2}\cos^2\theta^*\cdot\frac{1}{2}\left(\frac{d\Gamma_L^{11}}{dq^2}+\frac{d\Gamma_L^{22}}{dq^2}\right)$$
$$-\frac{3}{4}\sin^2\theta\cdot\cos 2\chi\cdot\frac{3}{4}\sin^2\theta^*\cdot\frac{1}{2}\left(\frac{d\Gamma_T^{11}}{dq^2}+\frac{d\Gamma_T^{22}}{dq^2}\right)$$
$$+\frac{9}{16}\sin 2\theta\cdot\cos\chi\cdot\sin 2\theta^*\cdot\frac{1}{2}\left(\frac{d\Gamma_I^{11}}{dq^2}+\frac{d\Gamma_I^{22}}{dq^2}\right)$$
$$+v\biggl[-\frac{3}{4}\cos\theta\cdot\frac{3}{4}\sin^2\theta^*\cdot\frac{d\Gamma_P^{12}}{dq^2}$$
$$-\frac{9}{8}\sin\theta\cdot\cos\chi\cdot\sin 2\theta^*\cdot\frac{1}{2}\left(\frac{d\Gamma_A^{12}}{dq^2}+\frac{d\Gamma_A^{21}}{dq^2}\right)$$
$$+\frac{9}{16}\sin\theta\cdot\sin\chi\cdot\sin 2\theta^*\cdot\left(\frac{d\Gamma_{II}^{12}}{dq^2}+\frac{d\Gamma_{II}^{21}}{dq^2}\right)\biggr]$$
$$-\frac{9}{32}\sin 2\theta\cdot\sin\chi\cdot\sin 2\theta^*\cdot\left(\frac{d\Gamma_{IA}^{11}}{dq^2}+\frac{d\Gamma_{IA}^{22}}{dq^2}\right)$$
$$+\frac{9}{32}\sin^2\theta\cdot\sin 2\chi\cdot\sin^2\theta^*\cdot\left(\frac{d\Gamma_{IT}^{11}}{dq^2}+\frac{d\Gamma_{IT}^{22}}{dq^2}\right)$$
$$+\frac{3}{4}\sin^2\theta\cdot\frac{3}{4}\sin^2\theta^*\cdot\frac{1}{2}\cdot\frac{d\tilde{\Gamma}_U^{11}}{dq^2}-\frac{3}{8}(1+\cos^2\theta)\cdot\frac{3}{4}\sin^2\theta^*\cdot\frac{d\tilde{\Gamma}_U^{22}}{dq^2}$$
$$+\frac{3}{2}\cos^2\theta\cdot\frac{3}{2}\cos^2\theta^*\cdot\frac{1}{2}\cdot\frac{d\tilde{\Gamma}_L^{11}}{dq^2}-\frac{3}{4}\sin^2\theta\cdot\frac{3}{2}\cos^2\theta^*\cdot\frac{d\tilde{\Gamma}_L^{22}}{dq^2}$$
$$+\frac{3}{4}\sin^2\theta\cdot\cos 2\chi\cdot\frac{3}{4}\sin^2\theta^*\cdot\left(\frac{d\tilde{\Gamma}_T^{11}}{dq^2}+\frac{d\tilde{\Gamma}_T^{22}}{dq^2}\right)$$

$$
-\frac{9}{8}\sin 2\theta\cdot\cos\chi\cdot\sin 2\theta^*\cdot\frac{1}{2}\left(\frac{d\tilde{\Gamma}_I^{11}}{dq^2}+\frac{d\tilde{\Gamma}_I^{22}}{dq^2}\right)+\frac{3}{2}\cos^2\theta^*\cdot\frac{1}{4}\frac{d\tilde{\Gamma}_S^{22}}{dq^2}
$$

$$
+\frac{9}{16}\sin 2\theta\cdot\sin\chi\cdot\sin 2\theta^*\cdot\left(\frac{d\Gamma_{IA}^{11}}{dq^2}+\frac{d\Gamma_{IA}^{22}}{dq^2}\right)
$$

$$
-\frac{9}{16}\sin^2\theta\cdot\sin 2\chi\cdot\sin^2\theta^*\cdot\left(\frac{d\Gamma_{IT}^{11}}{dq^2}+\frac{d\Gamma_{IT}^{22}}{dq^2}\right)\Bigg\}
$$

Integrating (33) over $\cos\theta^*$ and χ one recovers the two-fold $(q^2,\cos\theta)$ distribution of (25). Note that a similar four-fold distribution has also been obtained in [14,25–28] using, however, the zero lepton mass approximation. If there are sufficient data one can attempt to fit them to the full four-fold decay distribution and thereby extract the values of the coefficient functions $d\Gamma_X/dq^2$ and, in the case $l=\tau$ the coefficient functions $d\tilde{\Gamma}_X/dq^2$. Instead of considering the full four-fold decay distribution one can analyze single angle distributions by integrating out two of the remaining angles as done in [33].

6 Model Form Factors

We will employ the relativistic constituent quark model [18,19] to calculate the form factors relevant to the decays $B\to K\bar{l}l$ and $B_c\to D(D^*)\bar{l}l$. This model is based on an effective interaction Lagrangian which describes the coupling between hadrons and their constituent quarks.

For example, the coupling of the meson H to its constituent quarks q_1 and $\bar{q}_2$ is given by the Lagrangian

$$
\mathcal{L}_{\rm int}(x)=g_H H(x)\int dx_1\int dx_2 F_H(x,x_1,x_2)\bar{q}(x_1)\Gamma_H\lambda_H q(x_2)\,. \tag{34}
$$

Here, λ_H and Γ_H are Gell-Mann and Dirac matrices which entail the flavor and spin quantum numbers of the meson field $H(x)$. The function F_H is related to the scalar part of the Bethe-Salpeter amplitude and characterizes the finite size of the meson. The function F_H must be invariant under the translation $F_H(x+a,x_1+a,x_2+a)=F_H(x,x_1,x_2)$.

In our previous papers we have used the so-called impulse approximation for the evaluation of the Feynman diagrams. In the impulse approximation one omits a possible dependence of the vertex functions on external momenta. The impulse approximation therefore entails a certain dependence on how loop momenta are routed through the diagram at hand. This problem no longer exists in the present full treatment where the impulse approximation is no longer used. In the present calculation we will use a particular form of the vertex function given by

$$
F_H(x,x_1,x_2)=\delta\left(x-\frac{m_1x_1+m_2x_2}{m_1+m_2}\right)\Phi_H((x_1-x_2)^2). \tag{35}
$$

where m_1 and m_2 are the constituent quark masses. The vertex function F_H evidently satisfies the above translational invariance condition. As mentioned before we no longer use the impulse approximation in the present calculation.

The coupling constants g_H in (34) are determined by the so called *compositeness condition* proposed in [20] and extensively used in [21]. The compositeness condition means that the renormalization constant of the meson field is set equal to zero

$$Z_H = 1 - \frac{3g_H^2}{4\pi^2}\tilde{\Pi}'_H(m_H^2) = 0 \tag{36}$$

where $\tilde{\Pi}'_H$ is the derivative of the meson mass operator. For the pseudoscalar and vector mesons treated in this paper one has

$$\begin{aligned}\tilde{\Pi}'_P(p^2) &= \frac{1}{2p^2}\,p^\alpha\frac{d}{dp^\alpha}\int\frac{d^4k}{4\pi^2 i}\tilde{\Phi}_P^2(-k^2)\\ &\times \mathrm{tr}\Big[\gamma^5 S_1(\not k + w_{21}\not p)\gamma^5 S_2(\not k - w_{12}\not p)\Big]\\ \tilde{\Pi}'_V(p^2) &= \frac{1}{3}\left[g^{\mu\nu} - \frac{p^\mu p^\nu}{p^2}\right]\frac{1}{2p^2}\,p^\alpha\frac{d}{dp^\alpha}\int\frac{d^4k}{4\pi^2 i}\tilde{\Phi}_V^2(-k^2)\\ &\times \mathrm{tr}\Big[\gamma^\nu S_1(\not k + w_{21}\not p)\gamma^\mu S_2(\not k - w_{12}\not p)\Big]\end{aligned}$$

where $w_{ij} = m_j/(m_i+m_j)$.

The leptonic decay constant f_P is calculated from

$$\frac{3g_P}{4\pi^2}\int\frac{d^4k}{4\pi^2 i}\tilde{\Phi}_P(-k^2)\mathrm{tr}\Big[O^\mu S_1(\not k + w_{21}\not p)\gamma^5 S_2(\not k - w_{12}\not p)\Big] = f_P\,p^\mu. \tag{37}$$

The transition form factors $P(p_1) \to P(p_2), V(p_2)$ can be calculated from the Feynman integral corresponding to the diagram of Fig. 2:

$$\begin{aligned}\Lambda^{\Gamma^\mu}(p_1,p_2) &= \frac{3g_P g_{P'(V)}}{4\pi^2}\int\frac{d^4k}{4\pi^2 i}\tilde{\Phi}_P(-(k+w_{13}\,p_1)^2)\,\tilde{\Phi}_{P'(V)}(-(k+w_{23}\,p_2)^2)\\ &\times \mathrm{tr}\Big[S_2(\not k+\not p_2)\Gamma^\mu S_1(\not k+\not p_1)\gamma^5 S_3(\not k)\Gamma_{\mathrm{out}}\Big]\end{aligned} \tag{38}$$

where $\Gamma^\mu = \gamma^\mu,\ \gamma^\mu\gamma^5,\ i\sigma^{\mu\nu}q_\nu$, or $i\sigma^{\mu\nu}q_\nu\gamma^5$ and $\Gamma_{P',V} = \gamma^5,\ \gamma_\nu\epsilon_2^\nu$.

We use the local quark propagators

$$S_i(\not k) = \frac{1}{m_i - \not k}, \tag{39}$$

where m_i is the constituent quark mass. We do not introduce a new notation for constituent quark masses in order to distinguish them from the current quark masses used in the effective Hamiltonian and Wilson coefficients as described in Sect. 2 because it should always be clear from the context which set of masses is being referred to. As discussed in [18,19], we assume that

$$m_H < m_1 + m_2 \tag{40}$$

in order to avoid the appearance of imaginary parts in the physical amplitudes.

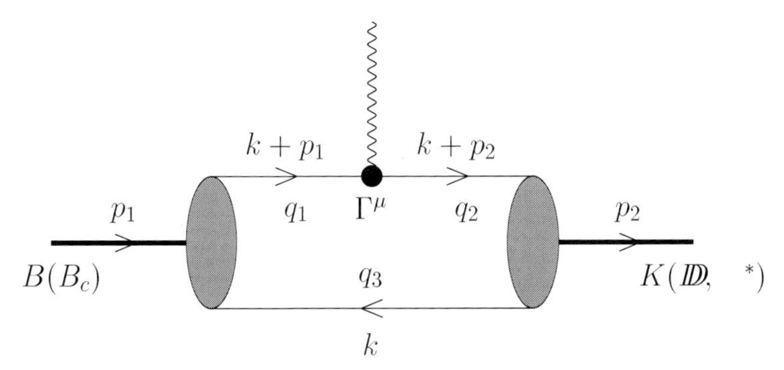

Fig. 2. Feynman diagram describing the form factors of the decay $B(B_c) \to K(D, D^*)\bar{l}l$

The fit values for the constituent quark masses are taken from our papers [18,19] and are given in (41).

$$\begin{array}{ccccc} m_u & m_s & m_c & m_b & \\ \hline 0.235 & 0.333 & 1.67 & 5.06 & \text{GeV} \end{array} \tag{41}$$

It is readily seen that the constraint (40) holds true for the low-lying flavored pseudoscalar mesons but is no longer true for the vector mesons. In the case of the heavy mesons D^* and B^* we will employ identical masses for the vector mesons and the pseudoscalar mesons for the calculation of matrix elements in (36),(37) and (38). It is a quite reliable approximation because of $(m_{D^*} - m_D)/m_D \sim 7\%$ and $(m_{B^*} - m_B)/m_B \sim 1\%$. In this vein, our model was successfully developed for the study of light hadrons (e.g., pion, kaon, baryon octet, Δ-resonance), heavy-light hadrons (e.g., D, D_s, B and B_s-mesons, Λ_Q, Σ_Q, Ξ_Q and Ω_Q-baryons) and double heavy hadrons (e.g, J/Ψ, Υ and B_c-mesons, Ξ_{QQ} and Ω_{QQ} baryons) [18,19]. To extend our approach to other hadrons we had to introduce extra model parameters or do some approximations, like, e.g., to introduce the cutoff parameter for external hadron momenta to guarantee the fulfillment of the mentioned above "threshold inequality". Therefore, at the present stage we can not apply our approach for the study of rare decays involving K^* mesons. Probably, it will be a subject of our future investigations.

We employ a Gaussian for the vertex function $\tilde{\Phi}_H(k_E^2/\Lambda_H^2) = \exp(-k_E^2/\Lambda_H^2)$ where k_E is the Euclidean momentum and determine the size parameters Λ_H by a fit to the experimental data, when available, or to lattice simulations for the leptonic decay constants. The quality of the fit can be seen from Table 3. The branching ratios of the semileptonic decays are shown in Table 4. The numerical values for Λ_H are $\Lambda_\pi = 1$ GeV, $\Lambda_K = 1.6$ GeV, $\Lambda_D = 2$ GeV and $\Lambda_B = 2.25$ GeV for all K, D and B partners, respectively.

We are now in a position to present our results for the $B(B_c) \to K(D, D^*)$ form factors. We have used the technique outlined in our previous papers [18,19] for the numerical evaluation of the Feynman integrals in (38). The results of our

Table 3. Leptonic decay constants f_H (MeV) used in the least-square fit. The values are taken either from PDG [30] or from the Lattice [31]: quenched (upper line) and unquenched (lower line).

Meson	This model	Expt/Lattice
π^+	131	$130.7 \pm 0.1 \pm 0.36$
K^+	161	$159.8 \pm 1.4 \pm 0.44$
D^+	211	$203\pm$ 14 $226\pm$ 15
D_s^+	222	$230\pm$ 14 $250\pm$ 30
B^+	180	$173\pm$ 23 $198\pm$ 30
B_s^0	196	$200\pm$ 20 $230\pm$ 30
B_c^+	398	

Table 4. Semileptonic decay branching ratios.

Meson	This model	Expt.
$\pi^+ \to \pi^0 l^+ \nu$	$1.03 \cdot 10^{-8}$	$(1.025 \pm 0.034) \cdot 10^{-8}$
$K^+ \to \pi^0 l^+ \nu$	$4.62 \cdot 10^{-2}$	$(4.82 \pm 0.06) \cdot 10^{-2}$
$B^+ \to \bar{D}^0 l^+ \nu$	$2.40 \cdot 10^{-2}$	$(2.15 \pm 0.22) \cdot 10^{-2}$
$B^+ \to \bar{D}^{*\,0} l^+ \nu$	$5.60 \cdot 10^{-2}$	$(5.3 \pm 0.8) \cdot 10^{-2}$
$B_c^+ \to D^0 l^+ \nu$	$2.05 \cdot 10^{-5}$	
$B_c^+ \to D^{*\,0} l^+ \nu$	$3.60 \cdot 10^{-5}$	

numerical calculations are well represented by the parametrization

$$F(s) = \frac{F(0)}{1 - as + bs^2}\,. \tag{42}$$

Using such a parametrization facilitates further integrations. The values of $F(0)$, a and b are listed in Tables 5.

At the end of this section we would like to discuss the impulse approximation used in our previous papers [18,19]. It was simply assumed that the vertex functions depend only on the loop momentum flowing through the vertex. The explicit translational invariant vertex function in (35) allows one to check the

Table 5. Parameter values for the approximated form factors $F(s) = F(0)/(1 - as + bs^2)$ $(s = q^2/m_B^2)$.

$B \to K\bar{l}l$	F_+	F_-	F_T
$F(0)$	0.357	-0.275	0.337
a	1.011	1.050	1.031
b	0.042	0.067	0.051

$B_c \to D(D^*)\bar{l}l$	F_+	F_-	F_T	A_0	A_+	A_-	V	a_0	a_+	g
$F(0)$	0.186	-0.190	0.275	0.279	0.156	-0.321	0.290	0.178	0.178	0.179
a	2.48	2.44	2.40	1.30	2.16	2.41	2.40	1.21	2.14	2.51
b	1.62	1.54	1.49	0.149	1.15	1.51	1.49	0.125	1.14	1.67

Table 6. Decay branching ratios without(with) long distance contributions.

Ref.	$\mathrm{Br}(B \to K\,\mu^+\mu^-)$	$\mathrm{Br}(B \to K\,\tau^+\tau^-)$	$\mathrm{Br}(B \to K\,\bar{\nu}\nu)$
[8]	$0.57 \cdot 10^{-6}$	$1.3 \cdot 10^{-7}$	
[4]	$(0.35 \pm 0.12) \cdot 10^{-6}$		
[11]	$0.44 \cdot 10^{-6}$	$1.0 \cdot 10^{-7}$	$5.6 \cdot 10^{-6}$
[32]	$0.5 \cdot 10^{-6}$	$1.3 \cdot 10^{-7}$	
our	$0.55\,(0.51) \cdot 10^{-6}$	$1.01\,(0.87) \cdot 10^{-7}$	$4.19 \cdot 10^{-6}$

	our	[17]
$\mathrm{Br}(B_c \to D_d\,\mu^+\mu^-)$	$0.44\,(0.38) \cdot 10^{-8}$	$0.41\,(0.33) \cdot 10^{-8}$
$\mathrm{Br}(B_c \to D_d^*\,\mu^+\mu^-)$	$0.71\,(0.58) \cdot 10^{-8}$	$1.01\,(0.78) \cdot 10^{-8}$
$\mathrm{Br}(B_c \to D_s\,\mu^+\mu^-)$	$0.97\,(0.86) \cdot 10^{-7}$	$1.36\,(1.12) \cdot 10^{-7}$
$\mathrm{Br}(B_c \to D_s^*\,\mu^+\mu^-)$	$1.76\,(1.41) \cdot 10^{-7}$	$4.09\,(3.14) \cdot 10^{-7}$
$\mathrm{Br}(B_c \to D_d\,\tau^+\tau^-)$	$0.11\,(0.09) \cdot 10^{-8}$	$0.13\,(0.11) \cdot 10^{-8}$
$\mathrm{Br}(B_c \to D_d^*\,\tau^+\tau^-)$	$0.11\,(0.08) \cdot 10^{-8}$	$0.18\,(0.13) \cdot 10^{-8}$
$\mathrm{Br}(B_c \to D_s\,\tau^+\tau^-)$	$0.22\,(0.18) \cdot 10^{-7}$	$0.34\,(0.27) \cdot 10^{-7}$
$\mathrm{Br}(B_c \to D_s^*\,\tau^+\tau^-)$	$0.22\,(0.15) \cdot 10^{-7}$	$0.51\,(0.34) \cdot 10^{-7}$
$\mathrm{Br}(B_c \to D_d\,\bar{\nu}\nu)$	$3.28 \cdot 10^{-8}$	
$\mathrm{Br}(B_c \to D_d^*\,\bar{\nu}\nu)$	$5.78 \cdot 10^{-8}$	
$\mathrm{Br}(B_c \to D_s\,\bar{\nu}\nu)$	$0.73 \cdot 10^{-6}$	
$\mathrm{Br}(B_c \to D_s^*\,\bar{\nu}\nu)$	$1.42 \cdot 10^{-6}$	

reliability of this approximation. We found that the results obtained with and without the impulse approximation are rather close to each other except for the heavy-to-light form factors. We consider the $B \to \pi$-transition as an example to illustrate this point. The calculated values of the $F_+^{B\pi}(q^2)$ form factor at $q^2 = 0$

are

$$F_+^{B\pi}(0) = \begin{cases} 0.27 \text{ exact} \\ \\ 0.48 \text{ impulse approximation} \end{cases}$$

One can see that the value of the form factor at $q^2 = 0$ calculated without the impulse approximation is considerably smaller than when calculated with the impulse approximation. Its value is close to the value of QCD SR estimates, see, for example, [29]: $F_+^{B\pi}(0) = 0.30$.

7 Numerical Results

We list our numerical results for the branching ratios in Table 6. When comparing the values of the branching ratios with those obtained in [8] and [17] one finds that they almost agree with each other.

References

1. R. Barate *et al.* [ALEPH Collaboration], Phys. Lett. B **429**, 169 (1998).
2. K. Abe *et al.* [Belle Collaboration], Phys. Lett. B **511**, 151 (2001).
3. S. Chen *et al.* [CLEO Collaboration], Phys. Rev. Lett. **87**, 251807 (2001).
4. A. Ali, E. Lunghi, C. Greub and G. Hiller, Phys. Rev. D **66**, 034002 (2002).
5. K. Abe *et al.* [BELLE Collaboration], Phys. Rev. Lett. **88**, 021801 (2002).
6. F. Abe *et al.* [CDF Collaboration], Phys. Rev. Lett. **81**, 2432 (1998); Phys. Rev. D **58**, 112004 (1998).
7. G. Buchalla, A. J. Buras and M. E. Lautenbacher, Rev. Mod. Phys. **68**, 1125 (1996). A. J. Buras and M. Munz, Phys. Rev. D **52**, 186 (1995).
8. A. Ali, P. Ball, L. T. Handoko and G. Hiller, Phys. Rev. D **61**, 074024 (2000).
9. C. Greub, A. Ioannisian and D. Wyler, Phys. Lett. B **346**, 149 (1995).
10. P. Colangelo, F. De Fazio, P. Santorelli and E. Scrimieri, Phys. Rev. D **53**, 3672 (1996) [Erratum-ibid. D **57**, 3186 (1998)].
11. D. Melikhov, N. Nikitin and S. Simula, Phys. Rev. D **57**, 6814 (1998).
12. T. M. Aliev, M. K. Cakmak, A. Ozpineci and M. Savci, Phys. Rev. D **64**, 055007 (2001); T. M. Aliev, M. Savci, A. Ozpineci and H. Koru, J. Phys. G **24**, 49 (1998).
13. Q. S. Yan, C. S. Huang, W. Liao and S. H. Zhu, Phys. Rev. D **62**, 094023 (2000).
14. F. Kruger, L. M. Sehgal, N. Sinha and R. Sinha, Phys. Rev. D **61**, 114028 (2000) [Erratum-ibid. D **63**, 019901 (2001)].
15. F. Kruger and L. M. Sehgal, Phys. Lett. B **380**, 199 (1996).
16. J. L. Hewett, Phys. Rev. D **53**, 4964 (1996)
17. C. Q. Geng, C. W. Hwang and C. C. Liu, Phys. Rev. D **65**, 094037 (2002).
18. M. A. Ivanov, M. P. Locher and V. E. Lyubovitskij, Few Body Syst. **21**, 131 (1996); M. A. Ivanov and V. E. Lyubovitskij, Phys. Lett. B **408**, 435 (1997); M. A. Ivanov, V. E. Lyubovitskij, J. G. Körner and P. Kroll, Phys. Rev. D **56**, 348 (1997); M. A. Ivanov, J. G. Körner, V. E. Lyubovitskij and A. G. Rusetsky, Phys. Rev. D **57**, 5632 (1998); D **60**, 094002 (1999); Phys. Lett. B **476**, 58 (2000); M. A. Ivanov, J. G. Körner and V. E. Lyubovitskij, Phys. Lett. B **448**, 143 (1999); M. A. Ivanov, J. G. Korner, V. E. Lyubovitskij, M. A. Pisarev and A. G. Rusetsky, Phys. Rev. D **61**, 114010 (2000).

19. M. A. Ivanov and P. Santorelli, Phys. Lett. B **456**, 248 (1999). M. A. Ivanov, J. G. Korner and P. Santorelli, Phys. Rev. D **63**, 074010 (2001). A. Faessler, T. Gutsche, M. A. Ivanov, J. G. Korner and V. E. Lyubovitskij, Phys. Lett. B **518**, 55 (2001).
20. A. Salam, Nuovo Cim. **25**, 224 (1962); S. Weinberg, Phys. Rev. **130**, 776 (1963); K. Hayashi *et al.*, Fort. der Phys. **15**, 625 (1967).
21. G. V. Efimov and M. A. Ivanov, *Bristol, UK: IOP (1993) 177 p*; Int. J. Mod. Phys. A **4**, 2031 (1989).
22. M. A. Ivanov, Y. L. Kalinovsky and C. D. Roberts, Phys. Rev. D **60**, 034018 (1999).
23. A. Ali, T. Mannel and T. Morozumi, Phys. Lett. B **273**, 505 (1991).
24. J. G. Korner and G. A. Schuler, Z. Phys. C **38**, 511 (1988) [Erratum-ibid. C **41**, 690 (1989)]; Z. Phys. C **46**, 93 (1990).
25. C. S. Kim, Y. G. Kim, C. D. Lu and T. Morozumi, Phys. Rev. D **62**, 034013 (2000).
26. A. Ali and A. S. Safir, arXiv:hep-ph/0205254.
27. C. H. Chen and C. Q. Geng, Nucl. Phys. B **636**, 338 (2002); Phys. Rev. D **63**, 114025 (2001).
28. D. Melikhov, N. Nikitin and S. Simula, Phys. Lett. B **442**, 381 (1998).
29. E. Bagan, P. Ball and V. M. Braun, Phys. Lett. B **417**, 154 (1998).
30. K. Hagiwara *et al.*, Phys. Rev. D **66**, 010001 (2002).
31. S. Ryan, Nucl. Phys. B (Proc. Suppl.) **106**, 86 (2002).
32. C. Q. Geng and C. P. Kao, Phys. Rev. D **57**, 4479 (1998); D **54**, 5636 (1996). [33]
33. A. Faessler, T. Gutsche, M. A. Ivanov, J. G. Korner and V. E. Lyubovitskij, Eur. Phys. J. directC **4**, 18 (2002) [arXiv:hep-ph/0205287].

Spectroscopy of Baryons Containing Two Heavy Quarks

I.M. Narodetskii and M.A. Trusov

Institute of Theoretical and Experimental Physics, Moscow 117259, Russia

Abstract. Doubly heavy baryons are considered within the Effective Hamiltonian (EH) approach by Simonov, which is derived from QCD. The EH has the form of the non relativistic three quark Hamiltonian with the perturbative Coulomb-like and non perturbative string interactions and the specific mass term. After outlining the approach methods of calculations of the baryon eigen energies and some simple applications are explained in details. A mass of 3620 MeV for the lightest *ccu* baryon is found by employing a variational solution.

1 Introduction

Doubly heavy baryons are baryons that contain two heavy quarks, either *cc*, *bc* or *bb*. Their existence is a natural consequence of the quark model of hadrons, and it would be surprising if they did not exist. In particular, data from the BaBar and Belle collaborations at the SLAC and KEK B-factories would be good places to look for doubly charmed baryons. Recently the SELEX, the charm hadroproduction experiment at Fermilab, reported a narrow state at 3519 ± 1 MeV decaying in $\Lambda_c^+ K^- \pi^+$, consistent with the weak decay of the doubly charged baryon Ξ_{cc}^+ [1]. The candidate is 6.3σ signal. The SELEX result was recently critically discussed in [2]. Whether or not the state that SELEX reports turns out to be the first observation of doubly charmed baryons, studying their properties is important for a full understanding of the strong interaction between quarks.

Estimations for the masses and spectra of the baryons containing two or more heavy quarks have been considered by many authors [3]. The purpose of this lecture is to present the results of the calculation [4] of the masses of the doubly-heavy baryons obtained in a simple approximation within the nonperturbative QCD.

2 The Effective Hamiltonian in QCD

The methodology of the approach has been reviewed recently [5] and so will be sketched here only briefly. The Y-shaped baryon wave function has the form:

$$B_X(x_1, x_2, x_3, X) = e_{\alpha\beta\gamma} q^\alpha(x_1, X) q^\beta(x_2, X) q^\gamma(x_3, X), \quad (1)$$

where Y is the equilibrium junction position (see below). This is the only gauge invariant configuration possible for baryons. The starting point of the approach

I.M. Narodetskii and M.A. Trusov, Spectroscopy of Baryons Containing Two Heavy Quarks, Lect. Notes Phys. **647**, 264–274 (2004)
http://www.springerlink.com/

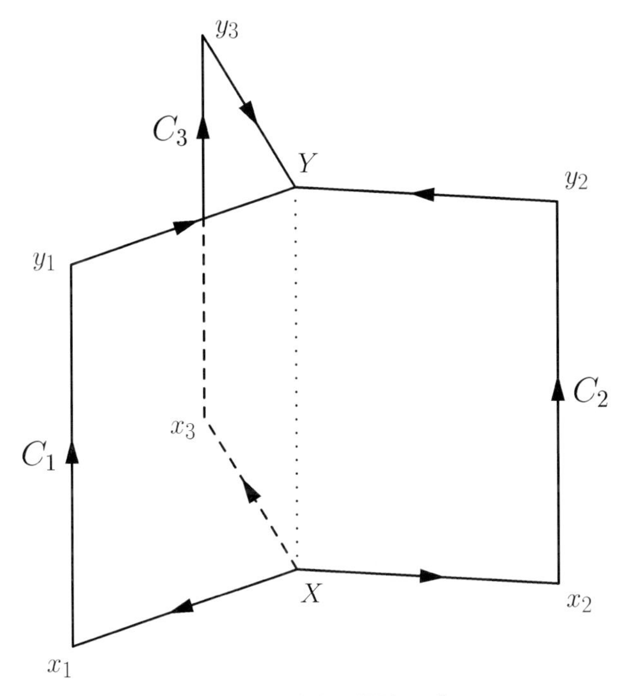

Fig. 1. Three lobes Wilson loop

is the Feynman-Schwinger representation for the Green function of the three quarks propagating in the nonperturbative confining background

$$G(x, y) = \sum_i \int_0^\infty ds_i \int Dz_i \exp(-K_i)\langle \mathcal{W} \rangle_B, \tag{2}$$

where $x = x_1, x_2, x_3$, $y = y_1, y_2, y_3$ and angular brackets mean averaging over background field. The quantities K_i are the kinetic energies of quarks, and all the dependence on the vacuum background field is contained in the generalized Wilson loop $\mathcal{W}$. The role of the time parameter along the trajectory of each quark is played by the Fock-Schwinger proper time s_i. Under the minimal area law assumption, the Wilson loop configuration takes the form

$$\langle \mathcal{W} \rangle_B \propto \exp(-\sigma(S_1 + S_2 + S_3)), \tag{3}$$

where S_i are the minimal areas inside the contours formed by quarks and the string junction trajectories, see Fig. 1, where the contours run over the classical trajectories of static quarks. The proper and real times for each quark are related via a new quantity that eventually plays the role of the dynamical quark mass. The final result is the derivation of the EH, see (5) below.

In contrast to the standard approach of the constituent quark model the dynamical masses m_i are no longer free parameters. They are expressed in terms of the running masses $m_i^{(0)}(Q^2)$ defined at the appropriate hadronic scale of Q^2 from the condition of the minimum of the baryon mass M_B as function of m_i:

$$\frac{\partial M_B(m_i)}{\partial m_i} = 0. \tag{4}$$

Technically, this has been done using the einbein (auxiliary fields) approach, which is proven to be rather accurate in various calculations for relativistic systems. Einbeins are treated as c number variational parameters: the eigenvalues of the EH are minimized with respect to einbeins to obtain the physical spectrum. Such procedure, first suggested in [6], provides the reasonable accuracy for the meson ground states [7].

This method was already applied to study baryon Regge trajectories [8] and very recently for computation of magnetic moments of light baryons [9]. The essential point adopted in [4] is that it is very reasonable that the same method should also hold for hadrons containing heavy quarks. As in [9] we take as the universal QCD parameter the string tension σ. We also include the perturbative Coulomb interaction with the frozen strong coupling constant α_s.

From experimental point of view, a detailed discussion of the excited $QQ'q$ states is probably premature. Therefore we consider the ground state baryons without radial and orbital excitations in which case tensor and spin-orbit forces do not contribute perturbatively. Then only the spin-spin interaction survives in the perturbative approximation. The EH has the following form

$$H = \sum_{i=1}^{3}\left(\frac{m_i^{(0)2}}{2m_i} + \frac{m_i}{2}\right) + H_0 + V, \tag{5}$$

where H_0 is the non-relativistic kinetic energy operator, and V is the sum of the perturbative one gluon exchange potentials V_c:

$$V_c = -\frac{2}{3}\alpha_s \cdot \sum_{i<j}\frac{1}{r_{ij}},$$

and the string potential V_{string}. The string potential calculated in [8] as the static energy of the three heavy quarks was shown to be consistent with that given by a minimum length configuration of the strings meeting in a Y-shaped configuration at a junction $\mathbf{X}$:

$$V_{\text{string}}(\mathbf{r}_1, \mathbf{r}_2, \mathbf{r}_3) = \sigma R_{\min}, \tag{6}$$

where $R_{\min}$ is the sum of the three distances $|\mathbf{r}_i|$ from the string junction point. The Y-shaped configuration was suggested long ago [10], and since then was used repeatedly in many dynamical calculations [11]. In a more refined analysis it was also shown that the Y-shaped static potential has a strong depletion at small distances due to a hole in string density at the string-junction position [12]. In what follow we shall neglect these complications.

3 Solving the Three Quark Equation

3.1 Jacobi Coordinates

The baryon wave function depends on the three-body Jacobi coordinates

$$\rho_{ij} = \sqrt{\frac{\mu_{ij}}{\mu}}(\mathbf{r}_i - \mathbf{r}_j), \tag{7}$$

$$\lambda_{ij} = \sqrt{\frac{\mu_{ij,k}}{\mu}}\left(\frac{m_i\mathbf{r}_i + m_j\mathbf{r}_j}{m_i + m_j} - \mathbf{r}_k\right) \tag{8}$$

$(i, j, k$ cyclic), where μ_{ij} and $\mu_{ij,k}$ are the appropriate reduced masses

$$\mu_{ij} = \frac{m_i m_j}{m_i + m_j}, \quad \mu_{ij,k} = \frac{(m_i + m_j)m_k}{m_i + m_j + m_k}, \tag{9}$$

and μ is an arbitrary parameter with the dimension of mass which drops off in the final expressions. The coordinate ρ_{ij} is proportional to the separation of quarks i and j and coordinate λ_{ij} is proportional to the separation of quarks i and j, and quark k.

In terms of the Jacobi coordinates the kinetic energy operator H_0 is written as

$$\begin{aligned} H_0 &= -\frac{1}{2\mu}\left(\frac{\partial^2}{\partial\rho^2} + \frac{\partial^2}{\partial\lambda^2}\right) \\ &= -\frac{1}{2\mu}\left(\frac{\partial^2}{\partial R^2} + \frac{5}{R}\frac{\partial}{\partial R} + \frac{K^2(\Omega)}{R^2}\right), \end{aligned} \tag{10}$$

where R is the six-dimensional hyper-radius

$$R^2 = \rho_{ij}^2 + \lambda_{ij}^2, \tag{11}$$

and $K^2(\Omega)$ is angular momentum operator whose eigen functions (the hyper spherical harmonics) are

$$K^2(\Omega)Y_{[K]} = -K(K+4)Y_{[K]}, \tag{12}$$

with K being the grand orbital momentum. In terms of $Y_{[K]}$ the wave function $\psi(\rho, \lambda)$ can be written in a symbolical shorthand as

$$\psi(\rho, \lambda) = \sum_K \psi_K(R)Y_{[K]}(\Omega).$$

In the hyper radial approximation which we shall use below $K = 0$ and $\psi = \psi(R)$. Since R^2 is exchange symmetric the baryon wave function is totally symmetric under exchange. Note that the centrifugal potential in the Schrödinger equation for the reduced radial function $\chi(R) = R^{5/2}\psi_K(R)$ with a given K

$$\frac{(K+2)^2 - 1/4}{R^2}$$

is not zero even for $K = 0$.

3.2 String Junction Point

The potential $V_{\text{string}}(\mathbf{r}_1, \mathbf{r}_2, \mathbf{r}_3)$ has rather complicated structure. Let θ_{ijk} be the angle between the line from quark i to quark j and that from quark j to quark k. If θ_{ijk} are all smaller than 120°, then the equilibrium junction position $\mathbf{X}$ coincides with the so-called Torrichelli point of the triangle in which vertices three quarks are situated. The geometrical construction of this point is presented on Fig. 2. Using this construction one can easy obtain an expression for a radius-vector of the Torrichelli point in terms of the lengths l_i of the segments between this point and the i-th quark, and the quark positions $\mathbf{r}_i$ [13] :

$$\mathbf{X} = \frac{l_2 l_3 \mathbf{r}_1 + l_1 l_3 \mathbf{r}_2 + l_1 l_2 \mathbf{r}_3}{l_2 l_3 + l_1 l_3 + l_1 l_2} \tag{13}$$

If θ_{ijk} is equal to or greater than 120°, the lowest energy configuration has the junction at the position of quark j.

An equivalent expression for $\mathbf{X}$ in terms of $\mathbf{R}_{\text{cm}}$, ρ and λ is [14]

$$\mathbf{X} = \mathbf{R}_{\text{cm}} + \alpha\rho + \beta\lambda, \tag{14}$$

with

$$\alpha = \frac{1}{2}\sqrt{\frac{\mu}{\mu_{ij}}}\left(\frac{m_j - m_i}{m_i + m_j} - \frac{1}{\sqrt{3}} \cdot \frac{4t + (3 - t^2)\cot\chi}{1 + t^2}\right),$$

$$\beta = \frac{\sqrt{\mu\mu_{ij,k}}}{m_i + m_j} + \sqrt{\frac{\mu}{3\mu_{ij}}} \cdot \frac{\rho}{2\lambda\sin\chi} \cdot \frac{3 - t^2}{1 + t^2},$$

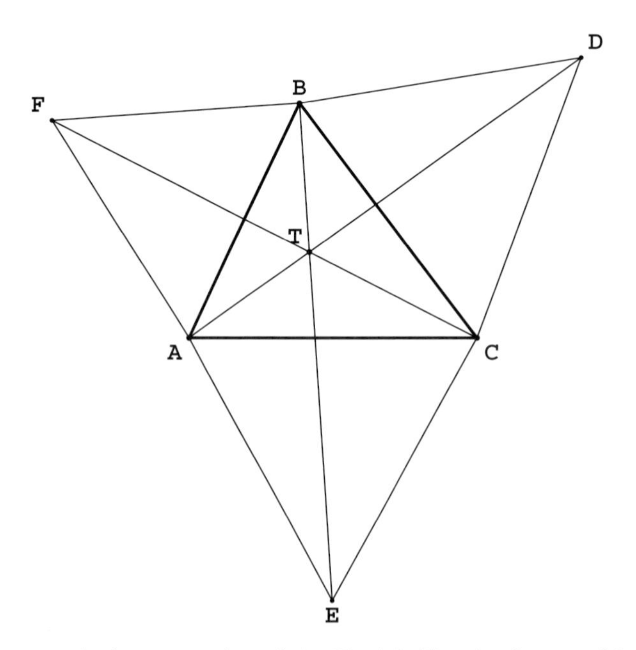

Fig. 2. The geometrical construction of the Torrichelli point for an arbitrary triangle △ABC. The triangles △AFB, △BDC, △CEA are equilateral.

where

$$t = \frac{2\lambda \sin\chi + \sqrt{\frac{3\mu_{ij,k}}{\mu_{ij}}}\rho}{2\lambda\cos\chi + \sqrt{\frac{3\mu_{ij,k}}{\mu_{ij}}} \cdot \frac{m_j - m_i}{m_i + m_j}\rho},$$

and χ is the angle between ρ and λ. It can be easily seen that dependence on m_i in (14) is apparent and $\mathbf{X}$ does not depend on quark masses as it should be.

In what follows the string junction point is chosen as coinciding with the center–of–mass coordinate. Accuracy of this approximation that greatly simplifies the calculations was discussed in [8]. In this case

$$V_{\text{string}} = \sigma \sum_{i<j} \frac{1}{m_k} \cdot \sqrt{\mu\mu_{ijk}} \cdot |\lambda_{ij}| \tag{15}$$

3.3 Hyper Radial Approximation

Introducing the variable $x = \sqrt{\mu}R$ and averaging the interaction $U = V_c + V_{\text{string}}$ over the six-dimensional sphere one obtains the Schrödinger equation for $\chi(x)$

$$\frac{d^2\chi(R)}{dx^2} + 2\left[E_0 + \frac{a}{x} - bx - \frac{15}{8x^2}\right]\chi(x) = 0, \tag{16}$$

where E_0 is the ground state eigenvalue and

$$\begin{aligned} a &= \tfrac{2\alpha_s}{3} \cdot \tfrac{16}{3\pi} \cdot \left(\sum_{i<j} \sqrt{\mu_{ij}}\right), \\ b &= \sigma \cdot \tfrac{32}{15\pi} \cdot \left(\sum_{i<j} \tfrac{\sqrt{\mu_{ij,k}}}{m_k}\right). \end{aligned} \tag{17}$$

Let us explain the numerical coefficients in (17) in more details. To this end we introduce the angles θ_{ij} such that

$$\rho_{ij} = R\sin\theta_{ij}, \quad \lambda_{ij} = R\cos\theta_{ij}, \quad 0 \le \theta_{ij} \le \frac{\pi}{2} \tag{18}$$

and write the phase space $d^3\rho_{ij}d^3\lambda_{ij}$ as

$$d^3\rho_{ij}\, d^3\lambda_{ij} = (4\pi)^2\rho_{ij}^2\lambda_{ij}^2 d\rho_{ij}d\lambda_{ij} = R^5\sin^2\theta_{ij}\cos^2\theta_{ij}dRd\theta_{ij} \tag{19}$$

The volume of the six-dimensional sphere is

$$\Omega_6 = (4\pi)^2 \int_0^{\frac{\pi}{2}} \sin^2\theta_{ij}\cos^2\theta_{ij}d\theta_{ij} = (4\pi)^2 \cdot \frac{\pi}{16} = \pi^3 \tag{20}$$

Then averaging the Coulomb and string terms yields

$$< \frac{1}{r_{ij}} >= \frac{1}{\pi^3} \cdot \sqrt{\frac{\mu_{ij}}{\mu}} \cdot \frac{1}{R} \cdot (4\pi)^2 \cdot \int_0^{\frac{\pi}{2}} \sin\theta_{ij}\cos^2\theta_{ij}\, d\theta_{ij} = \frac{16}{3\pi}\sqrt{\frac{\mu_{ij}}{\mu}} \cdot \frac{1}{R}, \tag{21}$$

$$< \lambda_{ij} >= \frac{1}{\pi^3} \cdot R \cdot (4\pi)^2 \cdot \int\limits_0^{\frac{\pi}{2}} \sin\theta_{ij}^2 \cos^3\theta_{ij}^3 \, d\theta_{ij} = \frac{32}{15\pi} \cdot R \tag{22}$$

Note that the wave function calculated in the hyper radial approximation shows the marginal diquark clustering in the doubly heavy baryons. This is principally kinematic effect related to the fact that in this approximation the difference between the various mean values $\bar{r}_{ij}$ in a baryon is due to the factor $\sqrt{1/\mu_{ij}}$ which varies between $\sqrt{2/m_i}$ for $m_i = m_j$ and $\sqrt{1/m_i}$ for $m_i \ll m_j$

3.4 Quasi Classical Solution

For the purpose of illustration, the problem is first solved quasi classically rather than using quantum mechanics. This approach is based on the well known fact that interplay between the centrifugal term and the confining potential produces a minimum of the effective potential specific for the three-body problem. The numerical solution of (16) for the ground state eigen energy may be reproduced on a per cent level of accuracy by using the parabolic approximation [15] for the effective potential

$$U(x) = V(x) + \frac{15}{8x^2}.$$

This approximation provides an analytical expression for the eigen energy. The potential $U(x)$ has the minimum at a point $x = x_0$, which is defined by the condition $U'(x_0) = 0$, *i.e.*:

$$bx_0^3 + ax_0 - 15/4 = 0\,. \tag{23}$$

Expanding $U(x)$ in the vicinity of the minimum one obtains

$$U(x) \approx U(x_0) + \frac{1}{2}U''(x_0)(x - x_0)^2,$$

i.e. the potential of the harmonic oscillator with the frequency $\omega = \sqrt{U''(x_0)}$. Therefore the ground state energy eigenvalue is

$$E_0 \approx U(x_0) + \frac{1}{2}\omega\,. \tag{24}$$

3.5 Variational Solution

Another method of solving (16) is the minimization of the baryon energy using a simple variational Ansätz

$$\chi(x) \sim x^{5/2} e^{-p^2x^2}, \tag{25}$$

where p is the variational parameter. Then the three-quark Hamiltonian admits explicit solutions for the energy and the ground state eigenfunction: $E_0 \approx \min\limits_p E_0(p)$, where

$$E_0(p) = \langle\chi|H|\chi\rangle = 3p^2 - a \cdot \frac{3}{4} \cdot \sqrt{\frac{\pi}{2}} \cdot p + b \cdot \frac{15}{16} \cdot \sqrt{\frac{\pi}{2}} \cdot \frac{1}{p}. \tag{26}$$

4 Quark Dynamical Masses

We first solve (4) for the dynamical masses m_i retaining only the string potential in the effective Hamiltonian (5). This procedure is in agreement with the strategy adopted in [9]. Then we add the perturbative Coulomb potential and solve (5) to obtain the ground state eigenvalues E_0. The masses m_i are then obtained from solving (4) with

$$M_B = \sum_{i=1}^{3} \left(\frac{m_i^{(0)2}}{2m_i} + \frac{m_i}{2} \right) + E_0(m_1, m_2, m_3) \tag{27}$$

We use the values of parameters $\sigma = 0.15$ GeV^2, $\alpha_s = 0.39$, $m_q^{(0)} = 0.009$ GeV, $m_s^{(0)} = 0.17$ GeV, $m_c^{(0)} = 1.4$ GeV, and $m_b^{(0)} = 4.8$ GeV, slightly different from those of [4]. The results for various baryons are given in Table 1. Note that there is no good theoretical reason why quark masses m_i need to be the same in different baryons. From the results of Table 1 we conclude that the masses of the light quarks (u, d or s) are increased by ~ 100 MeV when going from the light to heavy baryons.The dynamical masses of light quarks $m_q \sim \sqrt{\sigma} \sim 400 - 500$ MeV ($q = u, d, s$) qualitatively agree with the results of [16] obtained from the analysis of the heavy–light ground state mesons.

For the heavy quarks ($Q = c$ and b) the variation in the values of their masses m_Q is marginal. This is illustrated by the simple analytical results for Qud baryons [17]. These results were obtained from the approximate solution of equation

$$\frac{\partial E_0(m_1, m_2, m_3, p)}{\partial p} = 0 \tag{28}$$

Table 1. The constituent quark masses m_i and the ground state eigenvalues E_0 (in units of GeV) for the various baryon states.

baryon	m_1	m_2	m_3	E_0
(qqq)	0.372	0.372	0.372	1.426
(qqs)	0.377	0.377	0.415	1.398
(qss)	0.381	0.420	0.420	1.370
(sss)	0.424	0.424	0.424	1.343
(qqc)	0.424	0.424	1.464	1.171
(qsc)	0.427	0.465	1.467	1.146
(ssc)	0.468	0.468	1.469	1.121
(qqb)	0.446	0.446	4.819	1.085
(qsb)	0.448	0.487	4.820	1.059
(ssb)	0.490	0.490	4.821	1.033
(qcc)	0.459	1.498	1.498	0.904
(scc)	0.499	1.499	1.499	0.881
(qcb)	0.477	1.524	4.834	0.783
(scb)	0.517	1.525	4.834	0.759
(qbb)	0.495	4.854	4.854	0.593
(sbb)	0.534	4.855	4.855	0.570

where E_0 is given by (26) in the form of expansion in the small parameters

$$\xi = \frac{\sqrt{\sigma}}{m_Q^{(0)}} \quad \text{and} \quad \alpha_s. \tag{29}$$

Omitting the intermediate steps one has

$$E_0 = 3\sqrt{\sigma}\left(\frac{6}{\pi}\right)^{1/4}\left(1 + A\cdot\xi - \frac{5}{3}B\cdot\alpha_s + \ldots\right) \tag{30}$$

$$m_q = \sqrt{\sigma}\left(\frac{6}{\pi}\right)^{1/4}(1 - A\cdot\xi + B\cdot\alpha_s + \ldots), \tag{31}$$

$$m_Q = m_Q^{(0)}\left(1 + \mathcal{O}(\xi^2, \alpha_s^2, \alpha_s\xi) + \ldots\right), \tag{32}$$

where for the Gaussian variational Ansätz (25)

$$A = \frac{\sqrt{2}-1}{2}\left(\frac{6}{\pi}\right)^{1/4} \approx 0.24, \; B = \frac{4+\sqrt{2}}{18}\sqrt{\frac{6}{\pi}} \approx 0.42. \tag{33}$$

Note that the corrections of the first order in ξ and α_s are absent in the expression (32) for m_Q. Accuracy of this approximation is illustrated in Table 1 of [14].

5 Doubly Baryon Masses

To calculate hadron masses we, as in [8], first renormalize the string potential:

$$V_{\text{string}} \to V_{\text{string}} + \sum_i C_i, \tag{34}$$

where the constants C_i take into account the residual self-energy (RSE) of quarks [18]. In what follows we adjust the RSE constants C_i to reproduce the center-of-gravity for baryons with a given flavor. As a result we obtain $C_q = -0.30$ GeV, $C_s = -0.15$ GeV, $C_c \sim C_b \sim 0$.

We keep these parameters fixed to calculate the masses given in Table 2, namely the spin–averaged masses (computed without the spin–spin term) of the lowest double heavy baryons. It is interesting to compare results found here to the predictions of other models. In Table 2 we compare our predictions with the results obtained using the additive non–relativistic quark model with the power-law potential [19], relativistic quasipotential quark model [20], the Feynman-Hellmann theorem [21] and with the predictions obtained in the approximation of double heavy diquark [22].

The change of σ to 0.17 GeV^2 increases the mass of Ξ_{cc}^+ by $\sim$ 30 MeV. The hyperfine splitting with the spin $\frac{1}{2}$ states is calculated using the spin-spin interaction of the Fermi-Breit type [23]. It produces an additional shift of the Ξ_{cc}^+ mass ~ -20 MeV. Note that the mass of Ξ_{cc}^+ is rather sensitive to the value of the running c-quark mass $m_c^{(0)}$, see Fig. 3.

Table 2. Masses of doubly heavy baryons

State	this work	[19]	[20]	[21]	[22]
$\Xi\{qcc\}$	3.64	3.70	3.71	3.66	3.48
$\Omega\{scc\}$	3.82	3.80	3.76	3.74	3.58
$\Xi\{qcb\}$	6.93	6.99	6.95	7.04	6.82
$\Omega\{scb\}$	7.10	7.07	7.05	7.09	6.92
$\Xi\{qbb\}$	10.14	10.24	10.23	10.24	10.09
$\Omega\{sbb\}$	10.31	10.30	10.32	10.37	10.19

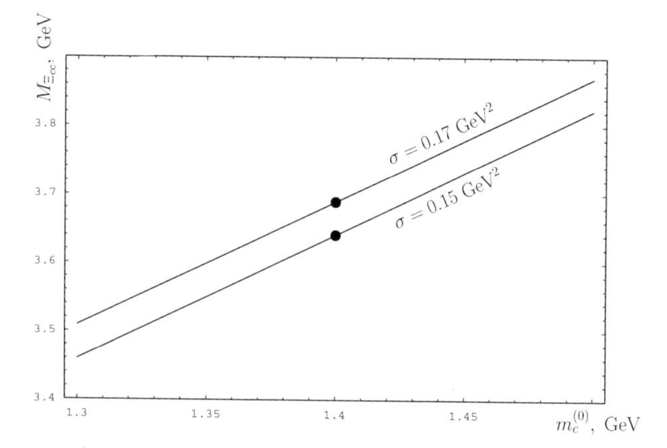

Fig. 3. Mass of Ξ_{cc}^{+} as a function of the running c–quark mass for $\sigma = 0.15$ GeV^2 and $\sigma = 0.17$ GeV^2. The masses are given in GeV. Bold points refer to the case $m_c^{(0)} = 1.4$ GeV

6 Conclusions

We have employed the general formalism for the baryons, which is based on nonperturbative QCD and where the only inputs are σ, α_s and two additive constants, C_q and C_s, the residual self–energies of the light quarks. Using this formalism we have also performed the calculations of the spin–averaged masses of baryons with two heavy quarks. One can see from Table 2 that our predictions are especially close to those obtained in [19] using a variant of the power–law potential adjusted to fit ground state baryons.

Acknowledgements

One of the authors (I.M.N.) would like to thank M.Ivanov and S.Nedelko for organizing an excellent School with a stimulating scientific program. This work was supported in part by RFBR grants ## 00-02-16363 and 00-15-96786.

References

1. M.Mattson *et al.*, Phys. Rev. Lett. **89**, 112001 (2002)
2. V.V.Kiselev and A.K.Likhoded, hep-ph/0208231
3. see references [111] - [124] in *B physics at the Tevatron: Run II and Beyond*, hep-ph/0201071
4. I.M.Narodetskii and M.A.Trusov, Phys. Atom. Nucl. **65**, 917 (2002) [hep-ph/0104019]
5. Yu.A.Simonov, hep-ph/0205331
6. A.Yu.Dubin, A.B.Kaidalov and Yu.A.Simonov, Phys. Lett. **B323**, 41 (1994)
7. V.L.Morgunov, A.V.Nefediev and Yu.A.Simonov, Phys. Lett. **B459**, 653 (1999)
8. M.Fabre de la Ripelle and Yu.A.Simonov, Ann. Phys. (N.Y.) **212**, 235 (1991)
9. B.O.Kerbikov, Yu.A.Simonov, Phys. Rev. D **62**, 093016 (2000)
10. X.Artru, Nucl. Phys. **B85**, 442 (1975); H.G.Dosch and V.Mueller, Nucl. Phys. **B116**, 470
11. J. Carlson, J. Kogut, and V.R. Pandharipande, Phys. Rev. D **27**, 233 (1983); N. Isgur and J. Paton, Phys. Rev. D **31**, 2910 (1985)
12. D.S.Kuzmenko, Yu.A.Simonov, hep-ph/0202277
13. S.Capstick and P.R.Page, LA UR-02-4082 [hep-ph/0207027]
14. I.M.Narodetskii and M.A.Trusov, in *Proceedings of the 5th International Conference BEACH2002, 25-29 June, 2002, Vancouver, BC, Canada* [hep-ph/0209044]
15. Yu.S.Kalashnikova, I.M.Narodetskii, and Yu.A.Simonov, Yad. Fiz. **46**, 1181 (1987)
16. Yu.S.Kalashnikova and A.Nefediev, Phys. Lett. B **492**, 91 (2000)
17. I.M.Narodetskii and M.A.Trusov, in *Proceedings of the 9th International Conference on the Structure of Baryons (Baryons-2002), 3-8 March, 2002, Newport News, VA, USA* [hep-ph/0304320]
18. Yu.A.Simonov, Phys. Lett. **B515**, 137 (2001)
19. E.Bagan *et al.* Z. Phys. C **64**, 57 (1994)
20. D.Ebert *et al.*, Z. Phys. C **76**, 111 (1997)
21. R.Roncaglia *et al.*, Phys. Rev. D **52**, 1248 (1995)
22. A.K.Likhoded and A.I.Onishchenko, hep-ph/9912425
23. A.De Rujula, H.Georgy and S.L.Glashow, Phys. Rev. D **12**, 147 (1975)

Extraction of the Strong Coupling Constant and Strange Quark Mass from Semileptonic τ Decays

Alexei A. Pivovarov

Institute for Nuclear Research of the Russian Academy of Science, 117312 Moscow, Russia

Abstract. In this lecture I present a pedagogical introduction to the low-energy phenomenology of light flavors. The renormalization scheme freedom in defining QCD parameters is discussed. I show in some details how one can extract an accurate numerical value for the strong coupling constant from the τ-lepton decay rate into hadrons. As a related topic I discuss some peculiarities of the precise definition of the quark mass in theories with confinement and describe the strange quark mass determination from data on τ-lepton decays employing contour resummation which is a modern technique of the precision analysis in perturbative QCD.

1 Definition of QCD Parameters α_s and m_s

Quantum ChromoDynamics (QCD) as a field theory for describing strong interactions is given by the Lagrangian

$$\mathcal{L}_{\rm QCD} = \sum_f \bar{q}_f(i\gamma_\mu\partial^\mu + g_s\gamma^\mu G^a_\mu t^a - m_f)q_f - \frac{1}{4}G^a_{\mu\nu}G^a_{\mu\nu}$$

where G^a_μ is a non-Abelian gluonic field and $G^a_{\mu\nu}$ is the field strength tensor [1]. There are six quark flavors $q = (u, d, s, c, b, t)$, three of which (u, d, s) are called light while the other three are heavy. There is a close analogy with QED – the Abelian gauge theory for describing the electromagnetic interaction of charged leptons. The QED Lagrangian for charged leptons reads

$$\mathcal{L}_{\rm QED} = \sum_l \bar{l}(i\gamma_\mu\partial^\mu + e\gamma^\mu A_\mu - m_l)l - \frac{1}{4}F_{\mu\nu}F_{\mu\nu}$$

where $F_{\mu\nu}$ is the electromagnetic field strength tensor. In the standard model there are three charged leptons $l = (e, \mu, \tau)$: electron and muon are very light at the hadronic mass scale of order 1 GeV while the τ lepton is rather heavy with a mass $M_\tau = 1.777$ GeV [2].

The interaction is given by a vertex in the Lagrangian and normalized to a coupling constant g_s at a tree level of perturbation theory (PT). A full theory (beyond the tree level) introduces a dressed vertex that eventually determines a coupling constant after renormalization. For the renormalizable models of quantum field theory as QED and QCD the PT dressing is straightforward and can in principle be done at any finite order of the expansion in the coupling constant. However, the dressing procedure is not unique because renormalization of

A.A. Pivovarov, Extraction of the Strong Coupling Constant and Strange Quark Mass from Semileptonic τ Decays, Lect. Notes Phys. **647**, 275–286 (2004)
http://www.springerlink.com/

loops can introduce some freedom in the choice of finite parameters through a particular definition of counterterms [3].

Leptons can be detected as asymptotic states in the scattering processes that allows one to relate the coupling – for instance, the fine structure constant α – and the particle mass m_l to observables very directly. Thus, the electron-photon scattering at low energies can be used to define the coupling constant (the fine structure constant, in fact) $\alpha = e^2/4\pi$ through the Thompson cross section. This is a natural definition directly through a physical observable. Lepton mass is also directly related to experiment: it can be defined as a physical mass of the asymptotic state (a position of the pole of the lepton propagator). In QCD there is a phenomenon of confinement and no asymptotic states of quarks and/or gluons can be observed. Only the colorless hadrons appear as the asymptotic states. Therefore, definitions of the coupling constant and quark masses in QCD are less direct than in QED. At the theoretical level of the given Lagrangian they are very similar though: a vertex for the coupling and the propagator for a mass. To determine numerical values for the coupling and quark masses one should turn to experiment. As there is no possibility to measure these quantities directly one should specify the research area as the definitions of the parameters can be adopted to specific experiments (basically to a corresponding energy scale). In this lecture I will talk about τ-lepton physics which is the area of low-energy hadron phenomenology: hadronic states have an energy $E < M_\tau$. The particles which can be observed, for instance, in the process of e^+e^- annihilation are ρ, ω, φ, $\pi\pi$. In τ decays $\tau \to \nu + hadrons$ one can in addition see π, $a_1(1260)$, K, $K^*(892)$.

The primary difficulty for extracting the QCD coupling from experimental data is that the qqg vertex cannot be "directly" measured. Indeed, a typical process (vertex) with hadrons at low energies is $\rho \to \pi\pi$ which is not directly expressed theoretically through the quark-gluon interaction vertex. Therefore, extracting α_s (defined in terms of quark-gluon vertex in the Lagrangian) from the experimental quantity as, for instance, the ρ-meson decay width $\Gamma(\rho \to \pi\pi)$ is highly nontrivial. One can also see a propagation of the pion but not that of a quark. No mass shell for the quark (or gluon) is seen in the experiment. Thus, in QCD there is no preferable definition of parameters related to experiment. Then the only guidance for the choice of a particular definition of the QCD parameters in PT is the technical convenience (and also some general requirements such as gauge invariance) [4].

Presently dimensional regularization is overwhelmingly used in many loop calculations. The renormalization procedure is usually a minimal one – subtraction of poles in $\varepsilon = (D-4)/2$ where D is space-time dimension (an arbitrary complex number formally introduced for the regularization purposes). This procedure is quite abstract and remote from experimental quantities. Thus, formally, the $\overline{\rm MS}$-scheme coupling constant is defined by $\alpha_s^{\overline{\rm MS}}(\mu) = Z_\alpha^{\overline{\rm MS}}(\mu)\alpha_s^B$ where α_s^B is a bare coupling constant. This definition of the renormalized coupling constant is not unique. In the momentum subtraction (MOM) scheme $\alpha_s^{\rm MOM}(\mu) \sim \Gamma_{\rm qqg}^{\rm ren}(p_1^2 = p_2^2 = p_3^2 = \mu^2)$. Since the definition is not unique the renormalization scheme freedom emerges. It is controlled by the renormalization group and can be conveniently parameterized by the coefficients $\beta_{2,3,\dots}$ of the β-function [5,6]

$$\mu\frac{\partial}{\partial\mu}\alpha_s(\mu) = 2\beta(\alpha_s), \quad \beta(\alpha_s) = -\beta_0\alpha_s^2 - \beta_1\alpha_s^3 - \beta_2\alpha_s^4 - \beta_3\alpha_s^5 + O(\alpha_s^6)$$

and, therefore, any particularly defined coupling constant α_s^{sch} generates an associated $\beta^{sch}(\alpha_s^{sch})$-function. The same is true for the definition of the quark mass. In the $\overline{\text{MS}}$-scheme one defines the renormalized mass as $m^{\overline{\text{MS}}}(\mu) = Z_m^{\overline{\text{MS}}}(\mu)m^B$ with an associated γ-function $(\mu\partial/\partial\mu)m(\mu) = 2\gamma(\alpha_s)m(\mu)$. Functions $\beta(\alpha_s)$ and $\gamma(\alpha_s)$ are known up to four-loop approximation in PT [7]. The pole mass is difficult to define for light quarks since numerically even the strange quark is very light $m_s \sim \Lambda_{\text{QCD}}$.

In fact, a QCD coupling constant can also be defined beyond PT through physical observables. For light flavor phenomenology it can directly be defined through the cross section of e^+e^- annihilation

$$1 + \alpha_s(s) \sim \frac{\sigma(e^+e^- \to \text{hadrons})}{\sigma(e^+e^- \to \mu\bar{\mu})}$$

while for heavy quark physics the definition based on the heavy-quark static potential $\alpha_V(\boldsymbol{q}^2) \sim V_{q\bar{q}}(\boldsymbol{q}^2)$ can be useful [8].

2 Kinematics of Semileptonic τ Decays

The differential decay rate of the τ lepton into an hadronic system $H(s)$ with a total squared energy s

$$\frac{d\sigma(\tau \to \nu H(s))}{ds} \sim \left(1 - \frac{s}{M_\tau^2}\right)^2 \left(1 + \frac{2s}{M_\tau^2}\right)\rho(s)$$

is determined by the hadronic spectral density $\rho(s)$ defined through the correlator of weak currents. For the (ud) current $j_\mu^W(x) = \bar{u}\gamma_\mu(1-\gamma_5)d$ one finds

$$i\int\langle Tj_\mu^W(x)j_\nu^{W+}(0)\rangle e^{iqx}dx = (q_\mu q_\nu - q^2 g_{\mu\nu})\Pi^{\text{had}}(q^2), \Pi^{\text{had}}(q^2) = \int\frac{\rho(s)ds}{s-q^2} \quad (1)$$

with $\rho(s) \sim \text{Im}\,\Pi^{\text{had}}(s+i0)$, $s = q^2$. The function $\Pi^{\text{had}}(Q^2)$ with $Q^2 = -q^2$ is calculable in pQCD far from the physical cut as a series in the running coupling constant $\alpha_s(Q^2)$.

NonPT effects (power corrections) are included using OPE for the correlator (1) at small distances as $x \to 0$ in Euclidean domain (that corresponds to large Q^2) through phenomenological characteristics of the vacuum such as gluon and quark condensates [9]. The lattice approximation for the evaluation of the correlator $\Pi^{\text{had}}(Q^2)$ beyond PT can also be used [10]. This is a basis for theoretical description of semileptonic τ decays in QCD.

Integrating the function $\Pi^{\rm had}(z)$ over a contour in the complex q^2 plane beyond the physical cut $s > 0$ one finds that for particular weight functions some integrals of the hadronic spectral density $\rho(s)$ can be reliably computed in PT [11,12]. Indeed, due to Cauchy theorem one gets

$$\oint_C \Pi(z)dz = \int_{\rm cut} \rho(s)ds \ .$$

Using the approximation $\Pi^{\rm had}(z)|_{z\in C} \approx \Pi^{\rm PT}(z)|_{z\in C}$ which is well justified sufficiently far from the physical cut one obtains

$$\oint_C \Pi^{\rm had}(z)dz = \int_{\rm cut} \rho(s)ds = \oint_C \Pi^{\rm PT}(z)dz$$

i.e. the integral over the hadronic spectrum can be evaluated in pQCD. The total decay rate of the τ lepton written in the form of an integral along the cut

$$R_{\tau S=0} = \frac{\Gamma(\tau \to H_{S=0}\nu)}{\Gamma(\tau \to l\bar{\nu}\nu)} \sim \int_{\rm cut} \left(1 - \frac{s}{M_\tau^2}\right)^2 \left(1 + \frac{2s}{M_\tau^2}\right) \rho(s)ds$$

is precisely the quantity that one can reliably compute in pQCD [13].

3 PT Analysis in QCD

For technical reasons (no overall UV divergence) a derivative of the correlator $\Pi^{\rm had}(Q^2)$ is often used for presenting results of PT evaluation

$$D(Q^2) = -Q^2 \frac{d}{dQ^2} \Pi^{\rm had}(Q^2) = Q^2 \int \frac{\rho(s)ds}{(s+Q^2)^2} \cdot$$

The PT expansion for the D-function in terms of $a_s(Q) = \alpha_s(Q)/\pi$ reads $D(Q^2) = 1 + a_s(Q) + k_1 a_s(Q)^2 + k_2 a_s(Q)^3 + k_3 a_s(Q)^4$. In the $\overline{\rm MS}$-scheme

$$k_1 = \frac{299}{24} - 9\zeta(3), \quad k_2 = \frac{58057}{288} - \frac{779}{4}\zeta(3) + \frac{75}{2}\zeta(5)$$

with ζ-function equal $\zeta(3) = 1.202...$ and $\zeta(5) = 1.037...$ The above numerical values for $k_{1,2}$ summarize the results of three and four loop PT calculations [14]. Numerically, one finds $D(Q^2) = 1 + a_s + 1.64a_s^2 + 6.37a_s^3 + k_3 a_s^4$. Coefficient k_3 is known only partly [15]. It is retained to obtain a feeling for the possible magnitude of the $O(\alpha_s^4)$ correction. The decay rate of the τ lepton into nonstrange hadrons is written in the form

$$R_{\tau S=0} = \frac{\Gamma(\tau \to H_{S=0}\nu)}{\Gamma(\tau \to l\bar{\nu}\nu)} = 3|V_{ud}|^2(1 + \delta_P + \delta_{NP}) \ .$$

Here the first term is the parton model result, the second term δ_P represents pQCD effects. NonPT effects are small, $\delta_{NP} \approx 0$, in the factorization approximation for the four-quark vacuum condensates which is quite accurate [13,16].

The experimental result $R^{\rm exp}_{\tau S=0} = 3.492 \pm 0.016$ leads to $\delta_P^{\rm exp} = 0.203 \pm 0.007$ [17]. In the $\overline{\rm MS}$-scheme the correction δ_P is given by the series

$$\delta_P^{\rm th} = a_s + 5.2023a_s^2 + 26.366a_s^3 + (78.003 + k_3)a_s^4 + O(a_s^5)$$

with a_s taken at the scale $\mu = M_\tau$. Usually one extracts a numerical value for $\alpha_s(M_\tau)$ by treating the first three terms of the expression as an exact function – the cubic polynomial $a_s + 5.2023a_s^2 + 26.366a_s^3 = \delta_P^{\rm exp}$. The solution reads $\pi a_s^{st}(M_\tau) \equiv \alpha_s^{st}(M_\tau) = 0.3404 \pm 0.0073_{exp}$. The error is due to the error of the input experimental value $\delta_P^{\rm exp}$. It is difficult to estimate the theoretical uncertainty of the approximation for the (asymptotic) series given by the cubic polynomial (higher order terms). One criterion is the pattern of convergence of the series

$$\delta_P^{\rm exp} = 0.203 = 0.108 + 0.061 + 0.034 + \ldots$$

The corrections provide a 100% change of the leading term. Another criterion is the order-by-order behavior of the extracted numerical value for the coupling constant. In consecutive orders of PT

$$\alpha_s^{st}(M_\tau)_{LO} = 0.6377, \quad \alpha_s^{st}(M_\tau)_{NLO} = 0.3882, \quad \alpha_s^{st}(M_\tau)_{NNLO} = 0.3404$$

that translates into a series for the coupling constant

$$\alpha_s^{st}(M_\tau)_{NNLO} = 0.6377 - 0.2495 - 0.0478 - \ldots$$

One can take a half(??) of the last term as an estimate of the theoretical uncertainty. It is only an indicative estimate. No rigorous justification can be given for such an assumption about the accuracy of the approximation without knowledge of the structure of the whole series. The uncertainty obtained in such a way $\Delta\alpha_s^{st}(M_\tau)_{th} = 0.0478/2 = 0.0239 \gg 0.0073_{exp}$ is much larger than that from experiment. This is a challenge for the theory: the accuracy of theoretical formulae cannot compete with experimental precision. Assuming this theoretical uncertainty one has

$$\alpha_s^{st}(M_\tau)_{NNLO} = 0.3404 \pm 0.0239_{th} \pm 0.0073_{exp}$$

Theory dominates the error. Still it is not the whole story. Now one can choose a different expansion parameter. The simplest way is to change the scale of the coupling along the RG trajectory $M_\tau \to 1$ GeV. In terms of $a_s(1\ {\rm GeV})$ one finds a series $a_s(1) + 2.615a_s(1)^2 + 1.54a_s(1)^3 = \delta_P^{\rm exp}$. The solution for the coupling constant is $\alpha_s(1\ {\rm GeV}) = 0.453$. The convergence pattern for the correction $\delta_P^{\rm exp}$ is

$$\delta_P^{\rm exp} = 0.203 = 0.144 + 0.054 + 0.005$$

and for the numerical value of the coupling constant

$$\alpha_s(1) = 0.453 = 0.638 - 0.177 - 0.008\,.$$

Should one conclude that now the accuracy is much better? What would be an invariant criterion for the precision of theoretical predictions obtained from

PT, i.e. finite number of terms of asymptotic series? Thus, one sees that the renormalization scheme dependence can strongly obscure the heuristic evaluation of the accuracy of theoretical formulae in the absence of any information on the structure of the whole PT series. The final result for the standard reference value of the coupling constant normalized to the Z boson mass M_Z reads [18]

$$\alpha_s(M_Z)_\tau = 0.1184 \pm 0.0007_{exp} \pm 0.0006_{hq\ mass} \pm 0.0010_{th=truncation}$$

with a theoretical uncertainty that mainly comes from the truncation of the PT series. The world average value given by the Particle Data Group reads [2]

$$\alpha_s(M_Z)^{\rm av}_{\rm PDG} = 0.1172 \pm 0.002\,.$$

To reduce a renormalization scheme dependence of the theoretical analysis one should use several observables simultaneously [19]. Such a possibility has recently emerged in study of τ decays since experimental data on Cabibbo suppressed ($S=1$) channel appeared [20]. For strange hadrons $H_{S=1}$ (us part of the weak current: Cabibbo suppressed decays) the decay rate becomes

$$R_{\tau S=1} = \frac{\Gamma(\tau \to H_{S=1}\nu)}{\Gamma(\tau \to l\bar{\nu}\nu)} = 3|V_{us}|^2(1+\delta'_P+\delta'_{NP})\,.$$

The first term ("1") is the parton model result, the second term δ'_P gives pQCD effects. Small s-quark mass effects for Cabibbo suppressed part of the rate are taken in PT at the leading order in the ratio m_s^2/M_τ^2 since $m_s \ll M_\tau$

$$\delta'_P(\alpha_s, m_s) = \delta_P(\alpha_s) + \frac{m_s^2}{M_\tau^2}\Delta_m(\alpha_s)$$

with $\delta_P(\alpha_s)$ being a correction in massless approximation for light quarks that is well justified for nonstrange decays (ud part) since u, d quarks are very light indeed $m_u + m_d = 14$ MeV [21,22]. The correlator of the weak charged strange current $j_\mu(x) = \bar{u}\gamma_\mu(1-\gamma_5)s$ with a finite s-quark mass is not transverse

$$i\int dx e^{iqx}\langle T j_\mu(x) j^\dagger_\nu(0)\rangle = q_\mu q_\nu \Pi_q(q^2) + g_{\mu\nu}\Pi_g(q^2)\,.$$

Retaining the first order term of expansion in the small ratio m_s^2/q^2 one finds the m_s^2 correction to the invariant functions $\Pi_{q,g}(q^2)$

$$\Pi_q(q^2) = \Pi(q^2) + 3\frac{m_s^2}{q^2}\Pi_{mq}(q^2), \quad \Pi_g(q^2) = -q^2\Pi(q^2) + \frac{3}{2}m_s^2\Pi_{mg}(q^2)$$

where $\Pi(q^2)$ is an invariant function for the mass zero case. The functions $\Pi_{q,g}(Q^2)$ are computable in QCD perturbation theory within operator product expansion for $Q^2 \to \infty$. Thus, the experimental data on τ lepton decays are theoretically described by three independent invariant functions (form factors) which can be analyzed simultaneously that may help to reduce uncertainties introduced by the renormalization scheme freedom.

In the actual analysis one can factor out the renormalization scheme freedom to large extent by introducing an effective scheme with definitions of effective quantities a, m_q^2, m_g^2 through the relations [23]

$$\begin{aligned}
-Q^2 \frac{d}{dQ^2} \Pi(Q^2) &= 1 + a(Q^2)\,, \\
-m_s^2(M_\tau^2) Q^2 \frac{d}{dQ^2} \Pi_{mg}(Q^2) &= m_g^2(M_\tau^2) C_g(Q^2)\,, \\
m_s^2(M_\tau^2) \Pi_{mq}(Q^2) &= m_q^2(M_\tau^2) C_q(Q^2)\,.
\end{aligned} \tag{2}$$

Here $C_{q,g}(Q^2)$ are coefficient functions of mass corrections. They are conveniently normalized by the requirement $C_{q,g}(M_\tau^2) = 1$. In terms of the $\overline{\text{MS}}$ scheme quantities $\alpha_s \equiv \alpha_s(M_\tau^2)$ and $m_s \equiv m_s(M_\tau^2)$ the effective parameters in (2) read

$$\begin{aligned}
a(M_\tau^2) &= \frac{\alpha_s}{\pi} + k_1 \left(\frac{\alpha_s}{\pi}\right)^2 + k_2 \left(\frac{\alpha_s}{\pi}\right)^3 + k_3 \left(\frac{\alpha_s}{\pi}\right)^4 + \mathcal{O}(\alpha_s^5)\,, \\
m_g^2(M_\tau^2) &= m_s^2(M_\tau^2)(1 + \frac{5}{3}\frac{\alpha_s}{\pi} + k_{g1} \left(\frac{\alpha_s}{\pi}\right)^2 + k_{g2} \left(\frac{\alpha_s}{\pi}\right)^3 + \mathcal{O}(\alpha_s^4))\,, \\
m_q^2(M_\tau^2) &= m_s^2(M_\tau^2)(1 + \frac{7}{3}\frac{\alpha_s}{\pi} + k_{q1} \left(\frac{\alpha_s}{\pi}\right)^2 + k_{q2} \left(\frac{\alpha_s}{\pi}\right)^3 + \mathcal{O}(\alpha_s^4))\,.
\end{aligned}$$

Numerical values for the coefficients k_3, k_{q2} are unknown though their estimates within various intuitive approaches can be found in the literature [24].

For confronting the experimental data obtained in τ decays with theory of strong interactions one uses special integrals of the hadronic spectral density $\rho(s)$ (called spectral moments) of the form

$$\mathcal{M}_{kl}^{\text{exp}} = \int_0^{M_\tau^2} \rho(s) \left(1 - \frac{s}{M_\tau^2}\right)^k \left(\frac{s}{M_\tau^2}\right)^l \frac{ds}{M_\tau^2}$$

which is suitable for experiment as the spectrum $\rho(s)$ is measured. Another representation of the moments is suitable for the perturbation theory evaluation in QCD. Due to analytic properties of the functions $\Pi_{q,g}(Q^2)$ the moments can be rewritten as contour integrals in the complex q^2 plane:

$$\mathcal{M}_{kl}^{\text{th}} = \frac{i}{2\pi} \oint_C \Pi_\#(z) \left(1 - \frac{z}{M_\tau^2}\right)^k \left(\frac{z}{M_\tau^2}\right)^l \frac{dz}{M_\tau^2}\,.$$

Technically it is convenient to choose a circular contour in the complex Q^2-plane with $Q^2 = M_\tau^2 e^{i\phi}$, $-\pi < \phi < \pi$ that converts all invariant amplitudes $\Pi_{q,g}(Q^2)$ to the certain functions of the angle ϕ. The evolution of the functions $\Pi_{q,g}(Q^2)$ along the contour in the complex plane is governed by the renormalization group [25]. For the massless case corresponding to the analysis of data in Cabibbo favored channel the only relevant quantity is the running "coupling constant" $a(Q^2) \to a(\phi)$ for $Q^2 = M_\tau^2 e^{i\phi}$ that serves as the expansion parameter for the function $\Pi(Q^2)$ along the contour. Note that there are no higher order corrections in the effective scheme by definition. The renormalization group evolution is

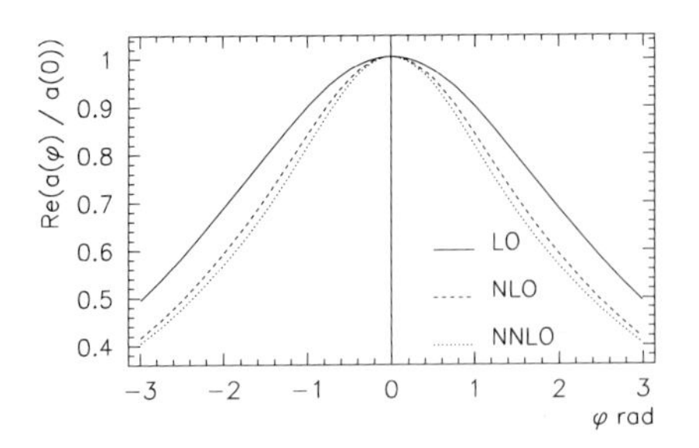

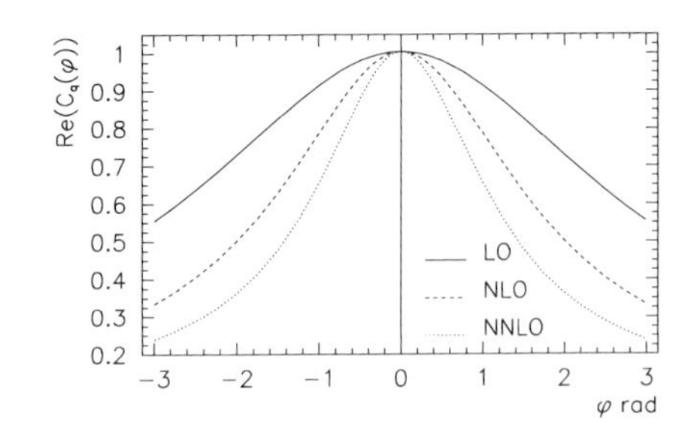

Fig. 1. Running of the functions $a(\phi)$ and $C_q(\phi)$ on a circular contour in the complex plane calculated at LO, NLO and NNLO (left: $a(\phi)$; right: $C_q(\phi)$; real parts only)

determined by the effective β function for the effective coupling constant [26,27]

$$-i\frac{d}{d\phi}a(\phi) = \beta(a(\phi)), \quad a(\phi = 0) = a(M_\tau^2),$$

and the anomalous dimension for the running mass

$$-i\frac{d}{d\phi}m_s(\phi) = \gamma_m[a(\phi)]m_s(\phi), \quad m_s(\phi = 0) = m_s .$$

The initial values for $a(\phi)$ and $m_s(\phi)$ are extracted from fit to data. Thus, we have three quantities $a(Q)$, $C_q(Q)$, $C_g(Q)$ associated with three invariant functions $\Pi(Q^2)$, $\Pi_{q,g}(Q^2)$ describing the τ system in the considered approximation.

The RG equations for the set of quantities $\{a(Q), C_q(Q), C_g(Q)\}$ are

$$Q^2\frac{d}{dQ^2}a(Q^2) = \beta(a), \quad Q^2\frac{d}{dQ^2}C_{g,q}(Q^2) = 2\gamma_{g,q}(a)C_{g,q}(Q^2).$$

The RG functions $\beta(a)$ and $\gamma_{g,q}(a)$ are given by the expressions [23]

$$\frac{-4\beta(a)}{9a^2} = 1 + 1.778a + 5.24a^2 + a^3(-34 + 2k_3);$$

$$\frac{-\gamma_g(a)}{a} = 1 + 4.03a + 17.45a^2 + a^3(249.59 - k_3);$$

$$\frac{-\gamma_q(a)}{a} = 1 + 4.78a + 32.99a^2 + a^3(-252 - k_3 + 3.4k_{q2}).$$

The solution of the renormalization group equation for the effective coupling constant $a(\phi)$ converges well when the higher order corrections of the β-function are included. The change from the next-to-leading order (NLO) solution to the next-to-next-to-leading order (NNLO) is small. The behavior of the coefficient function $C_g(\phi)$ (not shown) related to the contributions of spin one particles is rather similar to that of the coupling constant. However, the convergence pattern of the function $C_q(\phi)$ is much worse (see Fig. 1). It seems that the γ_q-function has already shown up an asymptotic growth in the next-to-next-to-leading order

Table 1. Coefficients of (3)

(k,l)	A_{kl}^{LO}	A_{kl}^{NLO}	A_{kl}^{NNLO}	B_{kl}^{LO}	B_{kl}^{NLO}	B_{kl}^{NNLO}
$(0,0)$	1.361	1.445	1.434	0.523	0.601	0.625
$(1,0)$	1.568	1.843	1.976	0.441	0.552	0.601
$(2,0)$	1.762	2.282	2.646	0.390	0.530	0.607

which will limit the precision of our results. Note that the fact that the function $C_q(\phi)$ can behave wilder in higher order of perturbation theory is expected since this function is more infrared sensitive than the coupling constant $a(\phi)$ and the function $C_g(\phi)$.

The numerical value for the s-quark mass m_s is extracted from the difference between moments of Cabibbo-favored (ud-type) and and Cabibbo-suppressed (us-type) decay rates

$$\delta R_\tau^{kl} = \frac{R_{\tau S=0}^{kl}}{|V_{ud}|^2} - \frac{R_{\tau S=1}^{kl}}{|V_{us}|^2}, \quad R_{\tau S=0,1}^{kl} = \int_0^{M_\tau^2} ds \left(1-\frac{s}{M_\tau^2}\right)^k \left(\frac{s}{M_\tau^2}\right)^l \frac{dR_{\tau S=0,1}}{ds}.$$

The theoretical expression for the m_s^2 corrections to the moments (k,l) corresponding to the above experimental quantity is given by the contour integral in the complex q^2 plane

$$\frac{3im_s^2}{\pi}\oint_C \left(1-\frac{z}{M_\tau^2}\right)^{2+k} \left(\frac{z}{M_\tau^2}\right)^l \left(\frac{\Pi_{mq}(z)}{z} - \frac{\Pi_{mg}(z)}{M_\tau^2}\right)\frac{dz}{M_\tau^2}.$$

In the theoretical expression for the difference δR_τ^{kl} we neglect terms of the order m_s^3/M_τ^3, set the u- and d-quark masses to zero, and retain only the most important term linear in m_s. Within operator product expansion the coefficient of this term is given by the quark condensate. The final result for the difference reads [23]

$$\delta R_\tau^{kl} = 3S_{EW}\left(6\frac{m_s^2}{M_\tau^2}(\omega_q A_{kl} + \omega_g B_{kl}) - 4\pi^2 \frac{m_s}{M_\tau}\frac{\langle \bar{s}s\rangle}{M_\tau^3}T_{kl}\right) \tag{3}$$

with $S_{EW} = 1.0194$ [28]. Here $m_{q,g}^2 = \omega_{q,g} m_s^2$ with $\omega_q = 1.73 \pm 0.04$, $\omega_g = 1.42 \pm 0.03$. We use the relation $\langle \bar{s}s\rangle = (0.8 \pm 0.2)\langle \bar{u}u\rangle$ and the numerical value $\langle \bar{u}u\rangle = -(0.23\ \mathrm{GeV})^3$ [29]. In the leading order approximation for the coefficient function the quantities T_{kl} multiplying the quark condensate are given by the expression

$$T_{kl} = 2\left(\delta_{l,0}(k+2) - \delta_{l,1}\right).$$

The numerical values for the first few coefficients T_{kl} read

$$T_{00} = 4, \quad T_{10} = 6, \quad T_{20} = 8, \quad T_{01} = -2, \quad T_{11} = -2.$$

The numerical values for the coefficients A_{kl} and B_{kl} are given in Table 1.

Table 2. Experimental moments and extracted mass

(k,l)	$(\delta R_\tau^{kl})^{\rm exp}$	$m_s(M_\tau^2)$ MeV
$(0,0)$	0.394 ± 0.137	$130 \pm \delta_{00}^{\rm th}(=6)$
$(1,0)$	0.383 ± 0.078	$111 \pm \delta_{10}^{\rm th}(=?)$
$(2,0)$	0.373 ± 0.054	$95 \pm \delta_{20}^{\rm th}(=?)$

Using experimental data one extracts m_s as given in Table 2. Theoretical prediction for the moment $(0,0)$ is the most reliable from PT point of view as $\delta_{20}^{\rm th}(=?) > \delta_{10}^{\rm th}(=?) > \delta_{00}^{\rm th} = 6$ MeV. The final result reads

$$m_s(M_\tau^2) = 130 \pm 27_{\rm exp} \pm 3_{\langle \bar{s}s\rangle} \pm 6_{\rm th} \text{ MeV}.$$

Normalization at 1 GeV gives

$$m_s(1 \text{ GeV}) = 176 \pm 37_{\rm exp} \pm 4_{\langle \bar{s}s\rangle} \pm 9_{\rm th} \text{ MeV}.$$

Only the moment $(0,0)$ is used for the m_s determination as the most reliable one from the PT point of view. The higher order moments with the weight function $(1 - s/M_\tau^2)^k$ for large k have an uncontrollable admixture of higher dimension condensates that makes them strongly nonperturbative and, therefore, unreliable for applications based on PT calculations [30]. The contributions of higher dimension condensates are unknown and from general considerations the errors $\delta_{20}^{\rm th}(=?) > \delta_{10}^{\rm th}(=?)$ are expected to be much larger than $\delta_{00}^{\rm th} = 6$ MeV. The value of the strange quark mass obtained by using the effective scheme approach as described in the present paper is in a reasonable agreement with other estimates [31–33]. It is a bit larger than the recent lattice determination [34].

To conclude, the experimental information on τ decays is a reliable source for the precision determination of the numerical values of important QCD parameters α_s and m_s.

Acknowledgements

I thank K.G. Chetyrkin, S. Groote, J.G. Körner, J.H. Kühn for useful discussions and fruitful collaboration. This work is partially supported by the Russian Fund for Basic Research under contracts 01-02-16171 and 02-01-00601 and by the Volkswagen grant.

References

1. C.N. Yang and R.L. Mills: Phys. Rev. **96**, 191 (1954); M. Gell-Mann: Phys. Lett. **8**, 214 (1964); M-Y. Han, Y. Nambu: Phys. Rev. B **139**, 1006 (1965); W.A. Bardeen, H. Fritzsch, M. Gell-Mann: In: *Scale and Conformal Symmetry in Hadron Physics*. ed. by R. Gatto (Wiley, New York 1973); H. Pagels: Phys. Rept. C **16**, 219 (1975)
2. D.E. Groom et al. (Particle Data Group): Eur. Phys. J. C **15**, 1 (2000)
3. N.N. Bogoliubov and D.V. Shirkov: *Quantum Fields* (Benjamin, New York 1983)

4. I.I. Bigi, M.A. Shifman, N.G. Uraltsev and A.I. Vainshtein: Phys. Rev. D **50**, 2234 (1994); M. Beneke and V.M. Braun: Nucl. Phys. B **426**, 301 (1994)
5. E. Stuekelberg and A. Peterman: Helv. Phys. Acta **26**, 499 (1953); M. Gell-Mann and E.E. Low: Phys. Rev. **95**, 1300 (1954); N.N. Bogoliubov and D.V. Shirkov: Dokl. Akad. Nauk **103**, 391 (1955)
6. P.M. Stevenson: Phys. Rev. D **23**, 2916 (1981)
7. T. van Ritbergen, J.A.M. Vermaseren and S.A. Larin: Phys. Lett. B **400**, 379 (1997); K.G. Chetyrkin: Phys. Lett. B **404**, 161 (1997); T. van Ritbergen, J.A.M. Vermaseren and S.A. Larin: Phys. Lett. B **405**, 327 (1997)
8. M. Jezabek, J.H. Kuhn, M. Peter, Y. Sumino and T. Teubner: Phys. Rev. D **58**, 014006 (1998); N.V. Krasnikov and A.A. Pivovarov: Mod. Phys. Lett. A **11**, 835 (1996); Phys. Atom. Nucl. **64**, 1500 (2001); M. Luscher, R. Sommer, P. Weisz and U. Wolff: Nucl. Phys. B **413**, 481 (1994); A.A. Pivovarov: "Heavy quark production near the threshold in QCD." [arXiv:hep-ph/0110398]
9. H.D. Politzer: Nucl. Phys. B **117**, 397 (1976); M.A. Shifman, A.I. Vainshtein and V.I. Zakharov: Nucl. Phys. B **147**, 385 (1979); B.V. Geshkenbein, B.L. Ioffe and K.N. Zyablyuk: Phys. Rev. D **64**, 093009 (2001)
10. A.C. Kalloniatis, D.B. Leinweber, W. Melnitchouk and A.G. Williams, *Lattice Hadron Physics*. Proceedings, (Cairns, Australia, July 9-18, 2001)
11. C. Bernard, A. Duncan, J. LoSecco and S. Weinberg: Phys. Rev. D **12**, 792 (1975); E. Poggio, H. Quinn and S. Weinberg: Phys. Rev. D **13**, 1958 (1976); R. Shankar: Phys. Rev. D **15**, 755 (1977); K.G. Chetyrkin, N.V. Krasnikov and A.N. Tavkhelidze: Phys. Lett. B **76**, 83 (1978)
12. N.V. Krasnikov, A.A. Pivovarov and N. Tavkhelidze: JETP Lett. **36**, 333 (1982), Z. Phys. C **19**, 301 (1983); N.V. Krasnikov and A.A. Pivovarov: Phys. Lett. B **112**, 397 (1982); A.A. Pivovarov: Phys. Atom. Nucl. **62**, 1924 (1999); A.A. Pivovarov and A.S. Zubov: Phys. Atom. Nucl. **63**, 1650 (2000)
13. K. Schilcher and M.D. Tran: Phys. Rev. D **29**, 570 (1984); E. Braaten: Phys. Rev. Lett. **60**, 1606 (1988); S. Narison and A. Pich: Phys. Lett. B **211**, 183 (1988); A.A. Pivovarov: Sov. J. Nucl. Phys. **54**, 676 (1991); E. Braaten, S. Narison and A. Pich: Nucl. Phys. B **373**, 581 (1992)
14. S.G. Gorishny, A.L. Kataev and S.A. Larin: Phys. Lett. B **259**, 144 (1991); L.R. Surguladze and M.A. Samuel: Phys. Rev. Lett. **66**, 560 (1991); Phys.Rev. D **44**, 1602 (1991); K.G. Chetyrkin: Phys. Lett. B **391**, 402 (1997)
15. P.A. Baikov, K.G. Chetyrkin and J.H. Kühn: Phys. Rev. Lett. **88**, 012001 (2002)
16. K.G. Chetyrkin and A.A. Pivovarov: Nuovo Cim. A **100**, 899 (1988); A.A. Pivovarov: JETP Lett. **55**, 6 (1992); Int. J. Mod. Phys. A **10**, 3125 (1995)
17. ALEPH collaboration: Z. Phys. C **76**, 15 (1997); Eur. Phys. J. C **4**, 409 (1998); *ibid* C **11**, 599 (1999); OPAL collaboration: Eur. Phys. J. C **7**, 571 (1999)
18. J.G. Körner, F. Krajewski and A.A. Pivovarov: Phys. Rev. D **63**, 036001 (2001)
19. S. Groote, J.G. Körner, A.A. Pivovarov and K. Schilcher: Phys. Rev. Lett. **79**, 2763 (1997); S.J. Brodsky, J.R. Pelaez and N. Toumbas: Phys. Rev. D **60**, 037501 (1999); J.G. Körner, F. Krajewski and A.A. Pivovarov: Eur. Phys. J. C **12**, 461 (2000); *ibid* C **14**, 123 (2000)
20. K.G. Chetyrkin, J.H. Kühn and A.A. Pivovarov: Nucl. Phys. B **533**, 473 (1998); A. Pich and J. Prades: JHEP **9806**, 013 (1998); *ibid* **9910**, 004 (1999)
21. S. Narison and E. de Rafael: Phys. Lett. B **103**, 57 (1981); J. Gasser and H. Leutwyler: Phys. Rept. **87**, 77 (1982)
22. A.L. Kataev, N.V. Krasnikov and A.A. Pivovarov: Phys. Lett. B **123**, 93 (1983); Nuovo Cim. A **76**, 723 (1983)

23. J.G. Körner, F. Krajewski and A.A. Pivovarov: Eur. Phys. J. C **20**, 259 (2001)
24. A.L. Kataev and V.V. Starshenko: Mod. Phys. Lett. A **10**, 235 (1995); K.G. Chetyrkin, B.A. Kniehl and A. Sirlin: Phys. Lett. B **402**, 359 (1997)
25. A.A. Pivovarov: Z. Phys. C **53**, 461 (1992); Nuovo Cim. A **105**, 813 (1992); F. Le Diberder and A. Pich: Phys. Lett. B **286**, 147 (1992); S. Groote, J.G. Körner and A.A. Pivovarov: Phys. Lett. B **407**, 66 (1997); Mod. Phys. Lett. A **13**, 637 (1998)
26. G. Grunberg: Phys. Lett. B **95**, 70 (1980)
27. N.V. Krasnikov: Nucl. Phys. B **192**, 497 (1981); A.L. Kataev, N.V. Krasnikov and A.A. Pivovarov: Phys. Lett. B **107**, 115 (1981); Nucl. Phys. B **198**, 508 (1982); A. Dhar and V. Gupta: Phys. Rev. D **29**, 2822 (1984)
28. W.J. Marciano and A. Sirlin: Phys. Rev. Lett. **61**, 1815 (1988); E. Braaten and C.S. Li: Phys. Rev. D **42**, 3888 (1990)
29. C. Becchi, S. Narison, E. de Rafael and F.J. Yndurain: Z. Phys. C **8**, 335 (1981); A.A. Ovchinnikov and A.A. Pivovarov: Phys. Lett. B **163**, 231 (1985); N.V. Krasnikov and A.A. Pivovarov: Nuovo Cim. A **81**, 680 (1984); Y. Chung et al.: Z. Phys. C **25**, 151 (1984); H.G. Dosch and S. Narison: Phys. Lett. B **417**, 173 (1998)
30. A.A. Pivovarov: "Spectrality, coupling constant analyticity and the renormalization group," to appear in Yad. Fiz. [arXiv:hep-ph/0104213]; S. Groote, J.G. Körner and A.A. Pivovarov: Phys. Rev. D **65**, 036001 (2002)
31. S.G. Gorishnii, A.L. Kataev and S.A. Larin: Phys. Lett. B **135**, 457 (1984); M. Jamin and M. Munz: Z. Phys. C **66**, 633 (1995); K.G. Chetyrkin, D. Pirjol and K. Schilcher: Phys. Lett. B **404**, 337 (1997)
32. K. Maltman and J. Kambor: Phys. Rev. D **65**, 074013 (2002); Phys. Lett. B **517**, 332 (2001); Nucl. Phys. A **684**, 348 (2001); Phys. Rev. D **62**, 093023 (2000); K. Maltman: Nucl. Phys. Proc. Suppl. A **109**, 178 (2002); arXiv:hep-ph/0209091
33. S. Chen, M. Davier, E. Gamiz, A. Hocker, A. Pich and J. Prades: Eur. Phys. J. C **22**, 31 (2001)
34. ALPHA and UKQCD Collaboration (J. Garden et al.): Nucl. Phys. B **571**, 237 (2000); ALPHA Collaboration (M. Guagnelli et al.): Nucl. Phys. B **560**, 465 (1999)

Exclusive Nonleptonic B Decays from QCD Light-Cone Sum Rules

Blaženka Melić[1,2,⋆]

[1] Institut für Theoretische Physik und Astrophysik, Julius-Maximilians-Universität Würzburg, 97074 Würzburg, Germany
[2] Institut für Physik, Johannes-Gutenberg Universität Mainz, 55099 Mainz, Germany

Abstract. We are going to review recent advances in the theory of exclusive nonleptonic B decays. The emphasis is going to be on the factorization hypothesis and the role of nonfactorizable contributions for nonleptonic B decays. In particular, we will discuss more in detail calculations of nonfactorizable contributions in the QCD light-cone sum rule approach and their implications to the $B \to \pi\pi$ and $B \to J/\psi K$ decays.

1 Exclusive Nonleptonic B Decays and Factorization

Exclusive nonleptonic decays represent a great challenge to theory. They are complicated by the hadronization of final states and strong-interaction effects between them. Today measurements have already reached sufficient precision to examine our knowledge of these effects. In order to make real use of data in the determination of fundamental parameters and in testing of the Standard Model, we are forced to provide a more accurate estimation of nonperturbative quantities, such as the matrix elements of weak operators.

At the first sight, the nonleptonic B meson decay seems to be simple, as far as we essentially consider this decay as a weak decay of heavy b quark. We are encouraged to use this argument by the facts that the b quark mass is heavy compared to the intrinsic scale of strong interactions and that the b quark decays fast enough to produce energetic constituents, which separate without interfering with each other. This naive picture was supported by the color-transparency argument [1] and natural application to nonleptonic two-body decays emerged under the name *the naive factorization* (discussed in detail below). However, although predictions from the naive factorization are in relatively good agreement with the data (apart from the color-suppressed decays), the naive factorization provides no insight into the dynamical background of exclusive nonleptonic decays.

The theoretical discussion of the nonleptonic decay starts with the effective weak Hamiltonian, which summarizes our knowledge of weak decays at low scales (for a review see [2]):

$$\mathcal{H}_{weak} = \frac{G_F}{\sqrt{2}} V_{Qq_1} V_{q_2 q_3} \left[C_1(\mu)\mathcal{O}_1 + C_2(\mu)\mathcal{O}_2 + ...\right] . \qquad (1)$$

⋆ Alexander von Humboldt fellow. On the leave of absence from Rudjer Bošković Institute, Zagreb, Croatia.

B. Melić, Exclusive Nonleptonic B Decays from QCD Light-Cone Sum Rules, Lect. Notes Phys. **647**, 287–301 (2004)
http://www.springerlink.com/

The Vs represent the Cabibbo-Kobayashi-Maskawa (CKM) matrix elements specified for the particular heavy-quark decay $Q \to q_1 q_2 \overline{q}_3$. Strong-interaction effects above some scale $\mu \sim m_b$ are retained in the Wilson coefficients $C_i(\mu)$. These coefficients are perturbatively calculable and therefore well known. Actually, the weak theory without strong corrections and QED effects knows only the operator $\mathcal{O}_1$, and in that case $C_1(M_W) = 1$ and $C_2(M_W) = 0$. The operator $\mathcal{O}_2$, defined in (2), emerges after taking the gluon exchange into account and therefore its contribution is suppressed as $C_2(\mu) \sim \ln(M_W)/\ln(\mu)$.

The main problem persists in the calculation of matrix elements of operators $\mathcal{O}_i$ in a particular process. In (1) we retain only the leading operators $\mathcal{O}_1$ and $\mathcal{O}_2$ and suppress explicitly so called penguin operators, $\mathcal{O}_{i=3,\ldots,10}$. Being multiplied by, in principle, small Wilson coefficients, the penguin operators usually can be neglected (except for the penguin-dominated decays), but could be extremely important for detection of CP violation in B decay [3–5].

The four-quark operators $\mathcal{O}_1$ and $\mathcal{O}_2$ differ only in their color structure:

$$\mathcal{O}_1 = (\overline{q}_{1i} \Gamma_\mu Q_i)(\overline{q}_{2j} \Gamma^\mu q_{3j}) \,, \qquad \mathcal{O}_2 = (\overline{q}_{1i} \Gamma_\mu Q_j)(\overline{q}_{2j} \Gamma^\mu q_{3i}) \,, \tag{2}$$

where i and j are color indices, and $\Gamma_\mu = \gamma_\mu(1-\gamma_5)$. The color-mismatched operator $\mathcal{O}_2$ can be projected to the color singlet state by using the relation $\delta_{ij}\delta_{kl} = 1/N_c \, \delta_{il}\delta_{jk} + 2\,(\lambda^a/2)_{il}\,(\lambda^a/2)_{jk}$, as

$$\mathcal{O}_2 = \frac{1}{N_c} \mathcal{O}_1 + 2\,\tilde{\mathcal{O}}_1 \,. \tag{3}$$

This projection, as can be seen from (3), results in a relative suppression of the $\mathcal{O}_2$ operator contribution of the order $1/N_c$ (N_c is the number of colors) and in the appearance of the new operator $\tilde{\mathcal{O}}_1$ with the explicit color SU(3) matrices λ^a:

$$\tilde{\mathcal{O}}_1 = (\overline{q}_{1i} \Gamma_\mu \frac{\lambda^a}{2} Q_i)(\overline{q}_{2j} \Gamma^\mu \frac{\lambda^a}{2} q_{3j}) \,. \tag{4}$$

Depending on the process involved, the operators $\mathcal{O}_1$ and $\mathcal{O}_2$ can exchange their roles, and then it is customary to define the effective parameters a_1 and a_2 as

$$a_1 = C_1(\mu) + \frac{1}{N_c} C_2(\mu) \,, \qquad a_2 = C_2(\mu) + \frac{1}{N_c} C_1(\mu) \,. \tag{5}$$

These parameters distinguish between three classes of decay topologies:
- *class-1* decay amplitude, where a charged meson is directly produced in the weak vertex; i.e. in the quark transition $b \to u d \overline{u}$ with $\mathcal{O}_1 = (\overline{d} \Gamma_\mu u)(\overline{u} \Gamma^\mu b)$:

$$\mathcal{A}(B \to \pi^+ \pi^-) \sim a_1 \langle O_1 \rangle \,, \tag{6}$$

- *class-2* decay amplitude, where a neutral meson is directly produced, i.e. in the quark transition $b \to c s \overline{c}$ with $\mathcal{O}_2 = (\overline{c} \Gamma_\mu c)(\overline{s} \Gamma^\mu b)$:

$$\mathcal{A}(B^+ \to J/\psi K^+) \sim a_2 \langle O_2 \rangle \,, \tag{7}$$

- *class-3* decay amplitude, where both cases are possible, but this amplitude is however connected by isospin symmetry with the class-1 and class-2 decays; i.e. in the quark transition $b \to cs\overline{u}$ with $\mathcal{O}_1 = (\overline{c}\Gamma_\mu u)(\overline{s}\Gamma^\mu b)$:

$$\mathcal{A}(B^- \to D^0 K^-) \sim (a_1 + x a_2)\langle O_1 \rangle \,, \tag{8}$$

where x denotes the nonperturbative factor being equal to one in the flavor-symmetry limit.

The effective parameters a_1 and a_2 are defined with respect to the naive factorization hypothesis, which assumes that the nonleptonic amplitude can be expressed as the product of matrix elements of two hadronic (bilinear) currents, for example:

$$\langle \pi^+\pi^- | (\overline{d}\Gamma_\mu u)(\overline{u}\Gamma^\mu b)|B\rangle \to \langle \pi^- | (\overline{d}\Gamma_\mu u)|0\rangle \langle \pi^+ | (\overline{u}\Gamma^\mu b)|B\rangle \tag{9}$$

and that there is no nonfactorizable exchange of gluons between the π^- and the $|\pi^+ B\rangle$ system. Effectively, that means that the 'nonfactorizable' matrix element of the $\tilde{\mathcal{O}}_1$ operator (4) vanishes, due to the projection of the colored current to the physical colorless state.

1.1 Nonfactorizable Contributions

The effective parameters a_1 and a_2 could be generalized to parametrize also *the nonfactorizable strong-interaction effects*, for example gluon exchanges between bilinear currents (i.e. in (9)) which introduce nonvanishing contribution from $\tilde{\mathcal{O}}$ operators. Schematically, in the large N_c limit,

$$\begin{aligned} a_1 &= C_1(\mu) + \frac{1}{N_c} C_2(\mu) + 2C_2(\mu)\xi_2^{nf}(\mu)\,, \\ a_2 &= C_2(\mu) + \frac{1}{N_c} C_1(\mu) + 2C_1(\mu)\xi_1^{nf}(\mu)\,, \end{aligned} \tag{10}$$

where we have explicitly indicated that the nonfactorizable contribution to the class-1 and class-2 decays $\xi^{nf}_{i=1,2}$, do not necessarily need to be the same, and also they can be process dependent quantities, which will be discussed later. Theoretically, nonfactorizable effects are desirable in order to cancel explicit the μ dependence of $C_i(\mu)$ and therefore of the a_i's. All physical quantities are μ independent, and because there is no explicit μ dependence of the matrix elements $\langle \mathcal{O}_i \rangle$ multiplying $C_i(\mu)$, there must be some underlying mechanism to cancel the explicit μ dependence of a_i's persisting in the factorization approach. In the calculation of the Wilson coefficients beyond the leading order, also the renormalization scheme dependence is presented [6]. Naturally, the parameter a_2 is more sensitive on the value of the factorization scale and on the renormalization scheme, due to the similar magnitude and different sign of the $C_2(\mu)$ and $1/N_c C_1(\mu)$ terms (calculated in the NDR scheme and for $\Lambda^{(5)}_{\overline{MS}} = 225\, GeV$, the Wilson coefficients have the following values: $C_1(m_b) = 1.082$ and $C_2(m_b) = -0.185$ [2]). This means also that a_2 is more sensitive to any additional nonperturbative long-distance contributions.

The global fit of a_1 and a_2 parameters to the B meson experimental data performed in [7], has shown that the a_1 coefficient, being essentially proportional to $C_1(\mu) \sim 1$, is in the expected theoretical range:

$$a_1 \sim 1.05 \pm 0.10 \,, \tag{11}$$

while a_2 has the fitted value of

$$a_2 \sim 0.25 \pm 0.05 \,. \tag{12}$$

Compared with the theoretical values calculated with the C_1 and C_2 stated above, we note that both fitted values show no explicit indication that there is a significant nonfactorizable contribution in B decays. This confirms the naive factorization picture, although the simple extrapolation of results in D decays to the B case would suggest that the a_2 coefficient could be negative, meaning a nontrivial cancellation of the $1/N_c$ terms and dominance of (negative) $C_2(\mu)$ in (10). The negative value of a_2 in D decays has found its confirmation in the large N_c hypothesis of neglecting the higher order $1/N_c$ terms, [8], and in the QCD sum rule calculation [9], where the cancellation of the $1/N_c$ part with the explicitly calculated nonfactorizable terms was verified.

However, there are additional indications that nonfactorizable contributions in B decays cannot be simply neglected and deserve to be investigated. New experimental data on B mesons indicate nonuniversality of the a_2 parameter and the strong final-state interaction phases in the color-suppressed class-2 decays being proportional to a_2 [10].

Therefore, the nonfactorizable contributions must play an important role in nonleptonic decays, particularly in the color-suppressed class-2 decays, such as the $B \to J/\psi K$ decay discussed in Sect.4.

1.2 Models for the Calculation of Nonfactorizable Contributions

Nowadays, there exist several approaches for the treatment of nonleptonic decays, which try to investigate the dynamical background and nonfactorizable contributions of such processes. The most exploited ones are *the QCD factorization*, [11], and *the PQCD approach* [12].

The PQCD approach claims the perturbativity of the two-body nonfactorizable amplitude if the Sudakov suppression is implemented into the calculation. The Sudakov form factor suppresses the configuration in which the soft gluon exchange could take place, and the amplitude is dominated by exchange of hard gluons and therefore perturbatively calculable.

A somewhat different method is applied in *the QCD factorization.* This method provides the factorization formula that separates soft and hard contribution on the basis of large m_b expansion. The perturbatively calculable nonleading terms of $1/m_b$ expansion can be then studied systematically, while soft (incalculable) contributions are suppressed by Λ_{QCD}/m_b. The method applies to class-1 decays and to class-2 decays under the assumption $m_c \ll m_b$.

None of these models take nonfactorizable soft $O(\Lambda_{QCD}/m_b)$ corrections into account. These corrections can be brought under control by using *the light-cone QCD sum rule method* [13]. This method is going to be discussed more in detail in what follows.

2 Light-Cone Sum Rules

All QCD sum rules are based on the general idea of calculating a relevant quark-current correlation function and relating it to the hadronic parameters of interest via a dispersion relation. Sum rules in hadron physics were already known before QCD was established (for a comprehensive introduction to sum rules see i.e. [14]), but have reached wide application in a calculation of various hadronic quantities in the form of so-called *SVZ sum rules* [15]. The other type of sum rules, *the light-cone QCD sum rules (LCSR)* were established for calculation of exclusive amplitudes and form factors ([16] and references therein).

2.1 Light-Cone Sum Rules *vs* SVZ Sum Rules

In order to illustrate application of the QCD sum rules and the main differences between SVZ sum rules and light-cone sum rules, we introduce an example.

One of the typical calculation using *the SVZ sum rules* is the estimation of the B meson decay constant f_B. The starting point is a correlation function defined as

$$F(q^2) = i \int d^4x e^{iqx} \langle 0|T\{m_b \overline{u} i\gamma_5 b(x), m_b \overline{b} i\gamma_5 u(0)\}|0\rangle \,. \tag{13}$$

In the Euclidean region of q momenta, $q^2 < 0$, we can perform a perturbative calculation in terms of quarks and gluons by applying the short-distance operator-product expansion (OPE) to the correlation function $F(q^2)$. The correlation function is then expressed via a dispersion relation in terms of the spectral function ρ^{OPE} representing the perturbative part, and the quark and gluon condensates, i.e. $\langle q\overline{q}\rangle$, $\langle GG\rangle$, etc (see for example [18]):

$$F^{OPE}(q^2) = \frac{1}{\pi}\int_0^\infty ds \frac{ImF^{OPE}(s)\, ds}{s-q^2} + A_i(s)\langle O|\Omega_i|0\rangle \,, \tag{14}$$

where

$$\frac{1}{\pi} ImF^{OPE}(s) = \rho^{OPE}(s) \tag{15}$$

and A_i are perturbative coefficients in front of the vacuum condensates of operators $\Omega_i = \overline{q}q\,, GG\,, \overline{q}\lambda^a/2\sigma \cdot Gq$, etc.

On the other hand, in the physical (Minkowskian) region, $q^2 > 0$, we insert the complete sum over hadronic states starting from the ground state B meson, and use a defining relation for f_B : $\langle m_b \overline{q} i\gamma_5 b|B\rangle = f_B m_B^2$. The correlation function $F(q^2)$ can then be written as

$$F^{hadron}(q^2) = \frac{m_B^4 f_B^2}{m_b^2 - q^2} + \int_{s_0^h}^\infty \frac{\rho^{hadron}(s)\, ds}{s-q^2} \,, \tag{16}$$

where the hadronic spectral density ρ^{hadron} contains all higher resonances and non-resonant states with the B meson quantum numbers.

By applying *the quark-hadron duality* to these higher hadronic (continuum) states, which means assuming that the continuum of hadronic states, described by the hadronic spectral function $\rho^{hadron}(s)$ via a dispersion relation, can be replaced by the spectral function $\rho^{OPE}(s)$ calculated perturbatively in the $q^2 < 0$ region, we match both sides, $F^{OPE}(q^2) = F^{hadron}(q^2)$, and extract the needed quantity f_B. The replacement is done for $s > s_0^B$, where s_0^B is an effective parameter of the order of the mass of the first excited B meson resonance squared.

In a practical calculation one performs a finite power expansion in $\rho^{OPE}(s)$. To improve the convergence of the expansion, *the Borel transform* of both sides, $F^{OPE}(q^2)$ and $F^{hadron}(q^2)$, is considered, defined by the following limiting procedure

$$\mathcal{B} = lim \frac{1}{(n-1)!}(-q^2)^n (\frac{d}{dq^2})^n \qquad |q^2|, n \to \infty \, , \frac{|q^2|}{n} = M^2 \;\; fixed. \tag{17}$$

M^2 is so called Borel parameter. It is determined by the search for stability criteria in a sense that, on the one hand, excited and continuum states are suppressed (asks for smaller M^2) and, on the other hand, the reliable perturbative calculation is enabled (asks for larger M^2).

The general procedure of QCD sum rules is depicted on Fig. 1.

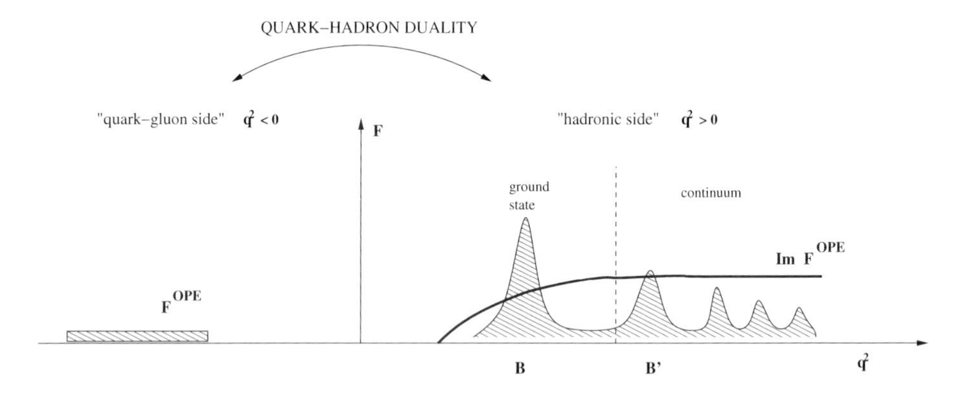

Fig. 1. An illustration of the matching procedure in QCD sum rules

For calculating quantities which involve hadron interactions, such as for example the $B \to \pi$ form factor, *the light-cone sum rules* are more suitable [17]. The correlation function is now defined as a vacuum-to-pion matrix element:

$$F_\mu = i \int d^4x e^{-ipx} \langle \pi(q) | T\{\overline{u}\gamma_\mu b(x), m_b \overline{b} i\gamma_5 d(0)\} | 0 \rangle \, . \tag{18}$$

The calculation follows by performing a light-cone OPE, an expansion in terms of the light-cone wave functions of increasing twist (twist = dimension - spin).

Physically, it means that one performs an expansion in the transverse quark distances in the infinite momentum frame, rather than a short-distance expansion [17]. Instead of dealing with the vacuum-to-vacuum quark and gluon condensates (numbers) like in the SVZ sum rules, we have now to know the pion distribution amplitude (wave function). The leading twist-2 pion distribution amplitude, ϕ_π is defined as

$$\langle \pi(q)|\overline{u}(x)\gamma_\mu\gamma_5 d(0)|0\rangle = -iq_\mu f_\pi \int_0^1 du e^{iuqx}\phi_\pi(u)\,. \tag{19}$$

Distribution amplitudes (DAs) describe distributions of the pion momentum over the pion constituents and u denotes the fraction of this momentum, $0 < u < 1$ (for a comprehensive paper on the exclusive decays and the light-cone DAs see [19]). The DAs represent a nonperturbative, noncalculable input and their form has to be determined by nonperturbative methods and/or somehow extracted from the experiment.

In the physical region of $(p-q)^2 > 0$ nothing changes in comparison to the SVZ sum rules. We insert the complete set of hadronic states with B meson quantum numbers as before, and extract the $B \to \pi$ form factor from the relation: $\langle \pi(q)|\overline{u}\gamma_\mu b|B(p+q)\rangle = 2f^+_{B\pi}(p^2)q_\mu + ...$ The matching procedure follows as described above.

3 Nonfactorizable Effects in the Light-Cone Sum Rules

Although the idea to apply QCD sum rules for calculating nonfactorizable contributions in nonleptonic B decays is not the new one, earlier applications were facing some problems which have caused unavoidable theoretical uncertainties in their results [13]. In the work [13], a new approach was introduced and we are going first to review its main ideas in the application to the $B \to \pi\pi$ decay.

3.1 Definitions

The correlator for the $B \to \pi\pi$ decay given in terms of two interpolating currents for the pion and the B meson, $J^{(\pi)}_{\nu 5} = \overline{u}\gamma_\nu\gamma_5 d$ and $J^{(B)}_5 = m_b\overline{b}i\gamma_5 d$ respectively, and relevant operators $\mathcal{O}_1 = (\overline{d}\Gamma_\mu u)(\overline{u}\Gamma^\mu b)$ and $\tilde{\mathcal{O}}_1 = (\overline{d}\Gamma_\mu\lambda^a/2u)(\overline{u}\Gamma^\mu\lambda^a/2b)$ looks like:

$$F_\nu(p,q,k) = \int d^4x e^{-i(p-q)x}\int d^4y e^{i(p-k)y}\langle 0|T\{J^{(\pi)}_{\nu 5}(y)\mathcal{O}_i(0)J^{(B)}_5(x)\}|\pi(q)\rangle\,. \tag{20}$$

The transition is defined again between a vacuum and an external pion state. The situation is illustrated in Fig. 2. One can note an unphysical momentum k coming out from the weak vertex. It was introduced in order to avoid the B meson four-momenta before ($p_B = (p-q)$), and after (P) the decay to be the same, Fig. 2. In such a way, it was prevented that the continuum of light states enters the dispersion relation of the B channel. States, like $D\overline{D}^*_s$ and $D^*\overline{D}_s$, have

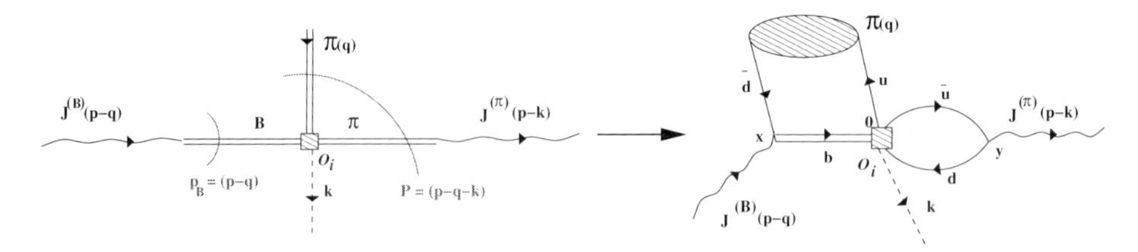

Fig. 2. $B \to \pi\pi$ decay in LCSR. The shaded oval region denotes nonperturbative input, π meson distribution amplitude. The other pion and the B meson are represented by the currents $J^{(\pi)}(p-k)$ and $J^{(B)}(p-q)$ respectively. The square stands for the four-quark operators $\mathcal{O}_i$.

masses smaller than the ground state B meson mass and spoil the extraction of the physical B meson. These 'parasitic' contributions have caused problems in the earlier application of the sum rules [13]. There are several other momenta involved into the decay. We take $p^2 = k^2 = q^2 = 0$ and consider region of large spacelike momenta

$$|(p-k)^2| \sim |(p-q)^2| \sim |P^2| \gg \Lambda_{QCD}^2 , \tag{21}$$

where the correlation function is explicitly calculable.

3.2 Procedure

The procedure which one performs is exhibited in Fig. 3. First, Fig. 3a, one makes a dispersion relation in a pion channel of momentum $(p-k)^2$ and applies the quark-hadron duality for this channel, as it was explained in Sect.2. Thereafter, to be able later to extract physical B meson state, one has to perform an analytical continuation of P momentum to its positive value, $P^2 = m_B^2$. This procedure is analogous to the one in the transition from the spacelike to the timelike pion form factor, Fig. 3b. Finally, Fig. 3c, a dispersion relation in the B channel of momentum $(p-q)^2$ has to be done, together with the application of the quark-hadron duality, now in the B channel. In such a way we arrive to the double dispersion relation. Apart from somewhat more complicated matching procedure, the calculation otherwise follows in a standard way.

3.3 Results and Implications in the $B \to \pi\pi$ Decay

In [13], first, the factorization of the $\mathcal{O}_1$ operator contribution in the $B \to \pi\pi$ decay was confirmed. The soft nonfactorizable contributions due to the $\tilde{\mathcal{O}}_1$ operator, which express the exchange of soft gluons between two pions were then calculated. Nonfactorizable soft contributions appear from the absorption of a soft gluon emerging from the light-quark loop $\overline{u}d$ in Fig. 2, by the distribution amplitude of the outcoming pion $\pi^+(-q)$ and there are of the higher, twist-3 and twist-4 order in comparison to the factorizable contributions.

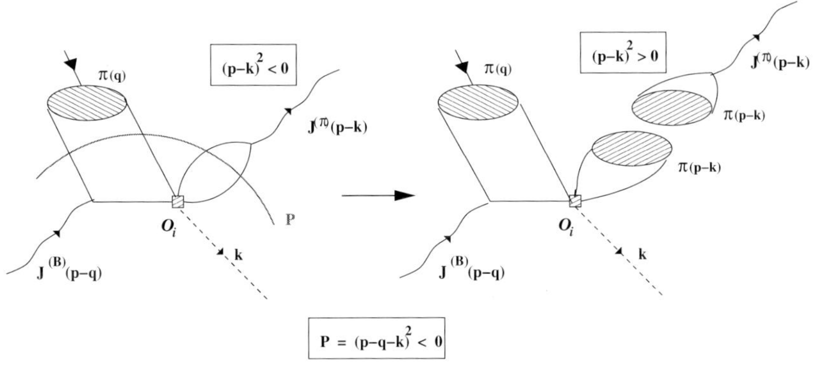

(**a**) Dispersion relation in the pion channel of momentum $(p-k)^2 < 0$

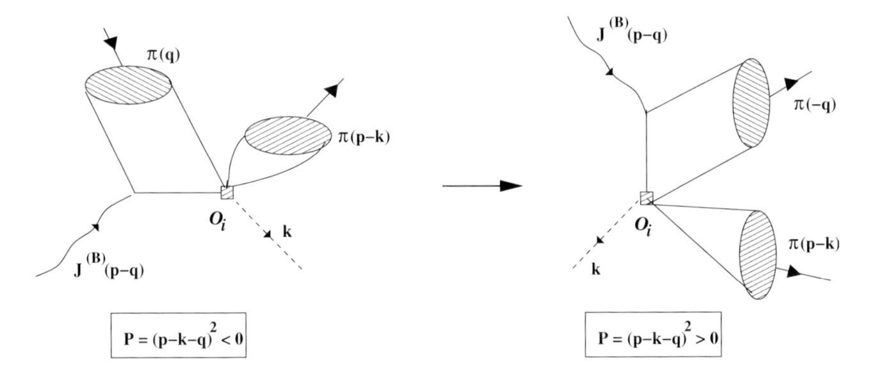

(**b**) Analytical continuation of P^2 to $P^2 = m_B^2$

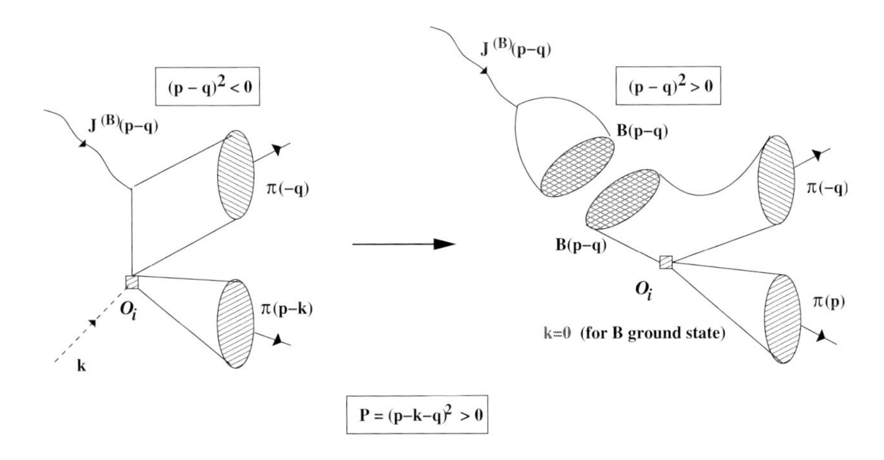

(**c**) Dispersion relation in the B meson channel of momentum $(p-q)^2 < 0$

Fig. 3. The light-cone sum rule procedure for exclusive decays

Nonfactorizable soft corrections appeared to be numerically small ($\sim 1\%$) and suppressed by $1/m_b$. Therefore, their impact on the complete decay amplitude was shown to be of the same order as that of hard nonfactorizable contributions calculated in the QCD factorization approach [11]. Also, the calculation has shown no imaginary phase from the soft contributions, whereas aforementioned hard nonfactorizable contributions get small complex phase because of the final state rescattering due to the hard gluon exchange.

4 Nonfactorizable Effects for $B \to J/\psi K$

The $B \to J/\psi K$ decay was considered in [20]. As it was emphasized at the beginning, this decay belongs to the color-suppressed class-2 decays in which one expects large nonfactorizable contributions. The confirmation of this assumption seems to be also found experimentally. Namely, there is a discrepancy between the experiment and the naive factorization prediction by at least a factor of 3 in the branching ratio. The Hamiltonian which describes the decay is given as

$$H_W = \frac{G_F}{\sqrt{2}} V_{cb} V_{cs}^* \left[(C_2(\mu) + \frac{1}{N_c} C_1(\mu)) \mathcal{O}_2 + 2\, C_1(\mu) \tilde{\mathcal{O}}_2 \right] , \tag{22}$$

with the operators $\mathcal{O}_2 = (\bar{c}\Gamma_\mu c)(\bar{s}\Gamma^\mu b)$ and $\tilde{\mathcal{O}}_2 = (\bar{c}\Gamma_\mu \lambda_a/2\, c)(\bar{s}\Gamma^\mu \lambda^a/2\, b)$. In the factorization approach, the matrix element of $\tilde{\mathcal{O}}_2$ vanishes, and the factorized matrix element of the operator $\mathcal{O}_2$ is given by

$$\begin{aligned} \langle J/\psi(p) K(q) | \mathcal{O}_2 | B(p+q) \rangle &= \langle J/\psi(p) | \bar{c}\Gamma_\mu c | 0 \rangle \langle K(q) | \bar{s}\Gamma^\mu b | B(p+q) \rangle \\ &= 2\epsilon \cdot q\, m_{J/\psi} f_{J/\psi} F^+_{BK}(m^2_{J/\psi}) \,. \end{aligned} \tag{23}$$

$F^+_{BK}(m^2_{J/\psi})$ is the $B \to K$ transition form factor calculated using the light-cone sum rules, in a way enlightened in Sect.2.1 on the example of $B \to \pi$ form factor calculation, and $f_{J/\psi}$ is the J/ψ decay constant. By evaluating numerically the $B \to J/\psi K$ branching ratio with the NLO Wilson coefficients used in Sec.1.1 and with the numerical input taken from [20], we arrive to

$$\mathcal{B}(B \to J/\psi K)^{fact} = 3.3 \cdot 10^{-4} \,. \tag{24}$$

This has to be compared with the recent measurements [21]

$$\begin{aligned} \mathcal{B}(B^+ \to J/\psi K^+) &= (10.1 \pm 0.3 \pm 0.5) \cdot 10^{-4} \,, \\ \mathcal{B}(B^0 \to J/\psi K^0) &= (8.3 \pm 0.4 \pm 0.5) \cdot 10^{-4} \,. \end{aligned} \tag{25}$$

It is clear that there a large discrepancy between the naive factorization prediction and the experiment.

To be able to discuss the impact of the nonfactorizable term $\tilde{\mathcal{O}}_2$, we parametrize the $\langle J/\psi K | H_W | B \rangle$ amplitude in terms of the a_2 parameter as

$$\langle J/\psi K | H_W | B \rangle = \sqrt{2}\, G_F\, V_{cb} V_{cs}^*\, \epsilon \cdot q\, m_{J/\psi} f_{J/\psi} F^+_{BK}(m^2_{J/\psi})\, a_2 \,, \tag{26}$$

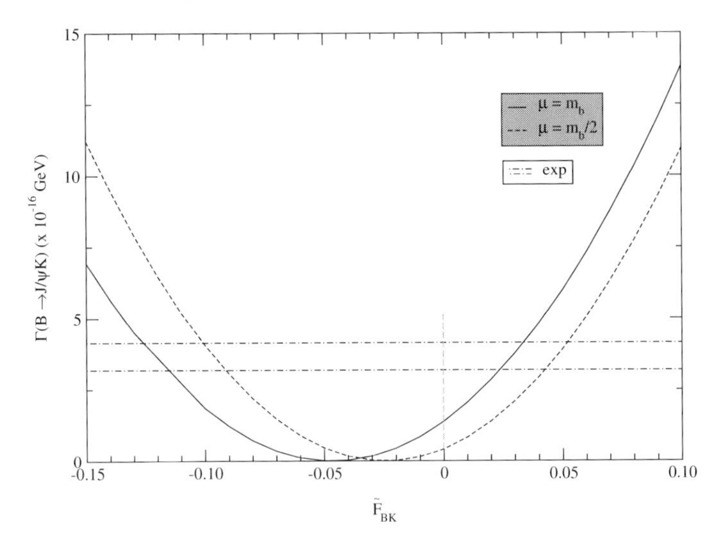

Fig. 4. The partial width for $B \to J/\psi K$ as a function of the nonfactorizable amplitude $\tilde{F}_{BK}$. The dashed-dotted lines denote the experimental region.

where

$$a_2 = C_2(\mu) + \frac{C_1(\mu)}{3} + 2C_1(\mu)\frac{\tilde{F}^+_{BK}(\mu)}{F^+_{BK}(m^2_{J/\psi})}\,. \tag{27}$$

The part proportional to $\tilde{F}^+_{BK}$ represents the contribution from the $\tilde{O}_2$ operator

$$\langle J/\psi K|\tilde{O}_2(\mu)|B\rangle = 2\epsilon \cdot q\, m_{J/\psi} f_{J/\psi} \tilde{F}^+_{BK}(\mu^2) \tag{28}$$

and $\tilde{F}^+_{BK} = 0$ corresponds to the naive factorization result, Eq. (23).

By using the parametrization (26) we can extract the a_2 coefficient from experiments (25). From the measurements (25) one obtains

$$|a_2^{exp}| = 0.29 \pm 0.03\,. \tag{29}$$

On the other hand, the naive factorization with the NLO Wilson coefficients [6] yields

$$a_{2,\,NLO}^{fact} = 0.176\,|_{\mu\simeq m_b}\,. \tag{30}$$

The value (30) is significantly below the value extracted from the experiment, although one should not forget a strong μ dependence of a_2^{fact}.

In Fig. 4 we show the partial width for $B \to J/\psi K$ as a function of the nonfactorizable amplitude $\tilde{F}_{B\to K}$. The zero value of $\tilde{F}_{BK}$ corresponds to the factorizable prediction. There exist two ways to satisfy the experimental demands on $\tilde{F}_{BK}$. Following the large $1/N_c$ rule [8], one can argue that there is a cancellation between $1/N_c$ piece of the factorizable part and the nonfactorizable contribution (27). This would ask for the relatively small and negative value of $\tilde{F}_{BK}$. The other possibility is to have even smaller, but positive values for $\tilde{F}_{BK}$, which then compensate the overall smallness of the factorizable part and bring the theoretical estimation for a_2 in accordance with experiment.

One can note significant μ dependence of the theoretical expectation for the partial width in Fig. 4, which brings an uncertainty in the prediction for $\tilde{F}_{BK}(\mu)$ in the order of 30%. This uncertainty is even more pronounced for the positive solutions of $\tilde{F}_{BK}(\mu)$. The values for $\tilde{F}^{+}_{BK}$ extracted from experiments

$$\tilde{F}^{+}_{BK}(m_b) = 0.028 \qquad \text{or} \qquad \tilde{F}^{+}_{BK}(m_b) = -0.120\,, \tag{31}$$

$$\tilde{F}^{+}_{BK}(m_b/2) = 0.046 \qquad \text{or} \qquad \tilde{F}^{+}_{BK}(m_b/2) = -0.095\,. \tag{32}$$

clearly illustrate the μ sensitivity of the nonfactorizable part.

In what follows we calculate the nonfactorizable contribution $\tilde{F}^{+}_{BK}$ which appears due to the exchange of soft gluons using the QCD light-cone sum rule method.

4.1 Light-Cone Sum Rule Calculation

The light-cone sum rule calculation starts by considering the correlator

$$F_\nu(p,q,k) = i^2 \int d^4x e^{-i(p+q)x} \int d^4y e^{i(p-k)y} \langle K(q)|T\{J_\nu^{(J/\psi)}(y)\mathcal{O}(0)J_5^{(B)}(x)\}|0\rangle \tag{33}$$

with the interpolating currents $J_\nu^{(J/\Psi)} = \bar{c}\gamma_\nu c$ and $J_5^{(B)} = m_b\bar{b}i\gamma_5 u$. The kinematics is the same as defined above in (21), with the exception that now $p^2 = m^2_{J/\psi}$. More explicitly the configuration is shown in Fig. 5.

The estimation of nonfactorizable contributions was performed for the exchange of soft gluons (shown by the dashed line in Fig. 5) and follows essentially steps of derivation explained in Sect. 3.2 for $B \to \pi\pi$ decay. Nonfactorizable contribution of the $\mathcal{O}_1$ operator appears first at $O(\alpha_s^2)$. The $\tilde{\mathcal{O}}_2$ operator gives nonvanishing result already at the one-gluon level and the leading contributions are given in terms of twist-3 and twist-4 kaon distribution amplitudes which contribute in the same order. Technical peculiarities of the calculation can be found in [20].

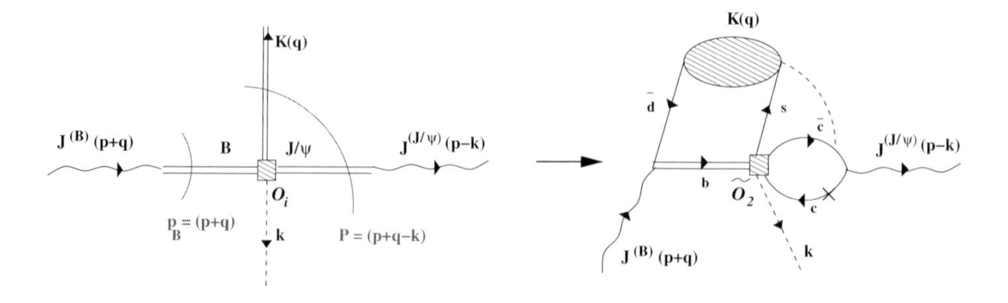

Fig. 5. $B \to J/\psi K$ decay in LCSR. The dashed line denote an exchange of a soft gluon for $\mathcal{O}_i = \tilde{\mathcal{O}}_2$ and the cross stands for other possible attachment of a soft gluon.

4.2 Results and Implications

The results can be summarized as follows. Soft nonfactorizable twist-3 and twist-4 contributions, expressed in terms of $\tilde{F}^+_{BK}$ are $\tilde{F}^+_{BK,tw3}(\mu_b) = 0.004 - 0.007$ and $\tilde{F}^+_{BK,tw4}(\mu_b) = 0.006 - 0.012$ and the final value is

$$\tilde{F}^+_{BK}(\mu_b) = 0.011 - 0.018\,. \tag{34}$$

where $\mu_b = \sqrt{M_B^2 - m_b^2} \simeq m_b/2$. The wide range prediction for $\tilde{F}^+_{BK}$ appears due to the variation of sum rule parameters.

First, we note that the nonfactorizable contribution (34) is much smaller than the $B \to K$ transition form factor $F^+_{BK} = 0.55 \pm 0.05$, which enters the factorization prediction (26). It is also significantly smaller than the value (32) extracted from experiments. Nevertheless, its influence on the final prediction for a_2 is significant, because of the large coefficient $2C_1$ multiplying it. Further, one has to emphasize that $\tilde{F}^+_{BK}$ is a positive quantity. Therefore, we do not find a theoretical support for the large N_c limit assumption discussed in Sect.4.1, that the factorizable part proportional to $C_1(\mu)/3$ should at least be partially canceled by the nonfactorizable part. Our result also contradicts the result of the earlier application of QCD sum rules to $B \to J/\psi K$ [23], where negative and somewhat larger value for $\tilde{F}^+_{BK}$ was found. However, earlier applications of QCD sum rules to exclusive B decays exhibit some deficiencies discussed in [13].

Using the same values for the NLO Wilson coefficients as in Sect.2, one gets from (34) the following value for the effective coefficient a_2:

$$a_2 \sim 0.15 - 0.18\,|_{\mu=\mu_b}\,. \tag{35}$$

Although the soft correction contributes in the order of $\sim 30\% - 70\%$, the net result (35) is still by approximately factor of two smaller than the experimentally determined value (29).

We would like to discuss our results for soft nonfactorizable contributions in comparison with the hard nonfactorizable effects calculated in the QCD factorization approach. The best thing would be to calculate both soft and hard contributions inside the same model. In principle, the light-cone sum rule approach presented here enables such a calculation, although the estimation of hard nonfactorizable contributions is technically very demanding, involving a calculation of two-loop diagrams. Therefore, we proceed with the QCD factorization estimations for the hard nonfactorizable contributions.

After including the hard nonfactorizable corrections, the a_2 parameter (35) is as follows

$$a_2 = \left[C_2(\mu) + \frac{C_1(\mu)}{3} + 2C_1(\mu)\left(\alpha_s F(\mu)^{hard} + \frac{\tilde{F}^+_{BK}(\mu)}{F^+_{BK}} \right) \right]\,. \tag{36}$$

The estimations done in the QCD factorization [24] show hard-gluon exchange corrections to the naive factorization result in the order of $\sim 25\%$, predicted by the LO calculation with the twist-2 kaon distribution amplitude. Unlikely large corrections are obtained by the inclusion of the twist-3 kaon distribution amplitude. Anyhow, due to the obvious dominance of soft contributions to the twist-3 part of the hard corrections in the BBNS approach [11], it is very likely that some double counting of soft effects could appear if we naively compare the results. Therefore, taking only the twist-2 hard nonfactorizable corrections from [24] into account, recalculated at the μ_b scale, our prediction (35) changes to

$$a_2 = 0.17 - 0.19 \,|_{\mu=\mu_b} \tag{37}$$

The prediction still remains too small to explain the data.

Nevertheless, there are several things which have to be stressed here in connection with the result. Soft nonfactorizable contributions are at least equally important as nonfactorizable contributions from the hard-gluon exchange, and can be even dominant. Soft nonfactorizable contributions are positive, and the same seems to be valid for hard corrections. While hard corrections have an imaginary part, in the soft contributions the annihilation and the penguin topologies as potential sources for the appearance of an imaginary part were not discussed. A comparison between the result (37) and the experimental value $|a_2| \sim 0.3$ for $B \to J/\psi K$ decay, with the recently deduced a_2 parameter from the color-suppressed $\overline{B}^0 \to D^{(*)0}\pi^0$ decays, $|a_2| \sim 0.4 - 0.5$ [10], provides clear evidence for the nonuniversality of the a_2 parameter in color-suppressed decays.

5 Conclusions

We have reviewed recent progress in the understanding of the underlying dynamics of exclusive nonleptonic decays, with the emphasis on the nonfactorizable corrections to the naive factorization approach. In the calculation of nonfactorizable contributions, we have focused to QCD light-cone sum rule approach and have shown results for $B \to \pi\pi$ [13] and $B \to J/\psi K$ [20] decays.

The QCD factorization method is reviewed in this volume by M. Neubert, [3].

Acknowledgment

I would like to thank R. Rückl for a collaboration on the subjects discussed in this lecture and A. Khodjamirian for numerous fruitful discussions and comments. The support by the Alexander von Humboldt Foundation is gratefully acknowledged. The work was also partially supported by the Ministry of Science and Technology of the Republic of Croatia under Contract No. 0098002.

References

1. J. B. Bjorken: Nucl. Phys. (Proc. Suppl.) **11** 321 (1989)
2. G. Buchalla, A.J. Buras, M.E. Lautenbacher: Rev. Mod. Phys. **68** 1125 (1996)
3. M. Neubert: Theory of Exclusive Hadronic B Decays, Lect. Notes Phys. **647**, 3–41 (2004)

4. R. Fleischer: B Physics and CP Violation, Lect. Notes Phys. **647**, 42–77 (2004)
5. T. Mannel: Theory of Rare B Decays: $b \to s\gamma$ and $b \to s\ell^+\ell^-$, Lect. Notes Phys. **647**, 78–99 (2004)
6. A. J. Buras: Nucl. Phys. B **434** 606 (1995)
7. M. Neubert, B. Stech: Adv. Ser. Direct. High Energy Phys. 15 (1998) 294 and hep-ph/9705292
8. A. J. Buras, J. M. Gerard, R. Rückl: Nucl. Phys.B **268** 16 (1986)
9. B.Y. Blok, M.A. Shifman: Sov. J. Nucl. Phys. **45** 135,307,522 (1987)
10. M. Neubert, A. A. Petrov: Phys. Lett. B **519** 50 (2001)
11. M. Beneke, G. Buchalla, M. Neubert, C. T. Sachrajda, Nucl. Phys. B 591 (2000) 313; Nucl. Phys. B 606 (2001) 245
12. Y.-Y. Keum, H.-n. Li and A.I. Sanda: Phys. Lett. B **504** 6 (2001); Phys. Rev. D **63** 054008 (2001)
13. A. Khodjamirian: Nucl. Phys. B **605** 558 (2001)
14. E. de Rafael: 'An Introduction to Sum Rules in QCD'. In: *Probing the Standard Model of Physical Interactions*, ed. by R. Gupta, A. Morel, E. De Rafael, F. David (Elsevier, Amsterdam, The Netherlands 1999) and hep-ph/9802448
15. M. A. Shifman, A.I. Vainstein and V.I. Zakharov: Nucl. Phys. B **147** 385, 448 (1979)
16. P. Colangelo, A. Khodjamirian: 'QCD Sum Rules, a Modern Perspective'. In: *At the Frontier of Particle Physics, Vol.3*, ed. M. Shifman (Singapore, World Scientific 2001) pp. 1495-1576 and hep-ph/0010175.
17. V.M. Braun: 'Light Cone Sum Rules'. In: *Rostock 1997, Progress in Heavy Quark Physics*, ed. by M. Beyer, T. Mannel, H. Schroderi (Rostock, Germany, University of Rostock 1998) pp. 105-118 and hep-ph/9801222
18. L.J. Reinders, H. Rubinstein: S. Yazaki: Phys. Rep. **127** 1 (1985)
19. G.P.Lepage and S. Brodsky: Phys. Rev. D **22** 2157 (1980)
20. B. Melić and R. Rückl: 'Nonfactorizable Effects in the $B \to J/\psi$ Decay', hep-ph/0212346, to appear in the Frascati Physics Series; B. Melić: 'Nonfactorizable corrections to $B \to J/\psi$ Decay', hep-ph/0303250, submitted to Phys. Rev. D
21. B. Aubert *et al.*: Phys. Rev. D **65** 032001 (2002)
22. J. Soares: Phys. Rev. D **51** 3518 (1995)
23. A. Khodjamirian, R. Rückl: 'Exclusive Nonleptonic Decays of Heavy Mesons in QCD'. In *Continuous Advances in QCD 1998*, ed. A.V. Smilga, (World Scientific, Singapore 1998), p. 287 and hep-ph/9807495.
24. H.-Y. Cheng, K.-Ch. Yang: Phys. Rev. D **63** 074011 (2001)

Part III

Production of Heavy Flavors

Heavy Flavor Production off Protons and in a Nuclear Environment

B.Z. Kopeliovich[1,2,3] and J. Raufeisen[4]

[1] Max-Planck Institut für Kernphysik, Postfach 103980, 69029 Heidelberg, Germany
[2] Institut für Theoretische Physik der Universität, 93040 Regensburg, Germany
[3] Joint Institute for Nuclear Research, Dubna, 141980 Moscow Region, Russia
[4] Los Alamos National Laboratory, MS H846, Los Alamos, NM 87545, USA

Abstract. These lectures present an overview of the current status of the QCD based phenomenology for open and hidden heavy flavor production at high energies. A unified description based on the light-cone color-dipole approach is employed in all cases. A good agreement with available data is achieved without fitting to the data to be explained, and nontrivial predictions for future experiments are made. The key phenomena under discussion are: (i) formation of the wave function of a heavy quarkonium; (ii) quantum interference and coherence length effects; (iii) Landau-Pomeranchuk suppression of gluon radiation leading to gluon shadowing and nuclear suppression of heavy flavors; (iv) higher twist shadowing related to the finite size of heavy quark dipoles; (v) higher twist corrections to the leading twist gluon shadowing making it process dependent.

1 Introduction

Reactions in which heavy flavors are produced involve a hard scale that allows one to employ perturbative QCD (pQCD). In particular, production off nuclei has always been an important topic in heavy quark physics. On the one hand, interest in this field is stimulated by demand to provide a proper interpretation for available and forthcoming data, especially from heavy ion collisions. On the other hand, nuclei have been traditionally employed as an analyzer for the dynamics and time scales of hadronic interactions.

In these lectures, we shall review the color dipole formulation of heavy flavor production, which was developed in [1–4]. The dipole approach to heavy flavor production expresses the cross section for production of open or hidden heavy flavor in terms of the cross section for scattering a color neutral quark-antiquark ($q\bar{q}$) pair off a nucleon. This approach is motivated by the need for a theoretical framework for the description of nuclear effects. Its main distinction from the conventional parton model is the possibility to actually calculate nuclear effects in the dipole formulation, rather than absorbing them into the initial conditions. In addition, the dipole approach is not restricted to the leading twist approximation. This is especially important for nuclear effects in J/ψ production, which are dominated by higher twists. Another advantage of the dipole approach is that it correctly describes the absolute normalization of cross sections in different processes, without introducing an arbitrary overall normalization factor

B.Z. Kopeliovich and J. Raufeisen, Heavy Flavor Production off Protons and in a Nuclear Environment, Lect. Notes Phys. **647**, 305–365 (2004)
http://www.springerlink.com/

("K-factor") [5,6]. However, the dipole approach is applicable only in the kinematical domain, where the heavy quark mass m_Q is much smaller than the center of mass (*cm.*) energy $\sqrt{s}$. The latter condition is fulfilled for charm production at the BNL Relativistic Heavy Ion Collider (RHIC) and for both, charm and bottom production, at the CERN Large Hadron Collider (LHC).

An alternative approach to heavy quark production that is designed especially for high energies and which is able to describe nuclear effects is desirable for a variety of reasons. At low x, the heavy quark pair is produced over large longitudinal distances, which can exceed the radius of a large nucleus by orders of magnitude. Indeed, even though the matrix element of a hard process is dominated by short distances, of the order of the inverse of the hard scale, the cross section of that process also depends on the phase space element. Due to gluon radiation, the latter becomes very large at high energies, and it is still a challenge how to resum the corresponding low-x logarithms. The dipole formulation allows for a simple phenomenological recipe to include these low-x logs. The large length scale in the problem leads to pronounced nuclear effects, giving one the possibility to use the nuclear medium as microscopic detectors to study the space-time evolution of heavy flavor production.

In addition, heavy quark production is of particular interest, because this process directly probes the gluon distributions of the colliding particles. Note that at the tremendous center of mass energies of RHIC and LHC, charm (and at LHC also bottom) decays will dominate the dilepton continuum [7]. Thus, a measurement of the heavy quark production cross section at RHIC and LHC will be relatively easy to accomplish and can yield invaluable information about the (nuclear) gluon density [8]. It is expected that at very low x, the growth of the gluon density will be slowed down by nonlinear terms in the QCD evolution equations [9]. The onset of this non-linear regime is controlled by the so-called saturation scale $Q_s(x, A)$, which is already of order of the charm quark mass at RHIC and LHC energies. Moreover, $Q_s(x, A) \propto A^{1/3}$ (A is the atomic mass of the nucleus), so that one can expect sizable higher twist corrections in AA collisions [4]. Note that saturation will lead to a breakdown of the twist expansion, since one cannot conclude any more that terms suppressed by powers of the heavy quark mass m_Q are small, $Q_s^n(x, A)/m_Q^n \in O(1)$ for any n [10]. Saturation effects are most naturally described in the dipole picture.

The most prominent motivation for a theoretical investigation of nuclear effects in heavy flavor production is perhaps the experimentally observed suppression of J/ψ mesons in nuclear collisions, see [11] for a review. Suppression of J/ψ production in nucleus-nucleus (AA) collisions has been proposed as signal of quark-gluon plasma (QGP) formation [12]. There are however several "mundane" nuclear effects that also lead to J/ψ suppression, such as gluon shadowing, final state absorption and breakup due to interactions with comovers. This issue has been widely and controversially discussed in the literature, and one clearly needs a reliable theoretical description of all these effects, before definite conclusions about any non-standard dynamics in heavy ion collisions can be drawn. Since the creation and study of the QGP is the main physics motivation for RHIC, the theoretical investigation of nuclear effects in J/ψ production is at the heart of the RHIC heavy ion program. Note that the production mechanism for heavy quarkonia itself is poorly understood, even in proton-proton (pp) collisions. However, since heavy flavors can be produced in many different reactions, and one may hope that a detailed experimental and theoretical study of quarkonium production in different environments and over a wide kinematical range will eventually clarify the underlying production mechanism.

2 The Foundations of the Color Dipole Approach to High Energy Scattering

It was first realized in [13] that at high energies color dipoles with a well defined transverse separation are the eigenstates of interaction at, i.e. can experience only diagonal transitions when interacting diffractively with a target. The eigenvalues of the amplitude operator are related to the cross section $\sigma_{q\bar{q}}(r)$ of interaction of a $q\bar{q}$ dipole with a nucleon. This dipole cross section is a universal, flavor independent quantity which depends only on transverse $q\bar{q}$ separation. Then, the total hadron (meson) nucleon cross section can be presented in a factorized form,

$$\sigma_{tot}^{hN} = \int d^2r\, |\Psi_{q\bar{q}}^h(\boldsymbol{r})|^2\, \sigma_{q\bar{q}}(r)\ , \tag{1}$$

where $\Psi_{q\bar{q}}^h(\boldsymbol{r})$ is the light-cone wave function of the $q\bar{q}$ component of the hadron. It was assumed in [13] that $|\Psi_{q\bar{q}}^h(\boldsymbol{r})|^2$ is integrated over longitudinal coordinate.

One of the advantages of this approach is simplicity of calculation of the effects of multiple interactions which have the eikonal form (exact for eigenstates),

$$\sigma_{tot}^{hA} = 2\int d^2b \int d^2r\, |\Psi_{q\bar{q}}^h(\boldsymbol{r})|^2 \left\{1 - \exp\left[-\frac{1}{2}\sigma_{q\bar{q}}(r)T_A(b)\right]\right\}\ , \tag{2}$$

where $T_A(b)$ is the nuclear thickness function which depends on impact parameter $\boldsymbol{b}$.

Later the energy dependence of the dipole cross section was taken into account [14], and was found that one can apply the same color-dipole formalism to variety of QCD processes, including radiation [15]. Here we review this approach applied to production of heavy flavors.

The simplest example for heavy flavor production is open heavy quark production in deep inelastic scattering (DIS) off a proton, which can be measured at the DESY ep collider HERA. We use this example to explain the basic ideas underlying the color dipole approach to high energy scattering.

The dipole approach is formulated in the target rest frame, where DIS looks like pair creation in the gluon field of the target, see Fig. 1. For further simplification, we consider only the case of a longitudinally polarized γ^* with virtuality Q^2 in this section, since the transverse polarization does not contain any qualitatively new physics. A straightforward calculation of the two Feynman diagrams in Fig. 1 yields for the transverse momentum ($\boldsymbol{\kappa}_\perp$) distribution of the heavy

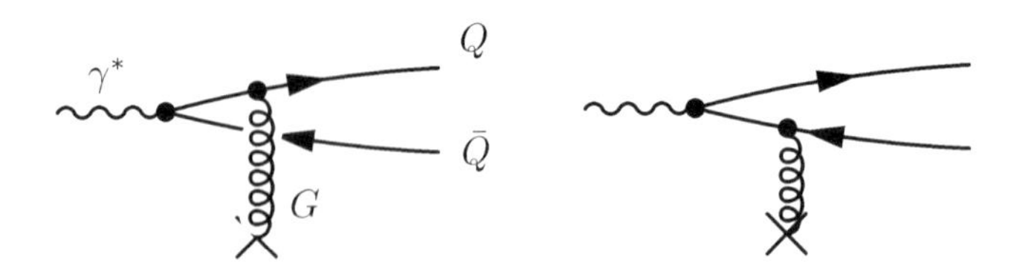

Fig. 1. Perturbative QCD graphs for heavy quark ($Q\bar{Q}$) production in DIS. The virtual photon (γ^*) fluctuates into a (virtual) $Q\bar{Q}$-pair far before the target. The interaction with the target, which is denoted by gluon (G) exchange, can put this fluctuation on mass shell.

quark [16],

$$\frac{d\sigma\left(\gamma_L^* p \to \{Q\bar{Q}\}X\right)}{d^2\kappa_\perp} = \frac{4\alpha_{\rm em}\alpha_S e_Q^2 Q^2}{\pi} \int d\alpha\, \alpha^2 (1-\alpha)^2 \times \int \frac{d^2k_T}{k_T^2} \left(\frac{1}{\kappa_\perp^2+\varepsilon^2} - \frac{1}{(\boldsymbol{\kappa}_\perp - \boldsymbol{k}_T)^2+\varepsilon^2}\right)^2 \mathcal{F}(x,k_T^2), \tag{3}$$

where α is the light-cone momentum fraction of the heavy quark and $\varepsilon^2 = \alpha(1-\alpha)Q^2 + m_Q^2$. The quark electric charge is denoted by e_Q; $\alpha_{\rm em} = 1/137$ and α_S are the electromagnetic and the strong coupling constants, respectively. The $Q\bar{Q}$-pair exchanges a gluon with transverse momentum $\boldsymbol{k}_T$ with the target. The latter is characterized by the unintegrated gluon density $\mathcal{F}(x,k_T^2)$. Note that (3) is also valid for light flavors.

In the dipole approach, a mixed representation is employed, that treats the longitudinal (γ^*) direction in momentum space, but the two transverse directions are described in coordinate (*i.e.* impact parameter) space. With help of the relation

$$\frac{1}{\kappa_\perp^2+\varepsilon^2} = \int \frac{d^2r}{2\pi} \mathrm{K}_0(\varepsilon r)\, \mathrm{e}^{-\mathrm{i}\boldsymbol{\kappa}_\perp\cdot\boldsymbol{r}}, \tag{4}$$

where K_0 is the zeroth order MacDonald function [17], one can Fourier transform (3) into impact parameter space,

$$\frac{d\sigma\left(\gamma_L^* p \to \{Q\bar{Q}\}X\right)}{d^2\kappa_\perp} = \frac{1}{(2\pi)^2}\int d\alpha \int d^2r_1 d^2r_2 \mathrm{e}^{\mathrm{i}\boldsymbol{\kappa}_\perp\cdot(\boldsymbol{r}_1-\boldsymbol{r}_2)} \times \Psi_{\gamma^*\to Q\bar{Q}}(\alpha,\boldsymbol{r}_1)\Psi^*_{\gamma^*\to Q\bar{Q}}(\alpha,\boldsymbol{r}_2) \times \frac{1}{2}\Big\{\sigma_{q\bar{q}}(r_1,x) + \sigma_{q\bar{q}}(r_2,x) - \sigma_{q\bar{q}}(\boldsymbol{r}_1-\boldsymbol{r}_2,x).\Big\} \tag{5}$$

Since the $\boldsymbol{r}_i$ are conjugate variables to $\boldsymbol{\kappa}_\perp$, one can interpret $\boldsymbol{r}_1$ as the transverse size of the $Q\bar{Q}$-pair in the amplitude and $\boldsymbol{r}_2$ as the size of the pair in the complex conjugate amplitude. An expression similar to (5) was also obtained in [18].

The light-cone (LC) wavefunctions for longitudinal (L) and for transverse (T) photons are given by

$$\Psi^{L}_{\gamma^*\to Q\bar{Q}}(\alpha,\boldsymbol{r}_1)\Psi^{*L}_{\gamma^*\to Q\bar{Q}}(\alpha,\boldsymbol{r}_2) = \frac{6\alpha_{\rm em}e_Q^2}{(2\pi)^2}4Q^2\alpha^2\left(1-\alpha\right)^2 \mathrm{K}_0(\varepsilon r_1)\mathrm{K}_0(\varepsilon r_2) \quad (6)$$

$$\Psi^{T}_{\gamma^*\to Q\bar{Q}}(\alpha,\boldsymbol{r}_1)\Psi^{*T}_{\gamma^*\to Q\bar{Q}}(\alpha,\boldsymbol{r}_2) = \frac{6\alpha_{\rm em}e_Q^2}{(2\pi)^2}\left\{m_Q^2\mathrm{K}_0(\varepsilon r_1)\mathrm{K}_0(\varepsilon r_2) + \left[\alpha^2+(1-\alpha)^2\right]\varepsilon^2\frac{\boldsymbol{r}_1\cdot\boldsymbol{r}_2}{r_1 r_2}\mathrm{K}_1(\varepsilon r_1)\mathrm{K}_1(\varepsilon r_2)\right\}. \quad (7)$$

The concept of LC wavefunction of a photon was first introduced in [19,20]. These wavefunctions are simply the $\gamma^*\to Q\bar{Q}$ vertex times the Feynman propagator for the quark line in Fig. 1, and can therefore be calculated in perturbation theory.

The flavor independent dipole cross section $\sigma_{q\bar{q}}$ in (5) carries all the information about the target. It is related to the unintegrated gluon density by [21]

$$\sigma_{q\bar{q}}(r,x) = \frac{4\pi}{3}\int\frac{d^2k_T}{k_T^2}\alpha_S\mathcal{F}(x,k_T)\left\{1-\mathrm{e}^{-\mathrm{i}\boldsymbol{k}_T\cdot\boldsymbol{r}}\right\}. \quad (8)$$

The color screening factor in the curly brackets in (8) ensures that $\sigma_{q\bar{q}}(r,x)$ vanishes $\propto r^2$ (modulo logs) at small separations. This seminal property of the dipole cross section is known as *color transparency* [13,22,23]. The dipole cross section cannot be calculated from first principles, but has to be determined from experimental data, see Sect. 3. In principle, the energy, *i.e.* x, dependence of $\sigma_{q\bar{q}}$ could be calculated in perturbative QCD. This has been attempted in the generalized BFKL approach of Nikolaev and Zakharov, see *e.g.* [21], by resumming higher orders in perturbation theory. However, the widely discussed next-to-leading order (NLO) corrections to the BFKL equation [24,25] has left the theory of low-x resummation in an unclear state. We shall account for higher order effects by using a phenomenological parameterization of the dipole cross section.

In the high energy limit, one can neglect the dependence of the gluon momentum fraction x on $\boldsymbol{\kappa}_\perp$ and integrate (5) over $\boldsymbol{\kappa}_\perp$ from 0 to ∞. One then obtains a particularly simple formula for the total cross section,

$$\sigma_{\rm tot}\left(\gamma^*p\to\{Q\bar{Q}\}X\right) = \sum_{T,L}\int d\alpha\int d^2r\left|\Psi^{T,L}_{\gamma^*\to Q\bar{Q}}(\alpha,\boldsymbol{r}_1)\right|^2\sigma_{q\bar{q}}(r,x). \quad (9)$$

It was argued in [26] that the dipole formulation is valid only in the leading $\log(x)$ approximation where (9) holds. Note however that (5) does not rely on any high energy approximation and is exactly equivalent the the k_T-factorized expression (3).

Equation (9) has an illustrative interpretation, which is the key to calculating nuclear effects in the dipole approach: The total cross section can be written in factorized form in impact parameter space, because partonic configurations with fixed transverse separations are eigenstates of the interaction [13,27], *i.e.* of the T matrix restricted to diffractive processes. Intuitively, the transverse size is frozen during the entire interaction because of time dilation. In the dipole approach, the projectile is expanded in these eigenstates,

$$|\gamma^*\rangle = \sum_k c_k^{\gamma^*} |\psi_k\rangle, \tag{10}$$

with

$$-\mathrm{i}T|\psi_k\rangle = \sigma_k|\psi_k\rangle. \tag{11}$$

Each eigenstate scatters independently off the target. According to the optical theorem, the total cross section is then given by

$$\sigma_{\mathrm{tot}} = \mathrm{Im}\langle\gamma^*|T|\gamma^*\rangle = \sum_k |c_k^{\gamma^*}|^2\sigma_k. \tag{12}$$

Comparing this expression, (12), with (3), one can identify $\sigma_{q\bar{q}}(r,x)$ as an eigenvalue of the T-matrix and the coefficients $c_k^{\gamma^*}$ as LC wavefunctions. The summation over the index k is replaced by the integrals over α and r.

Knowing the eigenstates of the interaction is a great advantage in calculating multiple scattering effects in nuclear collisions. Note however that the quark-antiquark pair is only the lowest Fock component of the virtual photon. There are also higher Fock states containing gluons, which are not taken into account by these simple considerations. These gluons cause the x-dependence of the dipole cross section and will be included in a phenomenological way, see Sect. 3. In addition, at lower energies color dipoles are no longer exact eigenstates. A dipole of size r_1 may evolve into a dipole of a different size r_2. This can be calculated from (5).

3 The Phenomenological Dipole Cross Section

The total cross sections for all hadrons and (virtual) photons are known to rise with energy. It is obvious that the energy dependence cannot originate from the hadronic wave functions, but only from the dipole cross section. In the approximation of two-gluon exchange used in [13] the dipole cross section is constant, the energy dependence originates from higher order corrections related to gluon radiation. Since no reliable way is known so far to sum up higher order corrections, especially in the semihard regime, we resort to phenomenology and employ a parameterization of $\sigma_{q\bar{q}}(r,x)$.

Few such parameterizations are available in the literature, we choose two of them which are simple, but quite successful in describing data and denote them by the initials of the authors as "GBW" [28] and "KST" [29].

We have

$$\text{“GBW”:} \qquad \sigma_{q\bar{q}}(r,x) = 23.03\left[1 - e^{-r^2/r_0^2(x)}\right]\ \mathrm{mb}\,, \tag{13}$$

$$r_0(x) = 0.4\left(\frac{x}{x_0}\right)^{0.144}\ \mathrm{fm}\,,$$

where $x_0 = 3.04 \cdot 10^{-4}$. The proton structure function calculated with this parameterization fits very well all available data at small x and for a wide range of Q^2 [28]. However, it obviously fails describing the hadronic total cross sections, since it never exceeds the value 23.03 mb. The x-dependence guarantees Bjorken scaling for DIS at high Q^2, however, Bjorken x is not a well defined quantity in the soft limit.

This problem as well as the difficulty with the definition of x have been fixed in [29], where the dipole cross section is treated as a function of the *c.m.* energy $\sqrt{s}$, rather than x, since $\sqrt{s}$ is more appropriate for hadronic processes. A similarly simple form for the dipole cross section is used

$$\text{“KST”:} \qquad \sigma_{q\bar{q}}(r,s) = \sigma_0(s)\left[1 - e^{-r^2/r_0^2(s)}\right] . \tag{14}$$

The values and energy dependence of hadronic cross sections is guaranteed by the choice of

$$\sigma_0(s) = 23.6\left(\frac{s}{s_0}\right)^{0.08}\left(1 + \frac{3}{8}\frac{r_0^2(s)}{\langle r_{ch}^2\rangle}\right) \text{mb} , \tag{15}$$

$$r_0(s) = 0.88\left(\frac{s}{s_0}\right)^{-0.14} \text{fm} . \tag{16}$$

The energy dependent radius $r_0(s)$ is fitted to data for the proton structure function $F_2^p(x,Q^2)$, $s_0 = 1000\,\text{GeV}^2$ and the mean square of the pion charge radius $\langle r_{ch}^2\rangle = 0.44\,\text{fm}^2$. The improvement at large separations leads to a somewhat worse description of the proton structure function at large Q^2. Apparently, the cross section dependent on energy, rather than x, cannot provide Bjorken scaling. Indeed, the parameterization (14) is successful only up to $Q^2 \approx 10\,\text{GeV}^2$.

In fact, the cases we are interested in, charmonium production and interaction, are just in between the regions where either of these parameterization is successful. Therefore, we suppose that the difference between predictions using (13) and (14) is a measure of the theoretical uncertainty which fortunately turns out to be rather small.

We demonstrate in Fig. 2 a few examples of r^2-dependence of the dipole cross section at different energies for both parameterization. Both, GBW and KST cross section, vanish $\propto r^2$ at small r, but deviate considerably from this simple behavior at large separations.

Quite often, the simplest parameterization ($\propto r^2$) is used for the dipole cross section. For the coefficient in front of r^2 we employ the expression obtained by the first term of the Taylor expansion of (14):

$$\text{“}r^2\text{”:} \qquad \sigma_{q\bar{q}}(r,s) = \frac{\sigma_0(s)}{r_0^2(s)} \cdot r^2 . \tag{17}$$

We shall refer to this form of the dipole cross section as r^2-approximation.

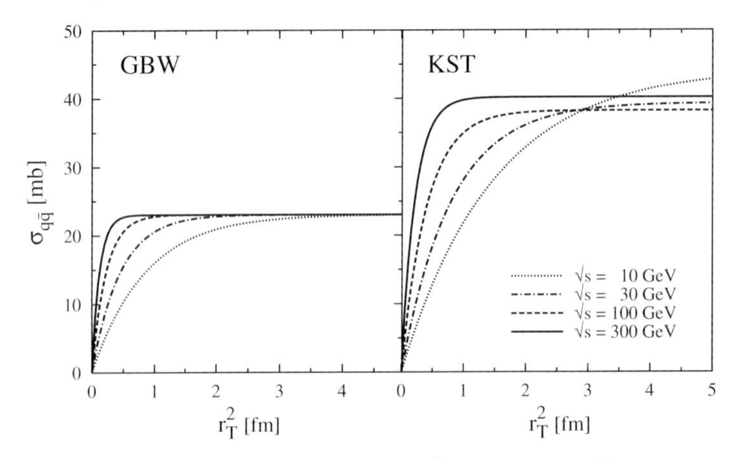

Fig. 2. The dipole cross section as function of r_T^2 at energies $\sqrt{s} = 10$, 30, 100 and 300 GeV for GBW (left) and KST (right) parameterizations. In the left panel, we used the prescription of [30], $x = (M_\psi^2 + Q^2)/s$, where M_ψ is the charmonium mass.

4 Diffractive Photoproduction of Charmonia off Protons

The dynamics of production and interaction of charmonia has drawn attention since their discovery back in 1974 [31]. As these heavy mesons have a small size it has been expected that hadronic cross sections may be calculated relying on perturbative QCD. The study of charmonium production became even more intense after charmonium suppression had been suggested as a probe for the creation and interaction of quark-gluon plasma in relativistic heavy ion collisions [12].

Since we will never have direct experimental information on charmonium-nucleon total cross sections one has to extract it from other data for example from elastic photoproduction of charmonia $\gamma p \to J/\psi(\psi')\ p$. The widespread believe that one can rely on the vector dominance model (VDM) is based on previous experience with photoproduction of ρ mesons. However, even a dispersion approach shows that this is quite a risky way, because the J/ψ pole in the complex Q^2 plane is nearly 20 times farther away from the physical region than the ρ pole. The multichannel analysis performed in [32] demonstrates that the corrections are huge, $\sigma_{tot}^{J/\psi\, p}$ turns out to be more that three times larger than the VDM prediction. Unfortunately, more exact predictions of the multichannel approach, especially for ψ', need knowledge of many diagonal and off-diagonal amplitudes which are easily summed only if one uses the oversimplified oscillator wave functions and a $q\bar{q}$-proton cross section of the form $\sigma_{q\bar{q}}(r) \propto r^2$, where r is the transverse $q\bar{q}$ separation.

Instead, one may switch to the quark basis, which should be equivalent to the hadronic basis because of completeness. In this representation the procedure of extracting $\sigma_{tot}^{J/\psi\, p}$ from photoproduction data cannot be realized directly, but has to be replaced by a different strategy. Namely, as soon as one has expressions for

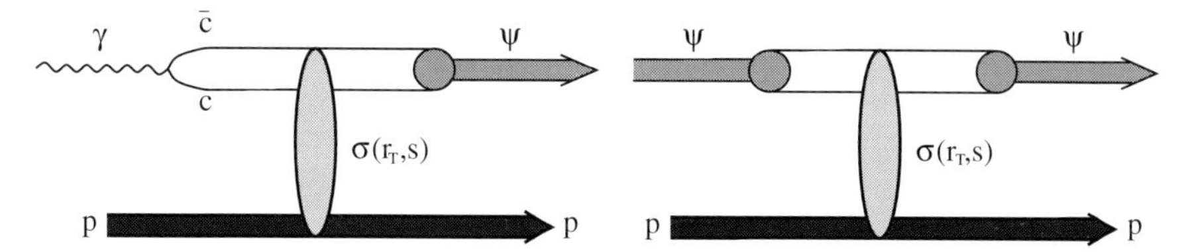

Fig. 3. Schematic representation of the amplitudes for the reactions $\gamma^* p \to \psi p$ (left) and $\psi\, p$ elastic scattering (right) in the rest frame of the proton. The $c\bar{c}$ fluctuation of the photon and the ψ with transverse separation r and c.m. energy $\sqrt{s}$ interact with the target proton via the cross section $\sigma(r,s)$ and produce a J/ψ or ψ'.

the wave functions of charmonia and the universal dipole cross section $\sigma_{q\bar{q}}(r,s)$, one can predict both, the experimentally known charmonium photoproduction cross sections and the unknown $\sigma_{tot}^{J/\psi(\psi')\,p}$. If the photoproduction data are well described one may have some confidence in the predictions for the $\sigma_{tot}^{J/\psi(\psi')p}$. Of course this procedure will be model dependent, but we believe that this is the best use of photoproduction data one can presently make. This program was performed for the first time in [33]. We do not propose a conceptually new scheme here, but calculate within a given approach as accurately as possible and without any free parameters. Wherever there is room for arbitrariness, like forms for the color dipole cross section and those for for charmonium wave functions, we use and compare other author's proposals, which have been tested on data different from those used here.

In the light-cone dipole approach the two processes, photoproduction and charmonium-nucleon elastic scattering look as shown in Fig. 3 [33]. The corresponding expressions for the forward amplitudes read

$$\mathcal{M}_{\gamma^* p}(s,Q^2) = \sum_{\mu,\bar{\mu}} \int_0^1 d\alpha \int d^2\boldsymbol{r}\, \Phi_{\psi}^{*(\mu,\bar{\mu})}(\alpha,\boldsymbol{r})\, \sigma_{q\bar{q}}(r,s)\, \Phi_{\gamma^*}^{(\mu,\bar{\mu})}(\alpha,\boldsymbol{r},Q^2)\ , \quad (18)$$

$$\mathcal{M}_{\psi\, p}(s) = \sum_{\mu,\bar{\mu}} \int_0^1 d\alpha \int d^2\boldsymbol{r}\, \Phi_{\psi}^{*(\mu,\bar{\mu})}(\alpha,\boldsymbol{r})\, \sigma_{q\bar{q}}(r,s)\, \Phi_{\psi}^{(\mu,\bar{\mu})}(\alpha,\boldsymbol{r})\ . \quad (19)$$

Here the summation runs over spin indexes μ, $\bar{\mu}$ of the c and $\bar{c}$ quarks, Q^2 is the photon virtuality, $\Phi_{\gamma^*}(\alpha,r,Q^2)$ is the light-cone distribution function of the photon for a $c\bar{c}$ fluctuation of separation r and relative fraction α of the photon light-cone momentum carried by c or $\bar{c}$. Correspondingly, $\Phi_{\psi}(\alpha,\boldsymbol{r})$ is the light-cone wave function of J/ψ, ψ' and χ (only in (19)). The dipole cross section $\sigma_{q\bar{q}}(r,s)$ mediates the transition (*cf* Fig. 3).

The light cone variable describing longitudinal motion which is invariant to Lorentz boosts is the fraction $\alpha = p_c^+/p_{\gamma^*}^+$ of the photon light-cone momentum $p_{\gamma^*}^+ = E_{\gamma^*} + p_{\gamma^*}$ carried by the quark or antiquark. In the nonrelativistic approximation (assuming no relative motion of c and $\bar{c}$) $\alpha = 1/2$ (e.g. [33]), otherwise one should integrate over α (see (18)). For transversely (T) and longitudinally

(L) polarized photons the perturbative photon-quark distribution function in (18) reads [19,20],

$$\Phi^{(\mu,\bar{\mu})}_{T,L}(\alpha,\boldsymbol{r},Q^2) = \frac{\sqrt{N_c\,\alpha_{em}}}{2\,\pi}\,Z_c\,\chi_c^{\mu\dagger}\,\widehat{O}_{T,L}\,\widetilde{\chi}_{\bar{c}}^{\bar{\mu}}\,\mathrm{K}_0(\varepsilon r)\;, \tag{20}$$

where

$$\widetilde{\chi}_{\bar{c}} = i\,\sigma_y\,\chi_{\bar{c}}^*\;; \tag{21}$$

χ and $\bar{\chi}$ are the spinors of the c-quark and antiquark respectively; $Z_c = 2/3$. $K_0(\epsilon r)$ is the modified Bessel function with

$$\varepsilon^2 = \alpha(1-\alpha)Q^2 + m_c^2\;. \tag{22}$$

The operators $\widehat{O}_{T,L}$ have the form:

$$\widehat{O}_T = m_c\,\boldsymbol{\sigma}\cdot\boldsymbol{e}_\gamma + i(1-2\alpha)\,(\boldsymbol{\sigma}\cdot\boldsymbol{n})\,(\boldsymbol{e}_\gamma\cdot\boldsymbol{\nabla}_r) + (\boldsymbol{n}\times\boldsymbol{e}_\gamma)\cdot\boldsymbol{\nabla}_r\;, \tag{23}$$

$$\widehat{O}_L = 2\,Q\,\alpha(1-\alpha)\,\boldsymbol{\sigma}\cdot\boldsymbol{n}\;, \tag{24}$$

where $\boldsymbol{n} = \boldsymbol{p}/p$ is a unit vector parallel to the photon momentum and $\boldsymbol{e}$ is the polarization vector of the photon. Effects of the non-perturbative interaction within the $q\bar{q}$ fluctuation are negligible for the heavy charmed quarks.

4.1 Charmonium Wave Functions

The spatial part of the $c\bar{c}$ pair wave function satisfying the Schrödinger equation

$$\left(-\frac{\Delta}{m_c} + V(R)\right)\,\Psi_{nlm}(\boldsymbol{R}) = E_{nl}\,\Psi_{nlm}(\boldsymbol{R}) \tag{25}$$

is represented in the form

$$\Psi(\boldsymbol{R}) = \Psi_{nl}(R)\cdot Y_{lm}(\theta,\varphi)\;, \tag{26}$$

where $\boldsymbol{R}$ is 3-dimensional $c\bar{c}$ separation (not to be confused with the 2-dimensional argument $\boldsymbol{r}$ of the dipole cross section), $\Psi_{nl}(R)$ and $Y_{lm}(\theta,\varphi)$ are the radial and orbital parts of the wave function. The equation for radial $\Psi(R)$ is solved with the help of the program [34]. The following four potentials $V(R)$ have been used:

- "COR": Cornell potential [35],

$$V(r) = -\frac{k}{R} + \frac{R}{a^2} \tag{27}$$

  with $k = 0.52$, $a = 2.34\,\mathrm{GeV}^{-1}$ and $m_c = 1.84\,\mathrm{GeV}$.
- "BT": Potential suggested by Buchmüller and Tye [36] with $m_c = 1.48\,\mathrm{GeV}$. It has a similar structure as the Cornell potential: linear string potential at large separations and Coulomb shape at short distances with some refinements, however.

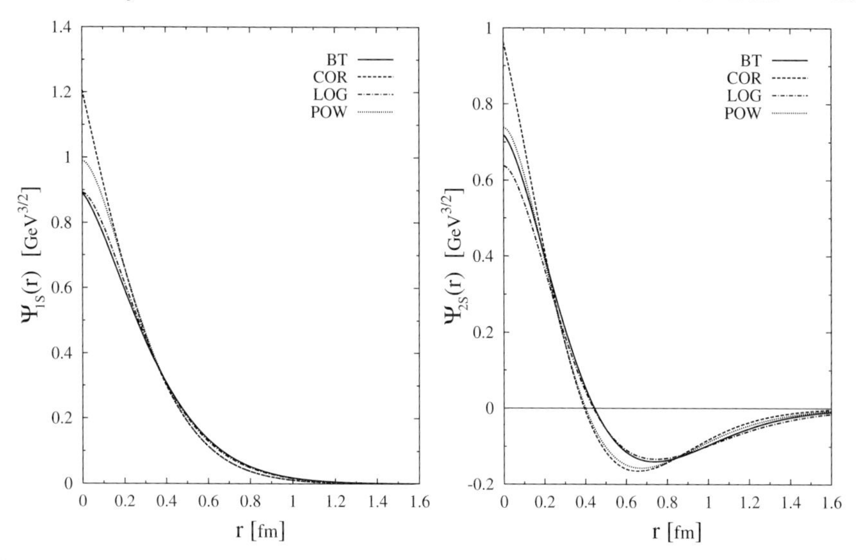

Fig. 4. The radial part of the wave function $\Psi_{nl}(r)$ for the $1S$ and $2S$ states calculated with four different potentials (see text). In this figure, $r = R$.

- "LOG": Logarithmic potential [37]

$$V(R) = -0.6635\,\mathrm{GeV} + (0.733\,\mathrm{GeV})\log(R \cdot 1\,\mathrm{GeV}) \tag{28}$$

with $m_c = 1.5\,\mathrm{GeV}$.
- "POW": Power-law potential [38]

$$V(R) = -8.064\,\mathrm{GeV} + (6.898\,\mathrm{GeV})(R \cdot 1\,\mathrm{GeV})^{0.1} \tag{29}$$

with $m_c = 1.8\,\mathrm{GeV}$.

The results of calculations for the radial part $\Psi_{nl}(R)$ of the $1S$ and $2S$ states are depicted in Fig. 4. For the ground state all the potentials provide a very similar behavior for $R > 0.3\,\mathrm{fm}$, while for small R the predictions are differ by up to 30%. The peculiar property of the $2S$ state wave function is the node at $R \approx 0.4\,\mathrm{fm}$ which causes strong cancelations in the matrix elements (18) and as a result, a suppression of photoproduction of ψ' relative to J/ψ [33,39].

Note that the lowest Fock component $|c\bar{c}\rangle$ in the infinite momentum frame is not related by simple Lorentz boost to the wave function of charmonium in the rest frame. This makes the problem of building the light-cone wave function for the lowest $|c\bar{c}\rangle$ component difficult, no unambiguous solution is yet known. There are only recipes in the literature, a simple one widely used [40], is the following. One applies a Fourier transformation from coordinate to momentum space to the known spatial part of the non-relativistic wave function (26), $\Psi(\boldsymbol{R}) \Rightarrow \Psi(\boldsymbol{p})$, which can be written as a function of the effective mass of the $c\bar{c}$, $M^2 = 4(p^2 + m_c^2)$, expressed in terms of light-cone variables

$$M^2(\alpha, p_T) = \frac{p_T^2 + m_c^2}{\alpha(1-\alpha)} \,. \tag{30}$$

In order to change the integration variable p_L to the light-cone variable α one relates them through M, namely $p_L = (\alpha - 1/2)M(p_T, \alpha)$. In this way the $c\bar{c}$ wave function acquires a kinematical factor

$$\Psi(\boldsymbol{p}) \Rightarrow \sqrt{2}\,\frac{(p^2+m_c^2)^{3/4}}{(p_T^2+m_c^2)^{1/2}}\cdot\Psi(\alpha,\boldsymbol{p}_T) \equiv \Phi_\psi(\alpha,\boldsymbol{p}_T)\ . \tag{31}$$

This procedure is used in [41] and the result is applied to calculation of the amplitudes (18). The result is discouraging, since the ψ' to J/ψ ratio of the photoproduction cross sections are far too low in comparison with data. However, the oversimplified dipole cross section $\sigma_{q\bar{q}}(r) \propto r^2$ has been used, and what is even more essential, the important ingredient of Lorentz transformations, the Melosh spin rotation, has been left out. The spin transformation has also been left out in the recent publication [42] which repeats the calculations of [41] with a more realistic dipole cross section which levels off at large separations. This leads to suppression of the node-effect (less cancelation) and enhancement of Ψ' photoproduction. Nevertheless, the calculated ψ' to J/ψ ratio is smaller than the data by a factor of two.

The 2-spinors χ_c and $\chi_{\bar{c}}$ describing c and $\bar{c}$ respectively in the infinite momentum frame are known to be related by Melosh rotation [43,40] to the spinors $\bar{\chi}_c$ and $\bar{\chi}_{\bar{c}}$ in the rest frame:

$$\begin{aligned}\overline{\chi}_{\mathbf{c}} &= \widehat{R}(\alpha,\boldsymbol{p}_T)\,\chi_c\ ,\\ \overline{\chi}_{\bar{\mathbf{c}}} &= \widehat{R}(1-\alpha,-\boldsymbol{p}_T)\,\chi_{\bar{c}}\ ,\end{aligned} \tag{32}$$

where the matrix $\widehat{R}(\alpha,\boldsymbol{p}_T)$ has the form:

$$\widehat{R}(\alpha,\boldsymbol{p}_T) = \frac{m_c + \alpha M - i\,[\boldsymbol{\sigma}\times\boldsymbol{n}]\,\boldsymbol{p}_T}{\sqrt{(m_c+\alpha M)^2 + p_T^2}}\ . \tag{33}$$

Since the potentials we use in Sect. 4.1 contain no spin-orbit term, the $c\bar{c}$ pair is in S-wave. In this case spatial and spin dependences in the wave function factorize and we arrive at the following light cone wave function of the $c\bar{c}$ in the infinite momentum frame

$$\Phi_\psi^{(\mu,\bar{\mu})}(\alpha,\boldsymbol{p}_T) = U^{(\mu,\bar{\mu})}(\alpha,\boldsymbol{p}_T)\cdot\Phi_\psi(\alpha,\boldsymbol{p}_T)\ , \tag{34}$$

where

$$U^{(\mu,\bar{\mu})}(\alpha,\boldsymbol{p}_T) = \chi_c^{\mu\dagger}\,\widehat{R}^\dagger(\alpha,\boldsymbol{p}_T)\,\boldsymbol{\sigma}\cdot\boldsymbol{e}_\psi\,\sigma_y\,\widehat{R}^*(1-\alpha,-\boldsymbol{p}_T)\,\sigma_y^{-1}\,\widetilde{\chi}_{\bar{c}}^{\bar{\mu}} \tag{35}$$

and $\widetilde{\chi}_{\bar{c}}$ is defined in (21).

Note that the wave function (34) is different from one used in [44–46] where it was assumed that the vertex $\psi \to c\bar{c}$ has the structure $\psi_\mu\,\bar{u}\,\gamma_\mu\,u$ like the for the photon $\gamma^* \to c\bar{c}$. The rest frame wave function corresponding to such a vertex contains S wave and D wave. The weight of the latter is dictated by the structure of the vertex and cannot be justified by any reasonable nonrelativistic potential model for the $c\bar{c}$ interaction.

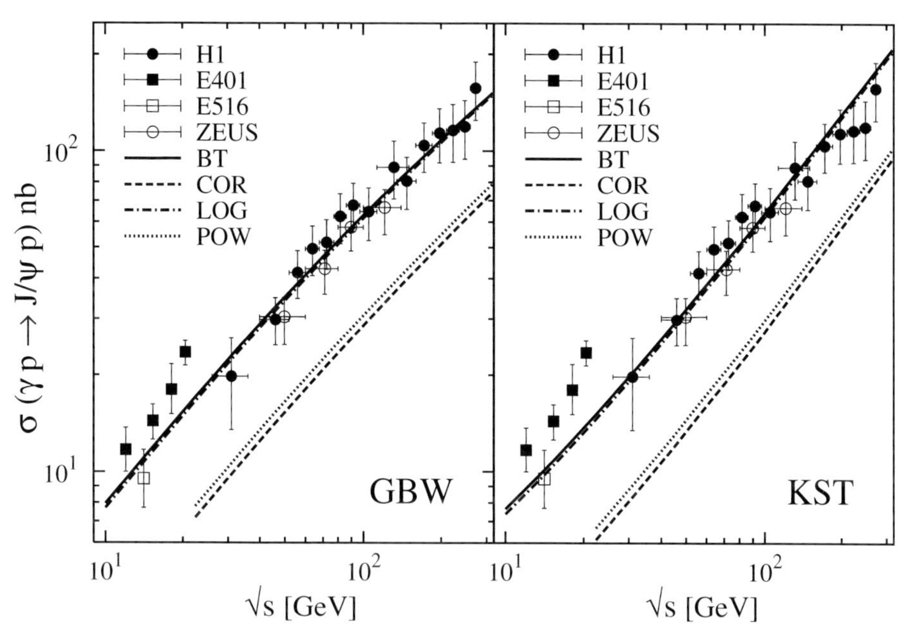

Fig. 5. Integrated cross section for elastic photoproduction $\gamma p \to J/\psi\, p$ with real photons ($Q^2 = 0$) as a function of the energy calculated with GBW and KST dipole cross sections and for four potentials to generate J/ψ wave functions. Experimental data points from the H1 [47], E401 [48], E516 [49] and ZEUS [50] experiments.

Now we can determine the light-cone wave function in the mixed longitudinal momentum - transverse coordinate representation:

$$\Phi_{\psi}^{(\mu,\bar{\mu})}(\alpha, \boldsymbol{r}) = \frac{1}{2\,\pi} \int d^2\boldsymbol{p}_T\, e^{-i\boldsymbol{p}_T\boldsymbol{r}}\, \Phi_{\psi}^{(\mu,\bar{\mu})}(\alpha, \boldsymbol{p}_T)\ . \tag{36}$$

4.2 Comparison with Data

Having the light-cone wave function of charmonium, we are now in the position to calculate the cross section of charmonium photoproduction using (18). The results for J/ψ are compared with the data in Fig. 5. Calculations are performed with GBW and KST parameterizations for the dipole cross section and for wave functions of the J/ψ calculated from BT, LOG, COR and POW potentials. One observes

- There are no major differences between different parameterizations [28,29] of the dipole cross section.
- The use of different potentials to generate the wave functions of the J/ψ leads to two distinctly different behaviors. The potentials labeled BT and LOG (see Sect. 4.1) describe the data very well, while the potentials COR and LOG underestimate them by a factor of two. The different behavior has been traced to the following origin: BT and LOG use $m_c \approx 1.5\,\text{GeV}$, but COR and POW $m_c \approx 1.8\,\text{GeV}$. While the bound state wave functions of J/ψ

Table 1. The photoproduction $\gamma p \to J/\psi p$ cross-section $\sigma(J/\psi)$ in nb and the ratio $\mathcal{R} = \sigma(\psi')/\sigma(J/\psi)$ for the four different types of potentials (BT, LOG, COR, POW) and the three parameterizations (GBW, KST, r^2) for the dipole cross section $\sigma(r,s)$ at $\sqrt{s} = 90$ GeV. The values in parentheses correspond to the case when the spin rotation is neglected. See [1] for a comparison with data.

		BT	LOG	COR	POW
σ	GBW	52.01 (37.77)	50.78 (36.63)	23.13 (17.07)	24.94 (18.64)
	KST	49.96 (35.87)	48.49 (34.57)	21.05 (15.42)	22.83 (16.92)
	r^2	66.67 (47.00)	64.07 (44.86)	25.81 (18.71)	28.23 (20.66)
$\mathcal{R}$	GBW	0.147 (0.075)	0.117 (0.060)	0.168 (0.099)	0.144 (0.085)
	KST	0.147 (0.068)	0.118 (0.054)	0.178 (0.099)	0.152 (0.084)
	r^2	0.101 (0.034)	0.081 (0.027)	0.144 (0.070)	0.121 (0.058)

are little affected by this difference (see Fig. 4), the photon wave function (20) depends sensitively on m_c via the argument (22) of the K_0 function.

4.3 Importance of Spin Effects for the ψ' to J/ψ Ratio

It turns out that the effects of spin rotation have a gross impact on the cross section of elastic photoproduction $\gamma p \to J/\psi(\psi')p$. To demonstrate these effects we present the results of our calculations at $\sqrt{s} = 90$ GeV in Table 1. The upper half of the table shows the photoproduction cross sections for J/ψ for different parameterizations of the dipole cross section (GBW, KST, "r^2") and potentials (BT, COR, LOG, POW). The numbers in parenthesis show what the cross section would be, if the spin rotation effects were neglected. We see that these effects add 30-40% to the J/ψ photoproduction cross section.

The spin rotation effects turn out to have a much more dramatic impact on ψ' increasing the photoproduction cross section by a factor 2-3. This is visible in the lower half of the table which shows the ratio $\mathcal{R} = \sigma(\psi')/\sigma(J/\psi)$ of photoproduction cross sections, where the number in parenthesis correspond to no spin rotation effects included. This spin effects explain the large values of the ratio $\mathcal{R}$ observed experimentally. Our results for $\mathcal{R}$ are about twice as large as evaluated in [42] and even more than in [41].

4.4 Charmonium-Nucleon Total Cross Sections

After the light-cone formalism has been checked with the data for virtual photoproduction we are in position to provide reliable predictions for charmonium-nucleon total cross sections. The corresponding expressions are given by (19)) (compare with [13]). The calculated J/ψ- and ψ'-nucleon total cross sections are plotted in Fig. 6 for for the GBW and KST forms of the dipole cross sections and all four types of the charmonium potentials.

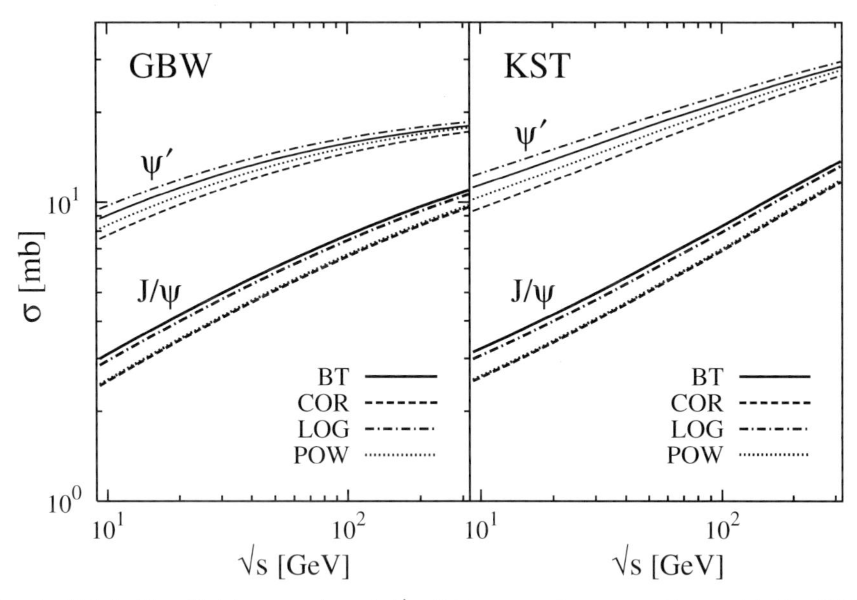

Fig. 6. Total $J/\psi\, p$ (thick curves) and $\psi'\, p$ (thin curves) cross sections with the GBW and KST parameterizations for the dipole cross section.

5 Nuclear Effects in Exclusive Leptoproduction of Charmonia

Charmonium production on nuclei can be exclusive, $\gamma^* A \to \Psi X$, where $X = A$ (coherent) or $X = A^*$ (incoherent), and inclusive when X includes pions. We skip the latter which is discussed in [51] and concentrate here on exclusive processes. In this case the following phenomena are to be expected: color filtering, *i.e.* inelastic interactions of the $c\bar{c}$ pair on its way through the nucleus is expected to lead to a suppression of Ψ production relative to $A\sigma_{\gamma^* p \to \Psi p}$. Since the dipole cross section $\sigma_{q\bar{q}}$ also depends on the gluon distribution in the target (p of A), nuclear shadowing of the gluon distribution is expected to reduce $\sigma_{q\bar{q}}$ in a nuclear reaction relative to the one on the proton. Production of a $c\bar{c}$ pair in a nucleus and its absorption are also determined by the values of the coherence length l_c and the formation length l_f [23].

Explicit calculations have been performed in [33] in the approximation of a short coherence (or production) length, when one can treat the creation of the colorless $c\bar{c}$ pair as instantaneous,

$$l_c = \frac{2\,\nu}{M_{c\bar{c}}^2} \approx \frac{2\,\nu}{M_{J/\psi}^2} \ll R_A, \tag{37}$$

where ν is the energy of the virtual photon in the rest frame of the nucleus. At the same time, the formation length may be long, comparable with the nuclear radius R_A,

$$l_f = \frac{2\,\nu}{M_{\psi'}^2 - M_{J/\psi}^2} \sim R_A\ . \tag{38}$$

In [33] the wave function formation is described by means of the light-cone Green function approach summing up all possible paths of the $c\bar{c}$ in the nucleus. The result has been unexpected. Contrary to naive expectation, based on the larger size of the ψ' compared to J/ψ, it has been found that ψ' is not more strongly absorbed than the J/ψ, but may even be enhanced by the nuclear medium. This is interpreted as an effect of filtering which is easy to understand in the limit of long coherence length, $l_c \gg R_A$. Indeed, the production rate of ψ' on a proton target is small due to strong cancelations in the projection of the produced $c\bar{c}$ wave packet onto the radial wave function of the ψ' which has a node. After propagation through nuclear matter the transverse size of a $c\bar{c}$ wave packet is squeezed by absorption and the projection of the ψ' wave function is enhanced [33,39] since the effect of the node is reduced (see another manifestation of the node in [52]).

However, the quantitative predictions of [33] are not trustable since the calculations have been oversimplified and quite some progress has been made on the form of the dipole cross section $\sigma_{q\bar{q}}$ and the light cone wave functions for the charmonia. Therefore we take the problem up again and provide more realistic calculations for nuclear effects in exclusive electroproduction of charmonia off nuclei relying on the successful parameter free calculations which have been performed recently in [1] for elastic virtual photoproduction of charmonia, $\gamma^*\,p \to \Psi\,p$ (see Sect. 4).

Whenever one deals with high-energy reactions on nuclei, one cannot avoid another problem of great importance: gluon shadowing. At small values of x, gluon clouds overlap in longitudinal direction and may fuse. As a result, the gluon density per one nucleon in a nucleus is expected to be reduced compared to a free proton. Parton shadowing, which leads to an additional nuclear suppression in various hard reactions (DIS, DY, heavy flavor, high-p_T hadrons, etc.) may be especially strong for exclusive vector meson production like charmonium production which needs at least two gluon exchange. Unfortunately, we have no experimental information for gluon shadowing in nuclei so far, and we have to rely on the available theoretical estimates, see *e.g.* [53,18,29,54].

5.1 Eikonal Shadowing Versus Absorption for $c\bar{c}$ Pairs in Nuclei

Exclusive charmonium production off nuclei, $\gamma^* A \to \Psi X$ is called coherent, when the nucleus remains intact, i.e. $X = A$, or incoherent, when X is an excited nuclear state which contains nucleons and nuclear fragments but no other hadrons. The cross sections depend on the polarization ϵ of the virtual photon (in all figures below we will imply $\epsilon = 1$),

$$\sigma^{\gamma^* A}(s, Q^2) = \sigma^{\gamma^*_T A}(s, Q^2) + \epsilon\,\sigma^{\gamma^*_L A}(s, Q^2)\ , \tag{39}$$

where the indexes T, L correspond to transversely or longitudinally polarized photons, respectively.

The cross section for exclusive production of charmonia off a nucleon target integrated over momentum transfer [13] is given by

$$\sigma_{inc}^{\gamma^*_{T,L}N}(s,Q^2) = \left|\left\langle \Psi \left|\sigma_{q\bar{q}}(r,s)\right| \gamma_{c\bar{c}}^{T,L} \right\rangle\right|^2 , \tag{40}$$

where $\Psi(\boldsymbol{r},\alpha)$ is the charmonium LC wave function which depends on the transverse $c\bar{c}$ separation $\boldsymbol{r}$ and on the relative sharing α of longitudinal momentum [1]. Both variables are involved in the integration in the matrix element (40). $\Psi(\boldsymbol{r},\alpha)$ is obtained by means of a Lorentz boost applied the solutions of the Schrödinger equation. This procedure involves the Melosh spin rotation [40,43] which produces sizable effects. In addition, $\gamma_{c\bar{c}}^{T,L}(\boldsymbol{r},\alpha,Q^2)$ is the LC wave function of the $c\bar{c}$ Fock component of the photon. It depends on the photon virtuality Q^2. One can find the details in [1] including the effects of a nonperturbative $q\bar{q}$ interaction.

The cross sections for coherent and incoherent production on nuclei will be derived under various conditions imposed by the coherence length (37). At high energies the coherence length (37) may substantially exceed the nuclear radius. In this case the transverse size of the $c\bar{c}$ wave packet is "frozen" by Lorentz time dilation, *i.e.* it does not fluctuate during propagation through the nucleus, and the expressions for the cross sections, incoherent (*inc*) or coherent (*coh*), are particularly simple [33],

$$\sigma_{inc}^{\gamma^*_{T,L}A}(s,Q^2) = \int d^2b\, T_A(b) \left|\left\langle \Psi \left|\sigma_{q\bar{q}}(r,s) \exp\left[-\frac{1}{2}\,\sigma_{q\bar{q}}(r,s)\,T_A(b)\right]\right| \gamma_{c\bar{c}}^{T,L} \right\rangle\right|^2 \tag{41}$$

$$\sigma_{coh}^{\gamma^*_{T,L}A}(s,Q^2) = \int d^2b \left|\left\langle \Psi \left|1 - \exp\left[-\frac{1}{2}\,\sigma_{q\bar{q}}(r,s)\,T_A(b)\right]\right| \gamma_{c\bar{c}}^{T,L} \right\rangle\right|^2 . \tag{42}$$

Here $T_A(b) = \int_{-\infty}^{\infty} dz\, \rho_A(b,z)$ is the nuclear thickness function given by the integral of the nuclear density along the trajectory at a given impact parameter b.

The nuclear suppression ratio for incoherent electroproduction of J/ψ and ψ' is shown in Fig. 7 as a function of $\sqrt{s}$. We use the GBW [28] and KST [29] parameterizations for the dipole cross section and show the results by solid and dashed curves, respectively. Differences are at most $10-20\,\%$. Analyzing the results shown in Fig. 7, we observe that nuclear suppression of J/ψ production becomes stronger with energy. This is an obvious consequence of the energy dependence of $\sigma_{q\bar{q}}(r,s)$, which rises with energy (see Sect. 3). For ψ' the suppression is rather similar to the J/ψ case. In particular we do not see any considerable nuclear enhancement of ψ' which has been found earlier [33,55], where the oversimplified form of the dipole cross section, $\sigma_{q\bar{q}}(r) \propto r^2$ and the oscillator form of the wave function had been used. Such a form of the cross section enhances the compensation between large and small distances in the wave function of ψ' in the process $\gamma^* p \to \psi' p$. Therefore, the color filtering effect which emphasizes the small distance part of the wave function leads to a strong enhancement of the ψ' production rate. This is why using the more realistic r-dependence of

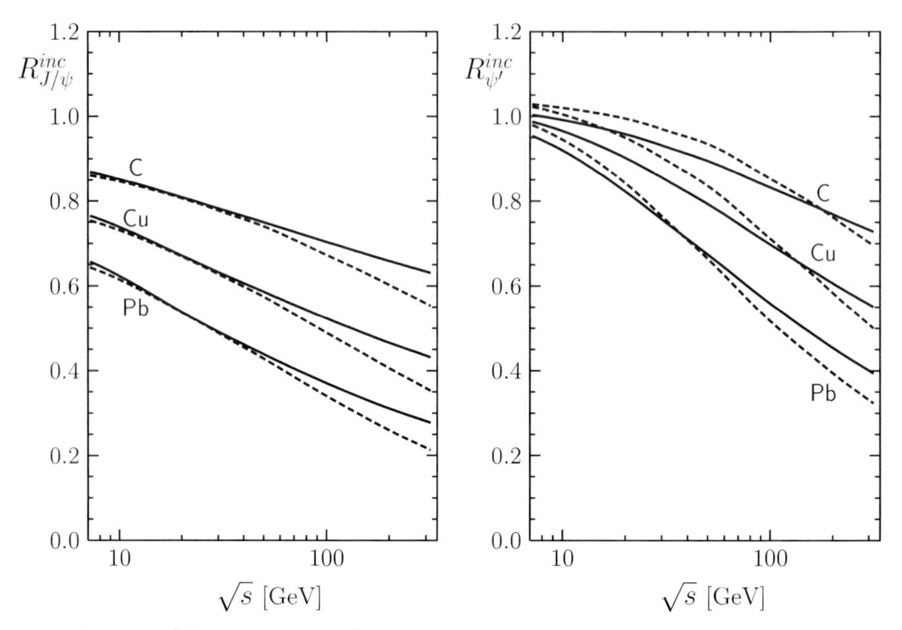

Fig. 7. Ratios R^{inc}_{Ψ} for J/ψ and ψ' incoherent production on carbon, copper and lead as function of $\sqrt{s}$ and at $Q^2 = 0$. The solid curves refer to the GBW parameterization of $\sigma_{q\bar{q}}$ and dashed one refer to the KST parameterization.

$\sigma_{q\bar{q}}(r)$ leveling off at large r leads to a weaker enhancement of the ψ'. This effect becomes even more pronounced at higher energies since the dipole cross section saturates starting at a value $r \sim r_0(s)$ where $r_0(s)$ decreases with energy. This observation probably explains why the ψ' is less enhanced at higher energies as one can see from Fig. 7.

Note that the "frozen" approximation is valid only for $l_c \gg R_A$ and can be used only at $\sqrt{s} > 20 - 30$ GeV. Therefore, the low-energy part of the curves depicted in Fig. 7 should be corrected for the effects related to the finiteness of l_c. This is done in [1].

One can change the effect of color filtering in nuclei in a controlled way by increasing the photon virtuality Q^2 thereby squeezing the transverse size of the $c\bar{c}$ fluctuation in the photon. For a narrower $c\bar{c}$ pair the cancelation which is caused by the node in the radial wave function of ψ' should be less effective. One expects that the ψ' to J/ψ ratio on a proton target increases with Q^2, as is observed both in experiment and calculation (Fig. 9 of [1]). A detailed investigation of the Q^2 dependence of nuclear effects is published in [3].

Cross sections for coherent production of charmonia on nuclei are calculated analogously using (42). The results for the energy dependence are depicted in Fig. 8.

It is not a surprise that the ratios exceed one. In the absence of $c\bar{c}$ attenuation the forward coherent production would be proportional to A^2, while integrated

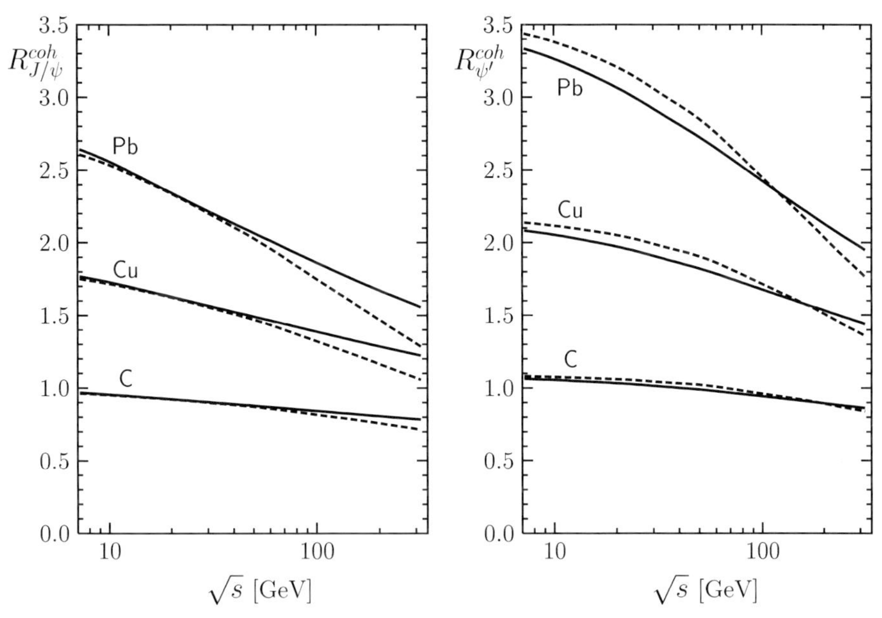

Fig. 8. The ratios $R^{coh}_{J/\psi}$ and $R^{coh}_{\psi'}$ for coherent production on nuclei as a function of $\sqrt{s}$. The meaning of the different lines is the same as in Fig. 7.

over momentum transfer, the one depicted in Fig. 8, behaves as $A^{4/3}$. This is a result of our definition that R^{coh}_{Ψ} exceeds one.

5.2 Gluon Shadowing

The gluon density in nuclei at small Bjorken x is expected to be suppressed compared to a free nucleon due to interferences. This phenomenon called gluon shadowing renormalizes the dipole cross section,

$$\sigma_{q\bar{q}}(r,x) \Rightarrow \sigma_{q\bar{q}}(r,x)\, R_G(x,Q^2,b)\,. \tag{43}$$

where the factor $R_G(x,Q^2,b)$ is the ratio of the gluon density at x and Q^2 in a nucleon of a nucleus to the gluon density in a free nucleon. No data are available so far which could provide direct information about gluon shadowing. Currently it can be evaluated only theoretically. In what follows we employ the technique developed in [29].

Note that the procedure (43) differs from the prescription in [45]. The latter is based on QCD factorization applied to a nuclear target and suggests to multiply by $R_G(x,Q^2,b)$ the whole nuclear cross section. This approximation should not be used for charmonium production which exposes according to above calculations a strong deviation from factorization. Besides, gluon shadowing is overestimated in [45] as is discussed in [29].

The interpretation of the phenomenon of gluon shadowing depends very much on the reference frame. It looks like glue-glue fusion in the infinite momentum frame of the nucleus: although the nucleus is Lorentz contracted, the bound nucleons are still well separated since they contract too. However, the gluon clouds of the nucleons are contracted less since they have a smaller momentum fraction $\sim x$. Therefore, they do overlap and interact at small x, and gluons originating from different nucleons can fuse leading to a reduction of the gluon density.

Although observables must be Lorentz invariant, the space-time interpretation of shadowing looks very different in the rest frame of the nucleus. Here it comes as a result of eikonalization of higher Fock components of the incident particles. Indeed, the nuclear effect included by eikonalization into (41)-(42) corresponds to the lowest $c\bar{c}$ Fock component of the photon. These expressions do not include any correction for gluon shadowing, but rather correspond to shadowing of sea quarks in nuclei, analogous to what is measured in deep-inelastic scattering. Although the phenomenological dipole cross section $\sigma_{q\bar{q}}(x,Q^2)$ includes all possible effects of gluon radiation, the eikonal expressions (41)-(42) assume that none of the radiated gluons takes part in multiple interaction in the nucleus. The leading order correction corresponding to gluon shadowing comes from eikonalization of the next Fock component which contains the $c\bar{c}$ pair plus a gluon. One can trace on Feynman graphs that this is exactly the same mechanism of gluon shadowing as glue-glue fusion in a different reference frame.

Note that (41)-(42) assume that for the coherence length $l_c \gg R_A$. Even if this condition is satisfied for a $c\bar{c}$ fluctuation, it can be broken for the $c\bar{c}G$ component which is heavier. Indeed, it was found in [58] that the coherence length for gluon shadowing as about an order of magnitude shorter than the one for shadowing of sea quarks. Therefore, one should not rely on the long coherence length approximation used in (41)-(42), but take into account the finiteness of l_c^G. This can be done by using the light-cone Green function approach developed in [56,29].

The factor $R_G(x,Q^2,b)$ has the form,

$$R_G(x,Q^2,b) = 1 - \frac{\Delta\sigma(\gamma^* A)}{T(b)\,\sigma(\gamma^* N)}\ , \tag{44}$$

where $\sigma(\gamma^* N)$ is the part of the total $\gamma^* N$ cross section related to a $c\bar{c}$ fluctuation in the photon,

$$\sigma(\gamma^* N) = \int d^2r \int\limits_0^1 d\alpha\ \left|\Psi_{\gamma^*\to c\bar{c}}(r,\alpha,Q^2)\right|^2\ \sigma_{q\bar{q}}(r,x)\ . \tag{45}$$

Here $\Psi_{\gamma^*\to c\bar{c}}(r,\alpha,Q^2)$ is the light-cone wave function of the $c\bar{c}$ pair with transverse separation $\boldsymbol{r}$ and relative sharing of the longitudinal momentum α and $1-\alpha$ (see details in Sect. 2). The numerator $\Delta\sigma(\gamma^* A)$ in (44) reads [29],

$$\Delta\sigma(\gamma^* A) = 8\pi\,\mathrm{Re}\int dM^2\ \left.\frac{d^2\sigma(\gamma^* N\to c\bar{c}GN)}{dM^2\,dq_T^2}\right|_{q_T=0} \tag{46}$$

$$\times \int_{-\infty}^{\infty} dz_1 \int_{-\infty}^{\infty} dz_2\, \Theta(z_2 - z_1)\, \rho_A(b, z_1)\, \rho_A(b, z_2)\, \exp\left[-i\, q_L\, (z_2 - z_1)\right] \, .$$

Here the invariant mass squared of the $c\bar{c}G$ system is given by,

$$M^2 = \sum_i \frac{m_i^2 + k_i^2}{\alpha_i} \, , \tag{47}$$

where the sum is taken over partons ($c\bar{c}G$) having mass m_i, transverse momentum $\boldsymbol{k}_i$ and fraction α_i of the full momentum. The $c\bar{c}G$ system is produced diffractively as an intermediate state in a double interaction in the nucleus. z_1 and z_2 are the longitudinal coordinates of the nucleons N_1 and N_2, respectively, participating in the diffractive transition $\gamma^*\, N_1 \to c\bar{c}G\, N_1$ and back $c\bar{c}G\, N_2 \to \gamma^*\, N_2$. The value of $\Delta\sigma$ is controlled by the longitudinal momentum transfer

$$q_L = \frac{Q^2 + M^2}{2\,\nu} \, , \tag{48}$$

which is related to the gluonic coherence length $l_c^G = 1/q_L$.

The Green function $G_{c\bar{c}G}(\boldsymbol{r}_2, \boldsymbol{\rho}_2, z_2; \boldsymbol{r}_1, \boldsymbol{\rho}_1 z_1)$ describes the propagation and interaction of the $c\bar{c}G$ system in the nuclear medium between the points z_1 and z_2. Here, $\boldsymbol{r}_{1,2}$ and $\boldsymbol{\rho}_{1,2}$ are the transverse separations between the c and $\bar{c}$ and between the $c\bar{c}$ pair and gluon at the point z_1 and destination z_2 respectively. Then the Fourier transform of the diffractive cross section in (46),

$$8\pi \int dM_X^2\, \left.\frac{d^2\sigma(\gamma^* N \to XN)}{dM_X^2\, dq_T^2}\right|_{q_T=0} \cos\left[q_L\, (z_2 - z_1)\right] \tag{49}$$

can be represented in the form,

$$\frac{1}{2}\,\mathrm{Re} \int d^2r_2 d^2\rho_2 d^2r_1 d^2\rho_1 \int d\alpha_q d\ln(\alpha_G) \tag{50}$$
$$\times\, F^\dagger_{\gamma^* \to c\bar{c}G}(\boldsymbol{r}_2, \boldsymbol{\rho}_2, \alpha_q, \alpha_G)\; G_{c\bar{c}G}(\boldsymbol{r}_2, \boldsymbol{\rho}_2, z_2; \boldsymbol{r}_1, \boldsymbol{\rho}_1, z_1)\; F_{\gamma^* \to c\bar{c}G}(\boldsymbol{r}_1, \boldsymbol{\rho}_1, \alpha_q, \alpha_G) \, .$$

Assuming that the momentum fraction taken by the gluon is small, $\alpha_G \ll 1$, and neglecting the $c\bar{c}$ separation $r \ll \rho$ we arrive at a factorized form of the three-body Green function,

$$G_{c\bar{c}G}(\boldsymbol{r}_2, \boldsymbol{\rho}_2, z_2; \boldsymbol{r}_1, \boldsymbol{\rho}_1, z_1) \Rightarrow G_{c\bar{c}}(\boldsymbol{r}_2, z_2; \boldsymbol{r}_1, z_1)\; G_{GG}(\boldsymbol{\rho}_2, z_2; \boldsymbol{\rho}_1, z_1) \, , \tag{51}$$

where $G_{GG}(\boldsymbol{\rho}_2, z_2; \boldsymbol{\rho}_1, z_1)$ describes propagation of the GG dipole (in fact the color-octet $c\bar{c}$ and gluon) in the nuclear medium. This Green function satisfies the two dimensional Schrödinger equation which includes the glue-glue nonperturbative interaction via the light-cone potential $V(\boldsymbol{\rho}, z)$, as well as interaction with the nuclear medium.

$$i\, \frac{d}{dz_2} G_{GG}(\boldsymbol{\rho}_2, z_2; \boldsymbol{\rho}_1, z_1) = \left[-\frac{\Delta(\boldsymbol{\rho}_2)}{2\,\nu\, \alpha_G (1 - \alpha_G)} + V(\boldsymbol{\rho}_2, z_2) \right] G_{GG}(\boldsymbol{\rho}_2, z_2; \boldsymbol{\rho}_1, z_1) \, , \tag{52}$$

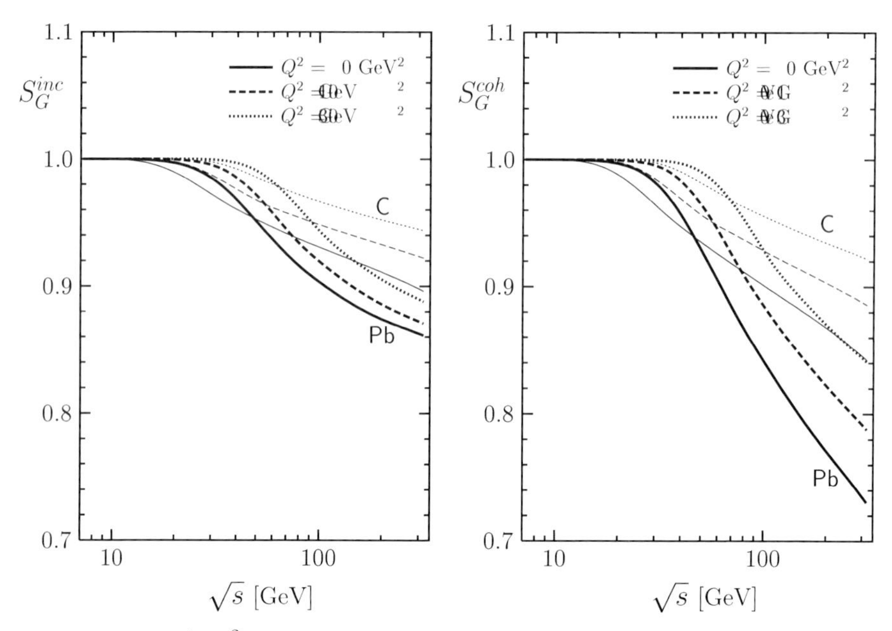

Fig. 9. Ratios $S_G(s, Q^2)$ of cross sections calculated with and without gluon shadowing for incoherent and coherent charmonia production. We only plot ratios for J/ψ production, because ratios for ψ' production are practically the same. All curves are calculated with the GBW parameterization of the dipole cross section $\sigma_{q\bar{q}}$.

where

$$2\,\mathrm{Im}\,V(\boldsymbol{\rho}, z) = -\sigma_{GG}(\boldsymbol{\rho})\,\rho_A(b, z)\ , \tag{53}$$

and the glue-glue dipole cross section is related to the $q\bar{q}$ one by the relation,

$$\sigma_{GG}(r, x) = \frac{9}{4}\,\sigma_{q\bar{q}}(r, x)\ . \tag{54}$$

Following [29] we assume that the real part of the potential has a form

$$\mathrm{Re}\,V(\boldsymbol{\rho}, z) = \frac{b_0^4\,\rho^2}{2\,\nu\,\alpha_G(1 - \alpha_G)}\ . \tag{55}$$

The parameter $b_0 = 0.65\,\mathrm{GeV}$ was fixed by the data on diffractive gluon radiation (the triple-Pomeron contribution in terms of Regge approach) which is an essential part of Gribov's inelastic shadowing [57]. The well known smallness of such a diffractive cross section explains why b_0 is so large, leading to a rather weak gluon shadowing. In other words, this strong interaction squeezes the glue-glue wave packet resulting in small nuclear attenuation due to color transparency.

Figure 9 shows the ratios of cross sections calculated with and without gluon shadowing for incoherent and coherent exclusive charmonium electroproduction. We see that the onset of gluon shadowing happens at a *c.m.* energy of few tens

GeV. This onset is controlled by the longitudinal nuclear formfactor

$$F_A(q_c^G, b) = \frac{1}{T_A(b)} \int\limits_{-\infty}^{\infty} dz\, \rho_A(b,z)\, e^{iq_c z} \tag{56}$$

where the longitudinal momentum transfer $q_c^G = 1/l_c^G$. For the onset of gluon shadowing $q_c^G R_A \gg 1$ one can keep only the double scattering shadowing correction,

$$S_G \approx 1 - \frac{1}{4}\sigma_{eff} \int d^2b\, T_A^2(b)\, F_A^2(q_c^G, b)\ , \tag{57}$$

where σ_{eff} is the effective cross section which depends on the dynamics of interaction of the $q\bar{q}G$ fluctuation with a nucleon.

It was found in [58] that the coherence length for gluon shadowing is rather short,

$$l_c^G \approx \frac{1}{10\, x\, m_N}\ , \tag{58}$$

where x in our case should be an effective one, $x = (Q^2 + M_\Psi^2)/2m_N\nu$. The onset of shadowing according to (56) and (57) should be expected at $q_c^2 \sim 3/(R_A^{ch})^2$ corresponding to

$$s_G \sim 10 m_N R_A^{ch} (Q^2 + M_\Psi^2)/\sqrt{3}\ , \tag{59}$$

where $(R_A^{ch})^2$ is the mean square of the nuclear charge radius. This estimate is in a good agreement with Fig. 9. Remarkably, the onset of shadowing is delayed with rising nuclear radius and Q^2. This follows directly from (57) and the fact that the formfactor is a steeper falling function of R_A for heavy than for light nuclei, provided that $q_c^G R_A \gg 1$.

6 Hadroproduction of Heavy Quarks

We now turn to open heavy flavor production in pp collisions [4,5]. The color dipole formulation of this process was first introduced in [59]. In the target rest frame, in which the dipole approach is formulated, heavy quark production looks like pair creation in the target color field, Figure 10. For a short time, a gluon G from the projectile hadron can develop a fluctuation which contains a heavy quark pair $(Q\bar{Q})$. Interaction with the color field of the target then may release these heavy quarks. Apparently, the mechanism depicted in Fig. 10 corresponds to the gluon-gluon fusion mechanism of heavy quark production in the leading order (LO) parton model. This can be verified by explicit an calculation [5]. The dipole formulation is therefore applicable only at low x_2, where the gluon density of the target is much larger than all quark densities[1].

[1] We use standard kinematical variables, $x_2 = 2P_{Q\bar{Q}} \cdot P_1/s$ and $x_1 = 2P_{Q\bar{Q}} \cdot P_2/s$, where P_1 (P_2) is the four-momentum of the projectile (target) hadron, and $P_{Q\bar{Q}}$ is the four-momentum of the heavy quark pair. In addition, $M_{Q\bar{Q}}$ is the invariant mass of the pair, and s is the hadronic center of mass energy squared.

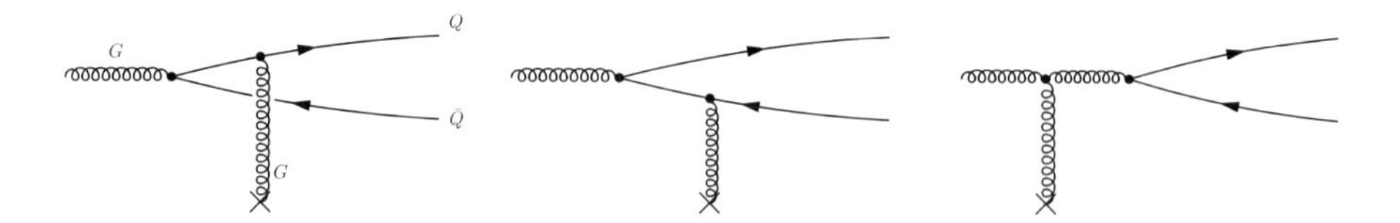

Fig. 10. The three lowest order graphs contributing to heavy quark production in the dipole approach. These graphs correspond to the gluon-gluon fusion mechanism of heavy quark production in the parton model.

The kinematical range where the dipole approach is valid can of course only be determined a posteriori. This is similar to determining the minimal value of Q^2 for which perturbative QCD still works. Note that while the mechanism for hadroproduction of heavy quark boundstates is still subject to active theoretical and experimental investigation, the mechanism for open heavy quark production is well established by now [60–62].

We shall now express the process depicted in Fig. 10 in terms of the cross section $\sigma_{q\bar{q}}(r)$ for scattering a color neutral $q\bar{q}$-pair with transverse size r off a nucleon. The $Q\bar{Q}$ pair can be produced in three different color and spin states. These states are orthogonal and do not interfere in the cross section. They include:

1. The color-singlet C-even $Q\bar{Q}$ state. The corresponding amplitude is odd (O) relative to simultaneous permutation of spatial and spin variables of the $Q\bar{Q}$ and has the form,

$$A^{\bar{\mu}\mu}_{ij,a}(\boldsymbol{\kappa},\boldsymbol{k}_T,\alpha)=\sum_{e=1}^{8}\frac{1}{6}\,\delta_{ae}\,\delta_{ij}\;O^{\bar{\mu}\mu}_{e}(\boldsymbol{\kappa},\boldsymbol{k}_T,\alpha)\;. \tag{60}$$

   Here $\boldsymbol{\kappa}$ and $\boldsymbol{k}_T$ are the relative and total transverse momenta of the $Q\bar{Q}$ pair respectively; $\mu,\bar{\mu}$ are spin indexes, a and i,j are color indexes of the gluon and produced quarks, respectively. We will classify such a state as 1^-, which means a color singlet with odd parity relative to index permutation. Note the 1^+ cannot be produced in the reaction shown in Fig. 10 .

2. Color-octet $Q\bar{Q}$ state with the production amplitude also antisymmetric relative simultaneous permutation of spatial and spin variables of the $Q\bar{Q}$ (8^-),

$$B^{\bar{\mu}\mu}_{ij,a}(\boldsymbol{\kappa},\boldsymbol{k}_T,\alpha)=\sum_{e,g=1}^{8}\frac{1}{2}\,d_{aeg}\,\tau_g(ij)\;O^{\bar{\mu}\mu}_{e}(\boldsymbol{\kappa},\boldsymbol{k}_T,\alpha)\;. \tag{61}$$

   Here $\lambda_g=\tau_g/2$ are the Gell-Mann matrices.

3. Color-octet $Q\bar{Q}$ with the amplitude symmetric relative permutation of quark variables (8^+),

$$C^{\bar{\mu}\mu}_{ij,a}(\boldsymbol{\kappa},\boldsymbol{k}_T,\alpha)=\sum_{e,g=1}^{8}\frac{i}{2}\,f_{aeg}\,\tau_g(ij)\;E^{\bar{\mu}\mu}_{e}(\boldsymbol{\kappa},\boldsymbol{k}_T,\alpha)\;. \tag{62}$$

The two amplitudes in (60) and (61) contain the common factor

$$O_e^{\bar{\mu}\mu}(\boldsymbol{\kappa},\boldsymbol{k}_T,\alpha) = \int d^2r\, d^2s\, \mathrm{e}^{\mathrm{i}\boldsymbol{\kappa}\cdot\boldsymbol{r}-\mathrm{i}\boldsymbol{k}_T\cdot\boldsymbol{s}}\, \Psi_{Q\bar{Q}}^{\bar{\mu}\mu}(\boldsymbol{r}) \left[\gamma^{(e)}(\boldsymbol{s}-\alpha\boldsymbol{r})-\gamma^{(e)}(\boldsymbol{s}+\bar{\alpha}\boldsymbol{r})\right], \quad (63)$$

which is odd (O) under permutation of the non-color variable of the quarks. Correspondingly, the even (E) factor in the amplitude (62) reads,

$$E_e^{\bar{\mu}\mu}(\boldsymbol{\kappa},\boldsymbol{k}_T,\alpha) =$$
$$\int d^2r\, d^2s\, \mathrm{e}^{\mathrm{i}\boldsymbol{\kappa}\cdot\boldsymbol{r}+\mathrm{i}\boldsymbol{k}_T\cdot\boldsymbol{s}}\, \Psi_{Q\bar{Q}}^{\bar{\mu}\mu}(\boldsymbol{r}) \left[\gamma^{(e)}(\boldsymbol{s}-\alpha\boldsymbol{r})+\gamma^{(e)}(\boldsymbol{s}+\bar{\alpha}\boldsymbol{r})-2\,\gamma^{(e)}(\boldsymbol{s})\right] \quad (64)$$

Here $\boldsymbol{s}$ and $\boldsymbol{r}$ is the position of the center of gravity and the relative transverse separation of the $Q\bar{Q}$ pair, respectively. It becomes evident from (63) and (64), that the *production* amplitude for the $Q\bar{Q}$-pair depends on the difference between the *interaction* amplitudes represented by the three graphs in Fig. 10. For example, the two terms in the square bracket in (63) correspond to the two first graphs in Fig. 10. If the Q and the $\bar{Q}$ would scatter at the same impact parameter, the production amplitude would vanish and nothing is produced. We stress that the interaction amplitudes represented by each of the three graphs in Fig. 10 is infrared divergent. This divergence, however, cancels in the production amplitude of the heavy quark pair, and therefore one can express the cross section for heavy flavor production in terms of color neutral quantities, such as the dipole cross section $\sigma_{q\bar{q}}$.

The LC wave function $\Psi_{Q\bar{Q}}^{\bar{\mu}\mu}(\boldsymbol{r})$ of the $Q\bar{Q}$ component of the incident gluon in (63)-(64) reads,

$$\Psi_{Q\bar{Q}}^{\bar{\mu}\mu}(\boldsymbol{r}) = \frac{\sqrt{2\,\alpha_s}}{4\pi}\,\xi^{\mu}\,\hat{\Gamma}\,\tilde{\xi}^{\bar{\mu}}\,\mathrm{K}_0(m_Q r)\;, \quad (65)$$

where the vertex operator has the form,

$$\hat{\Gamma} = m_Q\,\boldsymbol{\sigma}\cdot\boldsymbol{e} + \mathrm{i}(1-2\alpha)\,(\boldsymbol{\sigma}\cdot\boldsymbol{n})(\boldsymbol{e}\cdot\boldsymbol{\nabla}) + (\boldsymbol{n}\times\boldsymbol{e})\cdot\boldsymbol{\nabla}\;, \quad (66)$$

where $\boldsymbol{\nabla} = d/d\boldsymbol{r}$; α is the fraction of the gluon light-cone momentum carried by the quark Q and $\bar{\alpha}$ is the analogous quantity for the antiquark $\bar{Q}$; $\boldsymbol{e}$ is the polarization vector of the gluon and m_Q is the heavy quark mass.

The profile function $\gamma^{(e)}(\boldsymbol{s})$ in (63) –(64) is related by Fourier transformation to the amplitude $F^{(e)}(\boldsymbol{k}_T,\{X\})$, of absorption of a real gluon by a nucleon, $GN \to X$, which also can be treated as an "elastic" (color-exchange) gluon-nucleon scattering with momentum transfer $\boldsymbol{k}_T$,

$$\gamma^{(e)}(\boldsymbol{s}) = \frac{\sqrt{\alpha_s}}{2\pi\sqrt{6}} \int \frac{d^2k_T}{k_T^2+\lambda^2}\, \mathrm{e}^{-\mathrm{i}\boldsymbol{k}_T\cdot\boldsymbol{s}}\, F_{GN\to X}^{(e)}(\boldsymbol{k}_T,\{X\})\;, \quad (67)$$

where the upper index (e) shows the color polarization of the gluon, and the variables $\{X\}$ characterize the final state X including the color of the scattered gluon.

It is important for further consideration to relate the profile function (67) to the unintegrated gluon density $\mathcal{F}(k_T, x)$ and to the dipole cross section $\sigma_{q\bar{q}}(r, x)$ (*cf.* Sect. 2),

$$\int d^2b\, d\{X\} \sum_{e=1}^{8} \left|\gamma^{(e)}(\boldsymbol{s}+\boldsymbol{r}) - \gamma^{(e)}(\boldsymbol{s})\right|^2$$
$$= \frac{4\pi}{3}\,\alpha_s \int \frac{d^2k_T}{k_T^2}\left(1 - e^{i\boldsymbol{k}_T\cdot\boldsymbol{r}}\right) \mathcal{F}(k_T, x_2) = \sigma_{q\bar{q}}(r, x_2)\,. \tag{68}$$

Let us consider the production cross sections of a $Q\bar{Q}$ pair in each of three states listed above, (60)–(62). The cross section of a color-singlet $Q\bar{Q}$ pair, averaged over polarization and colors of the incident gluon reads,

$$\sigma^{(1)} = \frac{1}{(2\pi)^4} \sum_{\mu,\bar{\mu},i,j} \int_0^1 d\alpha \int d^2\kappa\, d^2k_T \overline{\left|A^{\bar{\mu}\mu}_{ij,a}(\boldsymbol{\kappa}, \boldsymbol{k}_T, \alpha)\right|^2} \tag{69}$$

Using (63), (65) and (68) this relation can be modified as,

$$\sigma^{(1)} = \sum_{\mu,\bar{\mu}} \int_0^1 d\alpha \int d^2r\, \sigma_1(r,\alpha) \left|\Psi^{\mu\bar{\mu}}(\boldsymbol{r},\alpha)\right|^2 , \tag{70}$$

where

$$\sigma_1(r,\alpha) = \frac{1}{8}\,\sigma_{q\bar{q}}(r, x_2)\,; \tag{71}$$

$$\sum_{\mu,\bar{\mu}} \left|\Psi^{\mu\bar{\mu}}(\boldsymbol{r},\alpha)\right|^2 = \frac{\alpha_s}{(2\pi)^2}\left[m_Q^2\, \mathrm{K}_0^2(m_Q r) + (\alpha^2 + \bar{\alpha}^2)\, m_Q^2\, \mathrm{K}_1^2(m_Q r)\right]. \tag{72}$$

One finds in a similar way that the cross sections of a color-octet $Q\bar{Q}$ pair production either in 8^- (Odd) or 8^+ (Even) states has the form,

$$\sigma^{(8)}_{O(E)} = \sum_{\mu,\bar{\mu}} \int_0^1 d\alpha \int d^2r\, \sigma^{(8)}_{O(E)}(r,\alpha) \left|\Psi^{\mu\bar{\mu}}(\boldsymbol{r},\alpha)\right|^2 , \tag{73}$$

where

$$\sigma^{(8)}_O(r,\alpha,x_2) = \frac{5}{16}\,\sigma_{q\bar{q}}(r,x_2)\,; \tag{74}$$

$$\sigma^{(8)}_E(r,\alpha,x_2) = \frac{9}{16}\left[2\sigma_{q\bar{q}}(\alpha r, x_2) + 2\sigma_{q\bar{q}}(\bar{\alpha} r, x_2) - \sigma_{q\bar{q}}(r, x_2)\right]. \tag{75}$$

After summation over all three color states in which the $Q\bar{Q}$ pair in Fig. 10 can be produced, one obtains for the partonic cross section [4],

$$\sigma(GN \to \{Q\bar{Q}\}X) = \int_0^1 d\alpha \int d^2r \left|\Psi_{G\to Q\bar{Q}}(\alpha, r)\right|^2 \sigma_{q\bar{q}G}(\alpha, r), \tag{76}$$

where $\sigma_{q\bar{q}G}$ is the cross section for scattering a color neutral quark-antiquark-gluon system on a nucleon [4],

$$\sigma_{q\bar{q}G}(\alpha, r) = \frac{9}{8}\left[\sigma_{q\bar{q}}(\alpha r) + \sigma_{q\bar{q}}(\bar{\alpha} r)\right] - \frac{1}{8}\sigma_{q\bar{q}}(r). \tag{77}$$

In order to simplify the notation, we do not explicitly write out the x_2 dependence of the dipole cross section.

The light-cone (LC) wavefunctions for the transition $G \to Q\bar{Q}$ can be calculated perturbatively and a very similar to the ones in leptoproduction, (34),

$$\begin{aligned}\Psi_{G\to Q\bar{Q}}(\alpha, \boldsymbol{r}_1)\Psi^*_{G\to Q\bar{Q}}(\alpha, \boldsymbol{r}_2) = \frac{\alpha_s(\mu_R)}{(2\pi)^2}\Bigg\{ & m_Q^2 \mathrm{K}_0(m_Q r_1)\mathrm{K}_0(m_Q r_2) \\ & + \left[\alpha^2 + \bar{\alpha}^2\right] m_Q^2 \frac{\boldsymbol{r}_1 \cdot \boldsymbol{r}_2}{r_1 r_2} \mathrm{K}_1(m_Q r_1)\mathrm{K}_1(m_Q r_2)\Bigg\},\end{aligned} \tag{78}$$

where $\alpha_s(\mu_R)$ is the strong coupling constant, which is probed at a renormalization scale $\mu_R \sim m_Q$.

Equation (76) is a special case of the general rule that at high energy, the cross section for the reaction $a + N \to \{b, c, \ldots\}X$ can be expressed as convolution of the LC wavefunction for the transition $a \to \{b, c, \ldots\}$ and the cross section for scattering the color neutral $\{\text{anti}-a, b, c\ldots\}$-system on the target nucleon N.

Note that although the dipole cross section is flavor independent, the integral (76) is not. Since the Bessel functions $\mathrm{K}_{1,0}$ decay exponentially for large arguments, the largest values of r which can contribute to the integral are of order $\sim 1/m_Q$. We point out, that as a consequence of color transparency [13,22], the dipole cross section vanishes $\propto r^2$ for small r. Therefore, the $Q\bar{Q}$ production cross section behaves roughly like $\propto 1/m_Q^2$ (modulo logs and saturation effects).

We can estimate the relative yield of the 1^-, 8^- and 8^+ states we can rely upon the approximation $\sigma_{q\bar{q}}(r) \propto r^2$ which is rather accurate in the case of a $Q\bar{Q}$ pair, since its separation $r \sim 1/m_Q$ is small. We then derive,

$$\sigma^{(1)} \ : \ \sigma_O^{(8)} \ : \ \sigma_E^{(8)} = 1 \ : \ \frac{5}{2} \ : \ \frac{117}{70} \ . \tag{79}$$

Thus, about 20% of the produced $Q\bar{Q}$ pairs are in a color-singlet state, the rest are color-octets.

In order to calculate the cross section for heavy quark pair production in pp collisions, (76) has to be weighted with the projectile gluon density,

$$\frac{d\sigma(pp \to \{QQ\}X)}{dy} = x_1 G\left(x_1, \mu_F\right) \sigma(GN \to \{Q\bar{Q}\}X), \tag{80}$$

where $y = \frac{1}{2}\ln(x_1/x_2)$ is the rapidity of the pair and $\mu_F \sim m_Q$. In analogy to the parton model, we call μ_F the factorization scale. Uncertainties arising from the choice of this scale will be investigated in section 6.1. Integrating over all kinematically allowed rapidities yields

$$\sigma_{\mathrm{tot}}(pp \to \{Q\bar{Q}\}X) = 2\int_0^{-\ln\left(\frac{2m_Q}{\sqrt{s}}\right)} dy\, x_1 G\left(x_1, \mu_F\right) \sigma(GN \to \{Q\bar{Q}\}X). \tag{81}$$

A word of caution is in order, regarding the limits of the α-integration in (76). Since the invariant mass of the $Q\bar{Q}$-pair is given by

$$M_{Q\bar{Q}}^2 = \frac{\kappa_\perp^2 + m_Q^2}{\alpha\bar{\alpha}}, \tag{82}$$

the endpoints of the α-integration include configurations corresponding to arbitrarily large invariant masses, eventually exceeding the total available *cm.* energy. However, since r and $\kappa_\perp$ (the single quark transverse momentum) are conjugate variables, the pair mass is not defined in the mixed representation, nor are the integration limits for α. Fortunately, this problem is present only at the very edge of the phase space and therefore numerically negligible.

6.1 Numerical Results for Hadroproduction of Heavy Quarks

Still the questions remain, how well does the dipole approach describe experimental data. Since there are not many data for the total cross section, we shall also compare predictions from the dipole approach to calculations in the NLO parton model [60–62][2] For $\sigma_{q\bar{q}}$, we use an improved version of the saturation model presented in [28], which now also includes DGLAP evolution [63]. This improvement has no effect on open charm, but is important for bottom production.

In the dipole approach, we use the one loop running coupling constant,

$$\alpha_s(\mu_R) = \frac{4\pi}{\left(11 - \frac{2}{3}N_f\right)\ln\left(\frac{\mu_R^2}{(200\,\mathrm{MeV})^2}\right)} \tag{83}$$

at a renormalization scale $\mu_R \sim m_Q$, and the number of light flavors is chosen to be $N_f = 3$ for open charm and $N_f = 4$ for open bottom production. Furthermore, we use the GRV98LO [64] gluon distribution to model the gluon density in the projectile. We use a leading order parton distribution function (PDF), because of its probabilistic interpretation. Note that one could attempt to calculate the projectile gluon distribution from the dipole cross section. However, the projectile distribution functions are needed mostly at large momentum fraction x_1, where the dipole cross section is not constrained by data.

Our results for the total charm pair cross section in proton-proton (pp) collisions is shown in Fig. 11 as function of center of mass energy. The left panel shows the uncertainties of both approaches by varying quark mass m_c and renormalization scale μ_R in the intervals $1.2\,\mathrm{GeV} \leq m_c \leq 1.8\,\mathrm{GeV}$ and $m_c \leq \mu_R \leq 2m_c$, respectively. The factorization scale is kept fixed at $\mu_F = 2m_c$, because in our opinion, the charm quark mass is too low for DGLAP evolution. A large fraction of the resulting uncertainty originates from different possible choices of the charm quark mass, since the total cross section behaves approximately like $\sigma_{\mathrm{tot}} \propto m_Q^{-2}$.

[2] A FORTRAN program for the NLO parton model calculation is available at http://n.home.cern.ch/n/nason/www/hvqlib.html.

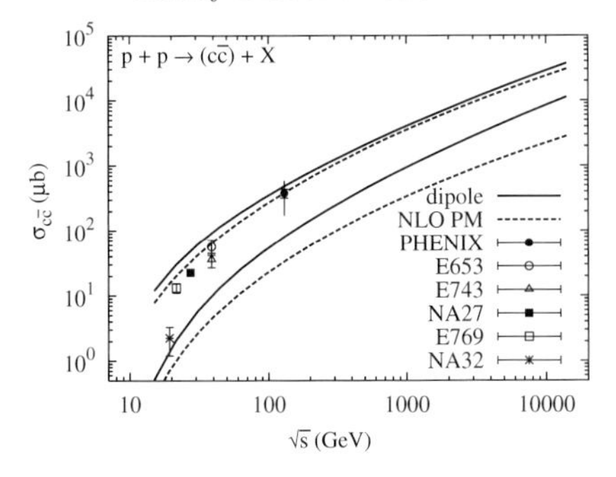

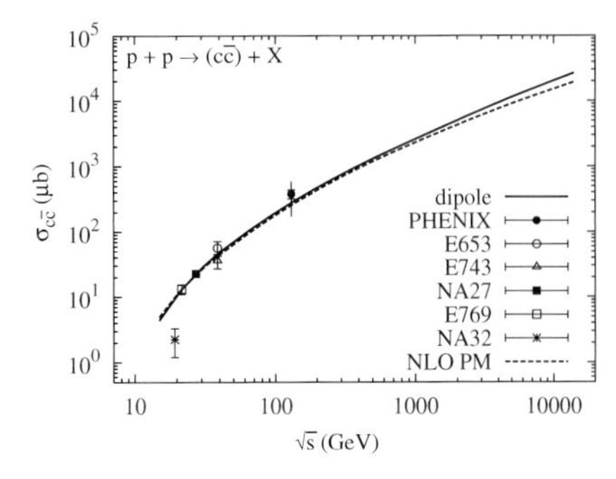

Fig. 11. Results for the total open charm pair cross section as function of cm. energy. Varying free parameters in dipole approach (solid lines) and in the parton model (dashed lines) gives rise to the uncertainties shown on the left. In the figure on the right, parameters in both models have been adjusted so that experimental data [65,66] are described.

Note that the mean value of x_2 increases with decreasing energy. At $\sqrt{s} = 130\,\mathrm{GeV}$ one has $x_2 \sim 0.01$. For lower energies, our calculation is an extrapolation of the saturation model. For the highest fixed target energies of $\sqrt{s} \approx 40\,\mathrm{GeV}$, values of $x_2 \sim 0.1$ become important. Unlike in the Drell-Yan case, which was studied in [6], the dipole approach to heavy quark production does not show any unphysical behavior when extrapolated to larger x_2. One reason for this is that the new saturation model [63] assumes a realistic behavior of the gluon density at large x_2. In addition, even at energies as low as $\sqrt{s} = 15\,\mathrm{GeV}$, the gluon-gluon fusion process is the dominant contribution to the cross section.

Because of the wide uncertainty bands, one can adjust m_c and μ_R in both approaches so that experimental data are reproduced. Then, dipole approach and NLO parton model yield almost identical results. However, the predictive power of the theory is rather small. In Fig. 11 (right), we used $m_c = 1.2\,\mathrm{GeV}$ and $\mu_R = 1.5m_c$ for the NLO parton model calculation and $m_c = 1.4\,\mathrm{GeV}$, $\mu_R = m_c$ in the dipole approach. The data points tend to lie at the upper edge of the uncertainty bands, so that rather small values of m_c are needed to describe them.

There are remaining uncertainties which are not shown in Fig. 11 (right), because different combinations of m_c and μ_R can also yield a good description of the data. In addition, different PDFs will lead to different values of the cross section at high energies, since the heavy quark cross section is very sensitive to the low-x gluon distribution. In [67], it was found that an uncertainty of a factor of ~ 2.3 remains at $\sqrt{s} = 14$ TeV (in the NLO parton model), even after all free parameters had been fixed to describe total cross section data at lower energies. It is interesting to see that 20–30% of the total pp cross section at LHC ($\sqrt{s} = 14$ TeV) goes into open charm [3].

[3] The Donnachie-Landshoff parameterization of the total pp cross section [68] predicts $\sigma_{\mathrm{tot}}^{pp}(\sqrt{s} = 14\,\mathrm{TeV}) = 100$ mb.

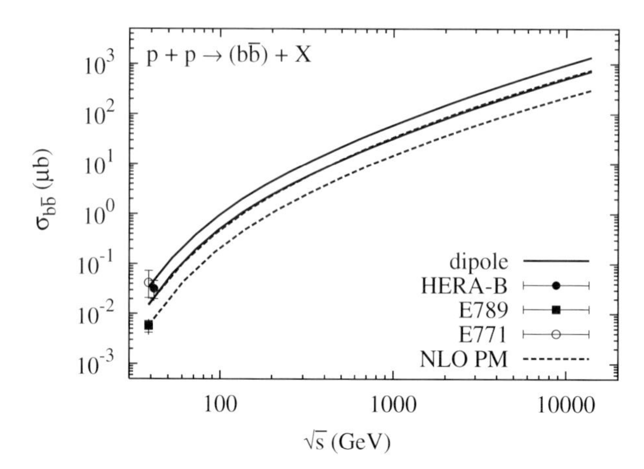

Fig. 12. Uncertainties of open $b\bar{b}$ pair production calculated in the dipole approach (solid) and in the NLO parton model (dashed). The dipole approach seems to provide a better description of the data, even though HERA-B energy is too low for the dipole approach.

Next, we calculate the total $b\bar{b}$-pair cross section as function of center of mass energy, see Fig. 12. In order to quantify the theoretical uncertainties, we vary the free parameters over the ranges $4.5\,\text{GeV} \leq m_b \leq 5\,\text{GeV}$ and $m_b \leq \mu_R, \mu_F \leq 2m_b$ Because of the large b-quark mass, uncertainties are much smaller than for open charm production. One can see that the dipole approach tends to predict higher values than the NLO parton model, even though the energy dependence expected in both approaches is very similar. In fact, the results calculated in the dipole approach with $m_b = 5\,\text{GeV}$ agree almost exactly with the NLO parton model calculation with $m_b = 4.5\,\text{GeV}$. For all other values of m_b, the uncertainty bands of the two approaches do not overlap, in contrast to the case for open charm production.

Three measurements of open $b\bar{b}$ production are published in the literature [69–71]. The two values of the open bottom cross section measured at Fermilab [69,70] at *cm.* energy $\sqrt{s} = 38.8\,\text{GeV}$ differ by almost three standard deviations. The HERA-B measurement at slightly larger *cm.* energy $\sqrt{s} = 41.6\,\text{GeV}$ [71] is consistent with the E771 [70] value. These two points seem to be better described by the dipole approach, though the NLO parton model (with $m_b = 4.5\,\text{GeV}$) still touches the HERA-B error bar. Note that also a different set of PDFs would not significantly pull up the parton model curve [67], as a lower value of the b-quark mass would do. With a resummation of terms from higher order corrections [72], however, the parton model can reproduce each of the three measurements within theoretical uncertainties, see [71]. On the other hand, typical values of x_2 which are important for $b\bar{b}$ production at HERA-B energy are of order $x_2 \sim 0.2$, while the parameterization [63] of the dipole cross section is constrained only by DIS data with $x_{Bj} \leq 0.01$.

While it is an advantage of the dipole formulation to provide very simple formulas that allow one to absorb much of the higher order corrections into a phenomenological parameterization of $\sigma_{q\bar{q}}(x_2, r)$, one cannot clarify the origin of the discrepancy in normalizations without a systematic calculation of higher orders in this approach.

7 Nuclear Effects in Hadroproduction of Open Charm

It is still unclear whether available data from fixed target experiments demonstrate any nuclear effects for open charm production [73–75]. Naively one might expect no effects at all, since a heavy quark should escape the nucleus without attenuation or reduction of its momentum. In fact, this is not correct even at low energies as is explained below. Moreover, at high energies one cannot specify any more initial or final state interactions. The process of heavy flavor production takes a time interval longer than the nuclear size, and the heavy quarks are produced coherently by many nucleons which compete with each other. As a result the cross section is reduced, and this phenomenon is called shadowing.

In terms of parton model the same effect is interpreted in the infinite momentum frame of the nucleus as reduction of the nuclear parton density due to overlap and fusion of partons at small Bjorken x. The kinematic condition for overlap is the same as for coherence in the nuclear rest frame. Thus, heavy quark via gluon fusion can be shadowed in the leading twist, if the gluon density in nuclei is reduced due to gluon shadowing.

There are well known examples of shadowing observed in hard reactions, like deep-inelastic scattering (DIS) [76] and the Drell-Yan process (DY) [77] demonstrating a sizable reduction of the density of light sea quarks in nuclei. Shadowing is expected also for gluons, although there is still no experimental evidence for that.

Shadowing for heavy quarks is a higher twist effect, and although its magnitude is unknown within the standard parton model approach, usually it is neglected for charm and beauty production. However, this correction is proportional to the gluon density in the proton and steeply rises with energy. Unavoidably, such a correction should become large at high energies. In some instances, like for charmonium production, this higher twist effect gains a large numerical factor and leads to a rather strong suppression even at energies of fixed target experiments (see below).

On the other hand, gluon shadowing which is a leading twist effect, is expected to be the main source of nuclear suppression for heavy flavor production at high energies. This is why this process is usually considered as a sensitive probe for the gluon density in hadrons and nuclei. If one neglects terms suppressed by a power of $1/m_Q^2$, the cross section of heavy $Q\bar{Q}$ production in pA collision is suppressed by the gluon shadowing factor R_A^G compared to the sum of A nucleon cross sections,

$$\sigma_{pA}^{Q\bar{Q}}(x_1, x_2) = R_A^G(x_1, x_2)\, A\, \sigma_{pN}^{Q\bar{Q}}(x_1, x_2) \,. \tag{84}$$

Here

$$R_A^G(x_1,x_2) = \frac{1}{A}\int d^2b\, R_A^G(x_1,x_2,b)\, T_A(b)\ , \tag{85}$$

where $R_A^G(x_1,x_2,b)$ is the (dimensional) gluon shadowing factor at impact parameter b; $T_A(b) = \int_{-\infty}^{\infty} dz\, \rho_A(b,z)$ is the nuclear thickness function, and x_1, x_2 are the Bjorken variables of the gluons participating in $Q\bar{Q}$ production from the colliding proton and nucleus.

The parton model cannot predict shadowing, but only its evolution at high Q^2, while the main contribution originates from the soft part of the interaction. The usual approach is to fit data at different values of x and Q^2 employing the DGLAP evolution and fitting the distributions of different parton species parametrized at some intermediate scale [78,79]. However, the present accuracy of data for DIS on nuclei do not allow to fix the magnitude of gluon shadowing, which is found to be compatible with zero [4]. Nevertheless, the data exclude some models with too strong gluon shadowing [80].

Another problem faced by the parton model is the impossibility to predict gluon shadowing effect in nucleus-nucleus collisions even if the shadowing factor (84) in each of the two nuclei was known. Indeed, the cross section of $Q\bar{Q}$ production in collision of nuclei A and B at impact parameter $\boldsymbol{b}$ reads,

$$\frac{d\sigma_{AB}^{Q\bar{Q}}(x_1,x_2)}{d^2b} = R_{AB}^G(x_1,x_2,b)\, AB\, \sigma_{NN}^{Q\bar{Q}}(x_1,x_2)\ , \tag{86}$$

where

$$R_{AB}^G(x_1,x_2,b) = \frac{1}{AB}\int d^2s\, R_A^G(x_1,\boldsymbol{s})\, T_A(s)\ R_B^G(x_2,\boldsymbol{b}-\boldsymbol{s})\, T_B(\boldsymbol{b}-\boldsymbol{s})\ . \tag{87}$$

In order to calculate the nuclear suppression factor (87) one needs to know the impact parameter dependence of gluon shadowing, $R_A^G(x_1,\boldsymbol{b})$, while only integrated nuclear shadowing (85) can be extracted from lepton- or hadron-nucleus data[5]. Note that the parton model prediction of shadowing effects for minimum bias events integrated over b suffers the same problem. Apparently, QCD factorization cannot be applied to heavy ion collisions even at large scales. The same is true for quark shadowing expected for Drell-Yan process in heavy ion collisions [15,54,81].

Nuclear shadowing can be predicted within the light-cone (LC) dipole approach which describes it via simple eikonalization of the dipole cross section. It was pointed out in [13] that quark configurations (dipoles) with fixed transverse separations are the eigenstates of interaction in QCD, therefore eikonalization is an exact procedure. In this way one effectively sums up the Gribov's inelastic corrections to all orders [13].

---

[4] Gluon shadowing was guessed in [78] to be the same as for $F_2(x,Q^2)$ at the semi-hard scale.

[5] One can get information on the impact parameter of particle-nucleus collision measuring multiplicity of produced particles or low energy protons (so called grey tracks). However, this is still a challenge for experiment.

The advantage of this formalism is that it does not need any K-factor. Indeed, it was demonstrated recently in [6] that the simple dipole formalism for Drell-Yan process [15,82,83] precisely reproduces the results of very complicated next-to-leading calculations at small x. The LC dipole approach also allows to keep under control deviations from QCD factorization. In particular, we found a substantial process-dependence of gluon shadowing due to the existence of a semi-hard scale imposed by the strong nonperturbative interaction of light-cone gluons [29]. For instance gluon shadowing for charmonium production off nuclei was found in [2] to be much stronger than in deep-inelastic scattering [29].

The LC dipole approach also provides effective tools for calculation of transverse momentum distribution of heavy quarks, like it was done for radiated gluons in [83,84], or Drell-Yan pairs in [54]. Nuclear broadening of transverse momenta of the heavy quarks also is an effective way to access the nuclear modification of the transverse momentum distribution of gluons, *i.e.* the so called phenomenon of color glass condensate or gluon saturation [85,23]. We consider only integrated quantities here.

In what follows we find sizable deviations from QCD factorization for heavy quark production off nuclei. First of all, for open charm production shadowing related to propagation of a $c\bar{c}$ pair through a nucleus is not negligible, especially at the high energies of RHIC and LHC, in spite of smallness of $c\bar{c}$ dipoles. Further, higher Fock components containing gluons lead to gluon shadowing which also deviates from factorization and depends on quantum numbers of the produced heavy pair $c\bar{c}$.

7.1 Higher Twist Shadowing for $c\bar{c}$ Production

An important advantage of the LC dipole approach is the simplicity of calculations of nuclear effects. Since partonic dipoles are the eigenstates of interaction one can simply eikonalize the cross section on a nucleon target [13] provided that the dipole size is "frozen" by Lorentz time dilation. Therefore, the cross section of a $c\bar{c}$ pair production off a nucleus has the form [59,4],

$$\sigma(GA \to c\bar{c}X) = 2 \sum_{\mu,\bar{\mu}} \int d^2b \int d^2r \int_0^1 d\alpha \left|\Psi^{\mu\bar{\mu}}(\boldsymbol{r},\alpha)\right|^2 \times \left\{1 - \exp\left[-\frac{1}{2}\,\sigma_{q\bar{q}G}(r,\alpha,x_2)\,T_A(b)\right]\right\} , \tag{88}$$

where $\sigma_{q\bar{q}G}(r,\alpha,x_2)$ is the cross section of interaction of a $c\bar{c}G$ three particle state with a nucleon, see Sect. 6.

Apparently, this expression leads to shadowing correction which is a higher twist effect and vanishes as $1/m_c^2$. Indeed, it was found in [59] that in the kinematic range of fixed target experiments at the Tevatron, Fermilab, $x_2 \sim 10^{-2}$, $x_F \sim 0.5$, the shadowing effects are rather weak even for heavy nuclei,

$$1 - R_A \lesssim 0.05 \ , \tag{89}$$

where R_A is defined in (84).

On the other hand, a substantial shadowing effect, several times stronger than in (89) was found in [2] for charmonium production (see below), although it is also a higher twist effect. In the case of open charm production there are additional cancelations which grossly diminish shadowing. The smallness of the effect maybe considered as a justification for the parton model prescription to neglect this correction as a higher twist effect. However, the dipole cross section $\sigma_{q\bar{q}}(r, x_2)$ steeply rises with $1/x_2$ especially at small r and and the shadowing corrections increase, reaching values of about 10% at $x_2 = 10^{-3}$, and about 30% at $x_2 = 10^{-5}$.

7.2 Process Dependent Gluon Shadowing

The phenomenological dipole cross section which enters the exponent in (88) is fitted to DIS data. Therefore it includes effects of gluon radiation which are in fact the source of rising energy $(1/x)$ dependence of the $\sigma_{q\bar{q}}(r, x)$. However, a simple eikonalization in (88) corresponds to the Bethe-Heitler approximation assuming that the whole spectrum of gluons is radiated in each interaction independently of other rescatterings. This is why the higher order terms in expansion of (88) contain powers of the dipole cross section. However, gluons radiated due to interaction with different bound nucleons can interfere leading to damping of gluon radiation similar to the Landau-Pomeranchuk [87] effect in QED. Therefore, the eikonal expression (88) needs corrections which are known as gluon shadowing.

Nuclear shadowing of gluons is a leading twist effect since the cloud of massless gluons has a larger size than the source which is a small size *barcc* pair. Gluon shadowing is treated by the parton model in the infinite momentum frame of the nucleus as a result of glue-glue fusion. On the other hand, in the nuclear rest frame the same phenomenon is expressed in terms of the Glauber like shadowing for the process of gluon radiation [88]. In impact parameter representation one can easily sum up all the multiple scattering corrections which have the simple eikonal form [13]. Besides, one can employ the well developed color dipole phenomenology with parameters fixed by data from DIS. Gluon shadowing was calculated employing the light-cone dipole approach for DIS [29] and production of charmonia [2], and a substantial deviation from QCD factorization was found. Here we calculate gluon shadowing for $c\bar{c}$ pair production.

First of all, one should develop a dipole approach for gluon radiation accompanying production of a $c\bar{c}$ pair in gluon-nucleon collision. Then nuclear effects can be easily calculated via simple eikonalization. This is done in [4].

According to the general prescription [15] the dipole cross section which enters the factorized formula for the process of parton a-nucleon collision leading to multiparton production, $a\,N \to b + c + \ldots + d\ X$, is the cross section for the colorless multiparton ensemble $|\bar{a}bc \ldots d\rangle$. The same multiparton dipole cross section is responsible for nuclear shadowing. Indeed, in the case of the process

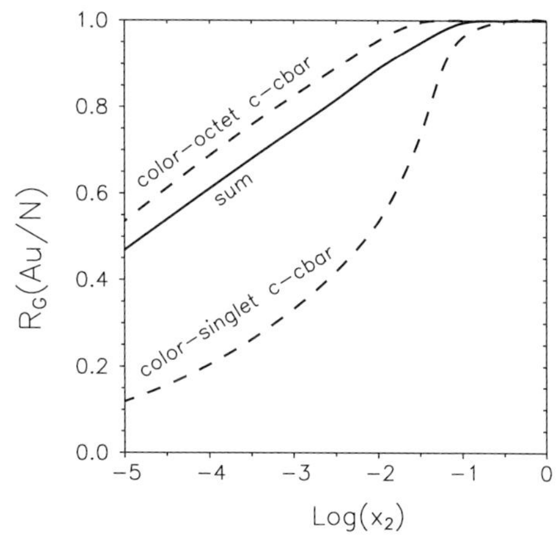

Fig. 13. Ratio of gluon densities $R_G(Au/p) = G_{Au}(x_2)/195\, G_p(x_2)$ for color octet-octet and singlet-octet states $(c\bar{c}) - G$ (dashed curves). The averaged gluon shadowing is depicted by the solid curve.

$GN \to c\bar{c}X$ it was the cross section $\sigma_{q\bar{q}G}$, (77), which correspond to a state $|c\bar{c}G\rangle$ interacting with a nucleon.

Correspondingly, in the case of additional gluon production, $G \to \bar{c}c\,G$, it is a 4-parton, $|c\bar{c}GG\rangle$, cross section $\sigma_4(\boldsymbol{r}, \boldsymbol{\rho}, \alpha_1, \alpha_2, \alpha_3)$. Here $\boldsymbol{r}$ and $\boldsymbol{\rho}$ are the transverse $c\bar{c}$ separation and the distance between the $c\bar{c}$ center of gravity and the final gluon, respectively. Correspondingly, $\alpha_1 = \alpha_c$, $\alpha_2 = \alpha_{\bar{c}}$, and $\alpha_3 = \alpha_G$. Treating the charm quark mass as a large scale, one can neglect $r \ll \rho$, then the complicated expression for σ_4 becomes rather simple. One can find details in [4].

One can treat partons as free only if their transverse momenta are sufficiently large, otherwise the nonperturbative interaction between partons may generate power corrections [29]. Apparently, the softer the process is, the more important are these corrections. In particular, diffraction and nuclear shadowing are very sensitive to these effects. Indeed, the cross section of diffractive dissociation to large masses (so called triple-Pomeron contribution) is proportional to the fourth power of the size of the partonic fluctuation. Therefore, the attractive nonperturbative interaction between the partons squeezes the fluctuation and can substantially reduce the diffractive cross section. Smallness of the transverse separation in the quark-gluon fluctuation is the only known explanation for the observed suppression of the diffractive cross section, which is also known as the problem of smallness of the triple-Pomeron coupling. While no data sensitive to gluon shadowing are available yet, a vast amount of high accuracy diffraction data can be used to fix the parameters of the nonperturbative interaction.

It turns out [4] that the color interaction between the gluon and the $c\bar{c}$ pair depends on color states of the latter. If the $c\bar{c}$ pair is in one of the two color octet states, the nonperturbative interaction between the pair and the gluon is strong, and the $|q\bar{q}G\rangle$ system cannot become larger than a typical constituent quark radius $\sim 0.3\,\mathrm{fm}$. For these small configurations, shadowing is rather small, see Fig. 13. If, on the other hand, the $c\bar{c}$ pair is in a color singlet state, the color charges of the c and the $\bar{c}$ screen each other, so that the pair cannot interact strongly with the radiated gluon, *i.e.* the value of b_0 (see Sect. 5.2) is much smaller than 0.65 GeV. The transverse size of these configurations is limited only by confinement, hence they can become as large as a typical hadron. Therefore, gluon shadowing is much stronger in the color singlet channel [2].

7.3 Numerical Results

To observe the shadowing effects in open charm production, one must access the kinematic region of sufficiently small $x_2 \lesssim 0.1$. With fixed targets it can be achieved at highest energies at Fermilab and in the experiment HERA-B at DESY. We apply the results of the previous section for gluon shadowing to $c\bar{c}$ pair production in proton-nucleus collisions. We assume that the $c\bar{c}$ is produced with Feynman x_F corresponding to $x_2 = (-x_F + \sqrt{x_F^2 + 4M_{c\bar{c}}^2/s})/2$, where we fix $M_{c\bar{c}} = 4\,\mathrm{GeV}$. The contribution of gluon shadowing to nuclear effects in proton-tungsten collision at $p_{lab} = 900\,\mathrm{GeV}$ is depicted by the dashed curve in Fig. 14.

The higher twist shadowing correction, which corresponds to the eikonalized dipole cross section $\sigma_{q\bar{q}G}$ in (88), is also a sizable effect and should be added. It

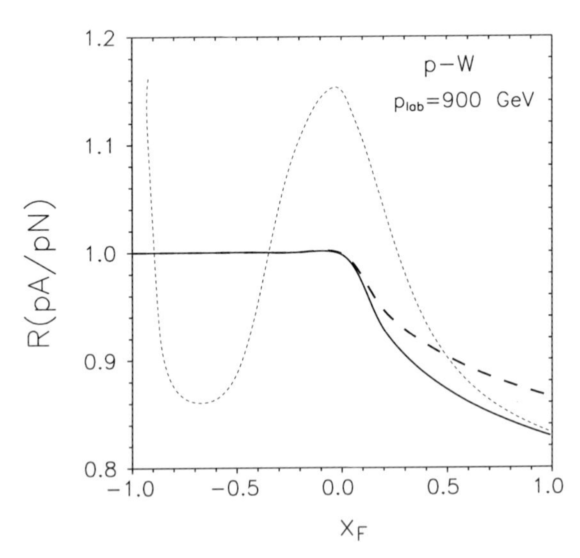

Fig. 14. Nuclear effects for open charm production in $p--W$ collisions at 900 GeV beam energy. The contribution of gluon shadowing is shown by dashed curve. The solid curve represents the full shadowing effect including the higher twist contribution given by (88). Possible medium effects including antishadowing and EMC-suppression for gluons [78] are also added and the result is represented by the dotted curve.

is diminished, however, due to the strong gluon shadowing which also reduces the amount of gluons available for multiple interactions compared to the eikonal approximation (88). We take this reduction into account multiplying $\sigma_{q\bar{q}G}$ in (88) by $R_G(x_2, M_{\bar{c}c})$. This procedure is justified at small transverse separations, since $\sigma(r,x) = (\pi^2/3)\,\alpha_s/\,r^2\,G(x,Q^2 \sim 1/r^2)$ [45]. For large separations see discussion in [2]. The summed shadowing suppression of $c\bar{c}$ production is depicted in Fig. 14 by the solid curve.

Besides shadowing, other nuclear effect are possible. The EMC effect, suppression of the nuclear structure function $F_2^A(x,Q^2)$ at large x, as well as the enhancement at $x \sim 0.1$ should also lead to similar modifications in the gluon distribution function $G^A(x,Q^2)$. These effects are different from shadowing which is a result of coherence. A plausible explanation relates them with medium effects, like swelling of bound nucleons [89]. To demonstrate a possible size of the medium effects on gluon distribution we parametrize and apply the effect of gluon enhancement and suppression at large x suggested in [78]. Although it is based on ad hoc gluon shadowing and underestimated shadowing for valence quarks (see discussion in [54]), it demonstrates the scale of possible effects missed in our analysis.

There are still other effects missed in our calculations. At this energy, the effect of energy loss due to initial state interactions [90] causes additional nuclear suppression at large x_F (compare with [2]). Another correction is related to the observation that detection of a charm hadron at large $|x_F|$ does not insure that it originates from a charm quark produced perturbatively with the same x_F. Lacking gluons with $x_{1,2} \to 1$ one can produce a fast charm hadron via a fast projectile (usually valence) quark which picks up a charm quark created at smaller $|x_F|$. This is actually the mechanism responsible for the observed $D/\bar{D}$ asymmetry. It provides a rapidity shift between the parent charm quark and the detected hadron. Therefore, it may reduce shadowing effects at largest $|x_F|$. We leave this problem open for further study.

To predict shadowing effects in heavy ion collisions we employ QCD factorization, which we apply only for a given impact parameter. For minimal bias events

$$R_{AB}(y) = R_A(x_1)\,R_B(x_2)\;, \tag{90}$$

where $y = \ln(x_1/x_2)/2$ is the rapidity of the $c\bar{c}$ pair. Our predictions for RHIC ($\sqrt{s} = 200\,\text{GeV}$) and LHC ($\sqrt{s} = 5500\,\text{GeV}$) are depicted in Fig. 15 separately for net gluon shadowing (dashed curves) and full effect including higher twist quark shadowing (solid curves). Although shadowing of charmed quarks is a higher twist effect, its contribution is about 10% at RHIC and rises with energy.

One might be surprised by the substantial magnitude of shadowing expected at the energy of RHIC. Indeed, the value of $x_{1,2} \approx 0.02$ at mid-rapidity is rather large, and no gluon shadowing would be expected for DIS [29]. However, the process of charm production demonstrates a precocious onset of gluon shadowing as was discussed above. Besides, the nuclear suppression is squared in AA collisions.

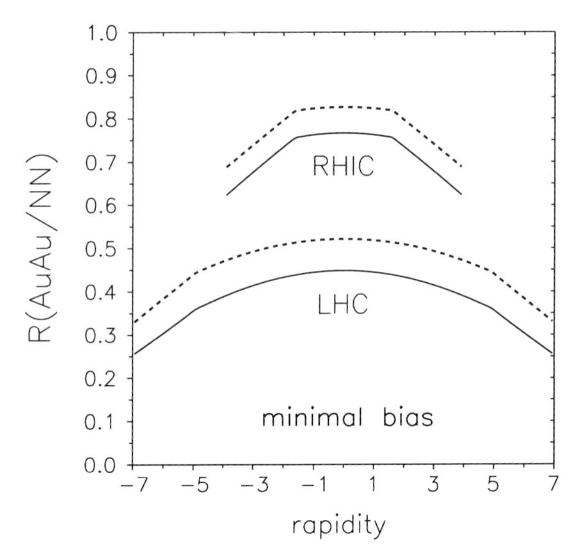

Fig. 15. Nuclear shadowing for open charm production in minimal bias gold-gold collision. Dotted curves show the net effect of gluon shadowing, while solid curves include both effects of gluon shadowing and the higher twist correction related to the non-zero separation of the $c\bar{c}$. The top (RHIC) and bottom (LHC) curves correspond to $\sqrt{s} = 200\,\text{GeV}$ and $5500\,\text{GeV}$ respectively.

Another interesting observation made in [4] is that shadowing is the same for central and minimal bias events. This has indeed been observed (within large error bars) by the PHENIX experiment [66].

8 The Light-Cone Dipole Formalism for Charmonium Production off a Nucleon

The important advantage of the light-cone (LC) dipole approach is its simplicity in the calculations of nuclear effects. It has been suggested two decades ago [13] that quark configurations (dipoles) with fixed transverse separations are the eigenstates of interaction in QCD. Therefore the amplitude of interaction with a nucleon is subject to eikonalization in the case of a nuclear target. In this way one effectively sums the Gribov's inelastic corrections in all orders.

Assuming that the produced $c\bar{c}$ pair is sufficiently small so that multigluon vertices can be neglected, we can write the cross section for $G\,N \to \chi\,X)$ as (see Fig. 16),

$$\sigma(GN \to \chi X) = \frac{\pi}{2(N_c^2-1)} \sum_{a,b} \int \frac{d^2k_T}{k_T^2}\, \alpha_s(k_T^2)\, \mathcal{F}(x,k_T^2) \left| M_{ab}(\boldsymbol{k}_T)\right|^2 , \quad (91)$$

where $\mathcal{F}(x,k_T^2) = \partial G(x,k_T^2)/\partial(k_T^2)$ is the unintegrated gluon density, $G(x,k_T^2) = x\,g(x,k_T^2)$ $(x = M_\chi^2/\hat{s})$; $M_{ab}(\boldsymbol{k}_T)$ is the fusion amplitude $G\,G \to \chi$ with a, b being the gluonic indices.

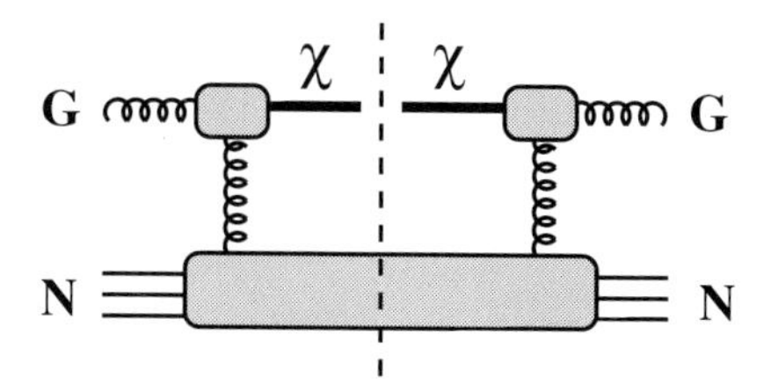

Fig. 16. Perturbative QCD mechanism of production of the χ states in a gluon-nucleon collision.

In the rest frame of the nucleon the amplitude can be represented in terms of the $c\bar{c}$ LC wave functions of the projectile gluon and ejectile charmonium,

$$M_{ab}(\boldsymbol{k}_T) = \frac{\delta_{ab}}{\sqrt{6}} \int\limits_0^1 d\alpha \int d^2r \sum_{\bar{\mu}\mu} \left(\Phi_\chi^{\bar{\mu}\mu}(\boldsymbol{r},\alpha)\right)^* \left[e^{i\boldsymbol{k}_T\cdot\boldsymbol{r}_1} - e^{i\boldsymbol{k}_T\cdot\boldsymbol{r}_2}\right] \Phi_G^{\bar{\mu}\mu}(\boldsymbol{r},\alpha) \,, \tag{92}$$

where

$$\boldsymbol{r}_1 = (1-\alpha)\,\boldsymbol{r}, \qquad \boldsymbol{r}_2 = -\boldsymbol{\alpha}\,\boldsymbol{r} \,. \tag{93}$$

For the sake of simplicity, we separate the color parts $\langle c\bar{c},\{8\}_a|$ and $\langle c\bar{c},\{1\}|$ from the LC wave function of the gluon and charmonium respectively, and calculate the matrix element,

$$\left\langle c\bar{c},\{8\}_a\left|\frac{1}{2}\,\lambda_b\right|c\bar{c},\{1\}\right\rangle = \frac{\delta_{ab}}{\sqrt{6}} \,, \tag{94}$$

which is shown explicitly in (92). Thus, the functions $\Phi^{\bar{\mu}\mu}_{G(\chi)}(\boldsymbol{r},\alpha)$ in (92) represent only the spin- and coordinate dependent parts of the corresponding full wave functions.

The gluon wave function differs only by a factor from the photon one,

$$\Phi_G^{\bar{\mu}\mu}(\boldsymbol{r},\alpha) = \frac{\sqrt{2}\,\alpha_s}{4\pi}\left(\xi_c^\mu\right)^\dagger \hat{O}\,\tilde{\xi}_{\bar{c}}^{\bar{\mu}}\,K_0(\epsilon\,r) \,, \tag{95}$$

where ξ_c^μ is the c-quark spinor, and

$$\begin{aligned}
\tilde{\xi}_{\bar{c}}^{\bar{\mu}} &= i\,\sigma_y\,\xi_{\bar{c}}^{\bar{\mu}^*} \,, && (96)\\
\hat{O} &= m_c\,\boldsymbol{\sigma}\cdot\boldsymbol{e} + i(1-2\alpha)\,(\boldsymbol{\sigma}\cdot\boldsymbol{n})\,(\boldsymbol{e}\cdot\boldsymbol{\nabla}) + (\boldsymbol{e}\times\boldsymbol{n})\boldsymbol{\nabla} \,, && (97)\\
\epsilon^2 &= Q^2\alpha(1-\alpha) + m_c^2 \,, && (98)\\
\boldsymbol{\nabla} &= \frac{d}{d\,\boldsymbol{r}} \,.
\end{aligned}$$

The gluon has virtuality Q^2 and polarization vector $\boldsymbol{e}$ and is moving along the unit vector $\boldsymbol{n}$ (in what follows we consider only transversely polarized gluons, $\boldsymbol{e}\cdot\boldsymbol{n}=0$).

The expression for the LC wave function of a charmonium and the wavefunction of the charmonium in its rest frame are related in a somewhat complicated way by Lorentz transformation, as discussed in Sect. 4. This complexity is a consequence of the nonlocal relation between the LC variables ($\boldsymbol{r}$, α) and the components of the 3-dimensional relative $c\bar{c}$ radius-vector in the rest frame of the charmonium. Also the Melosh spin rotation leads to a nontrivial relations between the two wave functions (see *e.g.* in [1]). This is a relativistic effect, it vanishes in the limit of small velocity $v \to 0$ of the quarks in the charmonium rest frame.

A word of caution is in order. In some cases the Melosh spin rotation is important even in the limit of vanishing quark velocity $v \to 0$. An example is the Landau-Yang theorem [91] which forbids production of the χ_1 state by two massless gluons. However, the LC approach leads to creation of the χ_1 even in the limit $v \to 0$ if the effect of spin rotation is neglected. It is demonstrated in [2] that the Landau-Yang theorem is restored only if the Melosh spin rotation is included. Such a cancelation of large values is a kind of fine tuning and is a good support for the procedure of Lorentz boosting which we apply to the charmonium wave functions.

Since the gluon LC wave function smoothly depends on α while the charmonium wave function peaks at $\alpha = 1/2$ with a tiny width estimated in [2], $\langle(\alpha - 1/2)^2\rangle = 0.01$, we can replace the charmonium wave function in the matrix element in (92) with

$$\Phi_\chi^{\bar{\mu}\mu}(r,\alpha) \approx \delta\left(\alpha - \frac{1}{2}\right) \int d\alpha\, \Phi_\chi^{\bar{\mu}\mu}(r,\alpha)\ . \tag{99}$$

It is convenient to expand the LC charmonium wave function in powers of v. The result depends on the total momentum J and its projection J_z on the direction $\boldsymbol{n}$. The charmonium wave function integrated over α has the form,

$$\int d\alpha\, \Phi_\chi^{\bar{\mu}\mu}(r,\alpha) = \left(\xi^\mu\right)^\dagger \left[\boldsymbol{\sigma}\cdot\boldsymbol{e}_\pm + \frac{1}{m}\left(\boldsymbol{e}_\pm \times \boldsymbol{n}\right)\cdot\boldsymbol{\nabla} - \frac{1}{2\,m_c^2}\left(\boldsymbol{e}_\pm\cdot\boldsymbol{\nabla}\right)\left(\boldsymbol{\sigma}\cdot\boldsymbol{\nabla}\right)\right] \tilde{\xi}^{\bar{\mu}}\, W + O(v^4)\ , \tag{100}$$

where

$$W = \frac{\boldsymbol{e}_\pm\cdot\boldsymbol{r}}{r}\left[R(r) + \frac{3}{4\,m_c^2}\,R''(r) + O(v^4)\right]\ , \tag{101}$$

and $R(r)$ is the radial part of the P-wave charmonium in its rest frame (see derivation in Appendix A of [2]). The new notations for the polarization vectors are,

$$\begin{aligned} \boldsymbol{e}_+ &= -\frac{\boldsymbol{e}_x + i\boldsymbol{e}_y}{\sqrt{2}}\ , \\ \boldsymbol{e}_- &= \frac{\boldsymbol{e}_x - i\boldsymbol{e}_y}{\sqrt{2}}\ . \end{aligned} \tag{102}$$

In what follows we use the LC wave functions of gluons and charmonium in order to calculate matrix elements of operators which depend only on the LC variables r and α. Therefore, for the sake of simplicity we can drop off the indexes $\mu, \bar{\mu}$ and summation over them, *i.e.* replace

$$\sum_{\mu\bar{\mu}} \left(\Phi_{\chi}^{\mu\bar{\mu}}(\boldsymbol{r},\alpha)\right)^* \Phi_{G}^{\mu\bar{\mu}}(\boldsymbol{r},\alpha) \Rightarrow \Phi_{\chi}^*(\boldsymbol{r},\alpha)\,\Phi_G(\boldsymbol{r},\alpha) \tag{103}$$

With this convention we can rewrite the cross section (92) as,

$$\sigma(GN \to \chi X) = \int\limits_0^1 d\alpha \int\limits_0^1 d\alpha' \int d^2r\, d^2r' \times \left\{\Phi_{\chi}^*(\boldsymbol{r},\alpha)\,\Phi_{\chi}(\boldsymbol{r}',\alpha')\,\Sigma^{tr}(\boldsymbol{r},\boldsymbol{r}',\alpha,\alpha')\,\Phi_G(\boldsymbol{r},\alpha)\,\Phi_G^*(\boldsymbol{r}',\alpha')\right\}, \tag{104}$$

where the transition cross section Σ^{tr} is a combination of dipole cross sections,

$$\Sigma^{tr}(\boldsymbol{r},\boldsymbol{r}',\alpha,\alpha') = \frac{1}{16}\left[\sigma_{\bar{q}q}(\boldsymbol{r}_1-\boldsymbol{r}_2') + \sigma_{q\bar{q}}(\boldsymbol{r}_2-\boldsymbol{r}_1') - \sigma_{q\bar{q}}(\boldsymbol{r}_1-\boldsymbol{r}_1') - \sigma_{q\bar{q}}(\boldsymbol{r}_2-\boldsymbol{r}_2')\right], \tag{105}$$

and $\boldsymbol{r}_1$, $\boldsymbol{r}_2'$ $\boldsymbol{r}_1'$ and $\boldsymbol{r}_2'$ are defined like in (93).

9 Charmonium Hadroproduction off Nuclei

Nuclear effects in charmonium production have drawn much attention during the last two decades since the NA3 experiment at CERN [92] has found a steep increase of nuclear suppression with rising Feynman x_F. This effect has been confirmed later in the same energy range [93], and at higher energy recently by the most precise experiment E866 at Fermilab [94]. No unambiguous explanation for these observations has been provided yet. With the advent of RHIC new data are expected soon in the unexplored energy range. Lacking a satisfactory understanding of nuclear effects for charmonium production in proton-nucleus collisions it is very difficult to provide a convincing interpretation of data from heavy ion collisions experiments [95,96] which are aimed to detect the creation of a quark-gluon plasma using charmonium as a sensitive probe. Many of existing analyses rely on an oversimplified dynamics of charmonium production which fail to explain even data for pA collisions, in particular the observed x_F dependence of J/Ψ suppression. Moreover, sometimes even predictions for RHIC employ those simple models. It is our purpose to demonstrate that the dynamics of charmonium suppression strikingly changes between the SPS and RHIC energies. We perform full QCD calculations of nuclear effects within the framework of the light-cone Green function approach aiming to explain observed nuclear effects without adjusting any parameters, and to provide realistic predictions for RHIC.

To avoid a confusion, we should make it clear that we will skip discussion of any mechanisms of charmonium suppression caused by the interaction with the produced comoving matter, although it should be an important effect in central heavy ion collisions. Instead, we consider suppression which originates from the production process and propagation of the $c\bar{c}$ pair through the nucleus. It serves as a baseline for search for new physics in heavy ion collisions.

We focus here on coherence phenomena which are still a rather small correction for charmonium production at the SPS, but whose onset has already been observed at Fermilab and which are expected to become a dominant effect at the energies of RHIC and LHC. One realizes the importance of the coherence effects treating charmonium production in an intuitive way as a hard $c\bar{c}$ fluctuation that loses coherence with the projectile ensemble of partons via interaction with the target, and is thus liberated. In spite of the hardness of the fluctuation, its lifetime in the target rest frame increases with energy and eventually exceeds the nucleus size. Apparently, in this case the $c\bar{c}$ pair is freed by interaction with the whole nucleus, rather than with an individual bound nucleon as it happens at low energies. Correspondingly, nuclear effects become stronger at high energies since the fluctuation propagates through the whole nucleus, and different nucleons compete with each other in freeing the $c\bar{c}$. In terms of the conventional Glauber approach it leads to shadowing. In terms of the parton model it is analogous to shadowing of c-quarks in the nuclear structure function. It turns out (see Sect. 9.3) that the fluctuations containing gluons in addition to the $c\bar{c}$ pair are subject to especially strong shadowing. Since at high energies the weight of such fluctuations rises, as well as the fluctuation lifetime, it becomes the main source of nuclear suppression of open and hidden charm at high energies, in particular at RHIC. In terms of the parton model, shadowing for such fluctuations containing gluons correspond to gluon shadowing.

The parton model interpretation of charmonium production contains no explicit coherence effects, but they are hidden in the gluon distribution function of the nucleus which is supposed to be subject to QCD factorization. There are, however, a few pitfalls on this way. First of all, factorization is exact only in the limit of a very hard scale. That means that one should neglect the effects of the order of the inverse c-quark mass, in particular the transverse $c\bar{c}$ separation $\langle r^2\rangle \sim 1/m_c^2$. However, shadowing and absorption of $c\bar{c}$ fluctuations is a source of a strong suppression which is nearly factor of 0.5 for heavy nuclei (see Fig. 18). QCD factorization misses this effect. Second of all, according to factorization gluon shadowing is supposed to be universal, *i.e.* one can borrow it from another process (although we still have no experimental information about gluons shadowing, it only can be calculated) and use to predict nuclear suppression of open or hidden charm. Again, factorization turns out to be dramatically violated at the scale of charm and gluon shadowing for charmonium production is much stronger than it is for open charm or deep-inelastic scattering (DIS) (compare gluon shadowing exposed in Fig. 21 with one calculated in [29] for DIS). All these important, sometimes dominant effects are missed by QCD factorization. This fact once again emphasizes the advantage of the light-cone dipole approach which does reproduce QCD factorization in the limit where it is expected to work, and which is also able to calculate the deviations from factorization in a parameter free way.

Unfortunately, none of the existing models for J/Ψ or Ψ' production in NN collisions is fully successful in describing all the features observed experimentally. In particular, the J/Ψ, Ψ' and χ_1 production cross sections in NN collisions come out too small by at least an order of magnitude [97]. Only data for production of χ_2 whose mechanism is rather simple seems to be in good accord with the theoretical expectation based on the color singlet mechanism (CSM) [98,99] treating χ_2 production via glue-glue fusion. The contribution of the color-octet mechanism is an order of magnitude less that of CSM [99], and is even more suppressed according to [100]. The simplicity of the production mechanism of χ_2 suggests to use this process as a basis for the study of nuclear effects. Besides, about 40% of the J/Ψs have their origin in χ decays. We drop the subscript of χ_2 in what follows unless otherwise specified.

9.1 Interplay of Formation and Coherence Time Scales and Related Phenomena

A lot of work has been done and considerable progress has been achieved in the understanding of many phenomena related to the dynamics of the charmonium production and nuclear suppression.

• Relative nuclear suppression of J/Ψ and Ψ' has attracted much attention. The Ψ' has twice as large a radius as the J/Ψ, therefore should attenuate in nuclear matter much stronger. However, formation of the wave function of the charmonia takes time, one cannot instantaneously distinguish between these two levels. This time interval or so called formation time (length) is enlarged at high energy E_Ψ by Lorentz time dilation,

$$t_f = \frac{2\,E_\Psi}{M_{\Psi'}^2 - M_{J/\Psi}^2}\,, \tag{106}$$

and may become comparable to or even longer than the nuclear radius. In this case neither J/Ψ, nor Ψ' propagates through the nuclear medium, but a pre-formed $c\bar{c}$ wave packet [23]. Intuitively, one might even expect a universal nuclear suppression, indeed supported by data [77,95,94]. However, a deeper insight shows that such a point of view is oversimplified, namely, the mean transverse size of the $c\bar{c}$ wave packet propagating through the nucleus varies depending on the wave function of the final meson on which the $c\bar{c}$ is projected. In particular, the nodal structure of the $2S$ state substantially enhances the yield of Ψ' [33,39] (see in [101,102] a complementary interpretation in the hadronic basis).

• The next phenomenon is related to the so called coherence time. Production of a heavy $c\bar{c}$ is associated with a longitudinal momentum transfer q_c which decreases with energy. Therefore the production amplitudes on different nucleons add up coherently and interfere if the production points are within the interval $l_c = 1/q_c$ called coherence length or time,

$$t_c = \frac{2\,E_\Psi}{M_{J/\Psi}^2}\,. \tag{107}$$

This time interval is much shorter than the formation time (106). One can also interpret it in terms of the uncertainty principle as the mean lifetime of a $c\bar{c}$ fluctuation. If the coherence time is long compared to the nuclear radius, $t_c \gtrsim R_A$, different nucleons compete with each other in producing the charmonium. Therefore, the amplitudes interfere destructively leading to an additional suppression called shadowing. Predicted in [33], this effect was confirmed by the NMC measurements of exclusive J/Ψ photoproduction off nuclei [76] (see also [39]). The recent precise data from the HERMES experiment [103] for electroproduction of ρ mesons also confirms the strong effect of coherence time [104].

Note that the coherence time (107) is relevant only for the lightest fluctuations $|c\bar{c}\rangle$. Heavier ones which contain additional gluons have shorter lifetime. However, at high energies they are also at work and become an important source of an extra suppression (see [29] and Sect. 9.3). They correspond to shadowing of gluons in terms of parton model. In terms of the dual parton model the higher Fock states contain additional $q\bar{q}$ pairs instead of gluons. Their contribution is enhanced on a nuclear target and leads to softening of the x_F distribution of the produced charmonium. This mechanism has been used in [105] to explain the x_F dependence of charmonium suppression. However the approach was phenomenological and data were fitted.

The first attempt to implement the coherence time effects into the dynamics of charmonium production off nuclei has been made in [106]. However, the approach still was phenomenological and data also were fitted. Besides, gluon shadowing (see Sect. 9.3) had been missed.

• The total J/Ψ-nucleon cross section steeply rises with energy, approximately as $s^{0.2}$. This behavior is suggested by the observation of a steep energy dependence of the cross section of J/Ψ photoproduction at HERA. This fact goes well along with observation of the strong correlation between x_{Bj} dependence of the proton structure function at small x_{Bj} and the photon virtuality Q^2: the larger is Q^2 (the smaller is its $q\bar{q}$ fluctuation), the steeper the rises $F_2(x_{Bj}, Q^2)$ with $1/x_{Bj}$. Apparently, the cross section of a small size charmonium must rise with energy faster than what is known for light hadrons. The J/Ψ-nucleon cross section has been calculated recently in [1] employing the light-cone dipole phenomenology, realistic charmonium wave functions and phenomenological dipole cross section fitted to data for $F_2(x, Q^2)$ from HERA. The results are in a good accord with data for the electroproduction cross sections of J/Ψ and Ψ' and also confirm the steep energy dependence of the charmonium-nucleon cross sections (see Sect. 4). Knowledge of these cross sections is very important for the understanding of nuclear effects in the production of charmonia. A new important observation made in [1] is a strong effect of spin rotation associated with boosting the $c\bar{c}$ system from its rest frame to the light cone. It substantially increases the J/Ψ and especially Ψ' photoproduction cross sections. The effect of spin rotation is also implemented in our calculations below and it is crucial for restoration of the Landau-Yang theorem [2,91].

• Initial state energy loss by partons traveling through the nucleus affects the x_F distribution of produced charmonia [107] especially at medium high energies. A shift in the effective value of x_1, which is the fraction of the incident momentum carried by the produced charmonium, and the steep x_1-dependence of the cross section of charmonium production off a nucleon lead to a dramatic nuclear suppression at large x_1 (or x_F) in a good agreement with data [92,93]. The recent analyses [90] of data from the E772 experiment for Drell-Yan process on nuclei reveals for the first time a nonzero and rather large energy loss.

9.2 Higher Twist Nuclear Effects

Nuclear effects in the production of a χ are controlled by the coherence and formation lengths which are defined in (106), (107). One can identify two limiting cases. The first one corresponds to the situation where both l_c and l_f are shorter that the mean spacing between bound nucleons. In this case one can treat the process classically, the charmonium is produced on one nucleon inside the nucleus and attenuates exponentially with an absorptive cross section which is the inelastic $\chi - N$ one. This simplest case is described in [107,11].

In the limit of a very long coherence length $l_c \gg R_A$ one can think about a $c\bar{c}$ fluctuation which emerges inside the incident hadron long before the interaction with the nucleus. Different bound nucleons compete and shadow each other in the process of liberation of this fluctuation. This causes an additional attenuation in addition to inelastic collisions of the produced color-singlet $c\bar{c}$ pair on its way out of the nucleus. Since $l_c \ll l_f$ an intermediate case is also possible where l_c is shorter than the mean internucleon separation, while l_f is of the order or longer than the nuclear radius.

The transition between the limits of very short and very long coherence lengths is performed using the prescription suggested in [51] for inelastic photoproduction of J/Ψ off nuclei. The amplitude of χ production off a nucleus can be represented in the form,

$$A^{(\lambda)}(b,z) = \int\limits_0^1 d\alpha \int d^2r \int d^2r'\, \Phi^*_\chi(\boldsymbol{r},\alpha)\, \hat{D}^{(\lambda)}(\boldsymbol{r},\boldsymbol{r}',\alpha;b,z)\, \Phi_G(\boldsymbol{r}',\alpha)\ , \quad (108)$$

where $\hat{D}^{(\lambda)}(\boldsymbol{r},\boldsymbol{r}',\alpha;b,z)$ is the amplitude of production of a colorless $c\bar{c}$ pair which reaches a separation $\boldsymbol{r}$ outside the nucleus. It is produced at the point $(\boldsymbol{b},z)$ by a color-octet $c\bar{c}$ with separation r'. The amplitude consists of two terms,

$$\hat{D}^{(\lambda)}(\boldsymbol{r},\boldsymbol{r}',\alpha;b,z) = \hat{D}_1^{(\lambda)}(\boldsymbol{r},\boldsymbol{r}',\alpha;b,z) + \hat{D}_2^{(\lambda)}(\boldsymbol{r},\boldsymbol{r}',\alpha;b,z)\ . \quad (109)$$

Here the first term reads,

$$\hat{D}_1^{(\lambda)}(\boldsymbol{r},\boldsymbol{r}',\alpha;b,z) = G^{(1)}_{c\bar{c}}(\boldsymbol{r},z_+;\boldsymbol{r}',z)\, \boldsymbol{e}^{(\lambda)}\cdot\boldsymbol{d}'\, e^{iq_L z}\ , \quad (110)$$

where $G^{(1)}_{c\bar{c}}(\boldsymbol{r},z_+;\boldsymbol{r}',z)$ is the color-singlet Green function describing evolution of a $c\bar{c}$ wave packet with initial separation $\boldsymbol{r}'$ at the point z up to the final separation $\boldsymbol{r}$ at $z_+ \to \infty$. This term is illustrated in Fig. 17a.

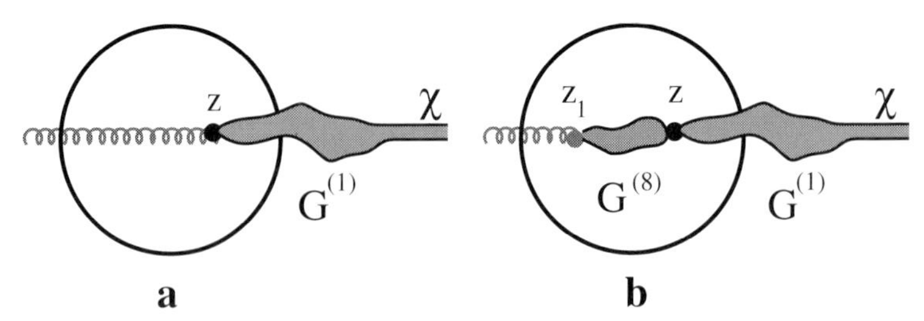

Fig. 17. The incident gluon can either produce the colorless $c\bar{c}$ pair with quantum numbers of χ at the point z (**a**), or it produces diffractively a color-octet $c\bar{c}$ with the quantum numbers of the gluon at the point z_1 which is then converted into a color singlet state at z (**b**). Propagation of a color-singlet or octet $c\bar{c}$ is described by the Green functions $G^{(1)}_{c\bar{c}}$ and $G^{(8)}_{c\bar{c}}$, respectively.

There is also a possibility for the projectile gluon to experience diffractive interaction with production of color-octet $c\bar{c}$ with the same quantum numbers of the gluon at the point z_1. This pair propagates from the point z_1 to z as is described by the corresponding color-octet Green function $G^{(8)}_{c\bar{c}}$ and produces the final colorless pair which propagation is described by the color-singlet Green function, as is illustrated in Fig. 17b. The corresponding second term in (109) reads,

$$\hat{D}^{(\lambda)}_2(\boldsymbol{r},\boldsymbol{r}',\alpha;b,z) = -\frac{1}{2}\int\limits_{-\infty}^{z} dz_1\, d^2r''\, G^{(1)}_{c\bar{c}}(\boldsymbol{r},z_+;r'',z) \times \boldsymbol{e}^{(\lambda)}\cdot\boldsymbol{d}''\, G^{(8)}_{c\bar{c}}(\boldsymbol{r}'',z;\boldsymbol{r}',z_1)\, e^{iq_L z_1}\, \sigma_{q\bar{q}G}(\boldsymbol{r}',\alpha)\, \rho_A(b,z_1)\,. \tag{111}$$

The singlet, $G^{(1)}_{c\bar{c}}$, and octet, $G^{(8)}_{c\bar{c}}$, Green functions describe the propagation of color-singlet and octet $c\bar{c}$, respectively, in the nuclear medium. They satisfy the Schrödinger equations,

$$i\,\frac{d}{d\,z}\,G^{(k)}_{c\bar{c}}(\boldsymbol{r},\boldsymbol{r}';z,z') = \left[\frac{m_c^2-\Delta_{\boldsymbol{r}}}{2\,E_G\,\alpha\,(1-\alpha)} + V^{(k)}(\boldsymbol{r},\alpha)\right] G^{(k)}_{c\bar{c}}(\boldsymbol{r},\boldsymbol{r}';z,z')\;, \tag{112}$$

with $k = 1,\ 8$ and boundary conditions

$$G^{(k)}_{c\bar{c}}(\boldsymbol{r},\boldsymbol{r}';z,z')\Big|_{z=z'} = \delta(\boldsymbol{r}-\boldsymbol{r}')\;. \tag{113}$$

The imaginary part of the LC potential $V^{(k)}$ is responsible for the attenuation in nuclear matter,

$$\mathrm{Im}\,V^{(k)}(\boldsymbol{r},\alpha) = -\frac{1}{2}\,\sigma^{(k)}(r,\alpha)\,\rho_A(b,z)\;, \tag{114}$$

where

$$\begin{aligned}\sigma^{(1)}(r,\alpha) &= \sigma_{q\bar{q}}(r)\;,\\ \sigma^{(8)}(r,\alpha) &= \sigma_3(r,\alpha)\;.\end{aligned} \tag{115}$$

The real part of the LC potential $V^{(k)}(\boldsymbol{r},\alpha)$ describes the interaction inside the $c\bar{c}$ system. For the singlet state $\mathrm{Re}\,V^{(1)}(\boldsymbol{r},\alpha)$ should be chosen to reproduce the charmonium mass spectrum. With a realistic potential (see *e.g.* [1]) one can solve (112) only numerically. Since we focus here on the principle problems of understanding of the dynamics of nuclear shadowing in charmonium production, we chose the oscillator form of the potential [29],

$$\mathrm{Re}\,V^{(1)}(\boldsymbol{r},\alpha) = \frac{a^4(\alpha)\,r^2}{2\,E_G\,\alpha\,(1-\alpha)}\;, \tag{116}$$

where

$$\begin{aligned} a(\alpha) &= 2\,\sqrt{\alpha(1-\alpha)\,\mu\,\omega}\;, \\ \mu &= \frac{m_c}{2}\;, \qquad \omega = 0.3\,GeV\;. \end{aligned} \tag{117}$$

The LC potential (116) corresponds to a choice of a potential,

$$U(\boldsymbol{R}) = \frac{1}{2}\,\mu\,\omega\,\boldsymbol{R}^2\;, \tag{118}$$

in the nonrelativistic Schrödinger equation,

$$\left[-\frac{\Delta}{2\mu} + U(\boldsymbol{R})\right]\,\Psi(\boldsymbol{R}) = E\,\Psi(\boldsymbol{r})\;, \tag{119}$$

which should describe the bound states of a colorless $c\bar{c}$ system. Of course this is an approximation we are forced to do in order to solve the evolution equation analytically.

To describe color-octet $c\bar{c}$ pairs we fix the corresponding potential at

$$\mathrm{Re}\,V^{(8)}(\boldsymbol{r},\alpha) = 0\;, \tag{120}$$

in order to reproduce the gluon wave function (95).

To keep calculations simple we use the r^2-approximation (17) for the dipole cross section which is reasonable for small-size heavy quark systems. Then, taking into account (114) - (116) we arrive at the final expressions,

$$V^{(k)}(r,\alpha) = \frac{1}{2}\,\kappa^{(k)}\,r^2\;, \tag{121}$$

$$\kappa^{(1)} = \frac{a^4(\alpha)}{\alpha(1-\alpha)\,E_G} - iC(s)\,\rho_A\;, \tag{122}$$

$$\kappa^{(8)} = -iC(s)\,\rho_A\left\{\frac{9}{8}\left[\alpha^2 + (1-\alpha)^2\right] - \frac{1}{8}\right\}\;. \tag{123}$$

Making use of this approximation and assuming a constant nuclear density $\rho_A(b,z) = \rho_A$ the Green functions can be obtained in an analytical form,

$$\begin{aligned} G_{c\bar{c}}^{(k)}(\boldsymbol{r},\boldsymbol{r}';z_2,z_1) &= \frac{b^{(k)}}{2\pi\,\sinh(\Omega^{(k)}\,\Delta z)} \\ &\times \exp\left\{-\frac{b^{(k)}}{2}\left[\frac{\boldsymbol{r}^2 + \boldsymbol{r}'^2}{\tanh(\Omega^{(k)}\,\Delta z)} - \frac{2\,\boldsymbol{r}\cdot\boldsymbol{r}'}{\sinh(\Omega^{(k)}\,\Delta z)}\right]\right\}\;, \end{aligned} \tag{124}$$

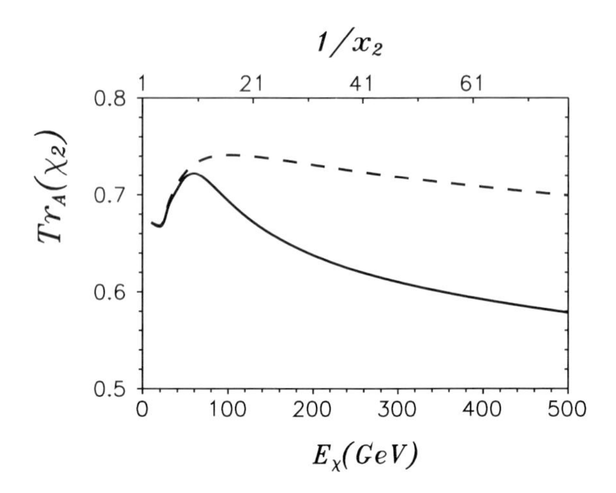

Fig. 18. Nuclear transparency for χ production off lead as function of energy of the charmonium, or x_2 (the upper scale). The solid curve includes both effects of coherence and formation, while the dashed curve corresponds to $l_c = 0$. Since transparency scales in x_2 according to (126), values of x_2 are shown on the top axis.

where

$$\begin{aligned} b^{(k)} &= \sqrt{\kappa^{(k)}\, E_G\, \alpha(1-\alpha)}\ , \\ \Omega^{(k)} &= \frac{b^{(k)}}{E_G\, \alpha(1-\alpha)}\ , \\ \Delta z &= z_2 - z_1\ . \end{aligned}$$

We define the nuclear transparency for χ production as

$$Tr_A(\chi) = \frac{\sigma(G\,A \to \chi\,X)}{A\,\sigma(G\,N \to \chi\,X)}\ . \tag{125}$$

It depends only on the χ or projectile gluon energy. We plot our predictions for lead in Fig. 18.

Transparency rises at low energy since the formation length increases and the effective absorption cross section becomes smaller. This behavior, assuming $l_c = 0$, is shown by dashed curve. However, at higher energies the coherence length is switched on and shadowing adds to absorption. As a result, transparency decreases, as is shown by the solid curve. On top of that, the energy dependence of the dipole cross section makes those both curves for $Tr_A(E_\chi)$ fall even faster.

Apparently, the nuclear transparency depends only on the χ energy, rather than the incident energy or x_1. It is interesting that this leads to x_2 scaling. Indeed, the χ energy

$$E_\chi = \frac{M_\chi^2}{2\,m_N\,x_2}\ , \tag{126}$$

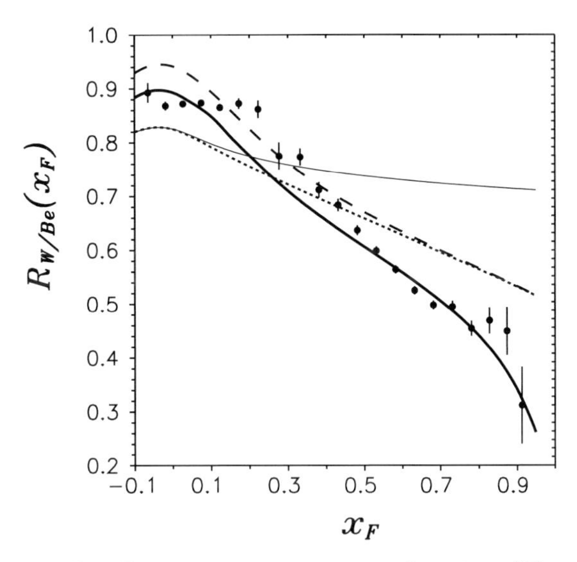

Fig. 19. Tungsten to beryllium cross section ratio as function of Feynman x_F for J/Ψ production at proton energy $800\,GeV$. The thin solid curve represents contribution of initial state quark shadowing and final state $c\bar{c}$ attenuation for χ production. The dotted curve includes also gluon shadowing. The dashed curve is corrected for gluon enhancement at large x_2 (small x_F) using the prescription from [78]. The final solid curve is also corrected for energy loss and for $\chi \to J/\Psi\gamma$ decay. Experimental points are from the E866 experiment [94].

depends only on x_2. We show the x_2 scale in Fig. 18 (top) along with energy dependence.

We also compare in Fig. 19 the contribution of quark shadowing and absorption (thin solid curve) with the nuclear suppression observed at $800\,GeV$ [94]. Since data are for W/Be ratio, and our constant density approximation should not be applied to beryllium, we assume for simplicity that all pA cross sections including pN obey the $A^{\alpha(x_F)}$. We see that the calculated contribution has quite a different shape from what is suggested by the data. It also leaves plenty of room for complementary mechanisms of suppression at large x_F (see below).

9.3 Leading Twist Gluon Shadowing

Previously we considered only the lowest $|c\bar{c}\rangle$ fluctuation of the gluon, which is apparently an approximation. The higher Fock components containing gluons should be also included. In fact they are already incorporated in the phenomenological dipole cross section we use, and give rise to the energy dependence of $\sigma_{q\bar{q}}$. However, they are still excluded from nuclear effects. Indeed, although we eikonalize the energy dependent dipole cross section the higher Fock components do not participate in that procedure, but they have to be eikonalized as well.

This corrections, as is demonstrated below, correspond to suppression of gluon density in nuclei at small x.

The gluon density at small x in nuclei is known to be shadowed, *i.e.* reduced compared to a free nucleons. The partonic interpretation of this phenomenon looks very different depending on the reference frame. In the infinite momentum frame, as was first suggested by Kancheli [108], the partonic clouds of nucleons are squeezed by the Lorentz transformation less at small than at large x. Therefore, while these clouds are well separated in longitudinal direction at large x, they overlap and can fuse at small x, resulting in a diminished parton density [108,9].

Different observables can probe this effect. Nuclear shadowing of the DIS inclusive cross section or Drell-Yan process demonstrate a reduction of the sea quark density at small x. Charmonium or open charm production is usually considered as a probe for gluon distribution.

Although observables are Lorentz invariant, partonic interpretations are not, and the mechanism of shadowing looks quite different in the rest frame of the nucleus where it should be treated as Gribov's inelastic shadowing. This approach seems to go better along with our intuition, besides, the interference or coherence length effects governing shadowing are under a better control. One can even calculate shadowing in this reference frame in a parameter free way (see [58,83,29]) employing the well developed phenomenology of color dipole representation suggested in [13]. On the other hand, within the parton model one can only calculate the Q^2 evolution of shadowing which is quite a weak effect. The main contribution to shadowing originates from the fitted to data input.

In the color dipole representation nuclear shadowing can be calculated via simple eikonalization of the elastic amplitude for each Fock component of the projectile light-cone wave function which are the eigenstates of interaction [13]. Different Fock components represent shadowing of different species of partons. The $|q\bar{q}\rangle$ component in DIS or $|q\gamma^*\rangle$ in Drell-Yan reaction should be used to calculate shadowing of sea quarks. The same components including also one or more gluons lead to gluon shadowing [88,29].

In the color dipole approach one can explicitly see deviations from QCD factorization, *i.e.* dependence of the measured parton distribution on the process measuring it. For example, the coherence length and nuclear shadowing in the Drell-Yan process vanish at minimal x_2 (at fixed energy) [90], while the factorization predicts maximal shadowing. Here we present even more striking deviation from factorization, namely, gluon shadowing for charmonium production turns out to be dramatically enhanced compared to DIS.

9.4 LC Dipole Representation for the Reaction $G\,N \to \chi\,G\,X$

In the case of charmonium production, different Fock components of the projectile gluon, $|(c\bar{c})_1 nG\rangle$ containing a colorless $c\bar{c}$ pair and n gluons ($n = 0, 1\ldots$) build up the cross section of charmonium production which steeply rises with energy (see [1]). The cross section is expected to factorize in impact parameter representation in analogy to the DIS and Drell Yan reaction. This representa-

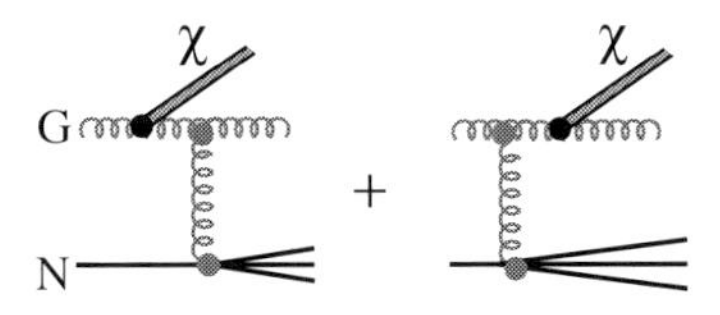

Fig. 20. The dominant Feynman diagrams contributing to χ production.

tion has the essential advantage in that nuclear effects can be easily calculated [15,83]. Feynman diagrams corresponding χ production associated with gluon radiation are shown in Appendix D of [2]. We treat the interaction of heavy quarks perturbatively in the lowest order approximation, while the interaction with the nucleon is soft and expressed in terms of the gluon distribution. The calculations are substantially simplified if the radiated gluon takes a vanishing fraction α_3 of the total light-cone momentum and the heavy quarkonium can be treated as a nonrelativistic system. In this case the amplitude of χG production has a simple form that corresponds to the "Drell-Yan" mechanism of χ production illustrated in Fig. 20.

Correspondingly, the cross section of χ production has the familiar factorized form similar to the Drell-Yan reaction [15,82,83],

$$\alpha_3 \frac{d\,\sigma(GN \to \chi GX)}{d\,\alpha_3} = \int d^2 s \left|\Psi_{G\chi}(s,\alpha_3)\right|^2 \sigma_{GG}\left[(1-\alpha_3)\boldsymbol{s}, x_2/\alpha_3\right] , \quad (127)$$

where $\sigma_{GG}(r,x) = 9/4\,\sigma_{q\bar{q}}(r,x)$ is the cross section of interaction of a GG dipole with a nucleon. $\Psi(s,\alpha_3)$ is the effective distribution amplitude for the $\chi - G$ fluctuation of a gluon, which is the analog to the $\gamma^* q$ fluctuation of a quark,

$$\begin{aligned}\Psi_{G\chi}(s,\alpha_3) = \sum_{\bar{\mu}\mu} \int d^2 r\, d\alpha\, \Phi_\chi^{\bar{\mu}\mu}(\boldsymbol{r},\alpha)\, \Phi_G^{\bar{\mu}\mu}(\boldsymbol{r},\alpha) \\ \times \left[\Phi_{cG}\left(\boldsymbol{s}+\frac{\boldsymbol{r}}{2},\frac{\alpha_3}{\alpha}\right) - \Phi_{cG}\left(\boldsymbol{s}-\frac{\boldsymbol{r}}{2},\frac{\alpha_3}{1-\alpha}\right)\right] .\end{aligned} \quad (128)$$

Here, $\Phi_\chi^{\bar{\mu}\mu}(\boldsymbol{r},\alpha)$ and $\Phi_G^{\bar{\mu}\mu}(\boldsymbol{r},\alpha)$ are the $q\bar{q}$ LC wave functions of the χ and gluon, respectively, which depend on transverse separation $\boldsymbol{r}$ and relative sharing α by the $q\bar{q}$ of the total LC momentum. $\Phi_{cG}(\boldsymbol{s},\alpha)$ is the LC wave function of a quark-gluon Fock component of a quark.

9.5 Gluon Shadowing for χ Production off Nuclei

The gluon density in nuclei is known to be modified, shadowed at small Bjorken x. Correspondingly, production of χ treated as gluon-gluon fusion must be additionally suppressed. In the rest frame of the nucleus, gluon shadowing appears as Gribov's inelastic shadowing [57], which is related to diffractive gluon radiation. The rest frame seems to be more convenient to calculate gluon shadowing, since techniques are better developed, and we use it in what follows. The process $GN \to \chi X$ considered in the previous section includes by default radiation of any number of gluons which give rise to the energy dependence of the dipole cross section.

Extending the analogy between the reactions of χG production by an incident gluon and heavy photon radiation by a quark to the case of nuclear target one can write an expression for the cross section of reaction $GA \to \chi GX$ in two limiting cases:

(i) the production occurs nearly instantaneously over a longitudinal distance which is much shorter than the mean free path of the χG pair in nuclear matter. In this case the cross sections on a nuclear and nucleon targets differ by a factor A independently of the dynamics of χG production.

(ii) The lifetime of the χG fluctuation,

$$t_c = \frac{2\,E_G}{M_{\chi G}^2}\,, \tag{129}$$

substantially exceeds the nuclear size. It is straightforward to replace the dipole cross section on a nucleon by a nuclear one [15,83], then (127) is modified to

$$\frac{d\,\sigma(GA \to \chi GX)}{d\,(\ln \alpha_3)} = 2 \int d^2b\, d^2s\, \left|\Psi_{G\chi}(\boldsymbol{s},\alpha_3)\right|^2 \times \left\{1 - \exp\left[-\frac{1}{2}\,\sigma_{GG}(\boldsymbol{s},x_2/\alpha_3)\,T_A(b)\right]\right\}\,. \tag{130}$$

In order to single out the net gluon shadowing we exclude here the size of the $c\bar{c}$ pair assuming that the cross section responsible for shadowing depends only on the transverse separation $\boldsymbol{s}$.

(iii) A general solution valid for any value of t_c is more complicated and must interpolate between the above limiting situations. In this case one can use the methods of the Landau-Pomeranchuk-Migdal (LPM) theory for photon bremsstrahlung in a medium generalized for targets of finite thickness in [109,83]. The general expressions for the cross section which reproduces the limiting cases $t_c \to 0$ (**i**) and $t_c \to \infty$ (**ii**) reads,

$$\frac{d\,\sigma(GA \to \chi GX)}{d^2b d\,(\ln \alpha_3)} = \left\{\int\limits_{-\infty}^{\infty} dz\,\rho_A(b,z) \times \int d^2s\, \left|\Psi_{G\chi}(s,\alpha_3)\right|^2 \sigma_{GG}\left[(1-\alpha_3)\boldsymbol{s},x_2/\alpha_3\right] - \frac{1}{2}\,\mathrm{Re}\int\limits_{-\infty}^{\infty} dz_2\,\rho_A(b,z_2) \int\limits_{-\infty}^{z_2} dz_1\,\rho_A(b,z_1)\,\widetilde{\Sigma}(z_2,z_1)\,e^{iq_L(z_2-z_1)}\right\}\,, \tag{131}$$

where

$$\widetilde{\Sigma}(z_2,z_1) = \int d^2s_1\, d^2s_2\, \Psi^*_{G\chi}(\boldsymbol{s}_2,\alpha_3)\,\sigma_{GG}(s_2,x_2/\alpha_3)\,G(\boldsymbol{s}_2,z_2;\boldsymbol{s}_1,z_1) \times \sigma_{GG}(s_1,x_2/\alpha_3)\,\Psi_{G\chi}(\boldsymbol{s}_1,\alpha_3)\,. \tag{132}$$

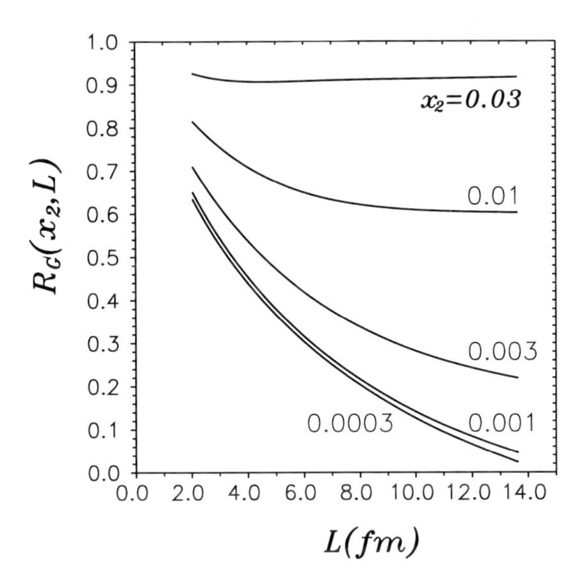

Fig. 21. Gluon suppression as function of thickness of nuclear matter with constant density $\rho_A = 0.16\, fm^{-3}$.

To single out the correction for gluon shadowing one should compare the cross section (131) with the impulse approximation term in which absorption is suppressed,

$$R_G(x_2) = \frac{G_A(x_2)}{A\, G_N(x_2)} = 1 - \frac{1}{A\, \sigma(GN \to \chi X)} \int\limits_{x_2}^{\alpha_{max}} d\alpha_3\, \frac{d\sigma(GA \to \chi GX)}{d\alpha_3}\,. \quad (133)$$

For further calculations and many other applications one needs to know gluon shadowing as function of impact parameter which is calculated as follows,

$$\begin{aligned} R_G(x_2, b) &= \frac{G_A(x_2, b)}{T_A(b)\, G_N(x_2)} \\ &= 1 - \frac{1}{T_A(b)\, \sigma(GN \to \chi X)} \int\limits_{x_2}^{\alpha_{max}} d\alpha_3\, \frac{d\sigma(GA \to \chi GX)}{d^2b\, d\alpha_3}\,. \end{aligned} \quad (134)$$

The results of calculations for the b-dependence of gluon shadowing (134) are depicted in Fig. 21 for different values of x_2 as function of thickness of nuclear matter, $L = \sqrt{R_A^2 - b^2}$.

The results confirm the obvious expectation that shadowing increases for smaller x_2 and for longer path in nuclear matter. One can see that for given thickness shadowing tends to saturate down to small x_2, what might be a result of one gluon approximation. Higher Fock components with larger number of gluons are switched on at very small x_2. At the same time, shadowing saturates at large lengths what one should have also expected as a manifestation of gluon saturation. Note that at large $x_2 = 0.03$ shadowing is even getting weaker at longer L. This is easy to understand, in the case of weak shadowing one can drop off the multiple scattering terms higher than two-fold one. Then the shadowing correction is controlled by the longitudinal formfactor of the nucleus which decreases with L (it is obvious for the Gaussian shape of the nuclear density, but is also true for the realistic Woods-Saxon distribution).

9.6 Antishadowing of Gluons

Nuclear modification of the gluon distribution is poorly known. There is still no experimental evidence for that. Nevertheless, the expectation of gluon shadowing at small x is very solid, and only its amount might be disputable. At the same time, some indications exist that gluons may be enhanced in nuclei at medium small $x_2 \sim 0.1$. The magnitude of gluon antishadowing has been estimated in [45] assuming that the total fraction of momentum carried by gluons is the same in nuclei and free nucleons (there is an experimental support for it). Such a momentum conservation sum rule leads to a gluon enhancement at medium x, since gluons are suppressed in nuclei at small x. The effect, up to $\sim 20\%$ antishadowing in heavy nuclei at $x \approx 0.1$, found in [45] is rather large, but it is a result of very strong shadowing which we believe has been grossly overestimated (see discussion in [29]).

Fit to DIS data based on evolution equations performed in [78] also provided an evidence for rather strong antishadowing effect at $x \approx 0.1$. However, the fit employed an ad hoc assumption that gluons are shadowed at the low scale Q_0^2 exactly as $F_2(x, Q^2)$ what might be true only by accident. Besides, in the x distribution of antishadowing was shaped ad hoc too.

A similar magnitude of antishadowing has been found in the analysis [110] of data on Q^2 dependence of nuclear to nucleon ratio of the structure functions, $F_2^A(x, Q^2)/F_2^N(x, Q^2)$. However it was based on the leading order QCD approximation which is not well justified at these values of Q^2.

Although neither of these results seem to be reliable, similarity of the scale of the predicted effect looks convincing, and we included the antishadowing of gluons in our calculations. We use the shape of x_2 dependence and magnitude of gluon enhancement from [78].

9.7 Comparison with Available Data and Predictions for Higher Energies

The dynamics of J/Ψ suppression at energy $800\,GeV$ is rather complicated and includes many effects. Now we can apply more corrections to the dotted curve in Fig. 19 which involves only quark and gluon shadowing. Namely, inclusion of the energy loss effect and decay $\chi \to J/\Psi\,\gamma$ leads to a stronger suppression depicted by the dashed curve. Eventually we correct this curve for gluon enhancement at $x_2 \sim 0.1$ (small x_F) and arrive at the final result shown by thick solid curve.

Since our calculation contains no free parameters we think that the results agree with the data amazingly well. Some difference in the shape of the maximum observed and calculated at small x_F may be a result of the used parameterization [78] for gluon antishadowing. We think that it gives only the scale of the effect, but neither the ad hoc shape, nor the magnitude should be taken literally. Besides, our calculations are relevant only for those J/Ψs which originate from χ decays which feed only about 40% of the observed ones.

At higher energies of RHIC and LHC, the effect of energy loss is completely gone and nuclear suppression must expose x_2 scaling. Much smaller x_2 can be reached at higher energies. Our predictions for proton-gold to proton-proton ratio is depicted in Fig. 22. One can see that at $x_F > 0.1$ shadowing suppresses charmonium production by nearly an order of magnitude. We can also estimate the effect of nuclear suppression in heavy ion collisions assuming factorization,

$$R_{AB}(x_F) = R_{pB}(x_F)\, R_{pA}(-x_F) \ . \tag{135}$$

Our predictions for gold-gold collisions at $\sqrt{s} = 200$ GeV are shown by the bottom curve in Fig. 22. Since factorization is violated this prediction should be verified.

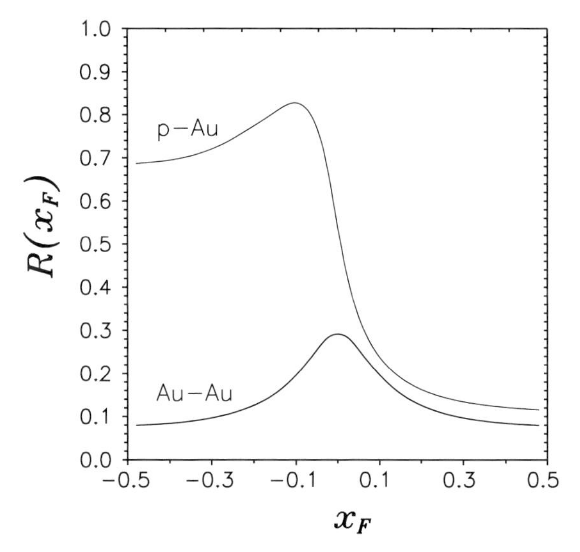

Fig. 22. Nuclear suppression of J/Ψ production in proton-gold collisions at $\sqrt{s} = 200\,GeV$ as function of x_F (the upper curve) and in gold-gold collisions (bottom curve). Effects of quark and gluon shadowing and gluon antishadowing are included.

10 Summary

In these lectures, we presented several QCD processes related to production of heavy quarks at high energies. We describe all these reactions within the same approach, a light-cone color-dipole formalism. We highlighted the main advantages of this approach in comparison with the alternative description based on the standard QCD parton model. The color-dipole formalism does not involve any uncertain K-factors, takes care and calculates higher twist corrections, predicts nuclear effects. At the same time, this approach is restricted to the small-x domain.

We started with the simplest process of diffractive production of charmonia off protons and nuclei. We made use of the best of our knowledge for charmonium wave functions and methods of their boosting to the light front. As a result, we arrived at a very good agreement with data, achieved without any adjustments. One of the key issues in reaching this agreement is inclusion of the Melosh spin rotation which substantially changes the production rates of J/Ψ and especially Ψ'. In particular, it solves the long standing problem of understanding photoproduction data for the Ψ' to J/Ψ ratio. These calculations also provided realistic predictions for charmonium-nucleon total cross sections.

We extended the study to charmonium photoproduction on nuclear targets aiming to study shadowing effects. Surprisingly, we found that higher twist shadowing which is neglected in parton model calculations, is the main source of shadowing, at least for foreseen energies. This shadowing effect is related to a nonzero separation of the produced $c\bar{c}$ and vanishes in the limit of very heavy quarks. The leading twist shadowing, so called gluon shadowing, is related to higher Fock components of the photon containing gluons. It depends only logarithmically on the quark mass, but its onset is delayed to very high energies.

The next important process is hadroproduction of open heavy flavor. The color-dipole approach turns out to be quite effective even for a proton target due to well developed phenomenology for the dipole cross section fitted to HERA data for $F_2^p(x, Q^2)$. Our predictions for open charm and beauty production well agrees with available data.

higher twist quark shadowing for open charm hadroproduction off nuclei is a quite weak effect. However, we find large higher twist corrections to the leading twist gluon shadowing, which makes the latter process dependent. Gluon shadowing in open charm production is found to be stronger than in photoproduction of charmonium (and in DIS), but somewhat weaker than in the case of hadroproduction of charmonium. Our predictions for RHIC will be tested soon.

The last and most complicated case is hadroproduction of charmonia. We restricted ourselves to the simplest case of production of the P-wave χ states. Like in photoproduction, we found a substantial higher twist shadowing, but much stronger leading twist gluon shadowing. For the first time we explained the steep x_F dependence of nuclear suppression of charmonia observed in the E772/866 experiments at Fermilab. Gluon shadowing is dramatically enhanced at the energies of RHIC and LHC and we predict very strong suppression of charmonia both for pA and AA collisions.

Acknowledgments

We are grateful to David Blaschke for his patience in encouraging us to write these lectures. Special thanks go to our collaborators Jörg Hüfner, Yuri Ivanov, Jen-Chieh Peng and Sasha Tarasov with whom the results presented here have been achieved. We thank the Institute for Nuclear Theory at the University of Washington for its hospitality during the workshop *The First Three Years of Heavy-ion Physics at RHIC*, where this work was completed. B.Z.K. is supported by the grant from the Gesellschaft für Schwerionenforschung Darmstadt (GSI), grant No. GSI-OR-SCH, and by the grant INTAS-97-OPEN-31696. J.R. is supported by the U.S. Department of Energy at Los Alamos National Laboratory under Contract No. W-7405-ENG-38.

References

1. J. Hüfner, Y. P. Ivanov, B. Z. Kopeliovich and A. V. Tarasov, Phys. Rev. D **62**, 094022 (2000) [arXiv:hep-ph/0007111].
2. B. Kopeliovich, A. Tarasov and J. Hüfner, Nucl. Phys. A **696**, 669 (2001) [arXiv:hep-ph/0104256].
3. Y. P. Ivanov, B. Z. Kopeliovich, A. V. Tarasov and J. Hüfner, Phys. Rev. C **66**, 024903 (2002) [arXiv:hep-ph/0202216].
4. B. Z. Kopeliovich and A. V. Tarasov, Nucl. Phys. A **710**, 180 (2002) [arXiv:hep-ph/0205151].
5. J. Raufeisen and J. C. Peng, Phys. Rev. D **67**, 054008 (2003) [arXiv:hep-ph/0211422].
6. J. Raufeisen, J. C. Peng and G. C. Nayak, Phys. Rev. D **66**, 034024 (2002) [arXiv:hep-ph/0204095];
   B. Z. Kopeliovich, J. Raufeisen and A. V. Tarasov, Phys. Lett. B **503**, 91 (2001) [arXiv:hep-ph/0012035].
7. S. Gavin, P. L. McGaughey, P. V. Ruuskanen and R. Vogt, Phys. Rev. C **54**, 2606 (1996).
8. K. J. Eskola, V. J. Kolhinen and R. Vogt, Nucl. Phys. A **696**, 729 (2001) [arXiv:hep-ph/0104124].
9. L. V. Gribov, E. M. Levin and M. G. Ryskin, Nucl. Phys. B **188**, 555 (1981); Phys. Rept. **100**, 1 (1983);
   A. H. Mueller and J. w. Qiu, Nucl. Phys. B **268**, 427 (1986).
10. D. E. Kharzeev and J. Raufeisen, AIP Conference Proceedings, PASI 2002, Campos do Jordao, *New states of matter in hadronic interactions* Vol. 631, Issue 1, pp. 27-69 [arXiv:nucl-th/0206073].
11. C. Gerschel and J. Hüfner, Ann. Rev. Nucl. Part. Sci. **49**, 255 (1999) [arXiv:hep-ph/9802245];
    R. Vogt, Phys. Rept. **310**, 197 (1999).
12. T. Matsui and H. Satz, Phys. Lett. B **178**, 416 (1986).
13. A. B. Zamolodchikov, B. Z. Kopeliovich and L. I. Lapidus, JETP Lett. **33**, 595 (1981) [Pisma Zh. Eksp. Teor. Fiz. **33**, 612 (1981)].
14. B. Blaettel, G. Baym, L. L. Frankfurt and M. Strikman, Phys. Rev. Lett. **70**, 896 (1993);
    L. L. Frankfurt, A. Radyushkin and M. Strikman, Phys. Rev. D **55**, 98 (1997) [hep-ph/9610274].

15. B. Z. Kopeliovich, proc. of the workshop *Dynamical Properties of Hadrons in Nuclear Matter,* Hirschegg, January 16 – 21, 1995, ed. by H. Feldmeyer and W. Nörenberg, Darmstadt, 1995, p. 102 (hep-ph/9609385).
16. S. Catani, M. Ciafaloni and F. Hautmann, Nucl. Phys. B **366**, 135 (1991);
J. Kwiecinski, A. D. Martin and A. M. Stasto, Phys. Rev. D **56**, 3991 (1997) [arXiv:hep-ph/9703445];
C. B. Mariotto, M. B. Gay Ducati and M. V. Machado, Phys. Rev. D **66**, 114013 (2002) [arXiv:hep-ph/0208155].
17. M. Abramowitz and I. A. Stegun (eds.), *Handbook of Mathematical Functions,* Dover Publications Inc., New York, NY, 1972.
18. A.H. Mueller, Nucl. Phys. B **558**, 285 (1999) [arXiv:hep-ph/9904404].
19. J. B. Kogut and D. E. Soper, Phys. Rev. D **1**, 2901 (1970).
20. J. D. Bjorken, J. B. Kogut and D. E. Soper, Phys. Rev. D **3**, 1382 (1971).
21. N. N. Nikolaev and B. G. Zakharov, J. Exp. Theor. Phys. **78**, 598 (1994) [Zh. Eksp. Teor. Fiz. **105**, 1117 (1994)].
22. G. Bertsch, S. J. Brodsky, A. S. Goldhaber and J. F. Gunion, Phys. Rev. Lett. **47**, 297 (1981).
23. S. J. Brodsky and A. H. Mueller, Phys. Lett. B **206**, 685 (1988).
24. E. A. Kuraev, L. N. Lipatov and V. S. Fadin, Sov. Phys. JETP **45**, 199 (1977) [Zh. Eksp. Teor. Fiz. **72**, 377 (1977)];
I. I. Balitsky and L. N. Lipatov, Sov. J. Nucl. Phys. **28**, 822 (1978) [Yad. Fiz. **28**, 1597 (1978)].
25. V. S. Fadin and L. N. Lipatov, Phys. Lett. B **429**, 127 (1998);
M. Ciafaloni and G. Camici, Phys. Lett. B **430**, 349 (1998).
26. A. Bialas, H. Navelet and R. Peschanski, Nucl. Phys. B **593**, 438 (2001) [arXiv:hep-ph/0009248].
27. H. I. Miettinen and J. Pumplin, Phys. Rev. D **18**, 1696 (1978).
28. K. Golec-Biernat and M. Wüsthoff, Phys. Rev. D **59**, 014017 (1999) [hep-ph/9807513]; Phys. Rev. D **60**, 114023 (1999) [hep-ph/9903358].
29. B. Z. Kopeliovich, A. Schäfer and A. V. Tarasov, Phys. Rev. D **62**, 054022 (2000) [arXiv:hep-ph/9908245].
30. M. G. Ryskin, R. G. Roberts, A. D. Martin and E. M. Levin, Z. Phys. C **76**, 231 (1997) [arXiv:hep-ph/9511228].
31. J. J. Aubert *et al.*, Phys. Rev. Lett. **33**, 1404 (1974);
J. E. Augustin *et al.*, Phys. Rev. Lett. **33**, 1406 (1974).
32. J. Hüfner and B. Z. Kopeliovich, Phys. Lett. B **445**, 223 (1998) [arXiv:hep-ph/9809300].
33. B. Z. Kopeliovich and B. G. Zakharov, Phys. Rev. D **44**, 3466 (1991).
34. W. Lucha and F. F. Schoberl, Int. J. Mod. Phys. C **10**, 607 (1999) [arXiv:hep-ph/9811453].
35. E. Eichten, K. Gottfried, T. Kinoshita, K. D. Lane and T. M. Yan, Phys. Rev. D **17**, 3090 (1978) [Erratum-ibid. D **21**, 313 (1980)];
E. Eichten, K. Gottfried, T. Kinoshita, K. D. Lane and T. M. Yan, Phys. Rev. D **21**, 203 (1980).
36. W. Buchmüller and S. H. Tye, Phys. Rev. D **24**, 132 (1981).
37. C. Quigg and J. L. Rosner, Phys. Lett. B **71**, 153 (1977).
38. A. Martin, Phys. Lett. B **93**, 338 (1980).
39. O. Benhar, B. Z. Kopeliovich, C. Mariotti, N. N. Nikolaev and B. G. Zakharov, Phys. Rev. Lett. **69**, 1156 (1992).
40. M. V. Terentev, Sov. J. Nucl. Phys. **24**, 106 (1976) [Yad. Fiz. **24**, 207 (1976)].

41. P. Hoyer and S. Peigne, Phys. Rev. D **61**, 031501 (2000).
42. K. Suzuki, A. Hayashigaki, K. Itakura, J. Alam and T. Hatsuda, Phys. Rev. D **62**, 031501 (2000) [arXiv:hep-ph/0005250].
43. H. J. Melosh, Phys. Rev. D **9**, 1095 (1974);
W. Jaus, Phys. Rev. D **41**, 3394 (1990).
44. M. G. Ryskin, Z. Phys. C **57**, 89 (1993).
45. S. J. Brodsky, L. Frankfurt, J. F. Gunion, A. H. Mueller and M. Strikman, Phys. Rev. D **50**, 3134 (1994) [arXiv:hep-ph/9402283];
L.L. Frankfurt, W. Koepf and M.I. Strikman, Phys. Rev. D **54** 3194 (1996).
46. J. Nemchik, N. N. Nikolaev, E. Predazzi and B. G. Zakharov, Z. Phys. C **75**, 71 (1997) [arXiv:hep-ph/9605231].
47. C. Adloff *et al.* [H1 Collaboration], Phys. Lett. B **483**, 23 (2000) [arXiv:hep-ex/0003020].
48. E401 Collab., M. Binkley *et al.*, Phys. Rev. Lett. **48** 73 (1982).
49. E516 Collab., B. H. Denby *et al.*, Phys. Rev. Lett. **52** 795 (1984).
50. J. Breitweg *et al.* [ZEUS Collaboration], Z. Phys. C **75**, 215 (1997) [arXiv:hep-ex/9704013].
51. J. Hüfner, B. Kopeliovich and A. B. Zamolodchikov, Z. Phys. A **357**, 113 (1997) [arXiv:nucl-th/9607033].
52. S. J. Brodsky and M. Karliner, Phys. Rev. Lett. **78**, 4682 (1997) [arXiv:hep-ph/9704379].
53. N. Armesto, A. Capella, A. B. Kaidalov, J. Lopez-Albacete and C. A. Salgado, arXiv:hep-ph/0304119.
54. B. Z. Kopeliovich, J. Raufeisen, A. V. Tarasov and M. B. Johnson, Phys. Rev. C **67**, 014903 (2003) [arXiv:hep-ph/0110221].
55. B. Z. Kopeliovich, J. Nemchik, N. N. Nikolaev and B. G. Zakharov, Phys. Lett. B **324**, 469 (1994) [arXiv:hep-ph/9311237].
56. B. Z. Kopeliovich, J. Raufeisen and A. V. Tarasov, Phys. Lett. B **440**, 151 (1998) [arXiv:hep-ph/9807211];
J. Raufeisen, A. V. Tarasov and O. O. Voskresenskaya, Eur. Phys. J. A **5**, 173 (1999) [arXiv:hep-ph/9812398].
57. V.N. Gribov, Sov. Phys. JETP **29**, 483 (1969) [Zh. Eksp. Teor. Fiz. **56**, 892 (1969)].
58. B. Z. Kopeliovich, J. Raufeisen and A. V. Tarasov, Phys. Rev. C **62**, 035204 (2000) [arXiv:hep-ph/0003136].
59. N. N. Nikolaev, G. Piller and B. G. Zakharov, J. Exp. Theor. Phys. **81** (1995) 851 [Zh. Eksp. Teor. Fiz. **108** (1995) 1554] [arXiv:hep-ph/9412344]; Z. Phys. A **354**, 99 (1996) [arXiv:hep-ph/9511384].
60. P. Nason, S. Dawson and R. K. Ellis, Nucl. Phys. B **303**, 607 (1988).
61. P. Nason, S. Dawson and R. K. Ellis, Nucl. Phys. B **327**, 49 (1989) [Erratum-ibid. B **335**, 260 (1990)].
62. M.L. Mangano, P. Nason and G. Ridolfi, Nucl. Phys. B **373**, 295 (1992).
63. J. Bartels, K. Golec-Biernat and H. Kowalski, Phys. Rev. D **66**, 014001 (2002) [arXiv:hep-ph/0203258].
64. M. Glück, E. Reya and A. Vogt, Eur. Phys. J. C **5**, 461 (1998) [arXiv:hep-ph/9806404].
65. K. Kodama *et al.* [Fermilab E653 Collaboration], Phys. Lett. B **263**, 573 (1991);
R. Ammar *et al.*, Phys. Rev. Lett. **61**, 2185 (1988);
M. Aguilar-Benitez *et al.* [LEBC-EHS Collaboration], Z. Phys. C **40**, 321 (1988);
G. A. Alves *et al.* [E769 Collaboration], Phys. Rev. Lett. **77**, 2388 (1996)

[Erratum-ibid. **81**, 1537 (1998)];
S. Barlag *et al.* [ACCMOR Collaboration], Z. Phys. C **39**, 451 (1988).
66. K. Adcox *et al.* [PHENIX Collaboration], Phys. Rev. Lett. **88**, 192303 (2002) [arXiv:nucl-ex/0202002].
67. R. Vogt, arXiv:hep-ph/0203151.
68. A. Donnachie and P. V. Landshoff, Phys. Lett. B **296**, 227 (1992) [arXiv:hep-ph/9209205].
69. D. M. Jansen *et al.* [E789 Collaboration], Phys. Rev. Lett. **74**, 3118 (1995).
70. T. Alexopoulos *et al.* [E771 Collaboration], Phys. Rev. Lett. **82**, 41 (1999).
71. I. Abt *et al.* [HERA-B Collaboration], arXiv:hep-ex/0205106.
72. R. Bonciani, S. Catani, M. L. Mangano and P. Nason, Nucl. Phys. B **529**, 424 (1998) [arXiv:hep-ph/9801375];
N. Kidonakis, E. Laenen, S. Moch and R. Vogt, Phys. Rev. D **64**, 114001 (2001) [arXiv:hep-ph/0105041].
73. E789 Collaboration, M.J. Leitch et al., Phys. Rev. Lett. **72** 2542 (1994).
74. E769 Collaboration, G.A. Alves et al., Phys. Rev. Lett. **70** 722 (1993).
75. WA82 Collaboration, M. Adamovich et al., Phys. Lett. B **284** 453 (1992).
76. NMC Coll., M. Arneodo et al., Nucl. Phys. B **481** 23 (1996).
77. The E772 Collaboration, D.M. Alde et al, Phys. Rev. Lett. **64** 2479 (1990).
78. K. J. Eskola, V. J. Kolhinen and P. V. Ruuskanen, Nucl. Phys. B **535**, 351 (1998) [arXiv:hep-ph/9802350];
K. J. Eskola, V. J. Kolhinen and C. A. Salgado, Eur. Phys. J. C **9**, 61 (1999) [arXiv:hep-ph/9807297].
79. M. Hirai, S. Kumano and M. Miyama, Phys. Rev. D **64**, 034003 (2001) [arXiv:hep-ph/0103208].
80. K.J. Eskola, H. Honkanen, V.J. Kolhinen, C.A. Salgado, hep-ph/0201256.
81. B. Z. Kopeliovich, J. Raufeisen and A. V. Tarasov, Phys. Lett. B **503**, 91 (2001) [arXiv:hep-ph/0012035];
M. A. Betemps, M. B. Gay Ducati and M. V. Machado, Phys. Rev. D **66**, 014018 (2002) [arXiv:hep-ph/0111473];
M. A. Betemps, M. B. Ducati, M. V. Machado and J. Raufeisen, arXiv:hep-ph/0303100.
82. S. J. Brodsky, A. Hebecker and E. Quack, Phys. Rev. D **55**, 2584 (1997) [arXiv:hep-ph/9609384].
83. B. Z. Kopeliovich, A. V. Tarasov and A. Schäfer, Phys. Rev. C **59**, 1609 (1999) [extended version in hep-ph/9808378].
84. M. B. Johnson, B. Z. Kopeliovich and A. V. Tarasov, Phys. Rev. C **63**, 035203 (2001) [arXiv:hep-ph/0006326];
J. Raufeisen, Phys. Lett. B **557**, 184 (2003) [arXiv:hep-ph/0301052].
85. L. McLerran and R. Venugopalan, Phys. Rev. D **49**, 2233 (1994); **49**, 3352 (1994); **49**, 2225 (1994).
86. A.H. Mueller, *Parton saturation: an overview*, hep-ph/0111244.
87. L.D.Landau, I.Ya.Pomeranchuk, *ZhETF* **24** (1953) 505,
L.D.Landau, I.Ya.Pomeranchuk, *Doklady AN SSSR* **92** (1953) 535, 735
E.L.Feinberg, I.Ya.Pomeranchuk, *Doklady AN SSSR* **93** (1953) 439,
I.Ya.Pomeranchuk, *Doklady AN SSSR* **96** (1954) 265,
I.Ya.Pomeranchuk, *Doklady AN SSSR* **96** (1954) 481,
E.L.Feinberg, I.Ya.Pomeranchuk, *Nuovo Cim. Suppl.* **4** (1956) 652.
88. A.H. Mueller, Nucl. Phys. **B335** (1990) 115; Nucl. Phys. B **558**, 285 (1999) [arXiv:hep-ph/9904404].

89. M. Arneodo, Phys. Rep. **240** 301 (1994).
90. M. B. Johnson *et al.* [FNAL E772 Collaboration], Phys. Rev. Lett. **86**, 4483 (2001) [arXiv:hep-ex/0010051];
M. B. Johnson *et al.*, Phys. Rev. C **65**, 025203 (2002) [arXiv:hep-ph/0105195].
91. V.B. Berestetski, E.M. Lifshitz, and L.P. Pitaevski, Relativistic Quantum Theory. Clarendon Press, Oxford, 1971.
92. The NA3 Collaboration, J. Badier et al., Z. Phys. **C20** 101 (1983).
93. The E537 Coll., S. Katsanevas et al., Phys. Rev. Lett., **60** 2121 (1988).
94. The E866 Collaboration, M.J. Leitch et al., Phys. Rev. Lett. **84** 3256 (2000).
95. The NA38 Collaboration, C. Baglin et al., Phys. Lett. B **345** 617 (1995).
96. The NA50 Collaboration, M.C. Abreu et al., Phys. Lett. B **477** 28 (2000).
97. M. Vaenttinen, P. Hoyer, S. J. Brodsky and W. K. Tang, Phys. Rev. D **51**, 3332 (1995) [arXiv:hep-ph/9410237].
98. E.L. Berger and D. Jones, Phys. Rev. **D23** 1521 (1981).
99. W. K. Tang and M. Vaenttinen, Phys. Rev. D **54**, 4349 (1996) [arXiv:hep-ph/9603266].
100. P. Hägler, R. Kirschner, A. Schäfer, L. Szymanowski and O. V. Teryaev, Phys. Rev. D **63**, 077501 (2001) [arXiv:hep-ph/0008316].
101. J. Hüfner and B. Kopeliovich, Phys. Rev. Lett. **76**, 192 (1996) [arXiv:hep-ph/9504379].
102. D. Kharzeev and H. Satz, Phys. Lett. B **356**, 365 (1995) [arXiv:hep-ph/9504397].
103. K. Ackerstaff *et al.* [HERMES Collaboration], Phys. Rev. Lett. **82**, 3025 (1999) [arXiv:hep-ex/9811011].
104. J. Hüfner, B. Kopeliovich and J. Nemchik, Phys. Lett. B **383**, 362 (1996) [arXiv:nucl-th/9605007].
105. K. Boreskov, A. Capella, A. Kaidalov and J. Tran Thanh Van, Phys. Rev. D **47** 919 (1993).
106. B.Z. Kopeliovich, *Dynamics and Phenomenology of Charmonium Production off Nuclei*, in proc. of the Workshop Hirschegg'97: QCD Phase Transitions', Hirschegg, Austria, January, 1997, ed. by H. Feldmeier, J. Knoll, W. Nörenberg and J. Wambach, Darmstadt, 1997, p. 281; hep-ph/9702365
107. B.Z. Kopeliovich and F. Niedermayer, *Nuclear screening in J/Ψ and Drell-Yan pair production*, JINR-E2-84-834, Dubna 1984, see the scanned version in KEK library:http://www-lib.kek.jp/cgi-bin/img_index?8504113
108. O.V. Kancheli, Sov. Phys. JETP Lett. **18** 274 (1973).
109. B. G. Zakharov, Phys. Atom. Nucl. **61**, 838 (1998) [Yad. Fiz. **61**, 924 (1998)]; [arXiv:hep-ph/9807540].
110. T. Gousset and H. J. Pirner, Phys. Lett. B **375**, 349 (1996) [arXiv:hep-ph/9601242].

Heavy Mesons and Impact Ionization of Heavy Quarkonia

David Blaschke[1,2], Yuri Kalinovsky[3], and Valery Yudichev[2]

[1] Department of Physics, University of Rostock, 18051 Rostock, Germany
[2] Bogoliubov Laboratory for Theoretical Physics, Joint Institute for Nuclear Research, 141980 Dubna, Russia
[3] Laboratory of Information Technologies, Joint Institute for Nuclear Research, 141980 Dubna, Russia

Abstract. At the chiral restoration/deconfinement transition, most hadrons undergo a Mott transition from being bound states in the confined phase to resonances in the deconfined phase. We investigate the consequences of this qualitative change in the hadron spectrum on final state interactions of charmonium in hot and dense matter, and show that the Mott effect for D-mesons leads to a critical enhancement of the J/ψ dissociation rate. Anomalous J/ψ suppression in the NA50 experiment is discussed as well as the role of the Mott effect for the heavy flavor kinetics in future experiments at the LHC. The status of our calculations of hadron-hadron cross sections using the quark interchange and chiral Lagrangian approaches is reviewed, and an Ansatz for a unification of these schemes is given.

1 Introduction

Heavy quarks as constituents of heavy quarkonia and heavy mesons play a decisive role in the diagnostics of hot and dense matter created in ultrarelativistic heavy-ion collisions. The most promiment example is the celebrated J/ψ suppression effect as a possible signal of the formation of a which was suggested by Matsui and Satz [1] in 1986 and has triggered experimental programs at CERN, Brookhaven and the future GSI facility as well as a broad spectrum of theoretical work ever since. In the present lecture, we will elucidate some aspects of the physics of heavy quarks in a hot and dense medium which demonstrate why J/ψ suppression is so difficult to interprete. We suggest that Υ may be a rather clean probe of the plasma state to be produced at CERN-LHC.

The production of heavy quark pairs is a hard process well separated from the soft physics governing the evolution of a quark gluon plasma in heavy-ion collisions. However, the formation of heavy quarkonia and heavy mesons as well as their final state interactions may be well modified by a surrounding hot medium. The basic principle of their use as indicators is similar to the situation in Astrophysics where information about the temperature, density, composition and other properties of stellar atmospheres is obtained from an analysis of the modification of emission and absorption spectra relative to terrestrial conditions [2].

The first and widely discussed probe in this context is the J/ψ. Contrary to naive expectations, Matsui and Satz suggested that with increasing cms energy

D. Blaschke, Y. Kalinovsky, and V. Yudichev, eavy Mesons and Impact Ionization of Heavy Quarkonia, Lect. Notes Phys. **647**, 366–375 (2004)
http://www.springerlink.com/

in a heavy-ion collision a suppression of the J/ψ production relative to the (Drell-Yan) continuum occurs due to the dissociation of the charmonium bound state in a dense medium. This is in analogy to the in semiconductor and plasma physics where under high pressure electrons become delocalized and a conduction band emerges, signaling the plasma state [3].

Correlators of quarkonia states are measured at finite temperature on the lattice [16–20]. Using the maximum entropy method [16], spectral functions can be obtained which give not only information about the mass but also about the spectral width of the states. No substantial modification of J/ψ spectral function is found for temperatures up to $T = 1.5T_c$. The χ_c and ψ' states may be dissolved at T_c. This new lattice analysis provides results alternative to those obtained earlier by solving the bound state Schrödinger equation for a (temperature-dependent) screened potential [22]. While the Mott-dissociation of χ_c and ψ' states is confirmed, the observation of a nearly unchanged J/ψ resonance well above the suspected Mott temperature $T_{J/\psi}^{\rm Mott} = 1.1 - 1.3\, T_c$ came as a surprise for simple potential models. The existence of above the Mott transition is, however, a well-known feature displayed, e.g., in analyses of exciton lines in semiconductor plasmas [3]. The picture which emerges from these lattice studies seems to be consistent with a modification of the effective interaction at finite temperatures so that excited states of the charmonium spectrum (χ_c and ψ') can be dissolved at T_c whereas the deeply bound ground states (η_c and J/ψ) are observed as rather narrow resonances well above that temperature. In order to interpret these findings a study of the spectral functions for screened Coulomb or Cornell-type potentials should be performed.

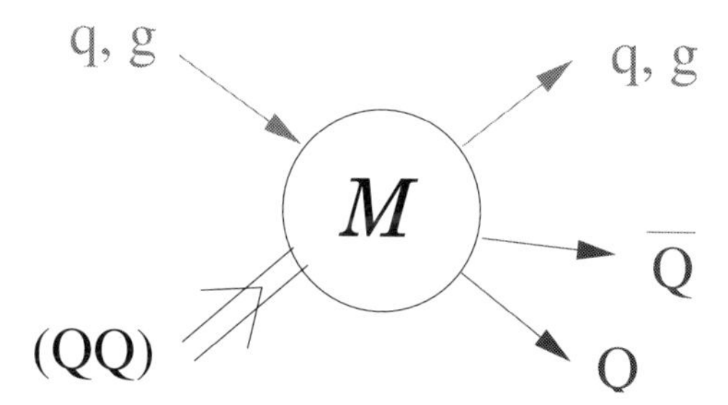

Fig. 1. Transition amplitude for a quarkonium breakup process in a fully developed quark-gluon plasma.

2 Quantum Kinetics for Quarkonium in a Plasma

The inverse lifetime of a state with 4-momentum p in a plasma is related to the imaginary part of its selfenergy by $\tau^{-1}(p) = \Gamma(p) = \Sigma^{>}(p) - \Sigma^{<}(p)$, where in the Born approximation for the quarkonium breakup by quark and gluon impact holds, see also Fig. 1

$$\Sigma^{\gtrless}(p) = \int_{p'} \int_{p_1} \cdots \int_{p_3} (2\pi)^4 \delta_{p+p',p_1+p_2+p_3} |\mathcal{M}|^2 G^<(p') G^>(p_1) G^>(p_2) G^>(p_3) ,$$

where the nonequilibrium Green functions can be expressed via distribution functions $f_i(p)$ and spectral functions $A_i(p)$ of the particle species i as $G_i^{\gtrless}(p) = [1 \pm f_i(p)]$, $G_i^<(p) = f_i(p)A_i(p)$. This quantum kinetic formulation includes not only the loss but also the reverse process of $Q\bar{Q}$ fusion in a consistent way. It can also be generalized to include strong correlations such as hadronic bound states or resonances in the visinity of the hadronization transition, see Fig. 2. First steps in this direction have been explored [23].

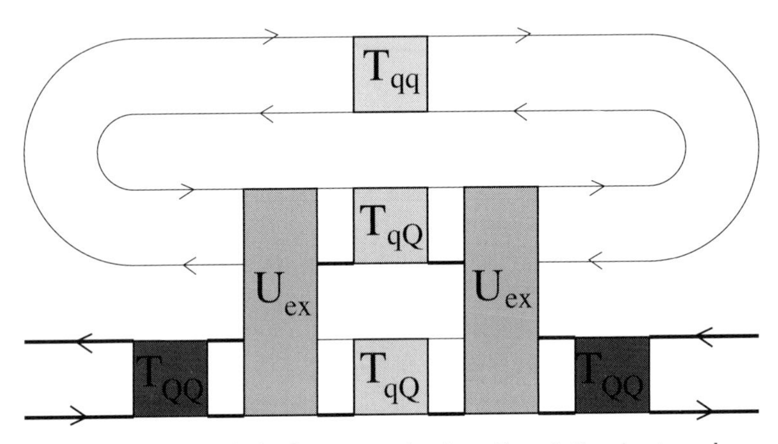

Fig. 2. Transition amplitude for heavy quarkonium dissociation in strongly correlated quark matter. The T-matrices stand for mesonic bound and/or scattering states.

3 Quarkonium Dissociation Cross Section in a Gluon Gas

In this section we give benchmark results for activation of Coulombic bound states by collisions in a medium. The medium will be represented as a gluon gas with thermal distribution function, $n_g(\omega) = g_g[\exp(\omega(p)/T) - 1]^{-1}$, where the degeneracy factor $g_g = 2(N_c^2 - 1)$..

The quarkonium breakup cross section by gluon impact can be estimated with the Bhanot/Peskin formula [4]

$$\sigma_{(Q\bar{Q})g}(\omega) = \frac{2^{11}}{3^4} \alpha_s \pi a_0^2 \frac{(\omega/\epsilon_0 - a(T))^{3/2}}{(\omega/\epsilon_0)^5} \Theta(\omega - \epsilon_0 a(T)) \tag{1}$$

with the binding energy ϵ_0 of the $1S$ quarkonium state with a Coulombic rms radius $\sqrt{\langle r^2 \rangle_{1S}} = \sqrt{3}a_0 = 2\sqrt{3}/(\alpha_s m_Q)$, and the heavy-quark mass m_Q, the energy of the impacting gluon is ω. With the parameter $0 < a(T) < 1$ we can

Table 1. Parameters of the heavy quarkonium systems: binding energy ε_0 and heavy quark mass m_Q from [5]. Left two columns refer to set(i), the right two columns to set(ii).

	set(i)		set(ii)	
$Q\bar{Q}$ system	ε_0[GeV]	m_Q[GeV]	ε_0 [GeV]	m_Q [GeV]
bottomonium	0.75	5.10	1.10	5.28
charmonium	0.78	1.94	0.62	1.86

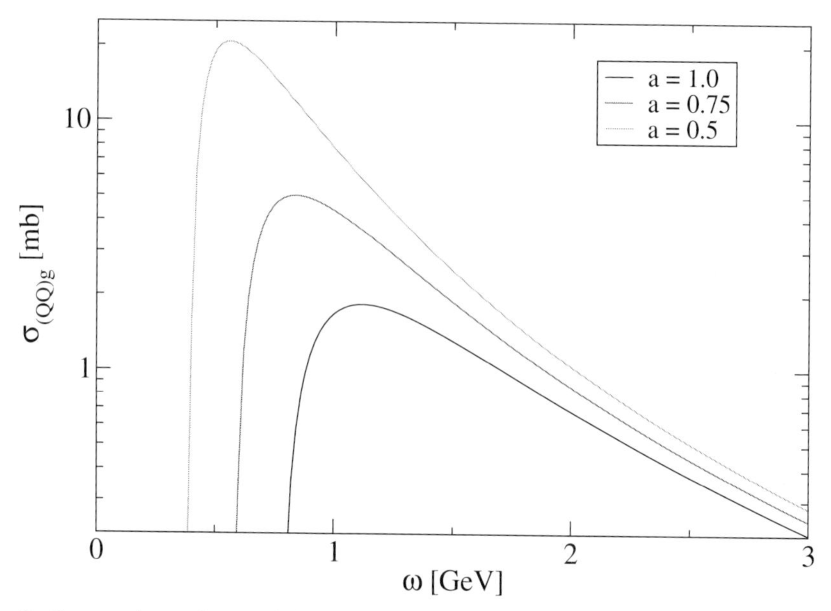

Fig. 3. Energy dependence of the quarkonium breakup cross section by gluon impact for different (constant) threshold depletion factors $a(T) = 1, 0.75, 0.5$.

vary the effective dissociation threshold from $\omega_{th}^{ideal} = \epsilon_0; a(T < T_c) = 1$ to $\omega_{th}^{Mott} = 0; a(T > T^{Mott}) = 0$. See Fig. 3 for numerical examples.

The dissociation rate for a heavy quarkonium $1S$ state at rest in a heat bath of massless gluons ($\omega = p$) at temparature T is

$$\frac{1}{\tau_{(Q\bar{Q})g}(T)} = \Gamma_{(Q\bar{Q})g}(T) = \langle \sigma_{(Q\bar{Q})g}^{ideal}(\omega) n_g(\omega) \rangle_T = \frac{1}{2\pi^2} \int_0^\infty \omega^2 d\omega \sigma_{(Q\bar{Q})g}^{ideal}(\omega) n_g(\omega) \tag{2}$$

and is shown in Fig. 4.

It is interesting to note that there is a discrepancy between the Lattice calculation of the charmonium decay with in a gluonic medium and the decay width of charmonium by gluon impact due to the gluonic E1 transition. It is conceivable that the Lattice simulation overestimates the thermal width of the charmonium states, but we would rather like to suggest that due to a lowering of the ionisation threshold in the gluonic medium the breakup rate will be anhanced. For an analytical estimate we adopt a temperature dependence of the strong coupling

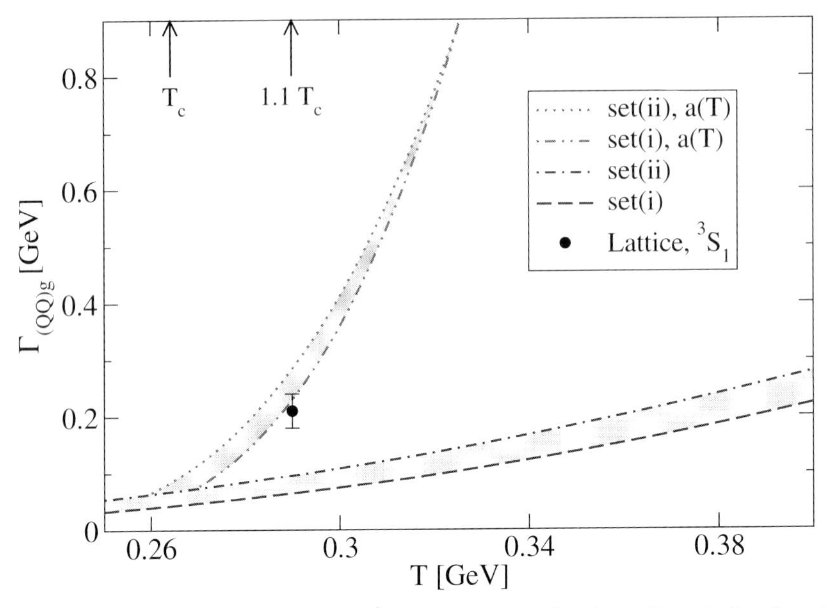

Fig. 4. Rate coefficient $\Gamma_{(Q\bar{Q})g} = \tau^{-1}$ for heavy quarkonium dissociation by gluon impact as a function of the gluon plasma temperature (no medium, $a(T) = 1$) parameters see Table 1. For comparison, the Breit-Wigner fit of the spectral width for $^1S_0$ charmonium from lattice simulations is shown. In-medium effects are estimated in two schemes: A)

constant by using the gluon momentum scale [6] $\langle p^2 \rangle = (3.2\ T)^2$ in the 1-loop β function such that

$$\frac{\varepsilon_0\ a(T)}{m_Q} = \left(\frac{3\pi}{22\ \ln(3.2\ T/\Lambda)} \right)^2 . \tag{3}$$

3.1 Quarkonia Abundances and Observable Signatures

In order to study observable signatures we will adopt here the Bjorken scenario [7] for the plasma evolution, i.e. longitudinal expansion with conserved entropy: $T^3\ t = T_0^3\ t_0 = \text{const}$. Parameters for the initial state are given in Table 2, where RHIC and LHC values are taken from [8], the SPS values are "canonical". As the quantity for comparison with experimental data of quarkonium production we consider the survival probability. We neglect here the hadronic comover and the nucleonic contributions, and also the effects due to the hadronisation phase

Table 2. Heavy-ion collision parameters from [8].

	LHC (3)	RHIC (3)	SPS
T_0 [GeV]	0.72	0.4	0.25
τ_0 [fm/c]	0.5	0.7	1.0

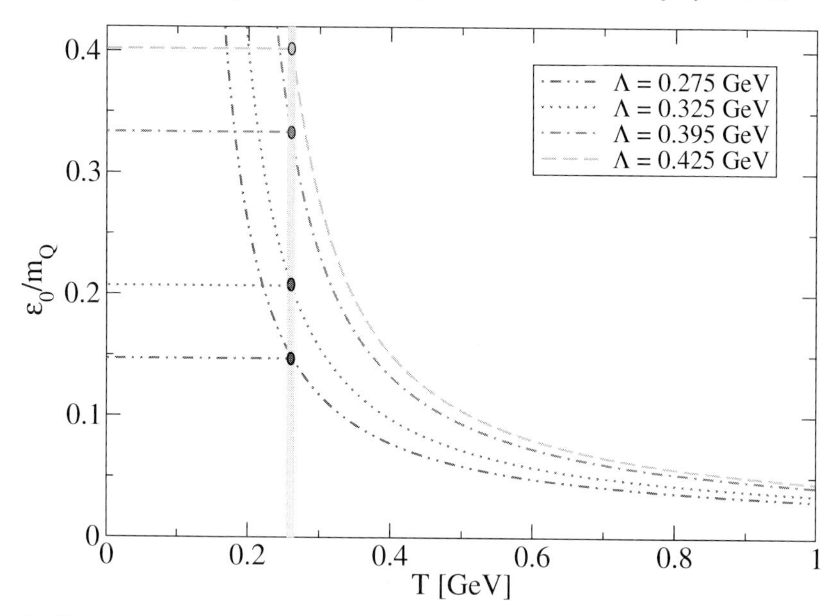

Fig. 5. Temperature dependence of the ionisation threshold due to the 1-loop running of the strong coupling constant (3).

transition

$$S(t_f) = \exp\left(-\int_{t_0}^{t_f} dt\ \tau^{-1}(T)\right) . \tag{4}$$

At the freeze-out time t_f the collisions stop to change the number of J/ψ (Υ). Using the Bjorken scaling it can be related into a freeze-out temperature. In Fig. 6 we give the survival probability for J/ψ (left panel) and for Υ (right panel) for the parameters of the Tables 1 and 2 due to gluon impact. For the LHC conditions, the simple estimates presented here don't give robust predictions for J/ψ. The Υ, however, shall be a good probe for the lifetime of the plasma state as well as for its temperature.

4 Quark Impact, $T > T_c$

The quarkonium breakup cross section by quark impact is estimated using the Bethe formula [9] for impact ionization of the $1S$ bound state of the Coulomb potential (the quark now plays the role of the impacting electron)

$$\sigma_{(Q\bar{Q})q}(\omega) = 2.5\pi a_0^2 \frac{\varepsilon_0}{\omega} \ln\left(\frac{\omega + \Delta(T)}{\varepsilon_0}\right) \Theta\left(\frac{\omega + \Delta(T)}{\varepsilon_0} - 1\right) . \tag{5}$$

Here we employ a generalization of Bethe's formula which takes into account the shift of the continuum edge of scattering states by $\Delta(T)$, which results in a temperature-dependent lowering of the ionisation threshold [15], $\varepsilon(T) =$

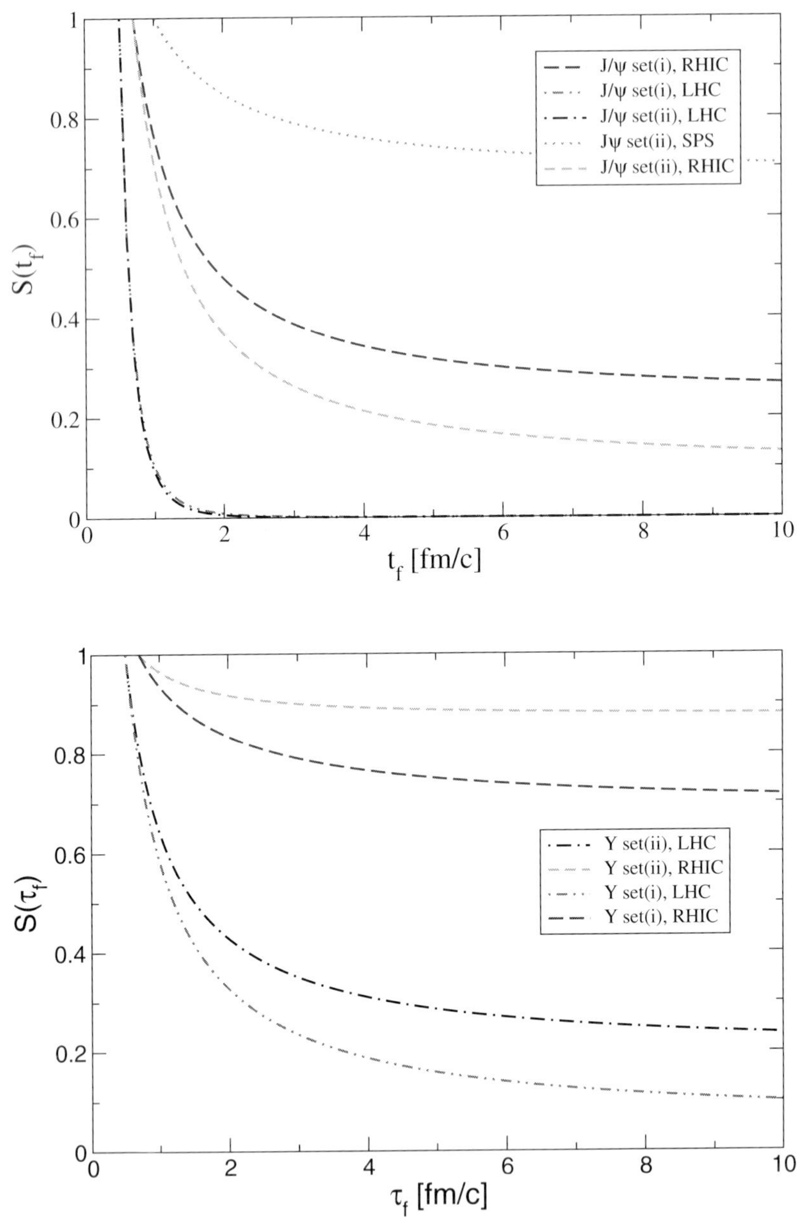

Fig. 6. Survival probability for heavy quarkonia states in a longitudinally expanding gluon plasma as a function of the plasma lifetime; ${}^1S_0$ charmonium (J/ψ, upper panel) and ${}^1S_0$ bottomonium (Υ, lower panel), parameters see Tables 1, 2.

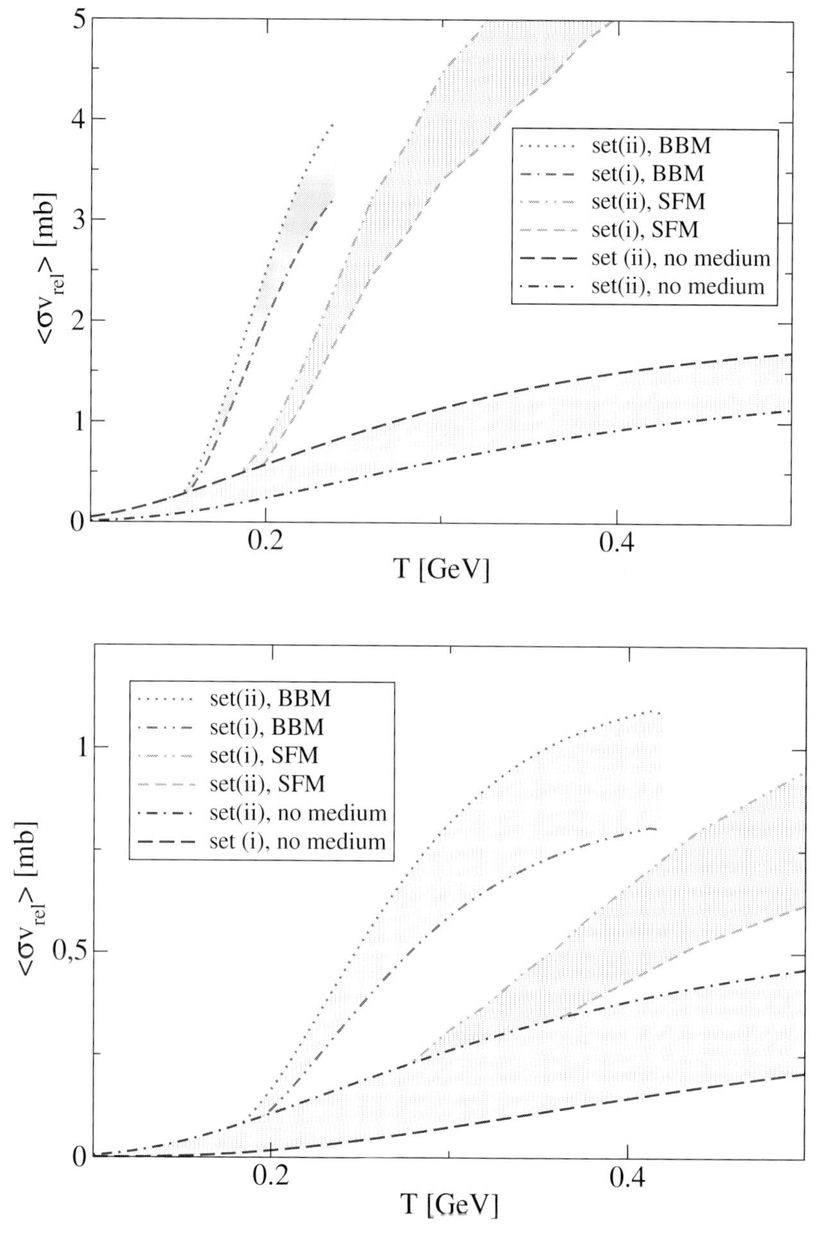

Fig. 7. Thermally averaged cross section for heavy quarkonium dissociation by quark impact as a function of the temperature: J/ψ (upper panel) and Υ (lower panel), parameters see Table 1.

$\varepsilon_0 - \Delta(T)$. This is named the Bethe-Born model (BBM). For quark matter systems, the energy shift $\Delta(T)$ has been obtained within the string-flip model (SFM), see [10,11]. Within this model also an effective cross section for heavy quarkonium breakup has been estimated [12,13]

$$\sigma(T) = \pi r_{Q\bar{Q}}^2(T) \exp(-\Delta(T)/T) \qquad (6)$$

It is remarkable that the enhancement of the breakup cross section due to the lowering of the threshold is comparable in magnitude in

5 Conclusions

In the QGP (and in the mixed phase), due to the presence of quasifree quarks and gluons, new channels for charmonium formation and dissociation exist which could drive chemical equilibration during the existence of the fireball formed in the heavy-ion collision. In this contribution we have given some benchmark estimates which have to be further elaborated. The role of partonic in-medium effects in the heavy quarkonium kinetics in a QGP, which has previously been discussed in the string-flip model of quark matter in the form of a modified mass action law [12] and dissociation rate [13], should be reconsidered by also including gluonic degrees of freedom which shall become dominant at RHIc and LHC conditions. We suggest that rate coefficients for the ionization and recombination of charm mesons could be described using an approach similar to methods used previously to study Coulomb plasmas [3,14,15]. It is suggested that the measurement of the Υ suppression at LHC can provide rather robust informations about the initial temperature (or lifetime) of the quark-gluon plasma.

Acknowledgement

V.Yu. acknowledges support by DFG under grant No. 436 RUS 17/../03.

References

1. T. Matsui and H. Satz, Phys. Lett. B **178** (1986) 416.
2. K. Kajantie, Quark Matter '84. Proceedings, 4th International Conference On Ultrarelativistic Nucleus-Nucleus Collisions, Helsinki, Finland, June 17-21, 1984,
3. R. Redmer, Phys. Rep. **282** (1997) 35
4. G. Bhanot and M.E. Peskin, Nucl. Phys. **B 156** (1979) 391
5. F. Arleo, P.-B. Gossiaux, T. Gousset, J. Aichelin, Phys. Rev. **D 65** (2001) 014005
6. J.I. Kapusta, Finite-temperature field theory, Cambridge University Press, Cambridge 1989
7. J. D. Bjorken, Phys. Rev. D **27** (1983) 140
8. X.-M. Xu, D. Kharzeev, H. Satz, X.-N. Wang, Phys. Rev. **C 53** (1996) 3051
9. H. Bethe, Ann. Phys. (Leipzig) **5** (1930) 325
10. D. Blaschke, F. Reinholz, G. Ropke and D. Kremp, Phys. Lett. B **151** (1985) 439.
11. G. Ropke, D. Blaschke and H. Schulz, Phys. Rev. D **34** (1986) 3499.

12. G. Röpke, D. Blaschke, and H. Schulz, *Phys. Lett. B* **202** (1988) 479.
13. G. Röpke, D. Blaschke and H. Schulz, Phys. Rev. **D 38** (1988) 3589
14. Th. Bornath and M. Schlanges *Physica A* **196** (1993) 427.
15. M. Schlanges, Th. Bornath, D. Kremp, Phys. Rev. **A 38** (1988) 2174.
16. Y. Nakahara, M. Asakawa and T. Hatsuda, Phys. Rev. D60 (1999) 091503
17. M. Asakawa, T. Hatsuda and Y. Nakahara, Prog. Part. Nucl. Phys. 46 (2001) 459
18. T. Umeda, R. Katayama, O. Miyamura and H. Matsufuru, Int. J. Mod. Phys. A 16 (2001) 2215
19. T. Umeda, K. Nomura and H. Matsufuru, hep-lat/0211003
20. S. Datta, F. Karsch, P. Petreczky and I. Wetzorke, hep-lat/0208012 (to be published in the proceedings of Lattice 2002)
21. M. Luescher et al, Nucl. Phys. B491 (1997) 344
22. F. Karsch, M.T. Mehr and H. Satz, Z. Phys. C (1988) 617
23. G. R. G. Burau, D. B. Blaschke and Y. L. Kalinovsky, Phys. Lett. B **506** (2001) 297 [arXiv:nucl-th/0012030].

Deep Inelastic J/ψ Production at HERA in the Colour Singlet Model with k_T-Factorization

A. Lipatov[1] and N. Zotov[2]

[1] Physical Department, M.V. Lomonosov Moscow State University, 119992 Moscow, Russia
[2] D.V. Skobeltsyn Institute of Nuclear Physics, M.V. Lomonosov Moscow State University, 119992 Moscow, Russia

Abstract. We consider J/ψ meson production in ep deep inelastic scattering in the colour singlet model using the k_T-factorization QCD approach. We investigate the z, Q^2, $\boldsymbol{p}_T^2$, y^* and W-dependences of inelastic J/ψ production on different forms of the unintegrated gluon distribution. The $\boldsymbol{p}_T^2$ and Q^2-dependences of the J/ψ spin aligment parameter α are presented also. We compare the theoretical results with recent experimental data taken by the H1 and ZEUS collaborations at HERA. It is shown that experimental study of the polarization J/ψ mesons at $Q^2 < 1\,\mathrm{GeV}^2$ is an additional test of BFKL gluon dynamics.

1 Introduction

It is known that from heavy quark and quarkonium production processes one can obtain unique information on gluon structure function of the proton because of the dominance of the photon-gluon or gluon-gluon fusion subprocess in the framework of QCD [1]. Studying gluon distributions at modern collider energy (such as HERA, Tevatron) is important for prediction of heavy quark and quarkonium production cross sections at future colliders (LHC, THERA). At the energies of HERA and LEP/LHC colliders heavy quark and quarkonium production processes are so called semihard processes [2]. In such processes by definition the hard scattering scale $\mu \sim m_Q$ is large compare to the Λ_{QCD} parameter but on the other hand μ is much less than the total center-of-mass energy: $\Lambda_{\mathrm{QCD}} \ll \mu \ll \sqrt{s}$. The last condition implies that the processes occur in small x region: $x \simeq m_Q/\sqrt{s} \ll 1$, and that the cross sections of heavy quark and quarkonium production processes are determined by the behavior of gluon distributions in the small x region.

It is also known that in the small x region the standard parton model (SPM) assumptions about factorization of gluon distribution functions and subprocess cross sections are broken because the subprocess cross sections and gluon structure functions depend on a gluon transverse momentum k_T [2]. So calculations of heavy quark production cross sections at HERA, Tevatron, LHC and other collider conditions are necessary to carry out in the so called k_T-factorization (or semihard) QCD approach, which is more preferable for the

A. Lipatov and N. Zotov, Deep Inelastic J/ψ Production at HERA in the Colour Singlet Model with k_T-Factorization, Lect. Notes Phys. **647**, 376–385 (2004)
http://www.springerlink.com/

small x region than SPM. The k_T-factorization QCD approach is based on Balitsky, Fadin, Kuraev, Lipatov (BFKL) [3] evolution equations. The resummation of the terms $\alpha_S^n \ln^n(\mu^2/\Lambda_{\rm QCD}^2)$, $\alpha_S^n \ln^n(\mu^2/\Lambda_{\rm QCD}^2) \ln^n(1/x)$ and $\alpha_S^n \ln^n(1/x)$ in the k_T-factorization approach leads to the unintegrated (dependent from $\boldsymbol{q}_T$) gluon distribution $\Phi(x, \boldsymbol{q}_T^2, \mu^2)$ which determine the probability to find a gluon carrying the longitudinal momentum fraction x and transverse momentum $\boldsymbol{q}_T$ at probing scale μ^2.

To calculate the cross section of a physical process the unintegrated gluon distributions have to be convoluted with off mass shell matrix elements corresponding to the relevant partonic subprocesses [2]. In the off mass shell matrix element the virtual gluon polarization tensor is taken in the BFKL form [2]:

$$L^{\mu\nu}(q) = \frac{q_T^\mu\, q_T^\nu}{\boldsymbol{q}_T^2}. \tag{1}$$

Nowadays, the significance of the k_T-factorization QCD approach becomes more and more commonly recognized [4]. It was already used for the description of a wide class heavy quark and quarkonium production processes [5–18]. It is notable that calculations in k_T-factorization approach provide results which are absent in other approaches, such as the fast growth of total cross sections in comparison with SPM, a broadening of the $\boldsymbol{p}_T$ spectra due to extra the transverse momentum of the colliding partons and other polarization properties of final particles in comparison with SPM.

We point out that heavy quark and quarkonium cross section calculations within the SPM in the fixed order of pQCD have some problems. For example, the very large discrepancy (by more than an order of magnitude) [19, 20] between the pQCD predictions for hadroproduction J/ψ and Υ mesons and experimental data at Tevatron was found. This fact has resulted in intensive theoretical investigations of such processes. In particular, it was required to use additional transition mechanism from $c\bar{c}$-pair to the J/ψ mesons, so-called the colour octet (CO) model [21], where $c\bar{c}$-pair is produced in the color octet state and transforms into final colour singlet (CS) state by help soft gluon radiation. The CO model was supposed to be applicable to heavy quarkonium hadro and electroproduction processes. However, the contributions from the CO mechanism to the J/ψ meson photoproduction contradict the experimental data at HERA for z-distribution [22–25].

Another difficulty of the CO model are the J/ψ polarization properties in $p\bar{p}$-interactions at the Tevatron. In the framework of the CO model, the J/ψ mesons should be transverse polarized at the large transverse momenta $\boldsymbol{p}_T$. However, this is in contradiction with the experimental data, too.

The CO model has been applied earlier in an analysis of the J/ψ inelastic production experimental data at HERA [26, 27]. However, the results do not agree with each other [27]. It is notable that results obtained within the usual collinear approach and CS model [28–31] underestimate experimental data by factor about 2.

Recently, first attempts to investigate the J/ψ polarization problem in $p\bar{p}$-interactions at Tevatron in k_T-factorization approach were made [17, 32, 33].

Also the theoretical prediction within semihard QCD approach [13] are stimulated the experimental analysis of J/ψ polarization properties at HERA conditions. However further theoretical and experimental studies of this problem are necessary.

Based on the above mentioned results we consider here deep inelastic J/ψ production at HERA in the framework of the CS model and k_T-factorization approach with emphasis of a role of the gluon distribution function. We investigate the Q^2, $\boldsymbol{p}_T^2$, z, y^* and W-dependences of J/ψ production on different forms of the unintegrated gluon distribution. Special attention is drawn to the unintegrated gluon distributions obtained from BFKL evolution equation which has been applied earlier in our previous papers [9–12]. For studying J/ψ meson polarization properties we calculate the $\boldsymbol{p}_T^2$ and Q^2-dependences of the spin aligment parameter α.

The outline of our paper is as follows. In Sect. 2 we give some details of the calculations. In Sect. 3 we compare the our numerical results with experimental data obtained by the H1 [34] and ZEUS [35] collaborations at HERA. Finally, in Sect. 4 we give some conclusions.

2 Details of the Calculations

In the k_T-factorization approach deep inelastic J/ψ meson production is determined by the contribution of six photon-gluon fusion diagrams with "unusual" properties of the gluons in proton. These gluons are off mass shell with the virtuality $q^2 = q_T^2 = -\boldsymbol{q}_T^2$, and their distribution in x and $\boldsymbol{q}_T^2$ in a proton is given by the unintegrated gluon structure function $\Phi(x, \boldsymbol{q}_T^2, \mu^2)$. The differential cross section for the deep inelastic J/ψ production differential cross section in the k_T-factorization approach has the following form [11]:

$$
\begin{aligned}
d\sigma(e\,p \to e'\,J/\psi\,X) &= \frac{1}{128\pi^3}\,\frac{\Phi(x_2,\,\boldsymbol{q}_{2T}^2,\,\mu^2)}{(x_2\,s)^2\,(1-x_1)}\,\frac{dz}{z\,(1-z)}\,dy_{J/\psi}\,\times \\
\times \sum |M|^2_{\mathrm{SHA}}(e\,g^* \to e'\,J/\psi\,g')\,&d\boldsymbol{p}_{J/\psi\,T}^2\,dQ^2\,d\boldsymbol{q}_{2T}^2\,\frac{d\phi_1}{2\pi}\,\frac{d\phi_2}{2\pi}\,\frac{d\phi_{J/\psi}}{2\pi}\,,
\end{aligned}
\tag{2}
$$

where $\boldsymbol{p}_{J/\psi\,T}$ and $\boldsymbol{q}_{2\,T}$ are transverse 4-momenta of the J/ψ meson and initial BFKL gluon, $y_{J/\psi}$ is the rapidity of J/ψ meson (in the ep c.m. frame), ϕ_1, ϕ_2 and $\phi_{J/\psi}$ are azimuthal angles of the incoming virtual photon, BFKL gluon and outgoing J/ψ meson, x_1 and x_2 are the photon and gluon momentum fraction.

The matrix element $\sum |M|^2_{\mathrm{SHA}}(e\,g^* \to e'\,J/\psi\,g')$ depends on the BFKL gluon virtuality $\boldsymbol{q}_{2\,T}^2$ and differs from the one of the usual parton model. We used here the results obtained in [11]. For studying J/ψ polarized production we used the 4-vector of the longitudinal polarization ε_L^μ as in [36].

There are several theoretical approximations for $\Phi(x, \boldsymbol{q}_T^2, \mu^2)$, which are based on solution of the BFKL evolution equations. As in our previous papers [9–12], we used here so called JB, KMS and GBW parametrizations (see also [4, 37] for the detail information).

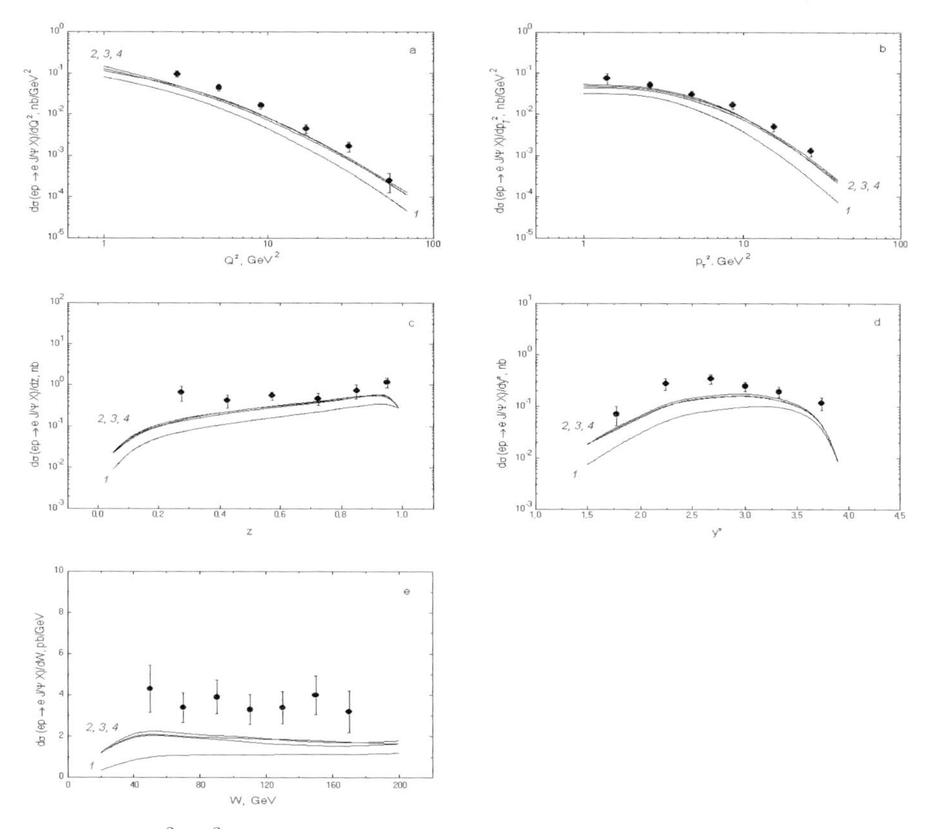

Fig. 1. The Q^2-, $\mathbf{p}^2_{J/\Psi\, T}$-, z-, $y^*_{J/\Psi}$- and W-distributions of the inelastic J/Ψ production.

3 Numerical Results and Discussion

There are three parameters which determine the common normalization factor of the cross section under consideration: J/ψ meson wave function at the origin $\psi(0)$, charmed quark mass m_c and factorization scale μ. The value of the J/ψ meson wave function at the origin may be calculated in a potential model or obtained from the well known experimental decay width $\Gamma(J/\psi \to \mu^+\,\mu^-)$. In our calculation we used $|\psi(0)|^2 = 0.0876\,\mathrm{GeV}^3$ as in [38]. Concerning a charmed quark mass, the situation is not clear: on the one hand, in the nonrelativistic approximation one has $m_c = m_{J/\psi}/2 = 1.55\,\mathrm{GeV}$, but on the other hand there are many examples when smaller value of a charm mass is used, for example, $m_c = 1.4\,\mathrm{GeV}$ [27, 39]. In the present paper we used both of charm mass values. Also the most significant theoretical uncertanties come from the choice of the factorization scale μ_F and renormalization one μ_R. One of them is related to the evolution of the gluon distributions $\Phi(x, \boldsymbol{q}_T^2, \mu_F^2)$, the other is responsible for strong coupling constant $\alpha_S(\mu_R^2)$. As often done in literature, we set $\mu_F = \mu_R = \mu$. In the present paper we used the following choice $\mu^2 = \boldsymbol{q}_{2T}^2$ as in [40].

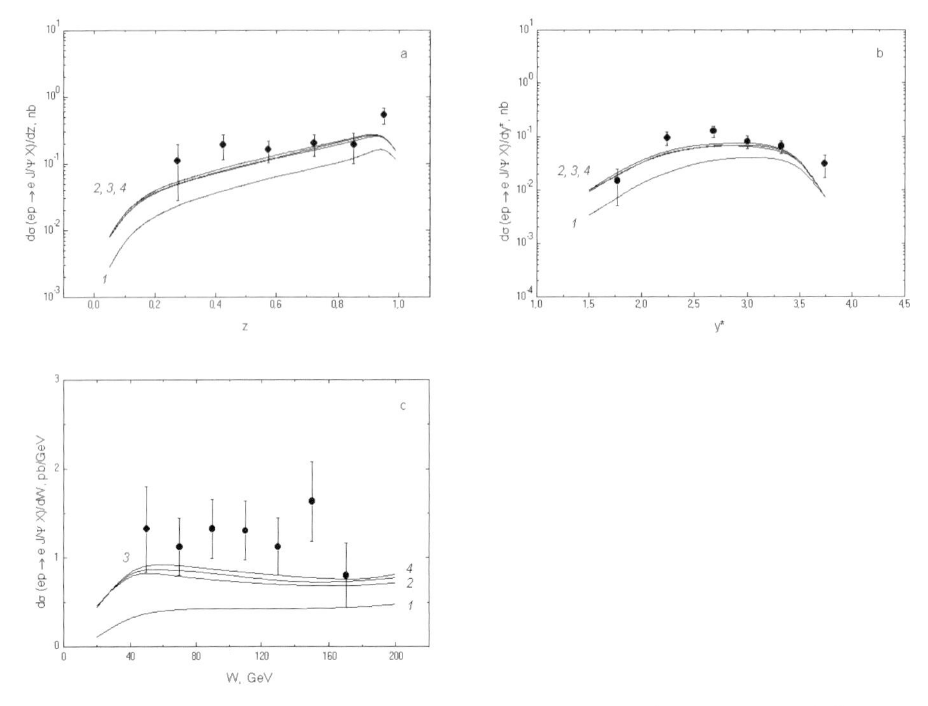

Fig. 2. The z-, $y^*_{J/\Psi}$- and W-distributions of the inelastic J/Ψ production ($\mathbf{p}^2_{J/\Psi\, T} \geq 4\,\mathrm{GeV}^2$).

The integration limits in (2) are taken as given by kinematical conditions of H1 experimental data [34]. One kinematical region is $2 \leq Q^2 \leq 80\,\mathrm{GeV}^2$, $40 \leq W \leq 180\,\mathrm{GeV}$, $z > 0.2$, $M_X \geq 10\,\mathrm{GeV}$ and other kinematical region is $\mathbf{p}^2_{J/\psi\, T} \geq 4\,\mathrm{GeV}^2$, $4 \leq Q^2 \leq 80\,\mathrm{GeV}^2$, $40 \leq W \leq 180\,\mathrm{GeV}$, $z > 0.2$, $M_X \geq 10\,\mathrm{GeV}$.

The results of our calculations are shown in Fig. 1–6. Figure 1 show the Q^2, z, $\mathbf{p}^2_{J/\psi\, T}$, $y^*_{J/\psi}$ [1] and W-distributions of the inelastic J/ψ meson production obtained in the first kinematical region at $\sqrt{s} = 314\,\mathrm{GeV}$, $m_c = 1.55\,\mathrm{GeV}$ and $\Lambda_{\mathrm{QCD}} = 250\,\mathrm{MeV}$. Curve *1* corresponds to the SPM calculations at the leading order approximation with GRV gluon density, curves *2*, *3* and *4* correspond to the k_T-factorization results with JB, KMS and GBW unintegrated gluon distributions. One can see that results obtained in the CS model with k_T-factorization (curves *2*–*4*) agree in shape but underestimate the H1 experimental data. The SPM calculation (curve *1*) are lower than the data by a factor 2.

Figure 2 show the z, $y^*_{J/\psi}$ and W-distributions of the inelastic J/ψ meson production obtained in the second kinematical region at $\sqrt{s} = 314\,\mathrm{GeV}$, $m_c = 1.55\,\mathrm{GeV}$ and $\Lambda_{\mathrm{QCD}} = 250\,\mathrm{MeV}$. Curves *1*–*4* are the same as in Fig. 1. One can

[1] The J/ψ meson rapidity in the $\gamma^* p$ c.m. frame.

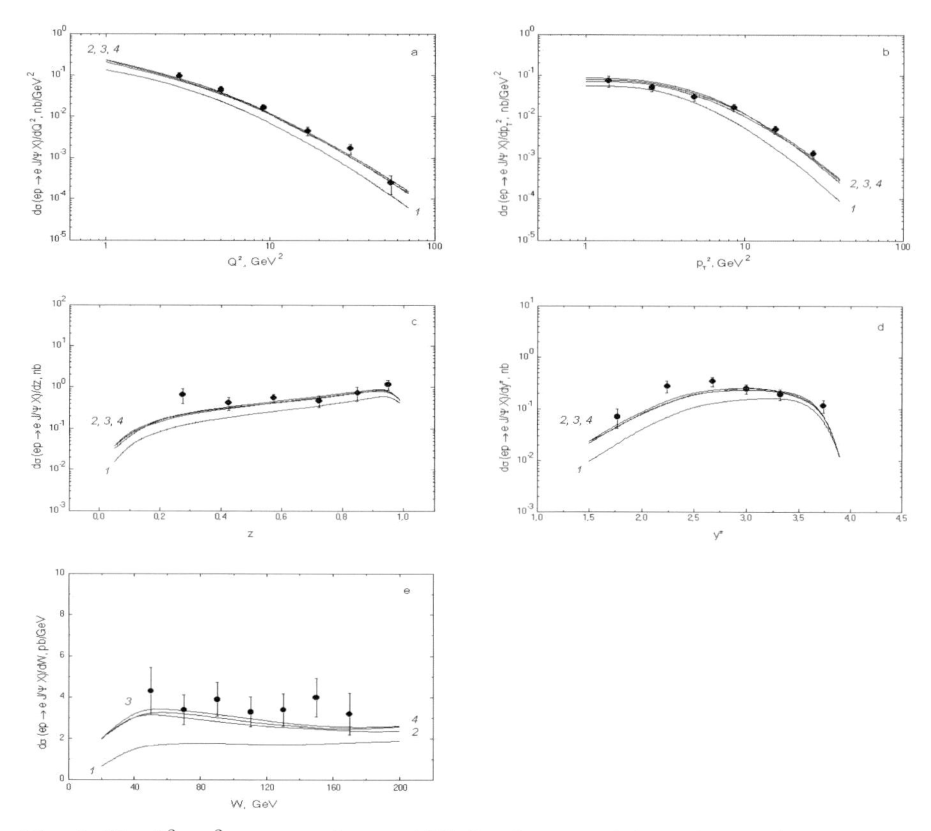

Fig. 3. The Q^2-, $\mathbf{p}^2_{J/\Psi\,T}$-, z-, $y^*_{J/\Psi}$- and W-distributions of the inelastic J/Ψ production ($m_c = 1.4\,\mathrm{GeV}$).

see that in this kinematical region the k_T-factorization approach (curves *2–4*) describe the data better than in the first one (without cut $\boldsymbol{p}^2_{J/\psi\,T} \geq 4\,\mathrm{GeV}^2$).

Figures 3, 4 show the Q^2, z, $\boldsymbol{p}^2_{J/\psi\,T}$, $y^*_{J/\psi}$ and W-distributions of the inelastic J/ψ meson production at $\sqrt{s} = 314\,\mathrm{GeV}$, $m_c = 1.4\,\mathrm{GeV}$ and $\Lambda_{\mathrm{QCD}} = 250\,\mathrm{MeV}$ obtained in the first kinematical region (Fig. 3) and in the second one (Fig. 4). Curves *1–4* are the same as in Fig. 1. One can see that a shift the charm quark mass down to $m_c = 1.4\,\mathrm{GeV}$ increases the cross section by a factor 1.5, and results obtained in the CS model with k_T factorization (curves *2–4*) agree with H1 experimental data.

The z-distributions are described at $z \geq 0.4$ only. It is because other J/ψ production mechanisms (such as resolved photon and/or colour octet contributions) may be impotant at $z < 0.4$ [41].

It is interesting to note that the saturation unintegrated gluon distribution (GBW parametrization) does not contradict the HERA experimental data.

Figures 1–4 show that k_T-factorization results with $m_c = 1.4\,\mathrm{GeV}$ (in contrast with SPM ones) for inelastic J/ψ electroproductionat HERA reasonably

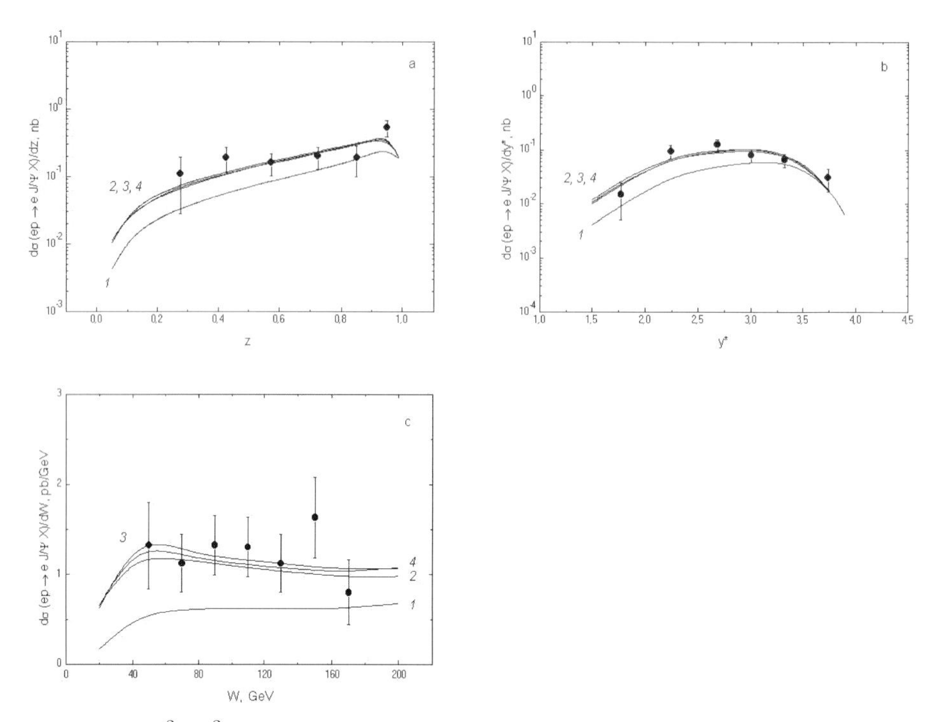

Fig. 4. The Q^2-, $\mathbf{p}^2_{J/\Psi\,T}$-, z-, $y^*_{J/\Psi}$- and W-distributions of the inelastic J/Ψ production ($\mathbf{p}^2_{J/\Psi\,T} \geq 4\,\mathrm{GeV}^2$, $m_c = 1.4\,\mathrm{GeV}$).

agree with H1 experimental data. Hovewer, at $m_c = 1.55\,\mathrm{GeV}$ they are lower than the data by a factor 1.5.

We analyzed specially the influence of charm quark mass, m_c, on the our theoretical results. We found that the main effect of change of the charm quark mass connects with final phase space of J/ψ meson. In the subprocess matrix elements this effect is neglectable. However the value of $m_c = 1.4\,\mathrm{GeV}$ corresponds to the unphysical phase space of J/ψ state[2].

For studying J/ψ meson polarization properties we calculate the $\boldsymbol{p}_T^2$ and Q^2-dependences of the spin aligment parameter α [11, 13, 42], which controls the angular distribution for leptons in the decay $J/\psi \to \mu^+\,\mu^-$ in the J/ψ meson rest frame. Figure 5 shows the parameter $\alpha(|\boldsymbol{p}_{J/\psi\,T}|)$, which is calculated in the region $0.4 < z < 0.9$ (a) and in the region $0.4 < z < 1$ (b) at $\sqrt{s} = 314\,\mathrm{GeV}$, $m_c = 1.4\,\mathrm{GeV}$ and $\Lambda_{\mathrm{QCD}} = 250\,\mathrm{MeV}$ in comparison with preliminary experimental data taken by the ZEUS [35] collaboration at HERA. Curve *1* corresponds to the SPM calculations at the leading order approximation with the GRV gluon density, curve *3* corresponds to the k_T-factorization QCD calculations with the KMS unintegrated gluon distribution. We note that it is impossible to make of exact conclusions about a BFKL gluon contribution to the polarized J/ψ

[2] We thank S. Baranov for the suggestion to study this problem.

production cross section because of large incertainties in the experimental data and large additional contribution from initial longitudinal polarization of virtual photons. However at low $Q^2 < 1\,\mathrm{GeV}^2$ such contributions are negligible. This fact should result in observable spin effects of final J/ψ mesons. As example, we have performed calculations for the spin parameter α as a function $\boldsymbol{p}^2_{J/\psi\,T}$ at fixed values of Q^2 for $40 \leq W \leq 180\,\mathrm{GeV}$, $z > 0.2$, $M_X \geq 10\,\mathrm{GeV}$ at $\sqrt{s} = 314\,\mathrm{GeV}$, $m_c = 1.4\,\mathrm{GeV}$ and $\Lambda_{\mathrm{QCD}} = 250\,\mathrm{MeV}$.

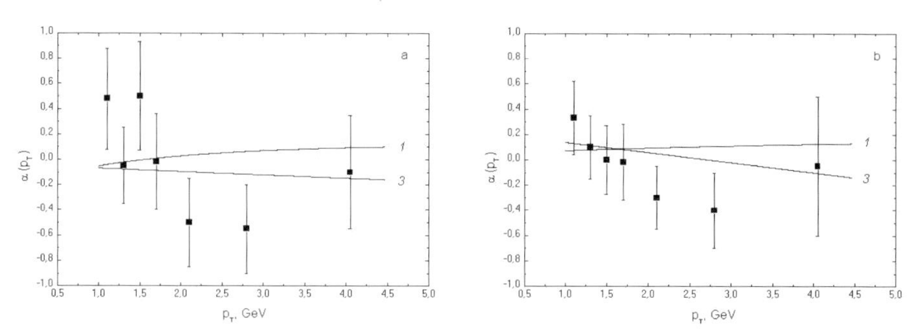

Fig. 5. The parameter $\alpha(\mathbf{p}_{J/\Psi\,T})$, which is calculated in the region $0.4 < z < 0.9$ (a) and in the region $0.4 < z < 1$ (b).

The results of our calculations at fixed values of $Q^2 = 0.1\,\mathrm{GeV}^2$, $Q^2 = 1\,\mathrm{GeV}^2$, $Q^2 = 5\,\mathrm{GeV}^2$ and $Q^2 = 10\,\mathrm{GeV}^2$ are shown in Fig. 6. Curves *1* and *3* are the same as in Fig. 5. We have large difference between predictions of the leading order of SPM and the k_T-factorization approach at low $Q^2 < 1\,\mathrm{GeV}^2$, as it is seen in Fig. 6. Therefore experimental measurement of polarization properties of the J/ψ mesons at low Q^2 will be an additional test of BFKL gluon dynamics.

4 Conclusions

In this paper we considered J/ψ meson production in ep deep inelastic scattering in the colour singlet model using the standard parton model in leading order in α_S and the k_T-factorization QCD approach. We investigated the z, Q^2, $\boldsymbol{p}^2_T$, y^*, W-dependences of inelastic J/ψ production on different forms of the unintegrated gluon distribution. The $\boldsymbol{p}^2_T$ and Q^2-dependences of the spin aligment parameter α presented also. We compared the theoretical results with available experimental data taken by the H1 and ZEUS collaborations at HERA. We have found that the k_T-factorization results (in contrast with the SPM) with the JB, KMS and GBW unintegrated gluon distributions reasonably agree with H1 experimental data at $m_c = 1.4\,\mathrm{GeV}$, $|\Psi(0)|^2 = 0.0876\,\mathrm{GeV}^3$ and $\Lambda_{\mathrm{QCD}} = 250\,\mathrm{MeV}$. At $m_c = 1.55\,\mathrm{GeV}$ they are lower than the data[3] by a factor 1.5. The saturation unintegrated gluon distribution (GBW parametrization) does not contradict the

[3] When the present paper has been prepared to publication, [43] appeared, in which experimental data taken by H1 collaboration at HERA were measured with increa-

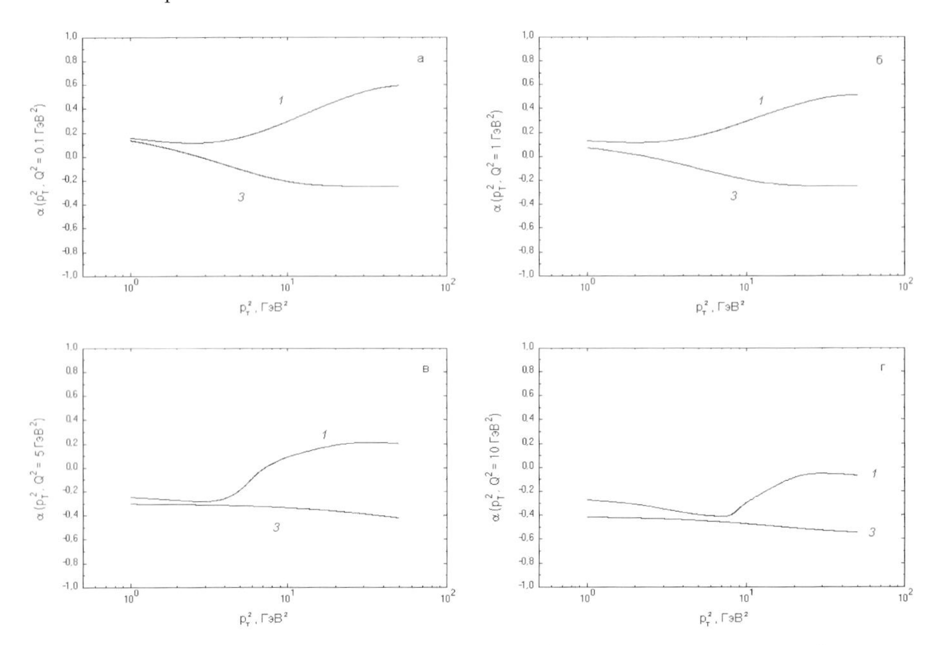

Fig. 6. The spin aligment parameter $\alpha(\mathbf{p}^2_{J/\Psi\, T})$ at fixed values of Q^2.

existing experimental data. The experimental study of a polarization of J/ψ meson at low $Q^2 < 1\,\mathrm{GeV}^2$ should be an additional test of BFKL gluon dynamics.

The authors would like to thank S. Baranov for encouraging interest and useful discussions, and H. Jung for careful reading manuscript of our paper and useful remarks. A.L. thanks also V. Saleev for the help on the initial stage of work. The study was supported in part by RFBR grant 02–02–17513 and INTAS grant YSF 2002–399.

References

1. E. Berger, D. Jones: Phys. Rev. D **23**, 1521 (1981);
   S. Gershtein, A. Likhoded, S. Slabospitsky: Sov. J. Nucl. Phys. **34**, 128 (1981);
   R. Baier, R. Ruckl: Nucl. Phys. B **218**, 289 (1983)
2. L. Gribov, E. Levin, M. Ryskin: Phys. Rep. **100**, 1 (1983);
   S. Catani, M. Ciafoloni, F. Hautmann: Nucl. Phys. B **366**, 135 (1991);
   J. Collins, R. Ellis: Nucl. Phys. B **360**, 3 (1991)
3. E. Kuraev, L. Lipatov, V. Fadin: Sov. Phys. JETP **44**, 443 (1976); **45**, 199 (1977);
   Yu. Balitsky, L. Lipatov: Sov. J. Nucl. Phys. **28**, 822 (1978)
4. B. Andersson *et al.*: DESY 02-041; hep-ph/0204115
5. M. Ryskin, Yu. Shabelski: Z. Phys. C **69**, 269 (1996)
6. M. Ryskin, Yu. Shabelski: Z. Phys. C **61**, 517 (1994); C **66**, 151 (1995)

---

sed statistics and precision as compared with [34]. We note that our preliminary results obtained in the colour singlet model and k_T-factorization approach agree with data [43] at $m_c = 1.55\,\mathrm{GeV}$.

7. V. Saleev, N. Zotov: Mod. Phys. Lett. A **9**, 151 (1994)
8. V. Saleev, N. Zotov: Mod. Phys. Lett. A **11**, 25 (1996)
9. A. Lipatov, N. Zotov: Mod. Phys. Lett. A **15**, 695 (2000)
10. A. Lipatov, V. Saleev, N. Zotov: Mod. Phys. Lett. A **15**, 1727 (2000)
11. A. Lipatov, N. Zotov: hep-ph/0208237
12. A. Kotikov, A. Lipatov, G. Parente, N. Zotov: hep-ph/0107135
13. S. Baranov: Phys. Lett. B **428**, 377 (1998)
14. S. Baranov, N. Zotov: Phys. Lett. B **458**, 389 (1999); B **491**, 111 (2000)
15. S. Baranov, M. Smizanska: Phys. Rev. D **62**, 014012 (2000)
16. H. Jung: Phys. Rev. D **65**, 034015 (2002); hep-ph/0110034
17. P. Hagler, R. Kirschner, A. Schefer *et al.*: Phys. Rev. D **62**, 071502 (2000); D **63**, 077501 (2001);
    F. Yuan, K.-T. Chao: Phys. Rev. D **63**, 034006 (2001); hep-ph/0009224
18. H. Jung: hep-ph/0110345
19. E. Braaten, S. Fleming: Phys. Rev. Lett. **74**, 3327 (1995);
    E. Braaten, T. Yuan: Phys. Rev. D **52**, 6627 (1995)
20. P. Cho, A. Leibovich: Phys. Rev. D **53**, 150 (1996); D **53**, 6203 (1996)
21. G. Bodwin, E. Braaten, G. Lepage: Phys. Rev. D **51**, 1125 (1995); D **55**, 5853 (1997)
22. M. Cacciari, M. Kramer: Phys. Rev. Lett. **76**, 4128 (1996)
23. P. Ko, J. Lee, H. Song: Phys. Rev. D **54**, 4312 (1996); D **60**, 119902 (1999)
24. S. Aid *et al.* (H1 Collab.): Nucl. Phys. B **472**, 3 (1996)
25. J. Breitweg *et al.* (ZEUS Collab.): Z. Phys. C **96**, 599 (1997)
26. S. Fleming, T. Mehen: Phys. Rev. D **57**, 1846 (1998)
27. B. Kniehl, L. Zwirner: Nucl. Phys. B **621**, 337 (2002); hep-ph/0112199
28. J. Korner, J. Cleymans, M. Kuroda, G. Gounaris: Phys. Lett. B **114**, 195 (1982)
29. J.-Ph. Guillet: Z. Phys. C **39**, 75 (1988)
30. H. Merabet, J. Mathiot, R. Mendez-Galain: Z. Phys. C **62**, 639 (1994)
31. F. Yuan, K.-T. Chao: Phys. Rev. D **63**, 034017 (2001)
32. F. Yuan, K.-T. Chao: Phys. Rev. Lett. D **87**, 022002-L (2001)
33. Ph. Hagler *et al.*: Phys. Rev. Lett. **86**, 1446 (2001)
34. C. Adloff *et al.* (H1 Collab.): DESY-99-026
35. R. Brugnera (ZEUS Collab.). In: *9th International Workshop on DIS and QCD (DIS'2001), Bologna, Italy, 2001*
36. M. Beneke, M. Kramer: Phys. Rev. D **55**, 5269 (1997)
37. S. Baranov, H. Jung, L. Jönsson *et al.*: Eur. Phys. J. C **24**, 425 (2002)
38. M. Kramer: Nucl. Phys. B **459**, 3 (1996)
39. P. Ball, M. Beneke, V. Braun: Phys. Rev. D **52**, 3929 (1995)
40. E. Levin, M. Ryskin, Yu. Shabelsky, A. Shuvaev: Yad. Fiz. **54**, 1420 (1991)
41. H1 Collab. In: *International Eur. Phys. Conference on High Energy Physics (HEP'99), Tampere, Finland, 1999*
42. S. Baranov, A. Lipatov, N. Zotov. In: *9th International Workshop on DIS and QCD (DIS'2001), Bologna, Italy, 2001*; hep-ph/0106229
43. H1 Collab.: DESY 02-060; hep-ex/0205065

The Structure Functions F_2^c, F_L and F_L^c in the Framework of the k_T Factorization

A.V. Kotikov[1], A.V. Lipatov[2], G. Parente[3], and N.P. Zotov[4]

[1] BLThPh, JINR, 141980 Dubna, Russia
[2] Dep. of Physics, MSU, 119899 Moscow, Russia
[3] Univ. de Santiago de Compostela, 15706 Santiago de Compostela, Spain
[4] INP, MSU, 119899 Moscow, Russia

Abstract. We present the perturbative parts of the structure functions F_2^c and F_L^c for a gluon target having nonzero transverse momentum squared at order α_s. The results of the double convolution (with respect to the Bjorken variable x and the transverse momentum) of the perturbative part and the unintegrated gluon densities are compared with HERA experimental data for F_2^c and F_L at low x values and with predictions of other approaches. The contribution from F_L^c structure function ranges $10 \div 30\%$ of that of F_2^c at the HERA kinematical range.

1 Introduction

The basic information on the internal structure of nucleons is extracted from the process of deep inelastic (lepton-hadron) scattering (DIS). Its differential cross-section has the form:

$$\frac{d^2\sigma}{dxdy} = \frac{2\pi\alpha_{em}^2}{xQ^4} \left[\left(1-y+y^2/2\right) F_2(x,Q^2) - \left(y^2/2\right) F_L(x,Q^2)\right],$$

where $F_2(x,Q^2)$ and $F_L(x,Q^2)$ are the transverse and longitudinal structure functions (SF), respectively, q^μ and p^μ are the photon and the hadron 4-momentums and $x = Q^2/(2pq)$ with $Q^2 = -q^2 > 0$.

In the lecture we will study only the charm part F_2^c of the transverse SF F_2 and the longitudinal SF F_L.

The study is related with the fact that recently there have been important new data on the charm SF F_2^c, of the proton from the H1 [1] and ZEUS [2] Collaborations at HERA, which have probed the small-x region down to $x = 8 \times 10^{-4}$ and $x = 2 \times 10^{-4}$, respectively. At these values of x, the charm contribution to the total proton SF, F_2, is found to be around 25%, which is a considerably larger fraction than that found by the European Muon Collaboration (EMC) at CERN [3] at larger x, where it was only $\sim 1\%$ of F_2. Extensive theoretical analyses in recent years have generally served to confirm that the F_2^c data can be described through perturbative generation of charm within QCD (see, for example, the review in [4] and references therein).

The second object of our study is the longitudinal SF $F_L(x,Q^2)$. It is a very sensitive QCD characteristic because it is equal to zero in the parton model

A.V. Kotikov, A.V. Lipatov, G. Parente, and N.P. Zotov, The Structure Functions F_2^c, F_L and F_L^c in the Framework of the k_T Factorization, Lect. Notes Phys. **647**, 386–400 (2004)
http://www.springerlink.com/

with spin$-1/2$ partons. Unfortunately, essentially at small values of x, the experimental extraction of F_L data is required a rather cumbersome procedure (see [5,6], for example). Moreover, the perturbative QCD leads to some controversial results in the case of SF F_L. The next-to-leading order (NLO) corrections to the longitudinal coefficient function, which are large and negative at small x [7,8], need a resummation procedure [1] that leads to coupling constant scale higher essentially then Q^2 (see [10,8,11]) [2].

Recently there have been important new data [17]-[22] of the longitudinal SF F_L, which have probed the small-x region down to $x \sim 10^{-2}$. Moreover, the SF F_L can be related at small x with SF F_2 and the derivation $dF_2/d\ln(Q^2)$ (see [23]). In this way most precise predictions based on data of F_2 and $dF_2/d\ln(Q^2)$ (see [19] and references therein) can be obtained for F_L. These predictions can be considered as indirect 'experimental data' for F_L.

We note, that perhaps more relevant analyses of the HERA data, where the x values are quite small, are those based on BFKL dynamics [16], because the leading $\ln(1/x)$ contributions are summed. The basic dynamical quantity in BFKL approach is the unintegrated gluon distribution $\varphi_g(x,k_\perp^2)$ [3] (f_g is the (integrated) gluon distribution multiplied by x and $k_\perp$ is the transverse momentum)

$$f_g(x,Q^2) = f_g(x,Q_0^2) + \int_{Q_0^2}^{Q^2} dk_\perp^2\, \varphi_g(x,k_\perp^2), \tag{1}$$

which satisfies the BFKL equation. The integral is divergent at the lower limit and it leads to the necessity to use the difference $f_g(x,Q^2) - f_g(x,Q_0^2)$ with some nonzero Q_0^2.

Then, in the BFKL-like approach (hereafter the k_t-factorization approach [24,25] is used) the SF $F_{2,L}^c(x,Q^2)$ and $F_L(x,Q^2)$ are driven at small x by gluons and are related in the following way to the unintegrated distribution $\varphi_g(x,k_\perp^2)$:

$$F_{2,L}^c(x,Q^2) = \int_x^1 \frac{dz}{z} \int dk_\perp^2\, e_c^2\, C_{2,L}^g(z,Q^2,m_c^2,k_\perp^2)\, \varphi_g(x/z,k_\perp^2), \tag{2}$$

$$\begin{aligned} F_L(x,Q^2) &- F_L^c(x,Q^2) \\ &= \int_x^1 \frac{dz}{z} \int dk_\perp^2 \sum_{i=u,d,s} e_i^2\, C_{2,L}^g(z,Q^2,0,k_\perp^2)\, \varphi_g(x/z,k_\perp^2), \end{aligned} \tag{3}$$

where e_i is the charge of the i-flavor quark.

The functions $C_{2,L}^g(x,Q^2,m_c^2,k_\perp^2)$ and $C_L^g(x,Q^2,0,k_\perp^2)$ may be regarded as the structure functions of the off-shell gluons with virtuality $k_\perp^2$ (hereafter we call

[1] Without a resummation the NLO approximation of SF F_L can be negative at low x and quite low Q^2 values (see [8,9]).

[2] Note that at low x a similar property has been observed also in the approaches [12–14] (see recent review [15] and discussions therein), which based on Balitsky-Fadin-Kuraev-Lipatov (BFKL) dynamics [16], where the leading $\ln(1/x)$ contributions are summed.

[3] Hereafter k^μ is the gluon 4-momentum.

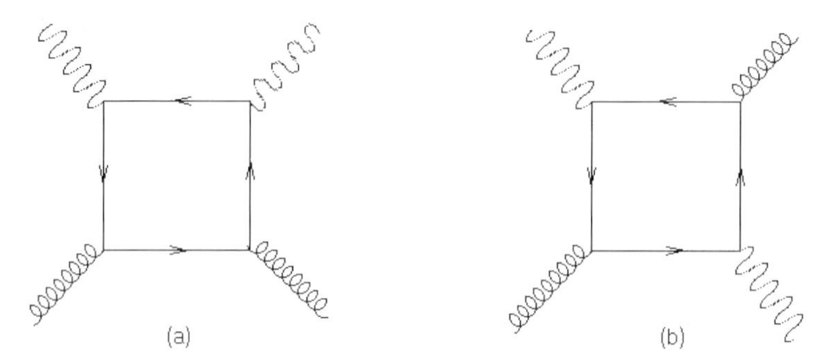

Fig. 1. The diagrams contributing to $T_{\mu\nu}$ for a gluon target (the upper and down wave lines on the diagram (a) correspond to photons and gluons, respectively). They should be multiplied by a factor of 2 because of the opposite direction of the fermion loop. The diagram (a) should be also doubled because of crossing symmetry.

them as *hard structure functions* [4]). They are described by the quark box (and crossed box) diagram contribution to the photon-gluon interaction (see Fig. 1.)

The purpose of the paper is to present the results of [26] for these hard SF $C^g_{2,L}(x, Q^2, m_c^2, k_\perp^2)$ and to analyze experimental data for $F_2^c(x, Q^2)$ and $F_L(x, Q^2)$ by applying (2) with different sets of unintegrated gluon densities (see [27,15]) and to give predictions for the longitudinal charm SF $F_L^c(x, Q^2)$.

It is instructive to note that the results should be similar to those of the photon-photon scattering process. The corresponding QED contributions have been calculated many years ago in [28] (see also the beautiful review in [29]). Our results have been calculated independently in [26] (based on approaches of [30,31]) and they are in full agreement with [28]. However, we hope that our formulas which are given in a simpler form could be useful for others.

2 Hard Structure Functions

The gluon polarization tensor (hereafter the indices α and β are connected with gluons), which gives the main contribution at high energy limit, has the form:

$$\hat{P}^{\alpha\beta}_{BFKLk} = \frac{k_\perp^\alpha k_\perp^\beta}{k_\perp^2} = \frac{x^2}{-k^2} p^\alpha p^\beta = -\frac{1}{2}\frac{1}{\tilde{\beta}^4}\left[\tilde{\beta}^2 g^{\alpha\beta} - 12bx^2 \frac{q^\alpha q^\beta}{Q^2}\right], \quad (4)$$

where $\tilde{\beta}^2 = 1 - 4bx^2$, $b = -k^2/Q^2 \equiv k_\perp^2/Q^2 > 0$.

Contracting the corresponding photon projectors, we have:

$$C_2^g(x) = \frac{\mathcal{K}}{\tilde{\beta}^2}\left[f^{(1)} + \frac{3a}{2\tilde{\beta}^2} f^{(2)}\right], \quad C_L^g(x) = \frac{\mathcal{K}}{\tilde{\beta}^2}\left[4bx^2 f^{(1)} + \frac{(1+2bx^2)}{\tilde{\beta}^2} f^{(2)}\right] \quad (5)$$

$$f^{(1)} = \frac{1}{\tilde{\beta}^4}\left[\tilde{\beta}^2 \hat{f}^{(1)} - 3bx^2 \tilde{f}^{(1)}\right], \quad f^{(2)} = \frac{1}{\tilde{\beta}^4}\left[\tilde{\beta}^2 \hat{f}^{(2)} - 3bx^2 \tilde{f}^{(2)}\right],$$

[4] This notation reflects the fact that SF $F^c_{2,L}$ and F_L connect with the functions $C^g_{2,L}$ and C^g_L at the same form as cross-sections connect with hard ones (see [24,25]).

where the normalization factor $\mathcal{K} = \alpha_s(Q^2)/(4\pi)\, x$ and

$$\hat{f}^{(1)} = -2\beta\Bigg[1 - \Big(1 - 2x(1+b-2a)\,[1 - x(1+b+2a)]\Big)\, f_1 + (2a-b)(1-2a)x^2\, f_2\Bigg], \tag{6}$$

$$\hat{f}^{(2)} = 8x\,\beta\Bigg[(1-(1+b)x) - 2x\Big(bx(1-(1+b)x)(1+b-2a) + a\tilde{\beta}^2\Big)\, f_1 + bx^2(1-(1+b)x)(2a-b)\, f_2\Bigg], \tag{7}$$

$$\tilde{f}^{(1)} = -\beta\Bigg[\frac{1-x(1+b)}{x} - 2\Big(x(1-x(1+b))(1+b-2a) + a\tilde{\beta}^2\Big)\, f_1 - x(1-x(1+b))(1-2a)\, f_2\Bigg], \tag{8}$$

$$\tilde{f}^{(2)} = 4\,\beta\,(1-(1+b)x)^2\Bigg[2 - (1+2bx^2)\, f_1 - bx^2\, f_2\Bigg], \tag{9}$$

with

$$\beta^2 = 1 - \frac{4ax}{(1-(1+b)x)}, \qquad f_1 = \frac{1}{\tilde{\beta}\beta}\ln\frac{1+\beta\tilde{\beta}}{1-\beta\tilde{\beta}}, \qquad f_2 = \frac{-4}{1-\beta^2\tilde{\beta}^2}$$

For the important regimes when $k^2 = 0$, $m_c^2 = 0$ and/or $Q^2 = 0$, the results are coincided with ones of [24,32]. Notice that our results in (5) should be also agree with those in [33] but the direct comparison is quite difficult because the structure of their results is quite cumbersome (see Appendix A in [33]). We have found numerical agreement in the case of $F_2(x, Q^2)$ for several types of unintegrated gluon distributions (see Fig. 4 in [26]).

3 Relations Between F_L, F_2 and Derivation of F_2 in the Case of Collinear Approximation

Another information about the SF F_L can be obtained in the collinear approximation (i.e. when $k_\perp^2 = 0$) in the following way (see also our study [34]).

In the framework of perturbative QCD, there is the possibility to connecting F_L to F_2 and its derivation $dF_2/d\ln Q^2$ due the fact that at small x the DIS structure functions depend only on two *independent* functions: the gluon distribution and singlet quark one (the nonsinglet quark density is negligible at small x), which in turn can be expressed in terms of measurable SF F_2 and its derivation $dF_2/d\ln Q^2$.

In this way, by analogy with the case of the gluon distribution function (see [35,36] and references therein), the behavior of $F_L(x,Q^2)$ has been studied in [23], using the HERA data [37,38] and the method [39] [5] of replacement of the Mellin convolution by ordinary products. Thus, the small x behavior of the SF $F_L(x,Q^2)$ can be extracted directly from the measured values of $F_2(x,Q^2)$ and its derivative without a cumbersome procedure (see [5,6]). These extracted values of F_L may be well considered as *new small x 'experimental data' of F_L*. The relations can be violated by nonperturbative corrections like higher twist ones (see [41,42]), which can be large exactly in the case of SF F_L [43,44].

Because k_T-factorization approach is one of popular perturbative approaches used at small x, it is very useful to compare its predictions with the results of [23] based on the relations between SF $F_L(x,Q^2)$, $F_2(x,Q^2)$ and $dF_2(x,Q^2)/d\ln Q^2$.

The k_T-factorization approach relates strongly to Regge-like behavior of parton distributions. So, we restrict our investigations to SF and parton distributions at the following form (hereafter $a=q,g$):

$$f_a(x,Q^2) \sim F_2(x,Q^2) \sim x^{-\delta(Q^2)} \tag{10}$$

Note that really the slopes of the sea quark and gluon distributions: δ_q and δ_g, respectively, and the slope δ_{F2} of SF F_2 are little different. The slopes have a familiar property $\delta_q < \delta_{F2} < \delta_g$ (see [45]–[50] and references therein). We will neglect, however, this difference and use in our investigations the experimental values of $\delta(Q^2) \equiv \delta_{F2}(Q^2)$ extracted by H1 Collaboration [6] (see [45] and references therein). We note that the Q^2-dependence is in very good agreement with perturbative QCD at $Q^2 \geq 2$ GeV^2 (see [52]). Moreover, the values of the slope $\delta(Q^2)$ are in agreement with recent phenomenological studies (see, for example, [12]) incorporating the NLO corrections [53] (see also [54]) in the framework of BFKL approach.

Thus, assuming the *Regge-like behavior* (10) for the gluon distribution and $F_2(x,Q^2)$ at $x^{-\delta} \gg 1$ and using the NLO approximation for collinear coefficient functions and anomalous dimensions of Wilson operators, the following approximate results for $\mathrm{F}_L(x,Q^2)$ has been obtained in [34]:

$$\begin{aligned} F_L(x,Q^2) = & \frac{r(1+\delta)(\xi(\delta))^{\delta}}{(1+30a_s(Q^2)[1/\tilde{\delta}-\frac{116}{45}\rho_1(\delta)])} \Biggl[\frac{dF_2(x\xi(\delta),Q^2)}{d\ln Q^2} \\ & + \frac{8}{3a}\,\rho_2(\delta)a_s(Q^2)F_2(x\xi(\delta),Q^2)\Biggr] + O(a_s^2,a_s x,x^2), \end{aligned} \tag{11}$$

[5] The method is based on previous investigations [5,40].

[6] Now the preliminary ZEUS data for the slope $d\ln F_2/d\ln(1/x)$ are available as some points on Figs. 8 and 9 in [51]. Moreover, the new preliminary H1 points have been presented on the Workshop DIS2002 (see [21]). Both the new points are shown quite similar properties to compare with H1 data [45]. Unfortunately, tables of the ZEUS data and the new H1 data are unavailable yet and, so, the points cannot be used here.

where

$$r(\delta) = \frac{4\delta}{2+\delta+\delta^2}, \quad \xi(\delta) = \frac{r(\delta)}{r(1+\delta)},$$
$$\rho_1(\delta) = 1+\delta+\delta^2/4, \quad \rho_2(\delta) = 1-2.39\delta+2.69\delta^2, \tag{12}$$

Here

$$\frac{1}{\tilde{\delta}} = \frac{1}{\delta}\left[1 - \frac{\Gamma(1-\delta)\Gamma(1+\nu)}{\Gamma(1-\delta+\nu)}x^{\delta}\right], \tag{13}$$

where Γ is the Rimmanian Γ-function. The value of ν comes (see [55]) from asymptotics of parton distributions $f_a(x)$ at $x \to 1$: $f_a \sim (1-x)^{\nu_a}$, and [7] $\nu \approx 4$ from quark account rules [56].

Note that the $1/\tilde{\delta}$ coincides approximately with $1/\delta$ when $\delta \neq 0$ and $x \to 0$. However, at $\delta \to 0$, the value of $1/\tilde{\delta}$ is not singular:

$$\frac{1}{\tilde{\delta}} \to \ln\left(\frac{1}{x}\right) - \left[\Psi(1+\nu) - \Psi(1)\right], \tag{14}$$

where Ψ-function is the logarithmic derivation of the Γ-function.

Note also that the results (11) are exact at $\delta = 0, 0.3$ and 0.5 values (see [34] and duscussions therein).

4 Comparison with F_2^c and F_L Experimental Data and Predictions for F_L^c

With the help of the results obtained in the previous sections we have analyzed HERA data [2] for SF F_2^c from ZEUS Collaboration and the data for SF F_L mostly from H1 Collaboration.

Notice that in [24] the $k_\perp^2$-integral in the r.h.s. of (2) has been evaluated using the BFKL results for the Mellin transform of the unintegrated gluon distribution and the Wilson coefficient functions have been calculated for the full perturbative series at asymptotically small x values. Since we would like to analyze experimental data for the SF F_2^c and F_L, we have an interest to obtain results at quite broad range of small x values. For the reason we need in a parameterization of unintegrated gluon distribution function.

To study F_2^c we consider two different parametrizations for the unintegrated gluon distribution (see [27]): the Ryskin-Shabelski (RS) one [57] and the Blumlein one [58].

In Fig. 2 we show the SF F_2^c as a function x for different values of Q^2 in comparison with ZEUS [2] experimental data. We see that at large Q^2 ($Q^2 \geq 10$ GeV^2) the SF F_2^c obtained in the k_T factorization approach is higher than the SF obtained in pure perturbative QCD with the GRV [59] gluon density at the

[7] In our formula (13) we have our interest mostly to gluons, so we can apply $\nu = \nu_g \approx 4$ below.

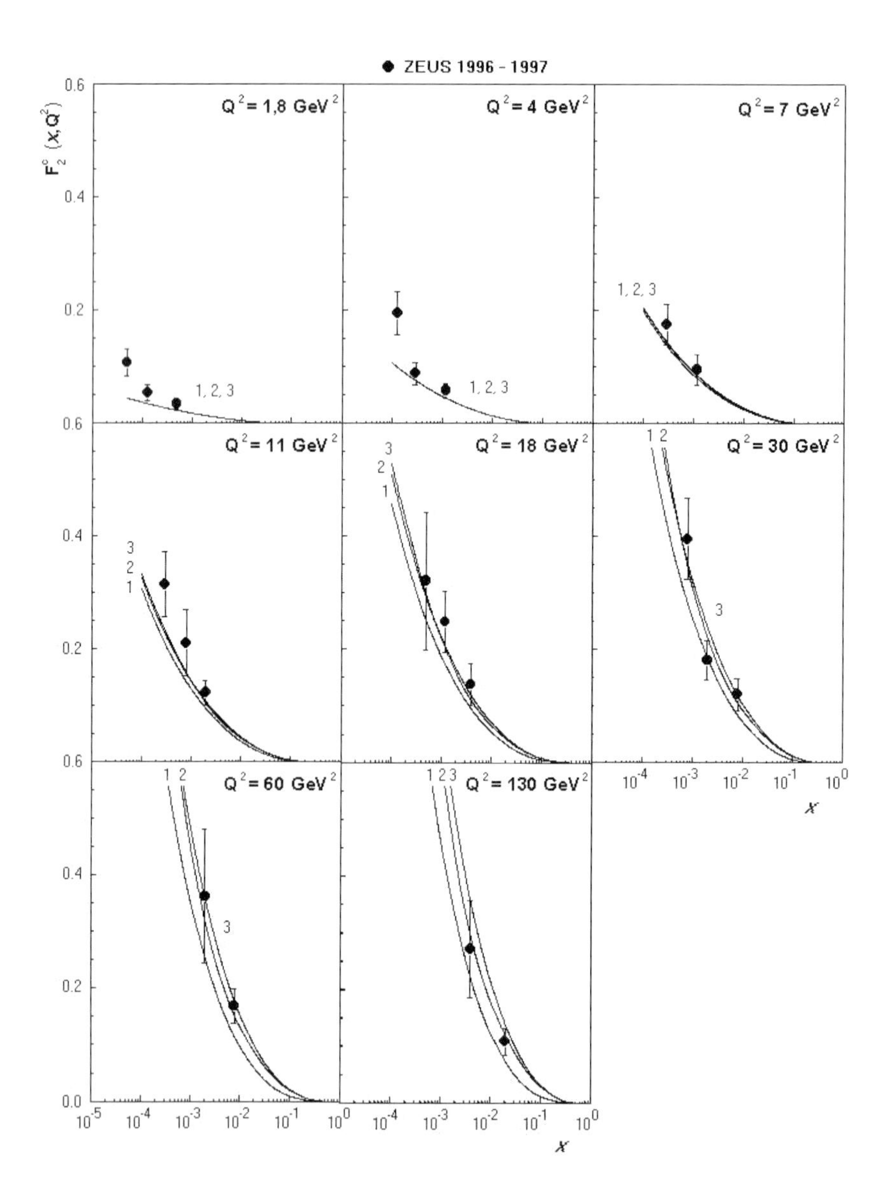

Fig. 2. The structure function $F_2^c(x, Q^2)$ as a function of x for different values of Q^2 compared to ZEUS data [2]. Curves 1, 2 and 3 correspond to SF obtained in pure perturbative QCD with the GRV [59] gluon density at the leading order approximation and to SF obtained in the k_T factorization approach with RS [57] and Blumlein (at $Q_0^2 = 4$ GeV^2) [58] parameterizations of unintegrated gluon distribution.

LO approximation (see curve 1) and has a more rapid growth in comparison with perturbative QCD results, especially at $Q^2 \sim 130$ GeV^2 [60]. At $Q^2 \leq 10$ GeV^2 the predictions from perturbative QCD and those based on the k_T factorization approach are very similar [8] and show a disagreement with data below $Q^2 = 7$ GeV^2 [9]. Unfortunately the available experimental data do not permit yet to distinguish the k_T factorization effects from those due to boundary conditions [57].

The results for the SF F_L^c obtained in perturbative QCD and from the k_T factorization approach are quite similar to the F_2^c case discussed above. The ratio $R^c = F_L^c/F_2^c$ is shown in Fig. 3. We see that $R^c \approx 0.1 \div 0.3$ in a wide region of Q^2. We would like to note that these values of R^c contradict the estimation obtained in [2]. So, the effect of the large R^c values should be considered in the extraction of F_2^c from the corresponding differential cross-section in future more precise measurements.

For the ratio R^c we found quite flat x-behavior at low x in the low Q^2 region (see Fig. 3), where approaches based on perturbative QCD and on k_T factorization give similar predictions (see Fig. 2). It is in agreement with the corresponding behaviour of the ratio $R = F_L/(F_2 - F_L)$ (see [23]) at quite large values of δ [10] ($\delta > 0.2 - 0.3$). The low x rise of R^c at high Q^2 disagrees with early calculations [23] in the framework of perturbative QCD. It could be due to the small x resummation, which is important at high Q^2 (see Fig. 2). We plan to study this effect in future.

In Fig. 4 we show the SF F_L as a function x for different values of Q^2 in comparison with H1 experimental data sets: old one of [17] (black triangles), last year one of [19] (black squares) and new preliminary one of [22] (black circles) and also with NMC [61] (white triangles), CCFR [62] (white circles) and BCDMS [63] data (white squares). For comparison with these data we present the results of the calculation with three different parameterizations for the unintegrated gluon distribution $\varphi_g(x, k_\perp^2, Q_0^2)$ at $Q_0^2 = 4$ GeV^2: Kwiecinski-Martin-Stasto (KMS) one [64], Blumlein one [58] and Golec-Biernat and Wusthoff (GBW) one [65].

We add also the 'experimental data' obtained using the relation between SF $F_L(x, Q^2)$, $F_2(x, Q^2)$ and $dF_2(x, Q^2)/d\ln Q^2$ (see Section 3) as black stars. Because the corresponding data for SF $F_2(x, Q^2)$ and $dF_2(x, Q^2)/d\ln Q^2$ essentially more precise (see [19]) to compare with the preliminary data [22] for F_L, the 'experimental data' have strongly suppressed uncertainties. As it is shown on Fig. 4 there is very good agreement between the new preliminary data [22], the

[8] This fact is also due to the quite large value of $Q_0^2 = 4$ GeV^2 chosen here.

[9] A similar disagreement with data at $Q^2 \leq 2$ GeV^2 has been observed for the complete structure function F_2 (see, for example, the discussion in [49] and reference therein). We note that the insertion of higher-twist corrections in the framework of usual perturbative QCD improves the agreement with data (see [41]) at quite low values of Q^2.

[10] The behaviour is in agreement with previous studies [55,23]. Note that at small values of δ, i.e. when $x^{-\delta} \sim Const$, the ratio R has the strong negative NLO corrections (see [7,8]) and tends to zero at $x \to 0$ after some resummation done in [8].

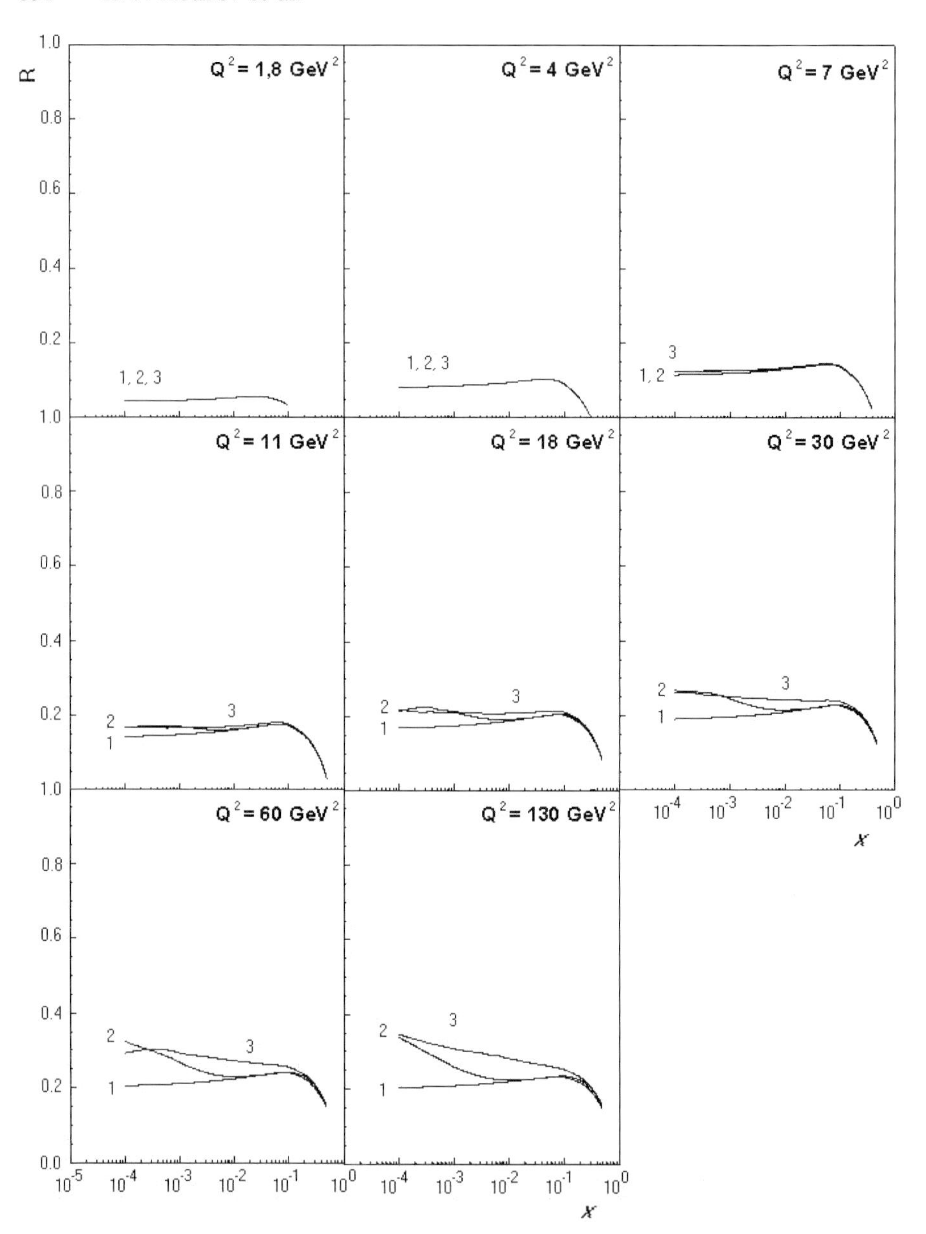

Fig. 3. The ratio $R^c = F_L^c(x, Q^2)/F_2^c$ as a function of x for different values of Q^2. Curves 1, 2 and 3 are as in Fig. 2.

'experimental data' and predictions of perturbative QCD and k_T-factorization approach.

The differences observed between the curves 2, 3 and 4 are due to the different behavior of the unintegrated gluon distribution as function x and $k_\perp$. We see that the SF F_L obtained in the k_T-factorization approach with KMS and Blumlein parameterizations is close each other and higher than the SF obtained in the pure perturbative QCD with the GRV gluon density at the leading order approximation. Otherwise, the k_T-factorization approach with GBW parameterization is very close to pure QCD predictions: it should be so because GBW model has deviations from perturbative QCD only at quite low Q^2 values. Thus, the predictions of perturbative QCD and ones based on k_T factorization approach are in agreement each other and with all data within modern experimental uncertainties. So, a possible high values of high-twist corrections to SF F_L predicted in [44] can be important only at low Q^2 values: $Q^2 \leq Q_0^2 = 4$ GeV^2.

Note that there are several other popular parameterizations (see, for example, Kimber-Martin-Ryskin (KMR) [66] and Jung-Salam (JS) [67]), which are not used in our study mostly because of technical difficulties. Note that all above parameterizations give quite similar results excepting, perhaps, the contributions from the small $k_\perp^2$-range: $k_\perp^2 \leq 1$ GeV^2 (see [15] and references therein). Because we use $Q_0^2 = 4$ GeV^2 in the study of SF F_2^c and F_L, our results depend very slightly on the the small $k_\perp^2$-range of the parameterizations. In the case RS, Blumlein, GBW and KMS sets this observation is supported below by our results and we expect that the application of KMR and JS sets should not strongly change our results.

5 Conclusions

We have presented the results for the perturbative parts of the SF F_2^c and F_L^c for a gluon target having nonzero momentum squared, in the process of photon-gluon fusion.

We have applied the results in the framework of k_T factorization approach to the analysis of present data for the SF F_L and for the charm contribution to F_2 and we have given the predictions for F_L^c. The analysis has been performed with several parameterizations of unintegrated gluon distributions, for comparison.

For SF F_2^c, we have found good agreement of our results with experimental HERA data, except at low Q^2 ($Q^2 \leq 7$ GeV^2). We have obtained also quite large contribution of the SF F_L^c at low x and high Q^2 ($Q^2 \geq 30$ GeV^2) and this effect should be considered for the extraction of F_2^c from the corresponding differential cross-section in future more precise measurements.

We would like to note the good agreement between our results for F_2^c and the ones obtained in [68] by Monte-Carlo studies. Moreover, we have also good agreement with fits of H1 and ZEUS data for F_2^c (see recent reviews in [69] and references therein) based on perturbative QCD calculations. But unlike to these fits, our analysis uses universal unintegrated gluon distribution, which gives in the simplest way the main contribution to the cross-section in the high-energy limit.

For SF F_L, we have found good agreement between all existing experimental data, the predictions for F_L obtained from the relation between SF $F_L(x,Q^2)$, $F_2(x,Q^2)$ and $dF_2(x,Q^2)/d\ln Q^2$ and the results obtained in the framework of

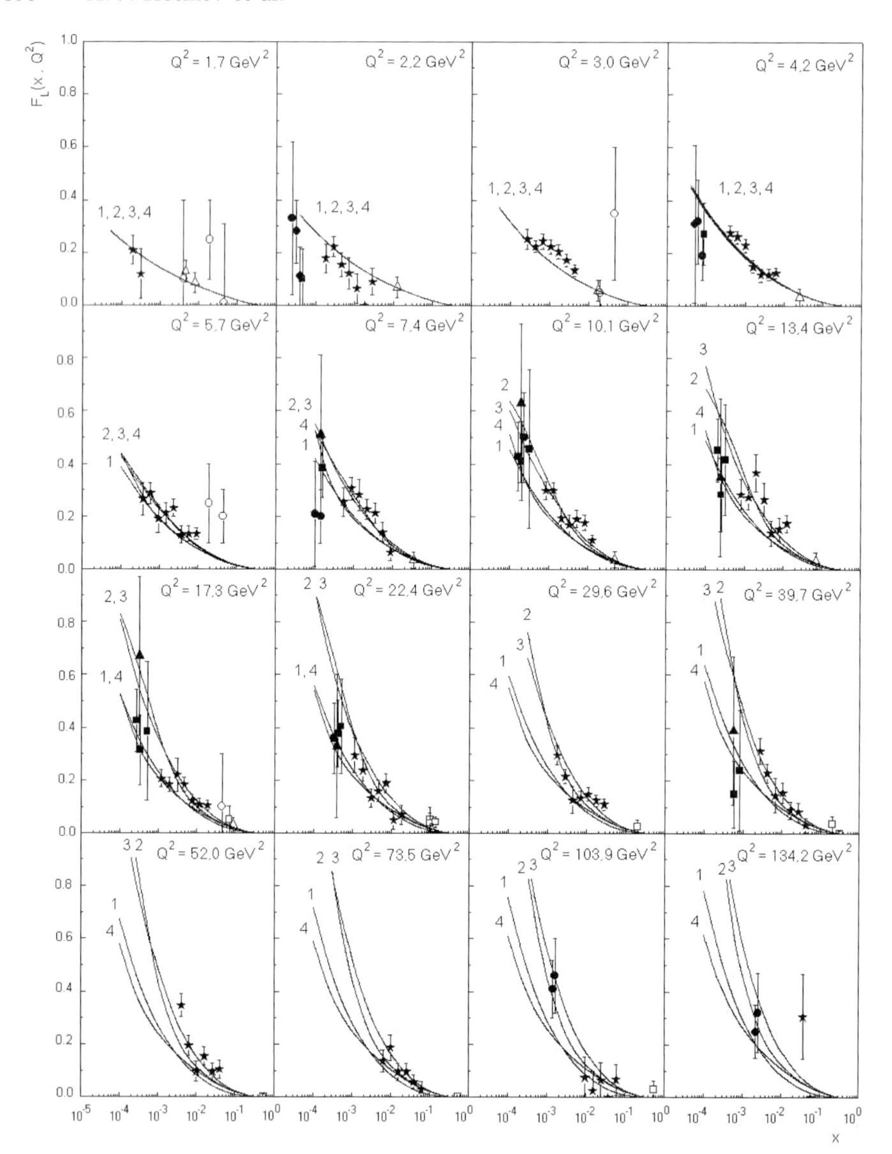

Fig. 4. The structure function $F_L(x, Q^2)$ as a function of x for different values of Q^2 compared to experimental data. The H1 data: the first 1997 ones [17], new 2001 ones [19] and preliminary ones [22] are shown as black triangles, squares and circles, respectively. The NM [61], CCFR [62] and BCDMS [63] data are shown as white triangles, circles and squares, respectively. The 'experimental data' are added as black stars. Curves 1, 2, 3 and 4 correspond to SF obtained in the perturbative QCD with the GRV [59] gluon density at the leading order approximation and to SF obtained in the k_T factorization approach with KMS [64], Blumlein (at $Q_0^2 = 4$ GeV^2) [58] and GBW [65] parametrizations of unintegrated gluon distribution.

perturbative QCD and ones based on k_T-factorization approach with the three different parameterizations of unintegrated gluon distributions.

It could be also very useful to evaluate the complete F_2 itself and the derivatives of F_2 with respect to the logarithms of $1/x$ and Q^2 with our expressions using the unintegrated gluons.

The consideration of the SF F_2 in the framework of the leading-twist approximation of perturbative QCD (i.e. for "pure" perturbative QCD) leads to very good agreement (see [49] and references therein) with HERA data at low x and $Q^2 \geq 1.5$ GeV^2. The agreement improves at lower Q^2 when higher twist terms are taken into account [41]. As it has been studied in [49,41], the SF F_2 at low Q^2 is sensitive to the small-x behavior of quark distributions. Thus, our future analysis of F_2 in a broader Q^2 range in the framework of k_T factorization should require the incorporation of parametrizations for unintegrated quark densities, introduced recently (see [66] and references therein).

Acknowledgments

Authors would like to express their sincerely thanks to the Organizing Committee of the International School "Heavy Quark Physics" for the kind invitation, the financial support and for fruitful discussions. We are grateful also to Professor Catani for useful discussions and comments.

The study is supported in part by the RFBR grant 02-02-17513. A.V.K. was supported in part by Alexander von Humboldt fellowship and INTAS grant N366. G.P. acknowledges the support of Galician research funds (PGIDT00 PX20615PR) and Spanish CICYT (FPA2002-01161). N.P.Z. also acknowledge the support of Royal Swedish Academy of Sciences.

References

1. H1 Coll.: S. Aid et al., Z. Phys. C **72**, 593 (1996); Nucl. Phys. B **545**, 21 (1999)
2. ZEUS Coll.: J. Breitweg et al., Phys. Lett. B **407**, 402 (1997); Eur.Phys.J. C **12** (2000) 35
3. EM Coll.: J.J. Aubert et al., Nucl. Phys. B **213**, 31 (1983); Phys. Lett. B **94**, 96 (1980); B **110**, 72 (1983)
4. A.M. Cooper-Sarkar et al., Int. J. Mod. Phys. A **13**, 3385 (1998)
5. A.M.Cooper-Sarkar et al., Z. Phys. C **39**, 281 (1988)
6. L. Bauerdick et al., in *Proc. of the Int. Workshop on Future Physics on HERA*, Hamburg, DESY (1996), p.77 (hep-ex/9609017)
7. S. Keller et al., Phys. Lett. B **270**, 61 (1990); L.H. Orr, W.J. Stirling, Phys. Rev. Lett. **66**, 1673 (1991); E. Berger, R. Meng, Phys. Lett. B **304**, 318 (1993)
8. A.V. Kotikov, JETP Lett. **59**, 1 (1994); Phys. Lett. B **338**, 349 (1994)
9. R.S. Thorne, in: *Proc. of the Int. Workshop on Deep Inelastic Scattering* (2002), Cracow
10. Yu.L. Dokshitzer, D.V. Shirkov, Z. Phys. C **67**, 449 (1995)
11. W.K. Wong, Phys. Rev. D **54**, 1094 (1996)

12. S.J. Brodsky et al., JETP. Lett. **70**, 155 (1999); V.T. Kim et al., in: *Proceedings of the VIIIth Blois Workshop at IHEP*, Protvino,Russia, 1999 (IITAP-99-013, hep-ph/9911228); in: *Proceedings of the Symposium on Multiparticle Dynamics (ISMD99)*, Providence, Rhode Island, 1999 (IITAP-99-014, hep-ph/9911242)
13. S.J. Brodsky et al., in: *Proceedings of the PHOTON2001*, Ascona, Switzerland, 2001 (CERN-TH/2001-341, SLAC-PUB-9069, hep-ph/0111390)
14. M. Ciafaloni et al., Phys. Rev. D **60**, 114036 (1999); JHEP **07**, 054 (2000); R.S. Thorne, Phys. Lett. B **474**, 372 (2000); Phys. Rev. D **60**, 054031 (1999); D **64**, 074005 (2001); G. Altarelli et al., Nucl. Phys. B **621**, 359 (2002)
15. Bo Andersson et al., Eur. Phys. J. C **25**, 77 (2002)
16. L.N. Lipatov, Sov. J. Nucl. Phys. **23**, 338 (1976); E.A. Kuraev et al., Sov. Phys. JETP **44**, 443 (1976); **45**, 199 (1977); Ya.Ya. Balitzki, L.N. Lipatov, Sov. J. Nucl. Phys. **28**, 822 (1978); L.N. Lipatov, JETP **63**, 904 (1986)
17. H1 Collab.: S. Aid et al., Phys. Lett. B **393**, 452 (1997)
18. R.S. Thorne, Phys. Lett. B **418**, 371 (1998)
19. H1 Collab.: S. Adloff et al., Eur. Phys. J. C **21**, 33 (2001)
20. H1 Collab.: D. Eckstein, in *Proc. Int. Workshop on Deep Inelastic Scattering*, DIS 2001 (2001), Bologna; M.Klein, in *Proc. of the 9th Int. Workshop on Deep Inelastic Scattering*, DIS 2001 (2001), Bologna
21. H1 Collab.: T. Lastovicka, in *Proc. of the 10th Int. Workshop on Deep Inelastic Scattering*, DIS 2002 (2002), Cracow; J. Gayler, in *Proc. of the 10th Int. Workshop on Deep Inelastic Scattering*, DIS 2002 (2002), Cracow
22. H1 Collab.: N. Gogitidze, J. Phys. G **28**, 751 (2002)
23. A.V. Kotikov, JETP **80**, 979 (1995); A.V. Kotikov, G. Parente, in *Proc. Int. Workshop on Deep Inelastic Scattering*, DIS96 (1996), Rome, p. 237 (hep-ph/9608409); Mod. Phys. Lett. A **12**, 963 (1997); JETP **85**, 17 (1997); hep-ph/9609439
24. S. Catani et al., Phys. Lett. B **242**, 97 (1990); Nucl. Phys. B **366**, 135 (1991)
25. J.C. Collins, R.K. Ellis, Nucl. Phys. B **360**, 3 (1991); E.M. Levin et al., Sov. J. Nucl. Phys. **53**, 657 (1991)
26. A.V. Kotikov et al., Eur. Phys. J. C **26**, 51 (2002)
27. A. V. Lipatov et al., Mod. Phys. Lett. A **15**, 695 (2000); A **15**, 1727 (2000)
28. V.N. Baier et al., JETP **50**, 156 (1966); V.G. Zima, Yad. Fiz. **16**, 1051 (1972)
29. V.M. Budnev et al., Phys. Rept. **15**, 181 (1975)
30. D.I. Kazakov, A.V. Kotikov, Theor. Math. Phys. **73**, 1264 (1987); Nucl. Phys. B **307**, 721 (1988); B **345**, 299 (1990)
31. A.V. Kotikov, Theor. Math. Phys. **78**, 134 (1989); Phys. Lett. B **375**, 240 (1996) 240; B **254**, 158 (1991); B **259**, 314 (1991); B **267**, 123 (1991); in: *Proc. of the XXXV Winter School*, Repino, S'Peterburg, 2001 (hep-ph/0112347).
32. E. Witten, Nucl. Phys. B **104** 445 (1976); M. Gluck, E. Reya, Phys. Lett. B **83**, 98 (1979); F.M. Steffens et al., Eur. Phys. J. C **11**, 673 (1999)
33. G. Bottazzi et al., JHEP **9812**, 011 (1998)
34. A.V. Kotikov et al., hep-ph/0207226, Eur. Phys. J. C in press
35. K. Prytz, Phys. Lett. B **311**, 286 (1993)
36. A.V. Kotikov, JETP Lett. **59**, 667 (1994); A.V. Kotikov, G. Parente, Phys. Lett. B **379**, 195 (1996)
37. H1 Collab.,T. Ahmed et al., Nucl. Phys. B **439**, 471 (1995)
38. ZEUS Collab., M. Derrick et al., Z. Phys. C **65**, 379 (1995)
39. A.V. Kotikov, Phys. Atom. Nucl. **57**, 133 (1994); Phys. Rev. D **49**, 5746 (1994)
40. F. Martin, Phys. Rev. D **19**, 1382 (1979); C. Lopez, F.I. Yndurain, Nucl. Phys. B **171**, 231 (1980); B **183**, 157 (1981)

41. A.V. Kotikov, G. Parente, in *Proc. Int. Seminar Relativistic Nuclear Physics and Quantum Chromodynamics* (2000), Dubna (hep-ph/0012299); in *Proc. of the 9th Int. Workshop on Deep Inelastic Scattering*, DIS 2001 (2001), Bologna (hep-ph/0106175)
42. V.G. Krivokhijine and A.V. Kotikov, JINR preprint E2-2001-190 (hep-ph/0108224); in: *Proc. of the XVIth International Workshop "High Energy Physics and Quantum Field Theory"* (2001), Moscow (hep-ph/0206221); in: *Proc. of the Int. Workshop "Renormalization Group 2002"* (2002), High Tatras, Slovakia (hep-ph/0207222).
43. S. Catani, F. Hautmann, Nucl. Phys. B **427**, 475 (1994); S. Catani, Preprint DFF 254-7-96 (hep-ph/9608310).
44. J. Bartels et al., Eur. Phys. J. C **17**, 121 (2000); J. Bartels, in: *Proc. of the Int. Workshop on Deep Inelastic Scattering* (2002), Cracow.
45. H1 Collab.: C. Adloff et al., Phys. Lett. B **520**, 183 (2001)
46. A.D. Martin et al., Phys. Lett. B **387**, 419 (1996)
47. M. Gluck et al., Eur. Phys. J. C **5**, 461 (1998)
48. A.D. Martin et al., Eur. Phys. J C **23**, 73 (2002); CTEQ Collab.: J. Pumplin et al., Preprint MSU-HEP-011101 (hep-ph/0201195)
49. A.V. Kotikov, G. Parente, Nucl. Phys. B **549**, 242 (1999); Nucl. Phys. (Proc. Suppl.) A**99**, 196 (2001); *in* Proc. of the Int. Conference PQFT98 (1998), Dubna (hep-ph/9810223); *in* Proc. of the 8th Int. Workshop on Deep Inelastic Scattering, DIS 2000 (2000), Liverpool, p. 198 (hep-ph/0006197)
50. A.V. Kotikov, Mod. Phys. Lett. A **11**, 103 (1996); Phys. At. Nucl. **59**, 2137 (1996)
51. ZEUS Collab., B. Surrow, in *Proc. of the Int. Europhysics Conference on High Energy Physics*, July 2001 (hep-ph/0201025)
52. A.V. Kotikov, G. Parente, Preprint US-FT/3-02 (hep-ph/0207276)
53. V.N. Fadin, L.N. Lipatov, Phys. Lett. B **429**, 127 (1998); M. Ciafaloni, G. Camici, Phys. Lett. B **430**, 349 (1998)
54. A.V. Kotikov, L.N. Lipatov, Nucl. Phys. B **582**, 19 (2000); in: *Proc. of the XXXV Winter School*, Repino, S'Peterburg, 2001 (hep-ph/0112346)
55. A.V. Kotikov et al., Theor. Math. Phys. **84**, 744 (1990); **111**, 442 (1997); A.V. Kotikov, Phys. Atom. Nucl. **56**, 1276 (1993); L. L. Jenkovszky et al., Sov. J. Nucl. Phys. **55**, 1224 (1992); JETP Lett. **58**, 163 (1993); Phys. Lett. B **314**, 421 (1993)
56. V.A. Matveev et al., Lett. Nuovo Cim. **7**, 719 (1973); S.J. Brodsky, G.R. Farrar, Phys. Rev. Lett. **31**, 1153 (1973); S.J. Brodsky et al., Phys. Rev. D **56**, 6980 (1997)
57. M.G. Ryskin, Yu.M. Shabelski, Z. Phys. C **61**, 517 (1994); C **66**, 151 (1995)
58. J. Blumlein, Preprints DESY 95-121 (hep-ph/9506403); DESY 95-125 (hep-ph/9506446)
59. M. Gluck et al., Z. Phys. C **67**, 433 (1995)
60. A.V. Lipatov, N.P. Zotov, *in* Proc. of the 8th Int. Workshop DIS 2000 (2000), World Scientific, p. 157
61. NM Collab., M. Arneodo et al., Nucl. Phys. B **483**, 3 (1997); A.J. Milsztajn, in *Proc. of the 4th Int. Workshop on Deep Inelastic Scattering*, DIS96 (1996), Rome, p.220
62. CCFR/NuTeV Collab.: U.K. Yang et al., Phys. Rev. Lett. **87**, 251802 (2001); CCFR/NuTeV Collab.: A. Bodek, in *Proc. of the 9th Int. Workshop on Deep Inelastic Scattering*, DIS 2001 (2001), Bologna (hep-ex/00105067)
63. BCDMS Collab., A.C. Benvenuti et al., Phys. Lett. B **223**, 485 (1989); B **237**, 592 (1990)
64. J. Kwiecinski et al., Phys. Rev. D **56**, 3991 (1997)

65. K. Golec-Biernat, M Wusthoff, Phys.Rev. D **59**, 014017 (1999); D **60**, 014015, 114023 (1999)
66. M.A. Kimber et al., Phys.Rev. D **63**, 114027 (2001)
67. H. Jung and G. Salam, Eur. Phys. J. C **19**, 351 (2001); H. Jung, hep-ph/9908497
68. H. Jung, Nucl. Phys. (Proc. Suppl.) **79**, 429 (1999) 429
69. G. Wolf, Preprint DESY 01-058 (hep-ex/0105055)

J/ψ and D^* Mesons Photoproduction at HERA

V.A.Saleev[1,2] and D.V.Vasin[1]

[1] Samara State University, Samara, 443011, Russia
[2] Samara Municipal Nayanova University, Samara, 443001, Russia

Abstract. In this report we compare the predictions of the collinear parton model and the k_T-factorization approach in J/Ψ and $D^\star$ meson photoproduction at HERA energies. It is shown that obtained $D^\star$ meson spectra over p_T and η are very similar in the parton model and k_T-factorization approach and they underestimate the experimental data. Opposite, the predictions of the both approaches for p_T- and z-spectra in the J/Ψ photoproduction are very different as well as the prediction obtained for the spin parameter $\alpha(p_T)$.

1 Hard Processes in the Parton Model and k_T-Factorization Approach

Nowadays, there are two approaches which are used in a study of the charmonia and charmed mesons photoproduction at high energies. In the conventional collinear parton model [1] it is suggested that hadronic cross section, for example $\sigma(\gamma p \to c\bar{c}X, s)$, and the relevant partonic cross section $\hat{\sigma}(\gamma g \to c\bar{c}, \hat{s})$ are connected as follows

$$\sigma^{PM}(\gamma p \to c\bar{c}X, s) = \int dx G(x, \mu^2)\hat{\sigma}(\gamma g \to c\bar{c}, \hat{s}), \tag{1}$$

where $\hat{s} = xs$, $G(x, \mu^2)$ is the collinear gluon distribution function in a proton, x is the fraction of a proton momentum, μ^2 is the typical scale of a hard process. The μ^2 evolution of the gluon distribution $G(x, \mu^2)$ is described by DGLAP evolution equation [2]. In the so-called k_T-factorization approach hadronic and partonic cross sections are related by the following condition [3–5]:

$$\sigma^{SHA}(\gamma p \to cX, s) = \int \frac{dx}{x} \int d\boldsymbol{k}_T^2 \int \frac{d\varphi}{2\pi} \Phi(x, \boldsymbol{k}_T^2, \mu^2)\hat{\sigma}(\gamma g^\star \to c\bar{c}, \hat{s}, \boldsymbol{k}_T^2), \tag{2}$$

where $\hat{\sigma}(\gamma g^\star \to c\bar{c}, \hat{s}, \boldsymbol{k}_T^2)$ is the $c\bar{c}$-pair photoproduction cross section on off mass shell gluon, $k^2 = {k_T}^2 = -\boldsymbol{k}_T^2$ is the gluon virtuality, $\hat{s} = xs - \boldsymbol{k}_T^2$, φ is the azimuthal angle in the transverse XOY plane between vector $\boldsymbol{k}_T$ and the fixed OX axis. The unintegrated gluon distribution function $\Phi(x, \boldsymbol{k}_T^2, \mu^2)$ can be related to the conventional gluon distribution by

$$xG(x, \mu^2) = \int_0^{\mu^2} \Phi(x, \boldsymbol{k}_T^2, \mu^2) d\boldsymbol{k}_T^2, \tag{3}$$

V.A. Saleev and D.V. Vasin, J/ψ and D^* Mesons Photoproduction at HERA, Lect. Notes Phys. **647**, 401–417 (2004)
http://www.springerlink.com/

where $\Phi(x, \boldsymbol{k}_T^2, \mu^2)$ satisfies the BFKL evolution equation [6]. In formulae (2) the four-vector of a gluon momentum is presented as follows:

$$k = xp_N + k_T, \tag{4}$$

where $k_T = (0, \boldsymbol{k}_T, 0)$, $p_N = (E_N, 0, 0, E_N)$ is the four-vector of a proton momentum. At the $x \ll 1$ the off mass-shell gluon has dominant longitudinal polarization along the proton momentum. Taking into account the gauge invariance of a total amplitude involving virtual gluon we can write the polarization four-vector in two different forms:

$$\varepsilon^\mu(k) = \frac{k_T^\mu}{|\boldsymbol{k}_T|} \tag{5}$$

or

$$\varepsilon^\mu(k) = -\frac{xp_N^\mu}{|\boldsymbol{k}_T|}. \tag{6}$$

As it will be shown above formulae (5) and (6) give the equal answers in calculating of squared amplitudes under consideration.

Our calculation in the parton model is down using the GRV LO [7] parameterization for the collinear gluon distribution function $G(x, \mu^2)$. In the case of the k_T-factorization approach we use the following parameterizations for an unintegrated gluon distribution function $\Phi(x, \boldsymbol{k}_T^2, \mu^2)$: JB by J. Bluemlein [8]; JS by H. Jung and G. Salam [9]; KMR by M.A. Kimber, A.D. Martin and M.G. Ryskin [10]. The detail analysis of the evolution equations lied in a basis of the different parameterizations is over our consideration. To compare different parameterizations we have plotted their as a function of x at the fixed $\boldsymbol{k}_T^2$ and μ^2 in Fig. 1 and as a function of $\boldsymbol{k}_T^2$ at the fixed x and μ^2 in Fig. 2.

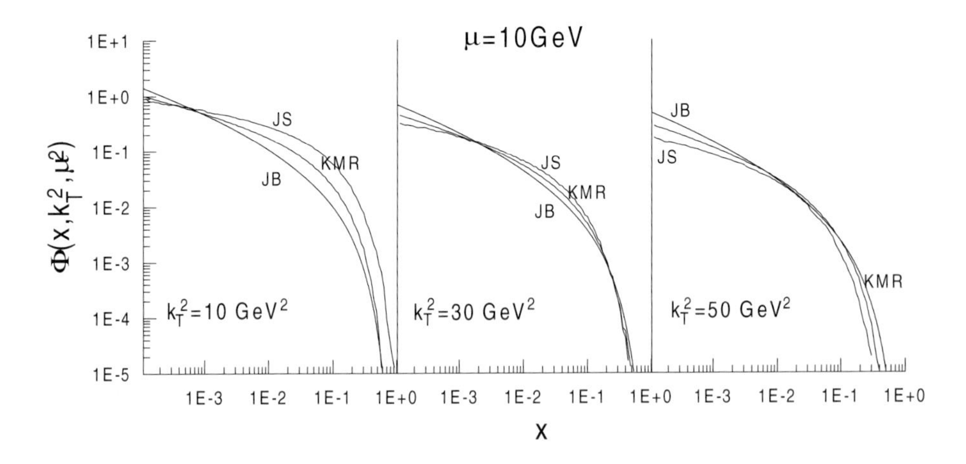

Fig. 1. The unintegrated gluon distribution function $\Phi(x, \boldsymbol{k}_T^2, \mu^2)$ versus x at the fixed values of μ and k_T^2.

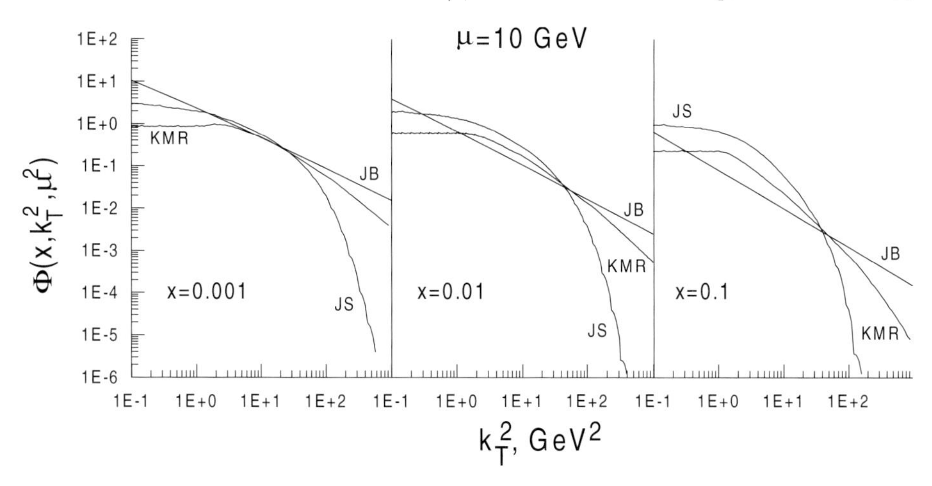

Fig. 2. The unintegrated gluon distribution function $\Phi(x, \boldsymbol{k}_T^2, \mu^2)$ versus k_T^2 at the fixed values of μ and x.

Note, that all parameterizations of an unintegrated gluon distribution function describe the data from HERA collider for the structure function $F_2(x, Q^2)$ well [8–10].

2 $D^\star$ Meson Photoproduction in LO QCD

The photoproduction of the $D^\star$ meson was studding experimentally by H1 and ZEUS collaborations at HERA ep-collider (E_e = 27.5 GeV, E_N = 820 GeV) [11,12]. Because of the large mass of a c-quark usually it is assumed that $D^\star$ meson production may be described in the fragmentation approach [1] [13], where

$$\sigma(\gamma p \to D^\star X, p) = \int D_{c\to D^\star}(z, \mu^2)\sigma(\gamma p \to cX, p_1 = p/z)dz \qquad (7)$$

and $D_{c\to D^\star}(z, \mu^2)$ is the universal fragmentation function of a c-quark into the $D^\star$ meson at the scale $\mu^2 = m_D^2 + p_T^2$. The fraction of the $D^\star$ produced by a c-quark as measured by OPAL Collaboration [15],

$$\omega_{c\to D^\star} = \int_0^1 D_{c\to D^\star}(z, \mu^2)dz = 0.222 \pm 0.014,$$

has been used in our LO QCD calculations to normalize the fragmentation function.

[1] The another approach based on recombination scenario was suggested recently in [14]

The Peterson [13] fragmentation function was used as a phenomenological factor:

$$D_{c\to D^\star}(z,\mu_0^2) = N\frac{z(1-z)^2}{[(1-z)^2+\epsilon z]^2}. \tag{8}$$

In the high energy limit or in the case of a massless quark one has following relation for the four-vectors $p = zp_1$, however in the discussed here process the $D^\star$ meson energy is not so large in compare to $M_{D^\star}$ and the following prescription was used

$$\boldsymbol{p} = z\boldsymbol{p}_1 \tag{9}$$

together with the mass-shell condition for the c-quark energy and momentum $E_1^2 = \boldsymbol{p}_1^2 + m_c^2$. We have used $\epsilon = 0.06$ as a middle value between two recent fits of $D^\star$ meson spectra in e^+e^--annihilation, which based on massive charm ($\epsilon = 0.036$)[16] and massless charm ($\epsilon = 0.116$)[17] calculations. The squared matrix element for the subprocess $\gamma g^\star \to c\bar{c}$ after summation over a gluon polarization accordingly (6) may be written as follows [4,5,18]:

$$\overline{|M|^2} = 16\pi^2 e_c^2\alpha_s\alpha\cdot(\hat{s}+\boldsymbol{k}_T^2)^2\left[\frac{\alpha_1^2+\alpha_2^2}{(\hat{t}-m_c^2)(\hat{u}-m_c^2)} - \frac{2m_c^2}{\boldsymbol{k}_T^2}\left(\frac{\alpha_1}{\hat{u}-m_c^2} - \frac{\alpha_2}{\hat{t}-m_c^2}\right)^2\right] \tag{10}$$

where $\hat{s}$, $\hat{t}$ and $\hat{u}$ are usual Mandelstam variables,

$$\alpha_1 = \frac{m_c^2+\boldsymbol{p}_{1T}^2}{m_c^2-\hat{t}},\ \alpha_2 = \frac{m_c^2+\boldsymbol{p}_{2T}^2}{m_c^2-\hat{u}},$$

$\boldsymbol{p}_{1T}$ and $\boldsymbol{p}_{2T}$ are the transverse momenta of c- and $\bar{c}$-quarks, $\boldsymbol{k}_T = \boldsymbol{p}_{1T} + \boldsymbol{p}_{2T}$.

Using formulas (5) for a BFKL gluon polarization four-vector we can rewrite (10) in the another form:

$$\begin{aligned}
\overline{|M|^2} = \ & \frac{16\pi^2 e_c^2\alpha_s\alpha}{(m_c^2-\hat{t})^2(m_c^2-\hat{u})^2}\Bigg[m_c^2\Big(-2m_c^6 - 4m_c^2\boldsymbol{p}_{1T}^2\boldsymbol{k}_T^2 + m_c^2\boldsymbol{k}_T^4 + 8\boldsymbol{p}_{1T}^2\boldsymbol{k}_T^4 + \\
& +3\boldsymbol{k}_T^6 + (4m_c^4 + 12\boldsymbol{p}_{1T}^2\boldsymbol{k}_T^2 + 5\boldsymbol{k}_T^4)\hat{s} - (3m_c^2 - 4\boldsymbol{p}_{1T}^2 - 3\boldsymbol{k}_T^2)\hat{s}^2 + \hat{s}^3\Big) + \\
& +\Big(8m_c^6 + 8m_c^2\boldsymbol{p}_{1T}^2\boldsymbol{k}_T^2 - 2m_c^2\boldsymbol{k}_T^4 - 4\boldsymbol{p}_{1T}^2\boldsymbol{k}_T^4 - \boldsymbol{k}_T^6 - 12m_c^4\hat{s} - 4\boldsymbol{p}_{1T}^2\boldsymbol{k}_T^2\hat{s} - \\
& -\boldsymbol{k}_T^4\hat{s} + 6m_c^2\hat{s}^2 - \boldsymbol{k}_T^2\hat{s}^2 - \hat{s}^3\Big)\hat{t} - \Big(4\boldsymbol{p}_{1T}^2\boldsymbol{k}_T^2 - \boldsymbol{k}_T^4 + 3(-2m_c^2+\hat{s})^2\Big)\hat{t}^2 + \\
& +4\Big(2m_c^2 - \hat{s}\Big)\hat{t}^3 - 2\hat{t}^4 - 4|\boldsymbol{p}_{1T}|\Big(|\boldsymbol{k}_T|\cos(\varphi)\Big(-2m_c^6 - \\
& -\Big(\boldsymbol{k}_T^2 - \hat{s} - 2\hat{t}\Big)\hat{t}\Big(2m_c^2 - \hat{u}\Big) + m_c^4\Big(\boldsymbol{k}_T^2 + 3\hat{s} + 6\hat{t}\Big) + m_c^2\Big(3\boldsymbol{k}_T^4 + \hat{s}^2 + \\
& +\boldsymbol{k}_T^2(4\hat{s} - 2\hat{t}) - 6\hat{s}\hat{t} - 6\hat{t}^2\Big)\Big) + |\boldsymbol{p}_{1T}|\cos(2\varphi)\Big(m_c^4\boldsymbol{k}_T^2 + \boldsymbol{k}_T^2\hat{t}\Big(2m_c^2 - \hat{u}\Big) - \\
& -m_c^2\Big(2\boldsymbol{k}_T^4 + \hat{s}^2 + \boldsymbol{k}_T^2(3\hat{s} + 2\hat{t})\Big)\Big)\Big)\Bigg],
\end{aligned} \tag{11}$$

where φ is the angle between $\boldsymbol{p}_{1T}$ and $\boldsymbol{k}_T$.

In the last case (11) it is easy to find the parton model limit:

$$\lim_{|\boldsymbol{k}_T|\to 0} \int_0^{2\pi} \frac{d\varphi}{2\pi} \overline{|M|^2} = \overline{|M_{PM}|^2}, \tag{12}$$

where

$$\boldsymbol{p}_{1T}^2 = \boldsymbol{p}_{2T}^2 = \frac{(\hat{u} - m_c^2)(\hat{t} - m_c^2)}{\hat{s}} - m_c^2,$$

and

$$\begin{aligned} \overline{|M_{PM}|^2} &= -\frac{16\pi^2 e_c^2 \alpha_s \alpha}{(9(m_c^2 - \hat{t})^2(-m_c^2 + \hat{s} + \hat{t})^2)} \Big[-2m_c^8 + 8m_c^6(\hat{s} + \hat{t}) - \\ &-\hat{t}(\hat{s}^3 + 3\hat{s}^2\hat{t} + 4\hat{s}\hat{t}^2 + 2\hat{t}^3) + m_c^2(\hat{s}^3 + 6\hat{s}^2\hat{t} + 8\hat{t}^3 + 4\hat{s}\hat{t}(3\hat{t} + \hat{u})) - \\ &-m_c^4(7\hat{s}^2 + 12\hat{t}^2 + 4\hat{s}(4\hat{t} + \hat{u})) \Big] \end{aligned} \tag{13}$$

3 $D^\star$ Meson Photoproduction at HERA

In this part we will compare our results obtained with leading order matrix elements for the partonic subprocess $\gamma g \to c\bar{c}$ in the conventional parton model as well as in the k_T-factorization approach with data from HERA ep-collider. The data under consideration taken by the ZEUS Collaboration [11]. Inclusive photoproduction of the $D^{\star\pm}$ mesons has been measured for the photon-proton center-of-mass energies in the range $130 < W < 280$ GeV and the photon virtuality $Q^2 < 1$ GeV^2. At low Q^2 the cross section for $ep \to eD^\star X$ are related to γp cross section using the equivalent photon approximation [19]:

$$d\sigma_{ep} = \int \sigma_{\gamma p} \cdot f_{\gamma/e}(y) dy,$$

where $f_{\gamma/e}(y)$ denotes the photon flux integrated over Q^2 from the kinematic limit of $Q^2_{min} = m_e^2 y^2/(1-y)$ to the upper limit $Q^2_{max} = 1$ GeV^2, $y = W^2/s$, $s = 4E_N E_e$, E_N and E_e are the proton and electron energies in the laboratory frame.

The exact formulas for $f_{\gamma/e}(y)$ is taken from [20]:

$$f_{\gamma/e}(y) = \frac{\alpha}{2\pi} \left[\frac{1 + (1-y)^2}{y} \log \frac{Q^2_{max}}{Q^2_{min}} + 2m_e^2 y \big(\frac{1}{Q^2_{min}} - \frac{1}{Q^2_{max}} \big) \right].$$

The limits of integration over y are $y_{\substack{max \\ min}} = W^2_{\substack{max \\ min}}/s$. In our calculations we used formulas for differential cross section in the following form:

$$\begin{aligned} \frac{d\sigma(ep \to eD^\star X)}{d\eta dp_T} &= \int dy f_{\gamma/e}(y) \int \frac{dz}{z} D_{c\to D^\star}(z, \mu^2) \int \frac{d\varphi}{2\pi} \int d\boldsymbol{k}_T^2 \frac{\Phi(x, \boldsymbol{k}_T^2, \mu^2)}{x} \times \\ &\times \frac{2|\boldsymbol{p}_1||\boldsymbol{p}_{1T}|}{E_1(2E_N(E_1 - p_{1z}) - W^2)} \cdot \frac{|\bar{M}|^2}{16\pi x W^2} \end{aligned} \tag{14}$$

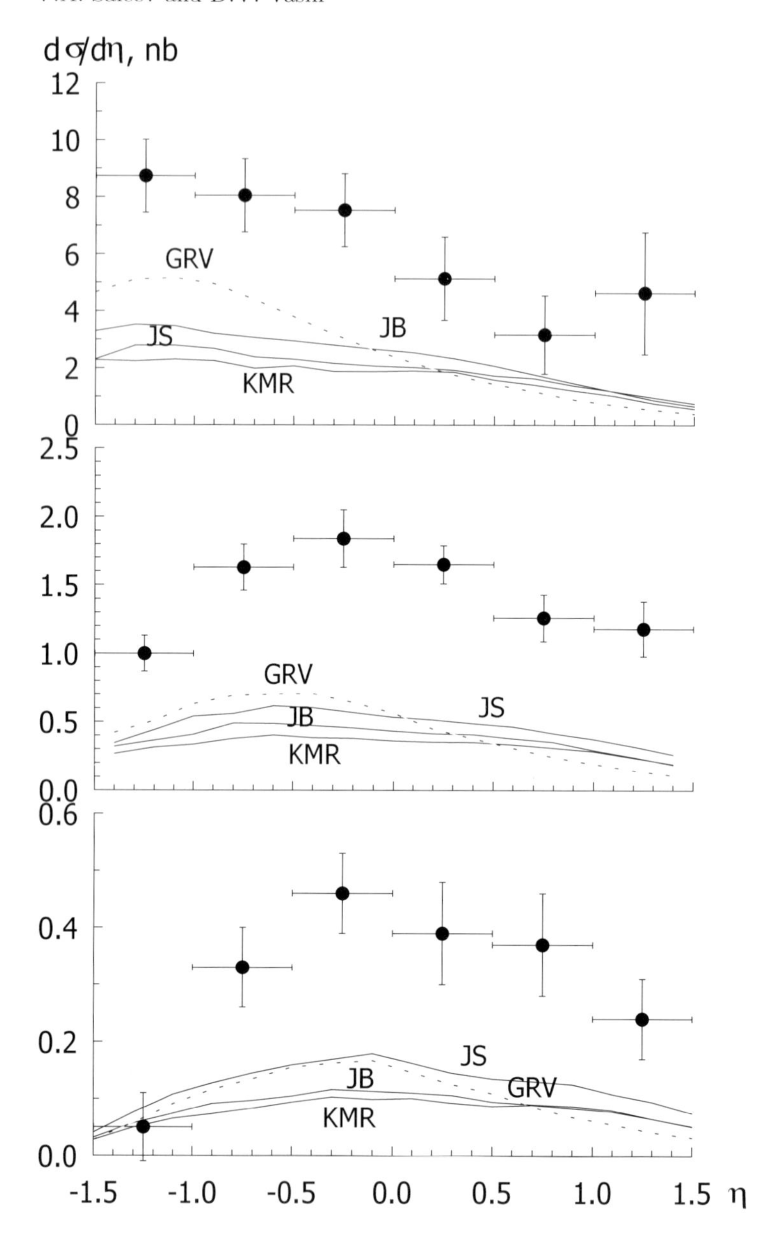

Fig. 3. The η spectra of the $D^{\star}$ meson at the various cut on a transverse momentum ($p_T > 2, 4, 6$ GeV, correspondingly from up to down) and $130 < W < 280$ GeV.

The differential cross section as a function of the $D^{\star}$ pseudorapidity, which is defined as $\eta = -\ln(\mathrm{tg}\,\frac{\theta}{2})$, where the polar angle θ is measured with respect to the proton beam direction, is shown in Fig. 3 where the kinematic ranges for

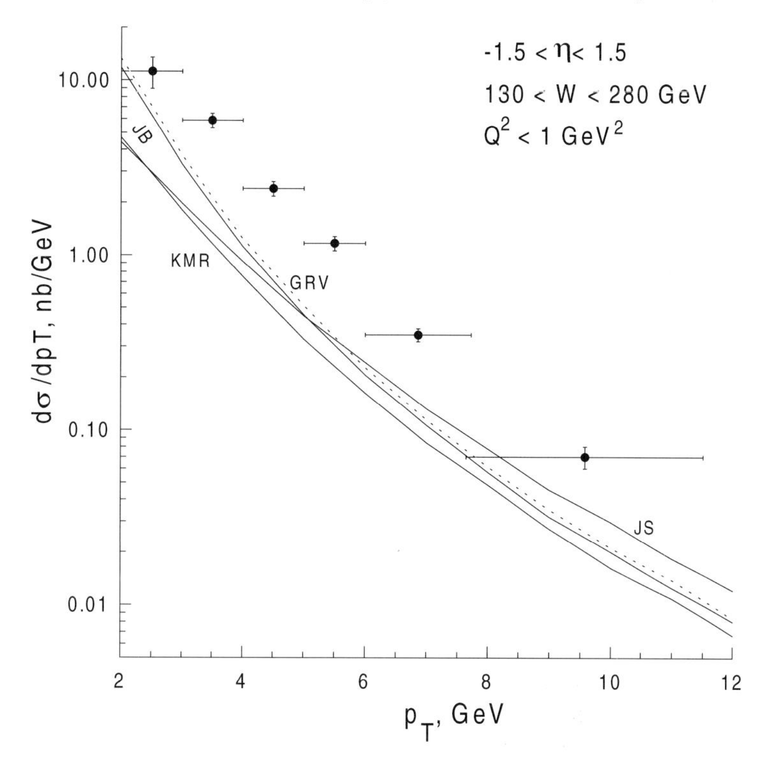

Fig. 4. The p_T-spectrum of $D^\star$ meson.

the $D^\star$ meson transverse momentum are $2 < p_T < 12$ GeV, $4 < p_T < 12$ GeV, $6 < p_T < 12$ GeV, correspondingly from up to down.

We see that the results of calculations performed in the collinear parton model as well as in the k_T-factorization approach with LO in α_s matrix elements need additional K-factor ($K \approx 2$) to describe the data. The value of this K-factor is usual for a heavy quark production cross section in the relevant energy range. Opposite the results obtained in k_T-factorization approach with JB parameterization in [21], where the strong enhance for the cross sections at all η in the k_T-factorization approach in compare to the collinear parton model was demonstrated, we see only deformation of the η-spectra. We have obtained that at low p_{Tmin} the maximum value of the cross section even higher in the collinear parton model and only at the large positive η the k_T-factorization approach gives more large values.

The p_T spectrum of $D^\star$ meson in photoproduction at $|\eta| < 1.5$ and $130 < W < 280$ GeV are shown in Fig. 4. All theoretical curves are under experimental points. As it was already mentioned typical value of the K-factor is equal 2.

Our results show that the introducing of a gluon transverse momentum k_T in the framework of the k_T-factorization approach does't increase the $D^\star$ meson photoproduction cross section at the large p_T as it is predicted for the J/Ψ

photoproduction [22,23] (see next part of the paper). We see the small effect in the case of JS parameterization [9] only.

The dependence of the $D^\star$ meson production cross section on a total photon-proton center-of-mass energy W is shown in Fig. 5. Nowadays the experimental data for $d\sigma/dW$ are absent. We see that the difference between the results obtained with the various parameterizations of a unintegrated gluon distribution function is about 50%. As well in the case of the η-spectra. At the small p_{Tmin} the cross section calculated in the parton model is larger than predictions obtained in the k_T-factorization approach.

The main uncertainties of our calculation come from the choice of a c-quark mass in the partonic matrix elements (10),(11) and (13), and from the choice of a parameter ϵ in the Peterson fragmentation function $D_{c\to D^\star}(z,\mu^2)$.

However, even at the very extremely choice ($m_c \approx 1.3$ GeV and $\epsilon \approx 0.02$) our theoretical predictions describe only shapes of the p_T- and η-spectra, but don't describe absolute values of the measured cross sections. This fact shows the famous role of the next to leading order corrections in the $D^\star$ meson photoproduction as in the parton model as in the k_T-factorization approach.

4 J/ψ Photoproduction in LO QCD

It is well known that in the processes of J/ψ meson photoproduction on protons at high energies the photon-gluon fusion partonic subprocess dominates [24]. In the framework of the general factorization approach of QCD the J/ψ photoproduction cross section depends on the gluon distribution function in a proton, the hard amplitude of $c\bar{c}$-pair production as well as the mechanism of a creation colorless final state with quantum numbers of the J/ψ meson. In such a way, we suppose that the soft interactions in the initial state are described by introducing a gluon distribution function, the hard partonic amplitude is calculated using perturbative theory of QCD at order in $\alpha_s(\mu^2)$, where $\mu \sim m_c$, and the soft process of the $c\bar{c}$-pair transition into the J/ψ meson is described in nonrelativistic approximation using series in the small parameters α_s and v (relative velocity of the quarks in the J/ψ meson). As is said in nonrelativistic QCD (NRQCD) [25], there are color singlet mechanism, in which the $c\bar{c}$-pair is hardly produced in the color singlet state, and color octet mechanism, in which the $c\bar{c}$-pair is produced in the color octet state and at a long distance it transforms into a final color singlet state in the soft process. However, as it was shown in papers [23,26], the data from the DESY ep-collider [27] in the wide region of p_T and z may be described well in the framework of the color singlet model and the color octet contribution is not needed. Based on the above mentioned result we will take into account in our analysis only the color singlet model contribution in the J/ψ meson photoproduction

[24]. We consider here the role of a proton gluon distribution function in the J/ψ photoproduction in the framework of the conventional parton model as well as in the framework of the k_T-factorization approach [3–5].

There are six Feynman diagrams (Fig. 6) which describe the partonic process $\gamma g \to J/\psi g$ at the leading order in α_s and α. In the framework of the color singlet

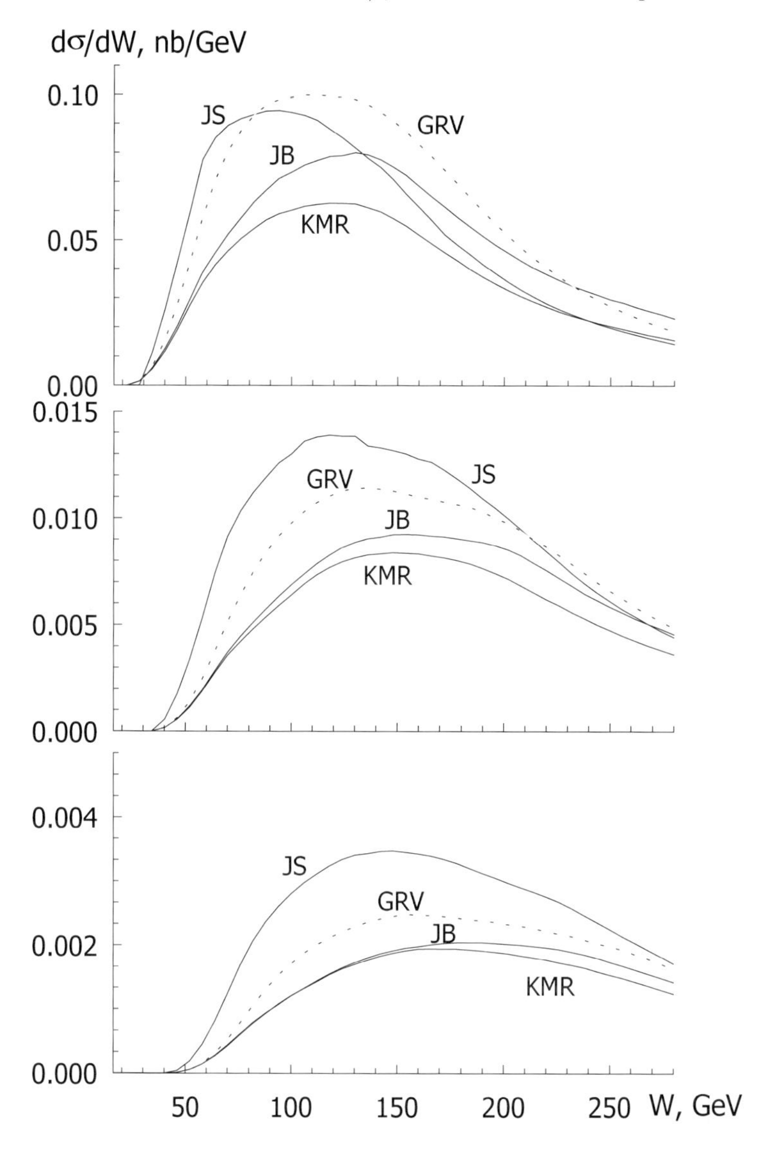

Fig. 5. The theoretical predictions for the W-spectra at the various cut on the $D^{\star}$ meson transverse momentum ($p_T > 2, 4, 6$ GeV, correspondingly from up to down) and $|\eta| < 1.5$.

model and nonrelativistic approximation the production of the J/ψ meson is considered as the production of a quark-antiquark system in the color singlet state with orbital momentum $L = 0$ and spin momentum $S = 1$. The binding energy and relative momentum of quarks in the J/ψ are neglected. In such a way $M = 2m_c$ and $p_c = p_{\bar{c}} = \dfrac{p}{2}$, where p is the 4-momentum of the J/ψ, p_c and

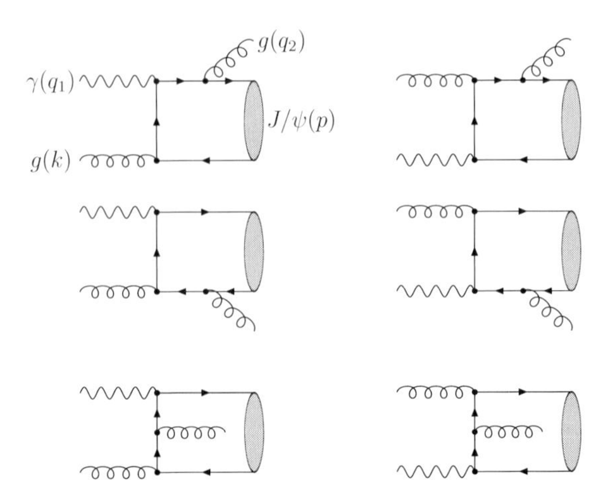

Fig. 6. Diagrams used for description partonic process $\gamma + g \to J/\psi + g$.

$p_{\bar{c}}$ are 4-momenta of quark and antiquark. Taking into account the formalism of the projection operator [28] the amplitude of the process $\gamma g \to J/\psi g$ may be obtained from the amplitude of the process $\gamma g \to \bar{c}cg$ after replacement:

$$V^i(p_{\bar{c}})\bar{U}^j(p_c) \to \frac{\Psi(0)}{2\sqrt{M}}\hat{\varepsilon}(p)(\hat{p}+M)\frac{\delta^{ij}}{\sqrt{3}}, \tag{15}$$

where $\hat{\varepsilon}(p) = \varepsilon_\mu(p)\gamma^\mu$, $\varepsilon_\mu(p)$ is a 4-vector of the J/ψ polarization, $\frac{\delta^{ij}}{\sqrt{3}}$ is the color factor, $\Psi(0)$ is the nonrelativistic meson wave function at the origin. The matrix elements of the process $\gamma g^\star \to J/\psi g$ may be presented as follows:

$$M_i = KC^{ab}\varepsilon_\alpha(q_1)\varepsilon^a_\mu(q)\varepsilon^b_\beta(q_2)\varepsilon_\nu(p)M_i^{\alpha\beta\mu\nu}, \tag{16}$$

$$M_1^{\alpha\beta\mu\nu} = \mathrm{Tr}\left[\gamma^\nu(\hat{p}+M)\gamma^\alpha\frac{\hat{p}_c-\hat{q}_1+m_c}{(p_c-q_1)^2-m_c^2}\gamma^\mu\frac{-\hat{p}_{\bar{c}}-\hat{q}_2+m_c}{(p_{\bar{c}}+q_2)^2-m_c^2}\gamma^\beta\right], \tag{17}$$

$$M_2^{\alpha\beta\mu\nu} = \mathrm{Tr}\left[\gamma^\nu(\hat{p}+M)\gamma^\beta\frac{\hat{p}_c+\hat{q}_2+m_c}{(p_c+q_2)^2-m_c^2}\gamma^\alpha\frac{\hat{k}-\hat{p}_{\bar{c}}+m_c}{(q-p_{\bar{c}})^2-m_c^2}\gamma^\mu\right], \tag{18}$$

$$M_3^{\alpha\beta\mu\nu} = \mathrm{Tr}\left[\gamma^\nu(\hat{p}+M)\gamma^\alpha\frac{\hat{p}_c-\hat{q}_1+m_c}{(p_c-q_1)^2-m_c^2}\gamma^\beta\frac{\hat{k}-\hat{p}_{\bar{c}}+m_c}{(q-p_{\bar{c}})^2-m_c^2}\gamma^\mu\right], \tag{19}$$

$$M_4^{\alpha\beta\mu\nu} = \mathrm{Tr}\left[\gamma^\nu(\hat{p}+M)\gamma^\mu\frac{\hat{p}_c-\hat{k}+m_c}{(p_c-q)^2-m_c^2}\gamma^\alpha\frac{-\hat{p}_{\bar{c}}-\hat{q}_2+m_c}{(q_2+p_{\bar{c}})^2-m_c^2}\gamma^\beta\right], \tag{20}$$

$$M_5^{\alpha\beta\mu\nu} = \mathrm{Tr}\left[\gamma^\nu(\hat{p}+M)\gamma^\beta\frac{\hat{p}_c+\hat{q}_2+m_c}{(p_c+q_2)^2-m_c^2}\gamma^\mu\frac{\hat{q}_1-\hat{p}_{\bar{c}}+m_c}{(q_1-p_{\bar{c}})^2-m_c^2}\gamma^\alpha\right], \tag{21}$$

$$M_6^{\alpha\beta\mu\nu} = \mathrm{Tr}\left[\gamma^\nu(\hat{p}+M)\gamma^\mu \frac{\hat{p}_c - \hat{k} + m_c}{(p_c-q)^2 - m_c^2}\gamma^\beta \frac{\hat{q}_1 - \hat{p}_{\bar{c}} + m_c}{(q_1 - p_{\bar{c}})^2 - m_c^2}\gamma^\alpha\right], \quad (22)$$

where q_1 is the 4-momentum of the photon, q is the 4-momentum of the initial gluon, q_2 is the 4-momentum of the final gluon,

$$K = e_c e g_s^2 \frac{\Psi(0)}{2\sqrt{M}}, \quad C^{ab} = \frac{1}{\sqrt{3}}\mathrm{Tr}[T^a T^b], \quad e_c = \frac{2}{3}, \quad e = \sqrt{4\pi\alpha}, \quad g_s = \sqrt{4\pi\alpha_s}\,.$$

The summation on the photon, the J/ψ meson and final gluon polarizations is carried out by covariant formulae:

$$\sum_{spin} \varepsilon_\alpha(q_1)\varepsilon_\beta(q_1) = -g_{\alpha\beta}, \quad (23)$$

$$\sum_{spin} \varepsilon_\alpha(q_2)\varepsilon_\beta(q_2) = -g_{\alpha\beta}, \quad (24)$$

$$\sum_{spin} \varepsilon_\mu(p)\varepsilon_\nu(p) = -g_{\mu\nu} + \frac{p_\mu p_\nu}{M^2}. \quad (25)$$

In case of the initial BFKL gluon we use the prescription (5). For studing J/ψ polarized photoproduction we introduce the 4-vector of the longitudinal polarization as follows:

$$\varepsilon_L^\mu(p) = \frac{p^\mu}{M} - \frac{M p_N^\mu}{(pp_N)}. \quad (26)$$

In the high energy limit of $s = 2(q_1 p_N) >> M^2$ the polarization 4-vector satisfies usual conditions $(\varepsilon_L \varepsilon_L) = -1$, $(\varepsilon_L p) = 0$.

Traditionally for a description of charmonium photoproduction processes the invariant variable $z = (pp_N)/(q_1 p_N)$ is used. In the rest frame of the proton one has $z = E_\psi / E_\gamma$. In the k_T-factorization approach the differential on p_T and z cross section of the J/ψ photoproduction may be written as follows:

$$\frac{d\sigma(\gamma p \to J/\Psi X)}{dp_T^2 dz} = \frac{1}{z(1-z)}\int_0^{2\pi}\frac{d\varphi}{2\pi}\int_0^{\mu^2} d\mathbf{k}_T^2 \Phi(x, \mathbf{k}_T^2, \mu^2)\frac{\overline{|M|^2}}{16\pi(xs)^2}. \quad (27)$$

The analytical calculation of the $\overline{|M|^2}$ is performed with help of REDUCE package and results are saved in the FORTRAN codes as a function of $\hat{s} = (q_1+q)^2$, $\hat{t} = (p-q_1)^2$, $\hat{u} = (p-q)^2$, $\mathbf{p}_T^2$, $\mathbf{k}_T^2$ and $\cos(\varphi)$. We directly have tested that

$$\lim_{\mathbf{k}_T^2 \to 0}\int_0^{2\pi}\frac{d\varphi}{2\pi}\overline{|M|^2} = \overline{|M_{PM}|^2}, \quad (28)$$

where $\mathbf{p}_T^2 = \frac{\hat{t}\hat{u}}{\hat{s}}$ in the $\overline{|M|^2}$ and $\overline{|M_{PM}|^2}$ is the square of the amplitude in the conventional parton model [24]. In the limit of $\mathbf{k}_T^2 = 0$ from formula (27) it is easy to find the differential cross section in the parton model, too:

$$\frac{d\sigma^{PM}(\gamma p \to J/\Psi X)}{dp_T^2 dz} = \frac{\overline{|M_{PM}|^2} xG(x,\mu^2)}{16\pi(xs)^2 z(1-z)}. \quad (29)$$

However, making calculations in the parton model we use formula (27), where integration over $\mathbf{k}_T^2$ and φ is performed numerically, instead of (29). This method fixes the common normalization factor for both approaches and gives a direct opportunity to study effects connected with virtuality of the initial BFKL gluon in the partonic amplitude.

5 J/ψ Photoproduction at HERA

After we fixed the selection of the gluon distribution functions $G(x,\mu^2)$ or $\Phi(x,\mathbf{k}_T^2,\mu^2)$ there are two parameters only, which values determine the common normalization factor of the cross section under consideration: $\Psi(0)$ and m_c. The value of the J/ψ meson wave function at the origin may be calculated in a potential model or obtained from experimental well known decay width $\Gamma(J/\psi \to \mu^+\mu^-)$. In our calculation we used the following choice $|\Psi(0)|^2 = 0.064$ GeV^3 which corresponds to NRQCD coefficient $< O^{J/\psi}, \mathbf{1}^3S_1 >= 1.12$ GeV^3 as the same as in [26]. Note, that this value is a little smaller (30 %) than the value which was used in our paper [23]. Concerning a charmed quark mass, the situation is not clear up to the end. From one hand, in the nonrelativistic approximation one has $m_c = \dfrac{M}{2}$, but there are many examples of taking smaller value of a c-quark mass in the amplitude of a hard process, for example $m_c = 1.4$ GeV. Taking into consideration above mentioned we perform calculations at $m_c = 1.5$ GeV. The cinematic region under consideration is determined by the following conditions: $Q^2 < 1$ GeV^2, $60 < W < 240$ GeV, $0.3 < z < 0.9$ and $p_T > 1$ GeV, which correspond to the H1 Collaboration data [29]. We assume that the contribution of the color octet mechanism is large at the $z > 0.9$ only. In the region of the small values of the $z < 0.2$ the contribution of the resolved photon processes [30] as well as the charm excitation processes [31] may be large, too. All of these contributions are not in our consideration.

Figures 7–10 show our results which were obtained as in the conventional parton model as well as in the k_T-factorization approach with the different parameterizations of the unintegrated gluon distribution function. The dependence of the results on selection of a hard scale parameter μ is much less than the dependence on selection of a c-quark mass and selection of a parameterization. We put $\mu^2 = M^2 + \mathbf{p}_\mathbf{T}^\mathbf{2}$ in a gluon distribution function and in a running constant $\alpha_s(\mu^2)$.

The count of a transverse momentum of the BFKL gluons in the k_T-factorization approach results in a flattening of the p_T-spectrum of the J/ψ as contrasted by predictions of the parton model. For the first time this effect was indicated in the [22], and later in [23]. Figure 7 shows the result of our calculation for the p_T spectrum of the J/ψ mesons. Using the k_T-factorization approach we have obtained the harder p_T-spectrum of the J/ψ than has been predicted in the LO parton model. It is visible that at large values of p_T only the k_T-factorization approach gives correct description of the data [29]. However, it is impossible to consider this visible effect as a direct indication on nontrivial developments of the small-x physics. In the article [26] was shown that the calculation in the

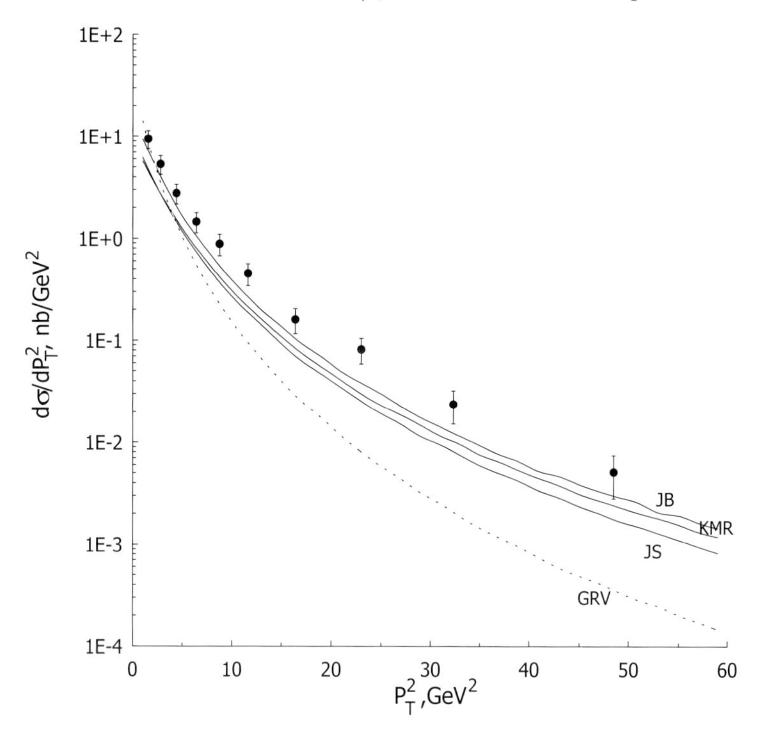

Fig. 7. The J/ψ spectrum on p_T^2 at the $60 < W < 240$ GeV and $0.3 < z < 0.9$.

NLO approximation gives a harder p_T spectrum of the J/ψ meson, too, which will agree with the data at the large p_T.

In the k_T-factorization approach JB parameterization [8] gives p_T-spectrum, which is very close to experimental data. From the another hand in the case of JS parameterization [9] the additional K-factor approximately equal 2 is needed.

The z spectra are shown in Fig. 8 at the various choice of the p_T cut: $p_T > 2$, 4 and 6 GeV, correspondingly. The relation between the theoretical predictions and experimental data is the same as in Fig. 7. The k_T factorization approach give more correct description of the data especially at large value of z where the curve obtained in the collinear parton model tends to zero.

Figure 9 shows the dependence of the total J/ψ photoproduction cross section on W at $0.3 < z < 0.8$ and $p_T > 1$ GeV. The shape of this dependence agrees well with the result obtained using JS parameterization [9] or KMR [10] parameterization. However, the predicted absolute value of the cross section $\sigma_{\gamma p}$ is smaller by factor 2 than obtained data [29]. The results of calculation using JB [8] or GRV [7] parameterizations are larger and coincide with the data [29] better.

As it was mentioned above, the main difference between the k_T-factorization approach and the conventional parton model is nontrivial polarization of the BFKL gluon. It is obvious, that such a spin condition of the initial gluon should

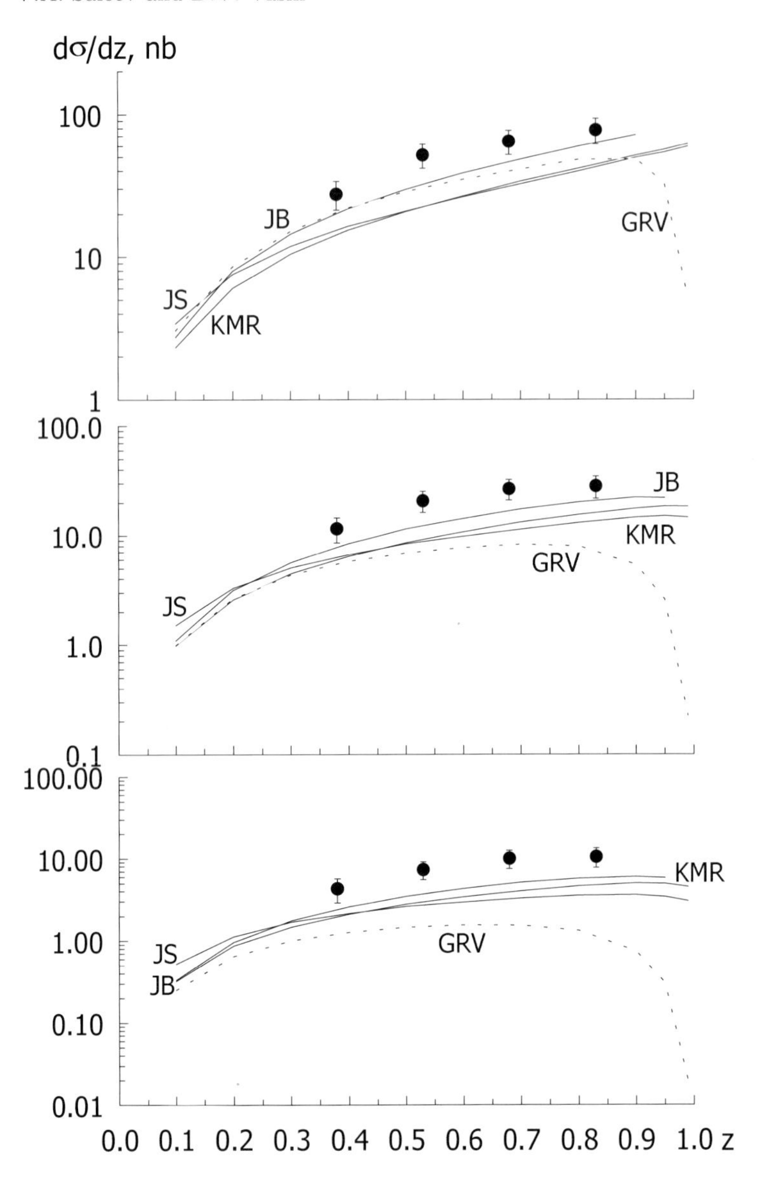

Fig. 8. The J/ψ spectrum on z at the $60 < W < 240$ GeV ($p_T > 1, 2, 3$ GeV, correspondingly from up to down).

result in observed spin effects during the birth of the polarized J/ψ meson. We have performed calculations for the spin parameter α as a function z or p_T in the conventional parton model and in the k_T-factorization approach :

$$\alpha(z) = \frac{\frac{d\sigma_{tot}}{dz} - 3\frac{d\sigma_L}{dz}}{\frac{d\sigma_{tot}}{dz} + \frac{d\sigma_L}{dz}}, \qquad \alpha(p_T) = \frac{\frac{d\sigma_{tot}}{dp_T} - 3\frac{d\sigma_L}{dp_T}}{\frac{d\sigma_{tot}}{dp_T} + \frac{d\sigma_L}{dp_T}} \tag{30}$$

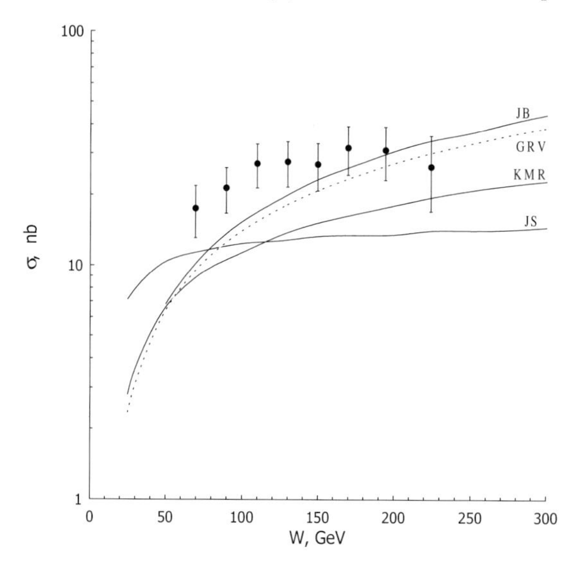

Fig. 9. The total J/ψ photoproduction cross section versus W at $0.3 < z < 0.8$ and $p_T > 1$ GeV.

Here $\sigma_{tot} = \sigma_L + \sigma_T$ is the total J/ψ production cross section, σ_L is the production cross section for the longitudinal polarized J/ψ mesons, σ_T is the production cross section for the transverse polarized J/ψ mesons. The parameter α controls the angle distribution for leptons in the decay $J/\psi \to l^+l^-$ in the J/ψ meson rest frame:

$$\frac{d\Gamma}{d\cos(\theta)} \sim 1 + \alpha\cos^2(\theta). \tag{31}$$

The theoretical results for the parameter $\alpha(z)$ are very close to each other irrespective of the choice of an approach or a gluon parameterization [23].

For the parameter $\alpha(p_T)$ we have found strongly opposite predictions in the parton model and in the k_T-factorization approach, as it is visible in Fig. 10. The parton model predicts that J/ψ mesons should have transverse polarizations at the large p_T ($\alpha(p_T) = 0.6$ at the $p_T = 6$ GeV), but k_T-factorization approach predicts that J/ψ mesons should be longitudinally polarized ($\alpha(p_T) = -0.4$ at the $p_T = 6$ GeV). The experimental points lie in the range $0 < p_T < 5$ GeV and they have the large errors. However, it is visible that $\alpha(p_T)$ decrease as p_T changes from 1 to 5 GeV. This fact coincide with theoretical prediction obtained in the k_T-factorization approach. Nowadays, a result of the NLO parton model calculation in the case of the polarized J/ψ meson photoproduction is unknown. It should be an interesting subject of future investigations. If the count of the NLO corrections will not change predictions of the LO parton model for $\alpha(p_T)$, the experimental measurement of this spin effect will be a direct signal about BFKL gluon dynamics.

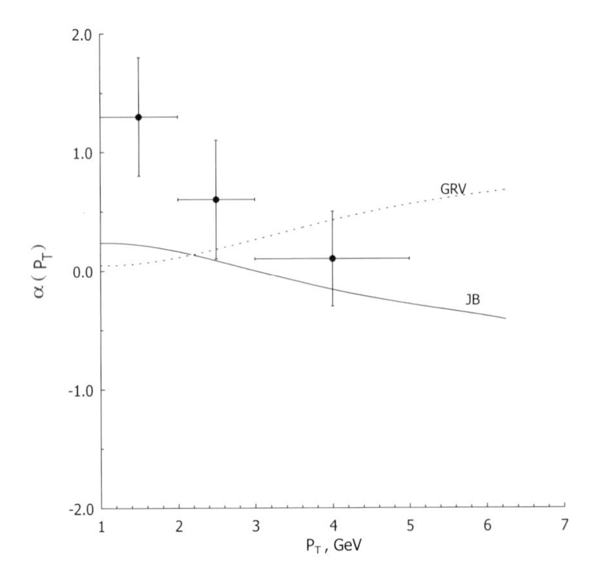

Fig. 10. Parameter α as a function of p_T at $0.3 < z < 0.9$, $p_T > 1$ GeV, $60 < W < 240$ GeV.

Nowadays, the experimental data on J/Ψ polarization in photoproduction at large p_T are absent. However there are similar data from CDF Collaboration [32], where J/ψ and ψ' p_T-spectra and polarizations have been measured. Opposite the case of J/ψ photoproduction, the hadroproduction dada needs to take into account the large color-octet contribution in order to explain J/ψ and ψ' production at Tevatron in the conventional collinear parton model. The relative weight of color-octet contribution may be smaller if we use k_T-factorization approach, as was shown recently in [33–36]. The predicted using collinear parton model transverse polarization of J/ψ at large p_T is not supported by the CDF data, which can be roughly explained by the k_T-factorization approach [34]. In conclusion, the number of theoretical uncertainties in the case of J/ψ meson hadroproduction is much more than in the case of photoproduction

and they need more complicated investigation, which is why the future experimental analysis of J/ψ photoproduction at THERA will be clean check of the collinear parton model and the k_T-factorization approach.

The authors would like to thank M. Ivanov and S. Nedelko for kind hospitality during workshop "Heavy quark - 2002" in Dubna, S. Baranov, A. Lipatov and O. Teryaev for discussions on the k_T-factorization approach of QCD and H. Jung for the valuable information on unintegrated gluon distribution functions. This work has been supported in part by the Russian Foundation for Basic Research under Grant 02-02-16253.

References

1. G. Sterman et al.: Rev. Mod. Phys. **59**, 158 (1995)
2. V.N. Gribov and L.N. Lipatov: Sov. J. Nucl. Phys. **15**, 438 (1972) Yu.A. Dokshitser: Sov. Phys. JETP. **46**, 641 (1977) G. Altarelli and G. Parisi: Nucl. Phys. **B126**, 298 (1977)
3. L.V. Gribov, E.M. Levin, M.G. Ryskin: Phys. Rep. **100**,1 (1983)
4. J.C. Collins and R.K. Ellis: Nucl. Phys. **360**, 3 (1991)
5. S. Catani,M. Ciafoloni and F. Hautmann: Nucl. Phys. **B366**, 135 (1991)
6. E. Kuraev, L. Lipatov, V. Fadin: Sov. Phys. JETP **44**, 443 (1976) Y. Balitskii and L. Lipatov: Sov. J. Nucl. Phys. **28**, 822 (1978)
7. M. Gluck, E. Reya and A. Vogt: Z. Phys. **C67**, 433 (1995)
8. J. Bluemlein: DESY 95-121.
9. H. Jung, G. Salam: Eur. Phys. J. **C19**, 351 (2001)
10. M.A. Kimber, A.D. Martin and M.G. Ryskin: Phys. Rev. **D63**, 114027 (2001)
11. J. Breitweg at al.[ZEUS Coll.]: Eur. Phys. J. **C6**, 67 (1999) Phys. Lett. **B481**, 213 (2000)
12. C. Adloff et al.[H1 Coll.]: Nucl. Phys. **B545**, 21 (1999)
13. C. Peterson et al.:Phys. Rev. **D27**, 105 (1983)
14. A.V. Berezhnoy, V.V. Kiselev and A.K. Likhoded: (hep-ph/9901333, hep-ph/9905555)
15. R. Akers et al.[OPAL Coll.]: Z. Phys. **C67**, 27 (1995)
16. P. Nason and C. Oleari: Phys. Lett. **B447**, 327 (1999)
17. B. Kniehe et al.: Z. Phys. **C76**, 689 (1997) J. Binnewics et al.: Phys. Rev. **D58**, 014014 (1998)
18. V.A. Saleev and N.P. Zotov: Mod. Phys. Lett. **A11**, 25 (1996)
19. V.M. Budnev et al.: Phys. Rep. **15**, 181 (1974)
20. S. Frixione et al.: Phys. Lett. **B319**, 339 (1993)
21. S. Baranov, N. Zotov: Phys. Lett. **B458**, 389 (1999)
22. V.A. Saleev, N.P. Zotov: Mod. Phys. Lett. **A9**, 151; 1517 (1994)
23. V.A. Saleev: Phys. Rev. **D65**, 054041 (2002)
24. E.L. Berger and D. Jones: Phys. Rev. **D23**, 1521 (1981) R. Baier and R. Ruckl: Phys. Lett. **B102**, 364 (1981) S.S. Gershtein, A.K. Likhoded, S.R. Slabospiskii: Sov. J. Nucl. Phys. **34**, 128 (1981)
25. G.T. Bodwin, E. Braaten, G.P. Lepage: Phys. Rev **D51**, 1125 (1995)
26. M. Kramer: Nucl. Phys. **B459**, 3 (1996)
27. Aid et al.[H1 Coll.]: Nucl. Phys. **B472**, 32 (1996) J. Breitweg et al.[ZEUS Coll.]: Z. Phys. **C76**, 599 (1997)
28. B. Guberina et al.: Nucl. Phys. **B174**, 317 (1980)
29. C. Adloff et al.[H1 Coll.]: DESY 02-059 (2002)
30. H. Jung, G.A. Schuler and J. Terron: DESY 92-028 (1992)
31. V.A. Saleev: Mod. Phys. Lett. **A9**, 1083 (1994)
32. T. Affolder et al.[CDF Coll.]: Phys. Rev. Lett. **85**, 2886 (2000)
33. F. Yuan, K-T. Chao: Phys. Rev. **D63**, 034006 (2001)
34. F. Yuan, K-T. Chao: Phys. Rev. Lett. **D87**, 022002-L (2001)
35. Ph. Hägler et al.: Phys. Rev. Lett. **86**, 1446 (2001)
36. Ph. Hägler et al.: Phys. Rev. **D63**, 077501 (2001)

Measurement of the $b\bar{b}$ Production Cross Section in 920 GeV Fixed-Target Proton-Nucleus Collisions at the HERA-B Detector

Alexander Lanyov for the HERA-B Collaboration

Particle Physics Laboratory, Joint Institute for Nuclear Research, 141980 Dubna, Moscow region, Russia

Abstract. The $b\bar{b}$ production cross section has been measured using the HERA-B detector in 920 GeV proton collisions on carbon and titanium targets. We used the data collected in a short period in summer 2000 during the commissioning of the detector and trigger before the HERA upgrade began. The $b\bar{b}$ events were identified via inclusive bottom quark decays producing a J/ψ by exploiting the longitudinal separation of the $J/\psi \to l^+l^-$ decay vertices from the primary proton-nucleus interactions, which benefits from the excellent resolution of the HERA-B silicon vertex detector. Both e^+e^- and $\mu^+\mu^-$ channels have been reconstructed and the combined analysis yields the cross section $\sigma(b\bar{b}) = 32^{+14}_{-12}(\text{stat})^{+6}_{-7}(\text{sys})$ nb/nucleon [1]. Our result is in good agreement with the latest QCD calculations beyond next-to-leading order and is compatible with the two previous measurements of the $b\bar{b}$ production cross section in proton-nucleus collisions.

1 Introduction

The question of a theoretical description of the $b\bar{b}$ cross section has attracted much attention during the last years due to the appearance of many new experimental papers which claim to disagree with the predictions of QCD at the next-to-leading order (NLO). One might refer to the CDF results [2] at $\sqrt{s} = 1800$ GeV where the ratio of the measured B^+ meson differential cross section to the NLO prediction is around 3, $\gamma\gamma$ reactions at LEP [3,4] where the measured $\sigma(e^+e^- \to e^+e^-b\bar{b})$ at $\sqrt{s} \approx 200$ GeV greatly exceeds the NLO predictions by a factor 3–4, and b photoproduction in the reaction $ep \to bX$ measured by H1 [5], where the results are larger than the NLO QCD predictions by more than a factor of 2. The following possible explanations of these discrepancies have been proposed: unintegrated k_T distributions and small-x resummations can be used [6], bottom fragmentation effects have been incorrectly accounted for [7], or there might be a signal from supersymmetry [8].

In a recent paper [9] by the CDF collaboration it was shown that the large b-quark cross section is not something that is specific to 1800 GeV data. Therefore, the measurement of $\sigma(b\bar{b})$ at smaller energies gains additional interest. In the region of proton-nucleon collisions at a total energy in the c.m. system of the order of 40 GeV, the situation cannot be considered satisfactory. Only two experimental measurements of $\sigma(b\bar{b})$ exist for this region: the E789 collaboration obtained $\sigma(b\bar{b}) = 5.7 \pm 1.5(\text{stat}) \pm 1.3(\text{sys})$ nb/nucleon in 800 GeV pAu

A. Lanyov, Measurement of the $b\bar{b}$ Production Cross Section in 920 GeV Fixed-Target Proton-Nucleus Collisions at the HERA-B Detector, Lect. Notes Phys. **647**, 418–426 (2004)
http://www.springerlink.com/

collisions using the decay chain $b \to J/\psi X \to \mu^+\mu^- X$ [10], while the E771 collaboration obtained $\sigma(b\bar{b}) = 43^{+27}_{-17}(\mathrm{stat}) \pm 7(\mathrm{sys})$ nb/nucleon in 800 GeV pSi collisions using μ semileptonic b decays [11]. These results differ from each other at the level of two standard deviations despite the large uncertainties. Therefore a new measurement of $\sigma(b\bar{b})$ is highly desirable in order to check the modern QCD predictions.

The HERA-B experiment is designed to identify B-meson decays in a dense hadronic environment, with a large geometrical coverage. The b-hadrons decaying into J/ψ are distinguished from the large prompt J/ψ background by exploiting the long b lifetime in a detached vertex analysis. The HERA-B detector can accomplish this task by using the decay chain $b \to J/\psi X$, $J/\psi \to l^+l^-$ not only in the $\mu^+\mu^-$ channel (which has been used before [10]), but also in the e^+e^- channel with a rather common experimental approach. The usage of both channels enhances the precision of the results and allows an additional cross-check with different trigger conditions.

2 The Experimental Setup

We used data collected over a short period during the summer 2000 while commissioning of the detector and trigger was being undertaken. The main parts of the detector which contributed to the measurements are the following. The target consisted of wires of different materials which were inserted into the halo of the HERA 920 GeV proton beam. In the data taken, carbon and titanium target wires were used at an interaction rate of ≈5 MHz. The Vertex Detector System (VDS) consisted of 64 double-sided silicon microstrip detectors organized in 8 superlayers and located in a Roman pot system in the vacuum vessel. The detector modules are movable in the radial and lateral directions by manipulators. The excellent work of the VDS is crucial for the analysis of the detached vertices which is the basis of the $b\bar{b}$ cross section measurement. The secondary vertex resolution along the beam direction z was $\sigma_z \approx 715\,\mu$m, while in the transverse direction accuracy $\sigma_{x,y} \approx 60$ μm was reached. When compared with the mean decay length of B mesons (which at HERA-B energies is ≈8 mm), a factor greater than one order of magnitude is achieved.

The track momenta are measured by a large-area Outer Tracker after the dipole magnet, with a field integral of 2.13 T·m. Muon identification is performed by the ring imaging Cherenkov hodoscope (RICH) and the muon detector (MUON), while the electrons are detected and identified by the electromagnetic calorimeter (ECAL). The trigger chain included a pretrigger, provided by ECAL or MUON, for the lepton candidate searching, and a first level trigger which selected dilepton events and forwarded them to the second level trigger (SLT), which confirmed the candidate tracks in OTR and VDS using a simplified Kalman filter algorithm. The SLT was a software trigger running on a PC-farm. A total of ≈450,000 dimuon and ≈900,000 dielectron candidates were recorded under these conditions. A more detailed description of the apparatus, trigger and software can be found in [1,12,13].

3 The Method of Measurement

Almost all J/ψ particles are produced directly on the target wire close to the primary interaction point. The $b \to J/\psi\, X$ events can be selected by using the long decay length of B mesons Δz which can be reliably measured. Additional impact parameter cuts make further suppression of the prompt J/ψ and other sources of background, e.g. from charm and bottom semileptonic decays.

In order to minimize the systematic errors related to detector and trigger efficiencies and to remove the dependence on the absolute luminosity determination, the measurement is performed relative to the known prompt J/ψ production cross section σ_{P}^{A} [14,15]. This measurement covers the J/ψ Feynman-x range $-0.25 \leq x_F \leq 0.15$. Within our acceptance, the b to prompt cross section ratio can be expressed as:

$$\frac{\Delta\sigma_{\mathrm{B}}^{A}}{\Delta\sigma_{\mathrm{P}}^{A}} = \frac{N_{\mathrm{B}}}{N_{\mathrm{P}}} \frac{1}{\varepsilon_{\mathrm{R}}\, \varepsilon_{\mathrm{B}}^{\Delta z}\, \mathrm{Br}(b\bar{b} \to J/\psi X)}\,, \tag{1}$$

where $\Delta\sigma_{\mathrm{B}}^{A}$ and $\Delta\sigma_{\mathrm{P}}^{A}$ are the $b \to J/\psi$ and prompt J/ψ cross sections in the detector x_F acceptance, respectively. N_{B} and N_{P} are the observed number of detached $b \to J/\psi$ and prompt J/ψ decays. ε_{R} is the relative detection efficiency of $b \to J/\psi$ with respect to prompt J/ψ, including contributions from the trigger, the dilepton reconstruction and the J/ψ vertexing. $\varepsilon_{\mathrm{B}}^{\Delta z}$ is the efficiency of the detached vertex selection. The branching ratio $\mathrm{Br}(b\bar{b} \to J/\psi X)$ was taken as $2 \cdot (1.16 \pm 0.10)\%$ [16].

The prompt J/ψ production cross section per nucleon, $\sigma(pN \to J/\psi X) = \sigma_{\mathrm{P}}^{A}/A^{\alpha}$, was previously measured by two fixed target experiments [14,15]. After correcting for the most recent measurement of the atomic number dependence ($\alpha = 0.955 \pm 0.005$ [17]) and rescaling [18] to the HERA-B c.m.s. energy, $\sqrt{s} = 41.6$ GeV, we obtain a reference prompt J/ψ cross section of $\sigma(pN \to J/\psi X) = (357 \pm 8(\mathrm{stat}) \pm 27(\mathrm{sys}))$ nb/nucleon. About 70% [14] of the J/ψ are produced in the kinematic range covered by our measurement. Since no nuclear suppression has been observed in D meson production [19], and as a similar behavior is expected in the b channel [20], we assume $\alpha = 1.0$ for the $b\bar{b}$ production cross section results presented here (i.e., $\sigma_{\mathrm{B}}^{A} = \sigma(pN \to b\bar{b}) \cdot A$).

To calculate the efficiencies in (1) and to evaluate the background contributions to the $b \to J/\psi$ channel from prompt J/ψ and semileptonic charm and b decays, a complete Monte Carlo (MC) simulation was performed. The heavy quark (Q) generation was done by simulating the basic process $pN \to Q\overline{Q}X$ using the Pythia event generator [21], and giving the remaining part of the process (X) as an input to the Fritiof package [22] to simulate further interactions inside the nucleus. The detector simulation, based on Geant 3.21 [23], included the description of the geometry, full digitization, and a realistic description of the working conditions. In order to describe the correct prompt J/ψ kinematics, measured p_T and x_F J/ψ distributions [14] were used to weight the generated events. For b-quark kinematics, the events generated by Pythia are weighted according to a model of M. Mangano et al. [24] using the most recent next-to-next-to-leading-logarithm (NNLL) MRST parton distribution functions [25].

Possible variations of the theoretical and experimental parameters have been included into the systematic errors [1].

4 Prompt J/ψ Selection

In order to calculate the number N_{P} of prompt J/ψ, the dilepton trigger tracks were reconstructed in the offline reprocessing with additional particle identification requirements. The $J/\psi \to l^+l^-$ selection and counting procedure differs between the muon and electron channels, due to differences in the background levels, shapes, and triggering conditions. To select $J/\psi \to \mu^+\mu^-$ decays the following criteria were used: a dimuon vertex requirement and muon identification cuts in both the MUON and the RICH systems. The result of the fit is shown in Fig. 1(a), with $N_{\mathrm{P}} = 2880 \pm 60$ prompt $J/\psi \to \mu^+\mu^-$ decays. The like-sign spectrum shown in Fig. 1(a) is obtained from the same set of triggered events: the small discrepancy in the number of reconstructed events in the background regions arises from the difference in trigger acceptance between the two cases and from physics contributions to the unlike-sign spectrum (Drell-Yan, open charm production).

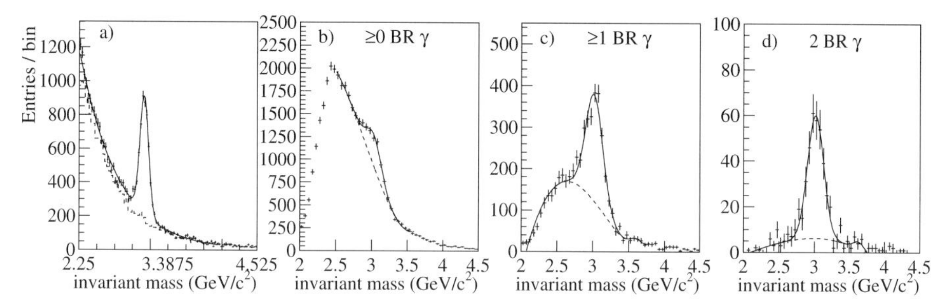

Fig. 1. (a) $\mu^+\mu^-$ invariant mass after the J/ψ selection cuts (*error bars*: $\mu^+\mu^-$; *solid line*: fit; *dashed line*: the like-sign background. **(b-c-d)** e^+e^- invariant mass with different requirements on number of identified bremsstrahlung (BR) photons (*error bars*: data with E/p cut at 1 σ level (9%); *solid line*: fit; *dashed line*: background). The bin size is 25 MeV/c^2 for $\mu^+\mu^-$ and 50 MeV/c^2 for e^+e^-.

To select $J/\psi \to e^+e^-$ decays, a dielectron vertex was required together with two additional electron identification cuts:

- Since electrons are fully absorbed in the ECAL, the ratio E/p should be around unity, where E is the energy of the ECAL cluster associated with the track and p is the track momentum measured by the tracking system. The E/p spectrum is compatible with a Gaussian distribution of mean 1.00 and width $\sigma \approx$9%.
- Bremsstrahlung photons emitted by electrons traversing the layers of material before the magnet keep the original electron direction. A cluster found in the ECAL at the position corresponding to the prolongation of the direction

measured by the VDS can be used to provide a clean electron signature and to correct the electron momentum at the vertex.

The e^+e^- invariant mass distribution is shown in Fig. 1(b,c,d), requiring only that E/p be within 1 σ from unity for each track. Figs. 1(c,d) show the improvements in signal significance obtained when the bremsstrahlung selection is added to the E/p requirement. A good understanding of the electron identification efficiencies [1] allowed us to infer the total number of prompt J/ψ in our sample with no bremsstrahlung tag requirement and with a 3σ E/p cut: $N_{\mathrm{P}} = 5710 \pm 380(\mathrm{stat}) \pm 280(\mathrm{sys})$.

5 Detached Vertex Analysis

To select $b \to J/\psi \to l^+l^-$ events from the prompt J/ψ background, we use the b-hadron long decay length Δz, defined as the distance along the beam axis between the J/ψ decay vertex and the closest target wire (primary production point). The r.m.s. resolution of the decay length is smaller than one-tenth of the mean decay length of triggered b-hadrons ($\approx$0.8 cm). Additionally, cuts on the minimum impact parameter of both leptons are used.

The $b \to J/\psi \to \mu^+\mu^-$ events were selected by requiring $\Delta z > 7.5\,\sigma_z$ where σ_z is the uncertainty on the secondary vertex position, a minimum track impact parameter to the assigned primary vertex of 160 μm and a minimum track impact parameter to the assigned wire of 45 μm. Both impact parameter cuts are needed, since the 2-dimensional vertex cut gives the best separation between the signal and the prompt background, whereas the target wire cut can additionally suppress background tracks originating from other potential primary vertices on the same wire.

For the electron channel the requirements are the following: a minimum decay length Δz of 0.5 cm, a minimum track impact parameter to the assigned wire of 200 μm or, alternatively, an isolation of the lepton candidate at the z position of the wire from any other track by a minimum distance of 250 μm. The "isolation cut" plays almost the same role as the vertex cut in the muon channel analysis, but with a less stringent efficiency requirement. The differences in types and values of the detached selection cuts between the muon and electron analysis are due to the very different background conditions. Due to the intrinsic "cleanness" of the muon sample, which benefits from the thick absorber before the muon detector, the combinatorial background is small for the muon channel, but significant for the electron one. Both channels have also a background from semileptonic charm and bottom decays, while the expected background from prompt J/ψ decays is negligible in both channels.

In order to extract the number of detached J/ψ events, an unbinned likelihood fit is performed on the invariant mass spectrum of the detached downstream l^+l^- candidates. For the $\mu^+\mu^-$ case, the shape of the signal is taken from the prompt J/ψ signal together with an exponential background contribution with a free slope. For the e^+e^- case, the shape of the signal is taken from simulated $b \to J/\psi$ decays, while the background shape is a combination

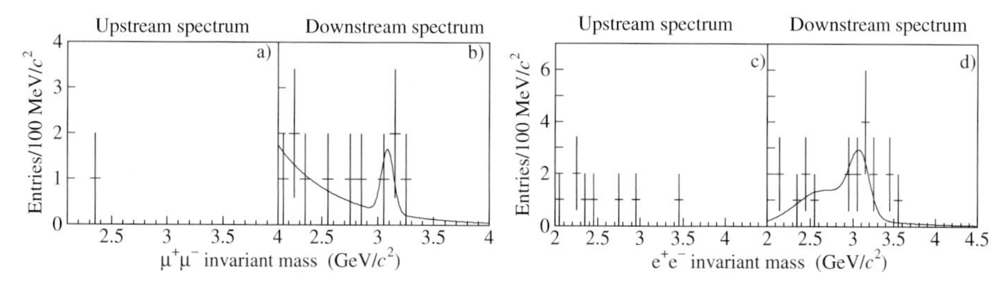

Fig. 2. The $\mu^+\mu^-$ upstream **(a)**, $\mu^+\mu^-$ downstream **(b)**, e^+e^- upstream **(c)**, e^+e^- downstream **(d)** invariant mass spectra after detached event selection. The downstream curves show the result of the unbinned likelihood fits, in which the yields of background and signal contributions, as defined in the text, were left as free parameters

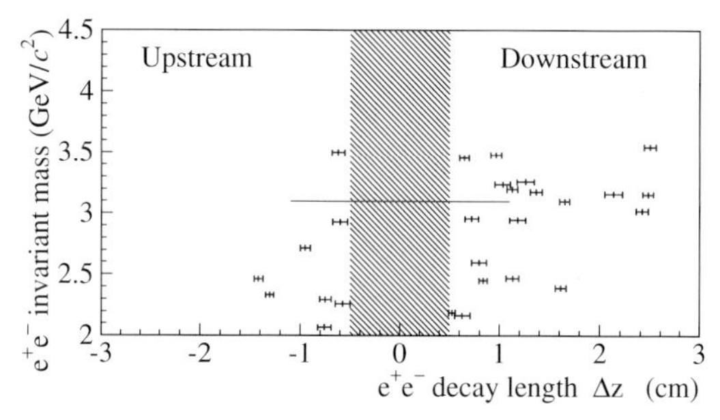

Fig. 3. The scatter plot of e^+e^- invariant masses versus the measured decay length Δz for the selected detached events. The shaded region is removed by the Δz cut. The horizontal line shows the mean J/ψ invariant mass value. A clear clustering of events around the J/ψ mass with large Δz is observed in the downstream sample

of the shapes obtained from simulated double semileptonic bottom quark decays and from pure combinatorial (upstream) events. The results of the likelihood fit are shown in Fig. 2, yielding $1.9^{+2.2}_{-1.5}$ $b \to J/\psi \to \mu^+\mu^-$ events and $8.6^{+3.9}_{-3.2}$ $b \to J/\psi \to e^+e^-$ events. The background levels are compatible with the MC predictions.

To determine $\Delta\sigma(b\bar{b})$ in the detector x_F acceptance, the prompt J/ψ and $b \to J/\psi$ MC events are submitted to the same analysis chain used for real data. From MC simulation, we obtain the efficiency terms entering in the cross section measurement (1): $\varepsilon_{\mathrm{R}} \cdot \varepsilon_{\mathrm{B}}^{\Delta z} = 0.44 \pm 0.02$ for the e^+e^- channel, and $\varepsilon_{\mathrm{R}} \cdot \varepsilon_{\mathrm{B}}^{\Delta z} = 0.41 \pm 0.01$ for the $\mu^+\mu^-$ channel. The corresponding $b\bar{b}$ cross sections are $\Delta\sigma(b\bar{b}) = 38^{+18}_{-15}$ and $\Delta\sigma(b\bar{b}) = 16^{+18}_{-12}$ nb/nucleon, respectively.

To confirm the b assignment, a detailed study of the decay length Δz has been performed. In Fig. 3, the selected detached events are displayed in a scatter plot of the invariant mass versus the measured decay length: a clustering is observed around the J/ψ invariant mass for large Δz values in the region downstream of the primary interaction. A mean decay length of 0.81 ± 0.03 cm on $b\bar{b}$ events is expected from MC. When an unbinned maximum likelihood fit is performed on Δz, we measure 1.0 ± 0.3 cm for the 10 downstream events in the

J/ψ region ($2.8\,\mathrm{GeV}/c^2 < m_{e^+e^-} < 3.3\,\mathrm{GeV}/c^2$), in good agreement with the $b\bar{b}$ interpretation, while the 8 upstream background events yield a mean decay length of 0.36 ± 0.13 cm (measured using $-\Delta z$).

We have verified that when the background shape used in the fit is replaced by a pure combinatorial background shape or by a pure double semileptonic $b\bar{b}$ background, the signal always stays above 2 standard deviations level (the corresponding variation is included in the systematic errors). Independent of the optimization criteria, a J/ψ signal with a significance greater than 2 σ is always observed in the downstream part of the spectrum, while a visible J/ψ signal is never present in the upstream part. The same behavior is observed in an analysis of the electron sample performed with the muon cuts: due to less stringent vertexing requirements, both the signal and the background contributions increased, resulting in the same final cross section value.

6 Combined Cross Section Measurement

In order to extract the maximum information on $b\bar{b}$ production cross section from our data, we performed a combined fit both for e^+e^- and $\mu^+\mu^-$ channels which gives the final result in our x_F range:

$$\Delta\sigma(b\bar{b}) = 30^{+13}_{-11}(\mathrm{stat})\ \mathrm{nb/nucleon},$$

where the quoted uncertainty has been estimated directly from the fit (see Fig. 4).

The main sources of systematic uncertainty in the present measurement, which are not related to the final $b\bar{b}$ statistics, are due to the prompt J/ψ cross section reference (11%), the branching ratio $\mathrm{Br}(b\bar{b} \to J/\psi X)$ (9%), the trigger and detector simulation (5%), the prompt J/ψ MC production models (3.5%), the $b\bar{b}$ MC production models (5%), the prompt $J/\psi \to e^+e^-$ counting (5%) and the carbon-titanium difference in efficiencies (1.7%). Other contributions are below the 1% level. For the uncertainties stemming from the background shapes used in the maximum likelihood fits on the invariant masses and from

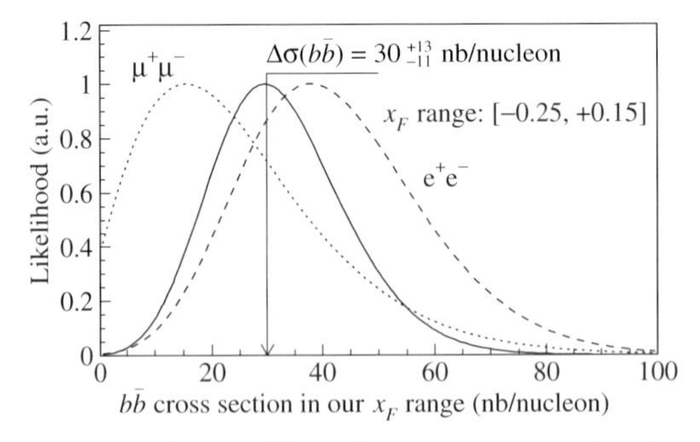

Fig. 4. The likelihood fits for the $b\bar{b}$ production cross section in our x_F range ($\Delta\sigma(b\bar{b})$) using the $\mu^+\mu^-$ and e^+e^- events separately (*dotted and dashed lines, respectively*) and in a combined analysis (*solid line*)

the cut values, we assign conservative uncertainties of $^{+10}_{-24}$% and 13% to the $\mu^+\mu^-$ and e^+e^- channels, respectively. The overall systematic uncertainty for our measurement, averaged over the muon and electron channels, is of $^{+20}_{-23}$%.

Extrapolating $\Delta\sigma(b\bar{b})$ to the full x_F range, we obtain the total $b\bar{b}$ production cross section:

$$\sigma(b\bar{b}) = 32^{+14}_{-12}(\text{stat})\,^{+6}_{-7}(\text{sys})\ \text{nb/nucleon}.$$

7 Conclusion

Analysing the data obtained in a short physics run during the HERA-B commissioning period in summer 2000, we identified $1.9^{+2.2}_{-1.5}$ candidates for $b \to J/\psi \to \mu^+\mu^-$ events, and $8.6^{+3.9}_{-3.2}$ for $b \to J/\psi \to e^+e^-$. From these candidates, we compute the $b\bar{b}$ production cross section by normalizing to the known prompt J/ψ cross section and extrapolating this measurement to the full J/ψ kinematic range. The combined result of the total $b\bar{b}$ production cross section measured by HERA-B at 920 GeV using pC and pTi interactions is [1]

$$\sigma(b\bar{b}) = 32^{+14}_{-12}(\text{stat})\,^{+6}_{-7}(\text{sys})\ \text{nb/nucleon}.$$

Figure 5 shows that this result is compatible with the E789 [10] and E771 [11] measurements. It also shows that the most recent QCD predictions beyond next-to-leading order: $\sigma(b\bar{b}) = 25^{+20}_{-13}$ nb/nucleon (value is based on [26], updated with the parton distribution function in [25]) and $\sigma(b\bar{b}) = 30 \pm 13$ nb/nucleon [27] are in good agreement with our measurement. An improved measurement with significantly enhanced statistics is foreseen in 2003 leading to a possibility to reduce both the statistical and systematic errors.

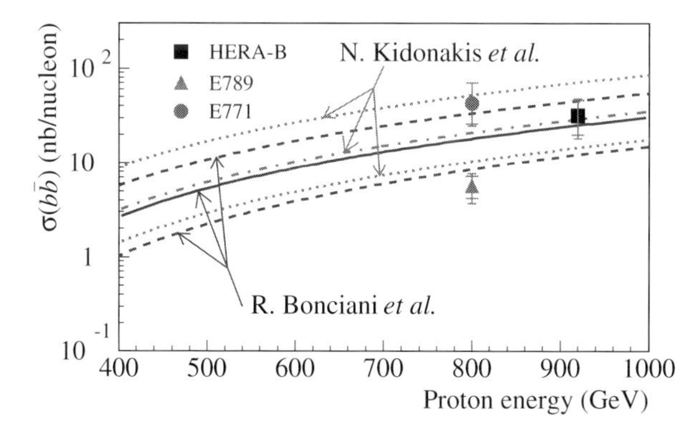

Fig. 5. The comparison of the HERA-B (2000) $\sigma(b\bar{b})$ value with other experiments [10,11] and the theoretical predictions of R. Bonciani et al. [26] updated with the NNLL parton distribution function in [25] (*solid line*: central value, *dashed lines*: upper and lower bounds) and N. Kidonakis et al. [27] (*dot-dashed line*: central value, *dotted lines*: upper and lower bounds)

References

1. The HERA-B Collaboration (I. Abt et al.): Preprint DESY 02-076 (2002), hep-ex/0205106, accepted to Eur. Phys. J. C
2. The CDF Collaboration (D. Acosta et al.): Phys. Rev. D **65**, 052005 (2002)
3. The OPAL Collaboration: OPAL Physics Note PN455 (2000)
4. The L3 Collaboration (M. Acciarri et al.): Phys. Lett. B **503**, 10 (2001)
5. The H1 Collaboration (C. Adloff et al.): Phys. Lett. B **467**, 156 (1999), erratum ibid. **518**, 331 (2001)
6. S. P. Baranov, M. Smižanská: Phys. Rev. D **62**, 014012 (2000);
   Ph. Hägler et al.: Phys. Rev. D **62**, 071502 (2000);
   A. V. Lipatov et al.: Preprint LUNFD6/(NFFL-7207) (2001), hep-ph/0112114;
   H. Jung: Phys. Rev. D **65**, 034015 (2002)
7. M. Cacciari, P. Nason: Phys. Rev. Lett. **89**, 122003 (2002)
8. E. L. Berger et al.: Phys. Rev.Lett. **86**, 4231 (2001)
9. The CDF Collaboration (D. Acosta et al.): Phys. Rev. D **66**, 032002 (2002)
10. The E789 Collaboration (D. M. Jansen et al.): Phys. Rev. Lett. **74**, 3118 (1995)
11. The E771 Collaboration (T. Alexopoulos et al.): Phys. Rev. Lett. **82**, 41 (1999)
12. E. Hartouni et al.: HERA-B Design Report, DESY-PRC-95-01 (1995)
13. The HERA-B Collaboration: HERA-B Status Report, DESY-PRC-00-04 (2000)
14. M. H. Schub et al.: Phys. Rev. D **52**, 1307 (1995)
15. T. Alexopoulos et al.: Phys. Rev. D **55**, 3927 (1997)
16. D. E. Groom et al.: Eur. Phys. J. C **15**, 1 (2000)
17. M. J. Leitch et al.: Phys. Rev. Lett. **84**, 3256 (2000)
18. T. Alexopoulos et al.: Phys. Lett. B **374**, 271 (1996)
19. M. J. Leitch et al.: Phys. Rev. Lett. **72**, 2542 (1994)
20. R. Vogt: Preprint LBNL-45350, hep-ph/0111271, to appear in the Proceedings of the Hard Probe Collaboration
21. T. Sjöstrand: Comp. Phys. Comm. **82**, 74 (1994)
22. H. Pi: Comp. Phys. Comm. **71**, 173 (1992)
23. R. Brun et al.: GEANT3, Internal Report CERN DD/EE/84-1 (1987)
24. M. Mangano, P. Nason, G. Ridolfi: Nucl. Phys. B **373**, 295 (1992);
    P. Nason, S. Dawson, R. K. Ellis: Nucl. Phys. B **327**, 49 (1988)
25. A. D. Martin, R. G. Roberts, W. J. Stirling, R. S. Thorne: Phys. Lett. B **531**, 216 (2002)
26. R. Bonciani et al.: Nucl. Phys. B **529**, 424 (1998)
27. N. Kidonakis et al.: Phys. Rev. D **64**, 114001 (2001)

Charmonium Production in 920 GeV Proton-Nucleus Interactions

Roman Mizuk for the HERA-B Collaboration

Institute for Theoretical and Experimental Physics, 117259 Moscow, Russia*

Abstract. Using data collected by the HERA-B experiment in the short commissioning run in the year 2000, we have studied the charmonium production in interactions of 920 GeV protons with carbon and titanium targets ($\sqrt{s} = 41.6 GeV$). We measured the fraction of J/ψ's produced via radiative χ_c decays $R_{\chi_c} = 0.32 \pm 0.06_{stat.} \pm 0.04_{syst.}$ averaged over proton-carbon and proton-titanium collisions [1]. The result is in agreement with previous measurements. It is compared with theoretical predictions. We present also preliminary results for J/ψ differential cross sections in x_F and p_T, which demonstrate the unique ability of HERA-B to study charmonium nuclear suppression in the backward hemisphere.

1 Introduction

Charmonium production in proton-nucleus collisions has recently attracted a considerable attention because of two problems. First, the mechanism of bound state formation from $c\bar{c}$ pair is not well understood. Second, the effects of nuclear matter on charmonium production are not understood, while a precise model of these effects is required for quark gluon plasma searches. In the following we discuss both problems in more details.

There are three models of charmonium production. The Colour Singlet Model (CSM) [2] requires that the $q\bar{q}$ pair be produced in a colour singlet state with the quantum numbers of the final meson. The Non-Relativistic QCD factorization approach (NRQCD) [3,4] assumes that a colour singlet or colour octet quark pair evolves towards the final bound state via exchange of soft gluons. The nonperturbative part of the process is described by long distance matrix elements which are extracted from data. Finally, the Colour Evaporation Model (CEM) [5,6] assumes the exchange of many soft gluons during the formation process such that the final meson carries no information about the production process of the $q\bar{q}$ pair. The CEM thus implies that quarkonium formation is independent of c.m.s. energy and the projectile/target parton density functions. There are many experimental observables which can be used to distinguish between the models. One of them is the dependence of the ratio of production cross sections for different states, e.g. the ratio of χ_c [1] and J/ψ production cross sections $\sigma(\chi_c)/\sigma(J/\psi)$, on $\sqrt{s}$. Experimentally, one reconstructs χ_c in the decay channel $\chi_c \to J/\psi\gamma$. Due

* supported in part by the Russian Fundamental Research Foundation under grant RFFI-00-15-96584 and the BMBF via the Max Planck Research Award

[1] In the following the notation χ_c indicates the sum of the two states χ_{c1} and χ_{c2}.

R. Mizuk, Charmonium Production in 920 GeV Proton-Nucleus Interactions, Lect. Notes Phys. **647**, 427–438 (2004)
http://www.springerlink.com/

to the small branching ratio of $\chi_{c0} \to J/\psi\gamma$, $(6.6 \pm 1.8) \times 10^{-3}$, the χ_{c0} contribution to the reconstructed signal can be neglected. In most experiments (and in HERA-B) the energy resolution is insufficient to resolve different χ_c states, so that one usually quotes the ratio

$$R_{\chi_c} = \frac{\sum_{i=1}^{2} \sigma(\chi_{ci}) Br(\chi_{ci} \to J/\psi\gamma)}{\sigma(J/\psi)} .$$

The experimental situation concerning R_{χ_c} is unclear, the uncertainties are large and the few existing measurements [8] differ strongly. We present here a measurement of R_{χ_c} for interactions of 920 GeV protons with carbon and titanium nuclei.

The suppression of J/ψ production in nuclear collisions is an important signature of quark gluon plasma [9]. The observation of suppressed J/ψ production in central Pb-Pb collisions by the NA50 collaboration [10] has been interpreted as possible evidence for the quark gluon plasma. However, alternative explanations are possible in terms of scattering from ordinary nuclear matter [11,12]. Measurements of charmonium suppression in proton-nucleus collisions, in which the nuclear state is almost certainly conventional, can firmly establish the properties with which to compare the results in nucleus-nucleus collisions. A variety of effects from nuclear targets have been considered in the literature [13]. However the available experimental data are not sufficient to understand the relative contribution of these effects. The charmonium suppression in proton-nucleus collisions is measured only for J/ψ and ψ' states and only for positive x_F [14]. To fix the nuclear suppression models one needs also the measurements at negative x_F [13] and for χ_c states [15]. The HERA-B experiment has a unique possibility to perform such measurements. We present here a preliminary result for differential cross section of inclusive J/ψ production in x_F and p_T, x_F values covering negative region. This result just demonstrates the capability of HERA-B. A new data sample with high statistics and improved detector will be collected during the year 2002/2003 run. A detailed study of the nuclear suppression for the different charmonium states will be performed on the forthcoming data.

2 Detector, Trigger and Data Sample

HERA-B is a fixed target experiment operating at the HERA storage ring at DESY. Charmonium and other heavy flavor states are produced in inelastic collisions of 920 GeV protons with fixed targets of different materials. The targets consist out of wires inserted in the halo of the proton beam circulating in HERA. The pN (N = p, n) c.m.s. energy is $\sqrt{s} = 41.6$ GeV. The detector is a magnetic spectrometer emphasizing vertexing, tracking and particle identification, with a dedicated J/ψ-trigger. The components of the HERA-B detector used for this analysis include a silicon strip vertex detector (VDS), honeycomb drift chambers (OTR), a large acceptance 2.13 T·m magnet, a finely segmented sampling electromagnetic calorimeter (ECAL), and a muon system (MUON) consisting of wire chambers interleaved with iron shielding which identifies muons with momenta larger than 5 GeV/c. The HERA-B detector allows an efficient reconstruction of particles with momenta larger than 1 GeV/c, including γ's and π^0's. A detailed description of the apparatus is given in [16].

The trigger selects $\mu^+\mu^-$ and e^+e^- pairs, the latter with an invariant mass larger than 2 GeV/c^2. For a muon candidate the trigger requires at least 3 MUON hits in coincidence with 9 OTR hits consistent with a particle track with a transverse momentum between 0.7 GeV/c and 2.5 GeV/c. The electron trigger requires that the transverse energy deposited in the calorimeter by the electron candidates be greater than 1.0 GeV and that at least 9 OTR hits confirm the track hypothesis. Both muon and electron candidates have to be confirmed by a track segment in the vertex detector with at least 6 hits.

The data presented here were collected during short commissioning run in the year 2000 before HERA luminosity upgrade shutdown began. Carbon and titanium target wires were used. The wires were separated by 3.3cm along the beam direction. The resolution of reconstructed dilepton vertices of 0.6mm along the beam direction [17] allowed a clear association of the interaction to a specific wire. The proton-nucleus interaction rate was approximately 5 MHz. About half of the data were taken with single carbon wire. The other half was taken with carbon and titanium wires simultaneously. The number of events recorded was 4.5×10^5 ($\mu^+\mu^-$) and 9.1×10^5 (e^+e^-). Another data sample consisting of 3.8×10^6 e^+e^- events collected without including the OTR information at the trigger level was used in addition for the J/ψ differential cross section measurement. The J/ψ acceptance was limited to $-0.25 < x_F < 0.15$. For more details concerning the trigger and the data sample of the year 2000 run see [18].

3 Monte Carlo Simulation

Since at present NRQCD is a favored approach, we use it to generate our Monte Carlo sample. The CSM is used to study model dependence. While CEM does not make any conclusive predictions for the differential charmonium cross sections, it is not used for simulation. The Monte-Carlo (MC) simulation of the events is done in three steps. First, PYTHIA 5.7 [19] is used to generate a $c\bar{c}$ pair and allow it to hadronize into charmonium. The transverse momentum and Feynman's x spectra are weighted according to NRQCD [4] or CSM [2] predictions. During the second step, the energy remaining after charmonium generation is used to simulate the rest of the interaction using FRITIOF 7.02 [20]. Finally, the J/ψ event is combined with n other inelastic interactions to simulate the event overlapping in the HERA-B conditions. The number n is distributed according to Poisson statistics with the mean 0.5. The detector response is simulated using GEANT 3.21 [21] and includes the measured hit resolution, the mapping of inefficient channels and the electronics noise. The simulated events are processed by the same trigger and reconstruction code as real data. For more details concerning MC simulation see [1].

4 Data Analysis

The measurement of R_{χ_c} is the main result of this work. The χ_c is observed in the decay $\chi_c \to J/\psi\gamma$ using the value ΔM which is the difference between the invariant mass of the $(l^+l^-\gamma)$ system and the invariant mass of the lepton pair l^+l^-:

$$\Delta M = M(l^+l^-\gamma) - M(l^+l^-)$$

Here, the uncertainty in the determination of the J/ψ mass essentially cancels. An excess of events with respect to the combinatorial background determines the number of χ_c candidates N_{χ_c} , from which the value of R_{χ_c} can be calculated as follows:

$$R_{\chi_c} = \frac{N_{\chi_c}}{N_{J/\psi} \cdot \epsilon_\gamma} \cdot \rho_\epsilon \tag{1}$$

where $N_{J/\psi}$ is the total number of reconstructed $J/\psi \to l^+l^-$ decays used for χ_c search, ϵ_γ is the photon detection efficiency and ρ_ϵ is the ratio of trigger and reconstruction efficiencies for J/ψ's from χ_c decays and for all J/ψ's:

$$\rho_\epsilon = \frac{\epsilon_{trig}^{J/\psi}\,\epsilon_{reco}^{J/\psi}}{\epsilon_{trig}^{\chi_c}\,\epsilon_{reco}^{\chi_c}}$$

Since the kinematics, triggering and reconstruction of direct J/ψ's and J/ψ's from χ_c decays are very similar, ρ_ϵ is close to unity.

The analysis consists of the reconstruction of the J/ψ events, the search of the photon candidates in the ECAL and the determination of the invariant mass of J/ψ and photons within the event. The analysis cuts for the χ_c selection were tuned to maximize the quantity $N_{J/\psi}\epsilon_\gamma/\sqrt{N}$, where $N_{J/\psi}$ is the number of J/ψ candidates above the background found in data, ϵ_γ is the photon efficiency determined from MC, N is the number of all events, including background, found in the measured ΔM distribution within a window of two standard deviations around the χ_c position.

4.1 J/ψ Selection

In the offline analysis, a track is accepted as a muon candidate if it passes a soft cut on the muon likelihood, derived from the ratio of expected and found hits in the MUON system. The requirement rejects random combinations of hits which satisfy trigger while keeping nearly all good tracks. For electron candidates, $|E/p - 1| < 0.3$ is required, where E is the energy deposited in the calorimeter and p is the track momentum. The cut on E/p corresponds to about 3.3 standard deviation of the electron E/p distribution. To further reject the background from hadrons, a presence of associated bremsstrahlung photons emitted in the region upstream or inside the magnet is required for both electron candidates. Thus an isolated electromagnetic cluster is searched for in the area where the bremsstrahlung would hit the ECAL. The total efficiency of this requirement is

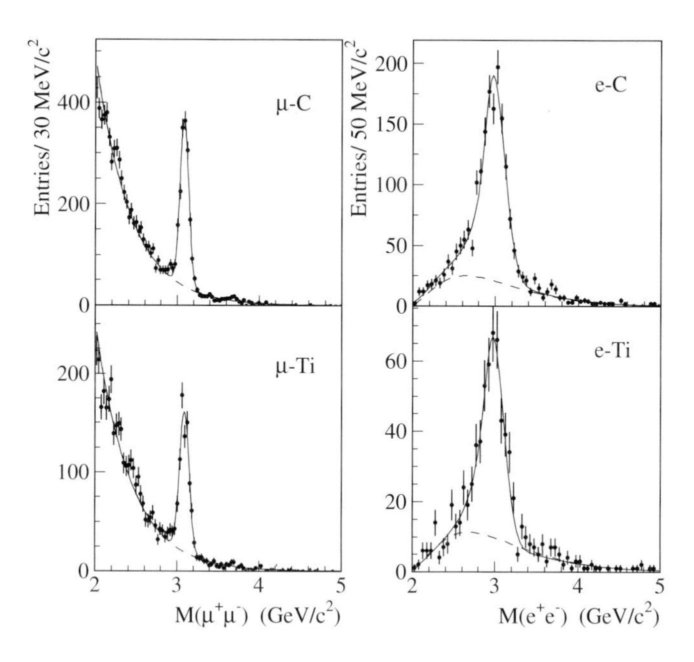

Fig. 1. Dilepton invariant mass spectrum for each of the four data samples ($\mu - C$, $\mu - Ti$, $e - C$ and $e - Ti$). The dashed line shows estimated background contribution. See text for the description of the fit (solid line).

20% (45% per electron candidate). The energy of the bremsstrahlung cluster is added to the energy of the electron candidate.

The lepton candidates are required to form a common vertex. Based on its position the lepton pair is assigned to the target wire. The data are divided into four separate subsamples: $\mu^+\mu^-$ or e^+e^- final state, each originating either from carbon or titanium targets (μ-C, e-C, μ-Ti, e-Ti). The invariant mass is calculated for each opposite-charge lepton pair. The resulting distributions are shown in Fig. 1. The signal observed for $J/\psi \to \mu^+\mu^-$ events is Gaussian while the $J/\psi \to e^+e^-$ signal has an asymmetric bremsstrahlung tail. In both cases the background underneath the signal is mainly combinatorial. The background shape is described either by an exponential distribution ($\mu^+\mu^-$) or by an exponential multiplied by a second order polynomial distribution (e^+e^-). The background shape was verified using the same sign pairs in case of muons and by fitting mass spectrum of all trigger candidates (mostly hadrons) in case of electrons. Only J/ψ candidates within a two standard deviations (2σ) window around J/ψ mass are considered for the analysis. In case of electrons σ is defined from the higher mass side of the signal. The numbers of J/ψ candidates obtained from the fit and corrected for the mass cut are shown in Table 1.

Using less stringent selection criteria we checked the x_F and p_T spectra of J/ψ. A fit was made to the individual mass distributions for each bin of x_F and p_T. The relative acceptance for each bin of x_F and p_T was determined from a de-

Table 1. The number of selected J/ψ events ($N_{J/\psi}$), the number of χ_c's observed (N_{χ_c}), photon detection efficiency (ϵ_γ) and the result for R_{χ_c}, for each of the four event samples. The quoted error on R_{χ_c} is statistical, except the contribution from systematic uncertainty in ϵ_γ.

	μ-C	e-C	μ-Ti	e-Ti
$N_{J/\psi}$	1510 ± 44	1180 ± 59	643 ± 29	382 ± 32
N_{χ_c}	159 ± 47	121 ± 38	59 ± 33	31 ± 27
ϵ_γ, %	27.3 ± 1.1	32.8 ± 1.5	24.4 ± 1.8	32.7 ± 2.6
R_{χ_c}	0.37 ± 0.11	0.30 ± 0.09	0.36 ± 0.20	0.23 ± 0.21

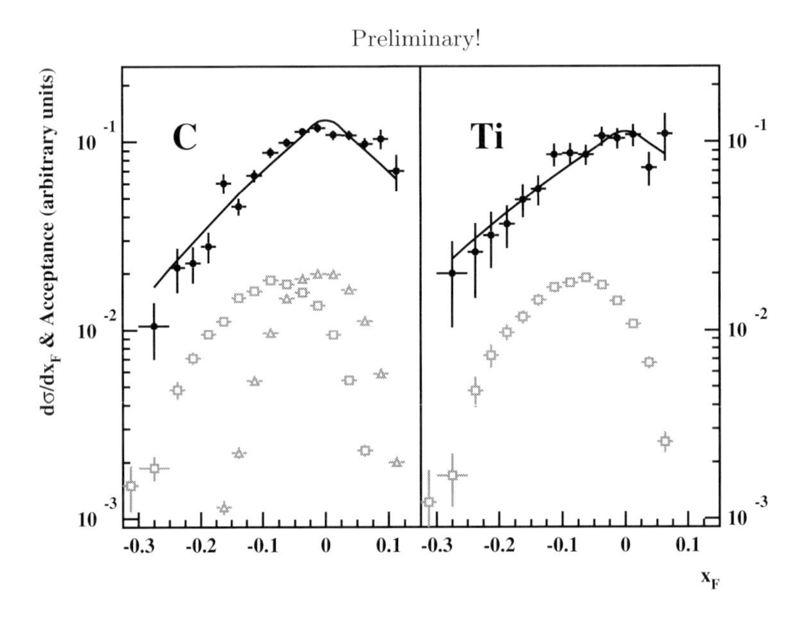

Fig. 2. Differential cross section (solid circles) in x_F for carbon and titanium targets. Acceptance is for J/ψ decay to e^+e^- (open triangles) and $\mu^+\mu^-$ (open squares). The solid curves are fits to the form $d\sigma/dx_F \propto (1-|x_F|)^c$. For titanium only $\mu^+\mu^-$ data are used.

tailed MC simulation. The resulting preliminary shapes of J/ψ differential cross section in x_F and p_T are shown in Figs. 2 and 3, for carbon and titanium targets. Only the shape of the differential cross section is measured while the absolute normalization is arbitrary. The acceptance is also shown in the bottom of the figures. It is flat in p_T but very nonuniform in x_F which makes the measurement of differential cross section in x_F more difficult than in p_T. The acceptance range is different for e^+e^- and for $\mu^+\mu^-$ because inner part of MUON and outer part of ECAL were not used in trigger in the year 2000 run. The errors include statistical and systematic contributions. The systematic is dominated by the uncertainty in the acceptance. The spectra are fit to the commonly used parametrisation

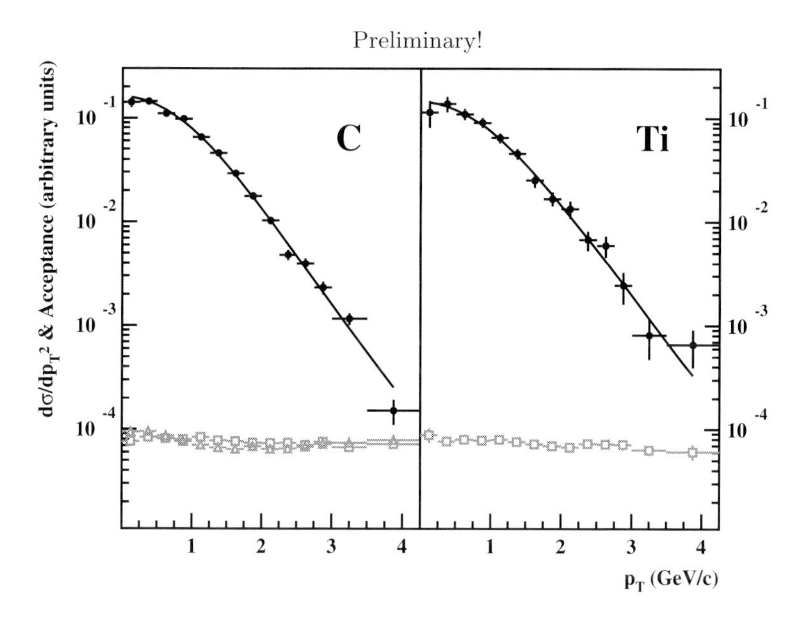

Fig. 3. Differential cross section (solid circles) in p_T for carbon and titanium targets. Acceptance is for J/ψ decay to e^+e^- (open triangles) and $\mu^+\mu^-$ (open squares). The solid curves are fits to the form $d\sigma/dp_T^2 \propto [1 + (p_T/p_0)^2]^{-6}$. For titanium only $\mu^+\mu^-$ data are used.

Table 2. Preliminary results for the shape parameters of the differential cross sections. Errors are in the order: statistical, systematic. The units for $\langle p_T \rangle$ are GeV/c.

Target	Parameter	Value	χ^2/N_{DF}
Carbon	c	6.32±0.26±0.32	39.3/14
Titanium	c	5.00±0.47±0.45	9.0/12
Carbon	$\langle p_T \rangle$	1.199±0.011±0.007	22.6/12
Titanium	$\langle p_T \rangle$	1.262±0.029±0.016	7.4/12

$d\sigma/dx_F \propto (1 - |x_F|)^c$ and $d\sigma/dp_T^2 \propto [1 + (p_T/p_0)^2]^{-6}$ where the parameters c and p_0 characterize the shapes in x_F and p_T, respectively. The parameter p_0 is analytically related to the mean p_T by $\langle p_T \rangle = (35\pi/256)p_0$. The preliminary results for the shape parameters are shown in Table 2. More information on the J/ψ differential cross section analysis can be found in [22].

4.2 χ_c Selection

To select χ_c events a cut on charged and neutral multiplicity is applied to throw away very busy events, which contribute more to the background than to the signal. This cut should not change the ratio $N_{\chi_c}/N_{J/\psi}$ since both types of events have similar multiplicities. For detailed discussion of this cut see [1].

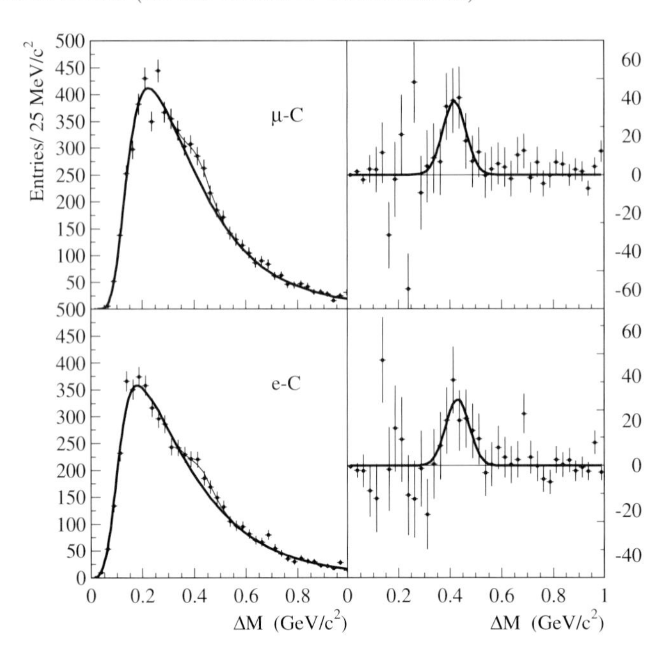

Fig. 4. The $\Delta M = M(l^+l^-\gamma) - M(l^+l^-)$ distribution for samples μ-C and e-C. In the left-most plots, the points represent data and the solid lines represent the combinatorial background estimated by event mixing. The right-most plots show the signal after background subtraction. See text and Table 1 for the details on the fit.

Each cluster in the ECAL with transverse energy $E_T > 0.1 GeV$ which is not associated with the leptons from the J/ψ, is considered as a photon candidate. Hadronic background is reduced by requiring that the ratio of the central cell energy to the total cluster energy (E_{centr}/E) be greater than 0.6. In order to suppress background due to soft secondary particles and noise clusters, an energy cut $E > 3.0 GeV$ is applied. A charged track veto is not applied, due to a high probability of the photon to convert in the detector material downstream of the magnet. The efficiency of photon detection is determined using MC simulation (see Table 1).

The ΔM distributions for all combinations of J/ψ and photon candidates for the carbon samples are shown in Fig. 4. The distributions show a χ_c signal. The shape of the dominantly combinatorial background in the ΔM distribution is obtained by combining J/ψ candidates with photon candidates from different events with similar multiplicity and applying the standard selection cuts. These "mixed events" reproduce the shape of the ΔM distribution everywhere except in the χ_c signal region (see solid line in Fig. 4, left panel). Similar results are obtained when events in the sidebands of the J/ψ mass region are combined with photon candidates. Since the experimental resolution is of the same order as the mass difference between χ_{c1} and χ_{c2} states and the statistics is limited, we use a single Gaussian to describe the signal. In the fit, the position and

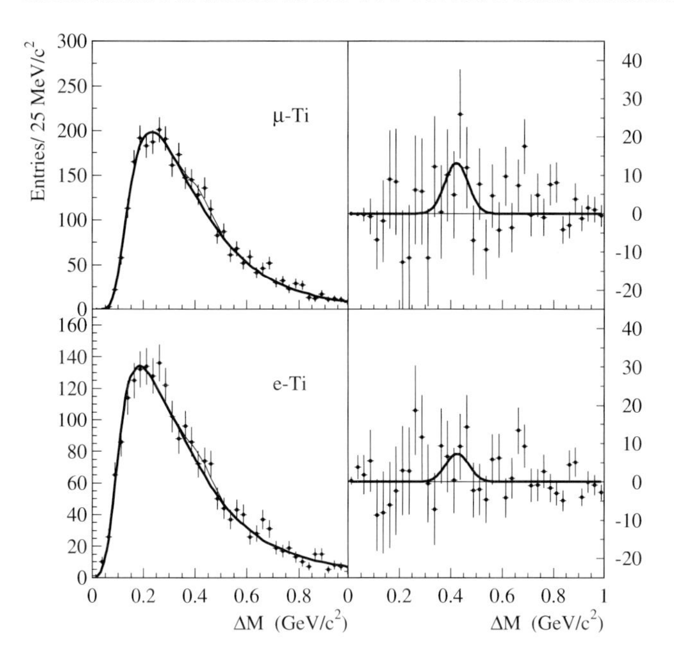

Fig. 5. Same as in Fig. 4 for μ-Ti and e-Ti samples.

normalization of the Gaussian, as well as the normalization of the background, are left free. The width of the Gaussian is fixed to the value predicted by MC. The position of the Gaussian agrees well with the value expected from MC. The background subtracted distributions are shown in the right panel of Fig. 4. The significances of both signals seen in the μ-C and e-C samples are about three standard deviations. The obtained number of χ_c events as well as the number of J/ψ events and the photon detection efficiency are summarized in Table 1. Taking into account the high background level and the ratio of N_{χ_c} to $N_{J/\psi}$ observed in the carbon samples, we do not expect to see a significant χ_c signal in the smaller titanium samples. The results obtained for the titanium sample with the same procedure are shown in Fig. 5. Although the signals are marginal, the R_{χ_c} values obtained from them are compatible with estimates from the carbon samples (see Table 1).

The following contributions to the systematic uncertainty are the most important: the uncertainty in the yield of J/ψ's due to background description (5%, only for electrons), the model dependence of efficiency factors in (1) (5%), the uncertainty in photon efficiency due to description of material in the detector (2%), the uncertainty in R_{χ_c} due to ECAL calibration (1%) and due to correlated noise in ECAL (3%), the uncertainty coming from fixed width of ΔM distribution, which depends upon the assumed ratio of χ_{c1} and χ_{c2} and detector resolution (6%), the variation of result with the cuts (6%), the finite MC statistics for determination of ϵ_γ (3%). All the uncertainties are assumed to be uncorrelated and the total systematic uncertainty on R_{χ_c} is estimated as 11%.

5 Result for R_{χ_c}

The values of R_{χ_c} (Table 1) for the two carbon samples agree with each other within the statistical errors. The results obtained from the titanium data are consistent with those obtained from the carbon data. Although nuclear dependence effects might be present in R_{χ_c} at the few percent level for the targets used here [15], they are beyond the statistical accuracy of the present measurement. We therefore average the results for the four samples obtaining:

$$R_{\chi_c} = 0.32 \pm 0.06_{stat} \pm 0.04_{sys}.$$

The first uncertainty listed is statistical only, whereas the second uncertainty is systematic.

Our result for R_{χ_c} is compatible with [8,23] and [24] as shown in Table 3 and Fig. 6. It is noteworthy that most of the measurements fall below the predictions of the CSM; our result confirms this behavior. In contrast, and due to the relatively large uncertainties, almost all data are compatible with a flat energy dependence as predicted by the CEM. On the other hand, the slope predicted by the NRQCD calculations is compatible with the pion data (except the point at lowest energy), whereas the uncertainties of the proton data do not allow to draw any conclusion. More precise measurements are needed to conclusively discriminate among these models.

Table 3. Previous πA [23], pA [8], $p\bar{p}$ [24] and HERA-B measurements of the R_{χ_c} value. The value quoted for exp. E771 has been calculated from the published cross sections [8] and branching ratios [7].

Exp.	coll.	$\sqrt{s}$ (GeV)	R_{χ_c}
IHEP140	$\pi^- p$	8.5	0.44 ± 0.16
WA11	π^-Be	18.7	0.30 ± 0.05
E610	π^-Be	18.9	0.31 ± 0.10
E673	$\pi^- H_2$, π^-Be	20.2	0.70 ± 0.28
E369	π^-Be	20.6	0.37 ± 0.09
E705	π^-Li	23.8	0.37 ± 0.03
E705	π^+Li	23.8	0.40 ± 0.04
E672/706	π^-Be	31.1	$0.443 \pm 0.041 \pm 0.035$
E610	pBe	19.4, 21.7	0.47 ± 0.23
E705	pLi	23.8	0.30 ± 0.04
E771	pSi	38.8	0.74 ± 0.17
R702	pp	52, 63	$0.15^{+0.10}_{-0.15}$
ISR	pp	62	0.47 ± 0.08
CDF	$p\bar{p}$	1800	$0.297 \pm 0.017 \pm 0.057$
HERA-B	pC, pTi	41.6	$0.32 \pm 0.06 \pm 0.04$

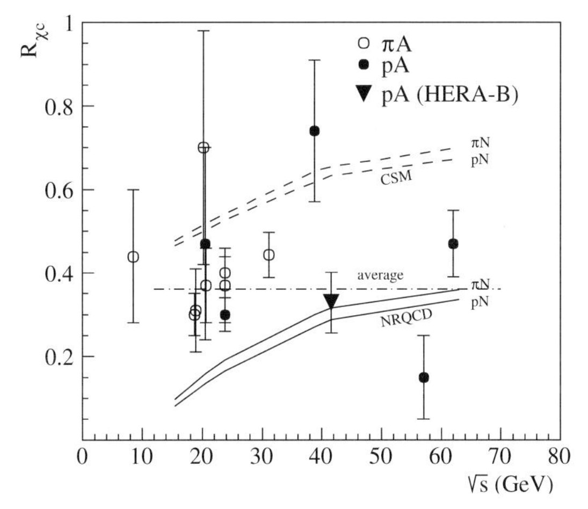

Fig. 6. Comparison of our measurement of R_{χ_c} (closed triangle) with those of other pp, pA [8] (closed circles) and πp, πA [23] (open circles) experiments. The CDF result [24] is not shown, since its kinematic acceptance differs strongly from the other experiments. The value quoted for exp. E771 has been calculated from the published cross sections [8] and branching ratios [7]. The error bars include statistical and systematic uncertainties. Also shown are predictions for pN and πN interactions based on the NRQCD [4] (solid), CSM [2] (dashed). The CEM [5,6] predicts a constant value. The dot-dashed line is the average of all measurements.

6 Conclusions

We present results from the HERA-B experiment based on data collected during short commissioning run in the year 2000. The fraction of J/ψ's produced from radiative decays of χ_{c1} and χ_{c2} states via $p-C$ and $p-Ti$ interactions is measured. For J/ψ's in the range $-0.25 < x_F < 0.15$ the value $R_{\chi_c} = 0.32 \pm 0.06_{stat} \pm 0.04_{sys}$ is obtained. Our result for R_{χ_c} agrees with most of previous measurements and with the predictions of NRQCD and CEM.

The preliminary results for J/ψ differential cross sections in x_F and p_T demonstrate the ability of the HERA-B experiment to study backward production of charmonium. A new data sample with high statistics and improved detector will be collected at the end of 2002 / beginning of 2003 year which will allow the measurement of nuclear suppression in the wide range of x_F including negative region ($-0.4 < x_F < 0.3$) for J/ψ, ψ' and χ_c states.

References

1. **HERA-B** collab., I. Abt et al., "J/ψ Production via χ_c Decays in 920 GeV pA Interactions", hep-ex/0211033, submitted to Phys. Lett. **B**
2. R. Baier, R. Rückl, Phys. Lett. **B102** (1981) 364; Z. Phys. C19 (1983) 251
3. G.T. Bodwin, E. Braaten, G.P. Lepage, Phys. Rev. **D51** (1995) 1125
4. P. Cho, A. Leibovich, Phys. Rev. **D53** (1996) 6203

5. H. Fritzsch, Phys. Lett. **B67** (1977) 217; F. Halzen, Phys. Lett. **B69** (1977) 105; F. Halzen, S. Matsuda, Phys. Rev. **D17** (1978) 1344; M. Glück, J. Owens, E. Reya, Phys. Rev. **D17** (1978) 2324; R. Gavai et al., Int. J. Mod. Phys. **A10** (1995) 3043
6. J. F. Amundson et al., Phys. Lett. **B372** (1996) 127; Phys. Lett. **B390** (1997) 323
7. C. Caso et al., Review of Particle Physics, Eur. Phys. Jour. **C15** (2000) 1
8. **ISR** collab., A.G. Clark et al., Nucl. Phys. **B142** (1978) 29; **E610** collab., D.A. Bauer et al., Phys. Rev. Lett. **54** (1985) 753; **E705** collab., L. Antoniazzi et al., Phys. Rev. Lett. **70** (1993) 383; **E771** collab., T. Alexopoulos et al., Phys. Rev. **D62** (2000) 032006
9. T. Matsui and H. Satz, Phys. Lett. **B178** (1986) 416
10. **NA50 collab.**, M.C. Abreu et al., Phys. Lett. **B499** (2001) 85
11. A. Capella et al., Phys. Lett. **B393** (1997) 431; S. Gavin and R. Vogt, Phys. Rev. Lett. **78** (1997) 1006
12. S. Gavin and R. Vogt, Phys. Rev. Lett. **78** (1997) 1006
13. R. Vogt, Phys. Rev. **C61** (2000) 035203
14. **E866/NuSea** collab., M.J. Leitch et al., Phys. Rev. Lett. **84** (2000) 3256
15. R. Vogt, Nucl. Phys. **A700** (2002) 539
16. HERA-B collab., E. Hartouni et al., HERA-B Design Report, DESY-PRC-95-01 (1995); **HERA-B** collab., HERA-B Status Report, DESY-PRC-00-04 (2000)
17. C. Bauer et al., Nucl. Instr. Methods **A453** (2000) 103
18. **HERA-B** collab., I. Abt et al., "Measurement of the $b\bar{b}$ Production Cross Section in 920 GeV Fixed-Target Proton-Nucleus Collisions", hep-ex/0205106, in press on Eur. Phys. J. C, DOI, 10.1140/epjc/s2002-01071-8
19. T. Sjöstrand, Comp. Phys. Comm. **82** (1994) 74
20. H. Pi, Comp. Phys. Comm. **71** (1992) 173
21. R. Brun et al., GEANT3, CERN-DD-EE-84-1 (1987)
22. **HERA-B** collab., I. Abt et al., "Backward production of J/ψ mesons in proton-nucleus collisions at $\sqrt{s}$41.6 GeV", in preparation
23. **IHEP140** collab., F. Binon et al., Nucl. Phys. **B239** (1984) 311; **WA11** collab., Y. Lemoigne et al., Phys. Lett. **B113** (1982) 509; **E610** collab., D.A. Bauer et al., Phys. Rev. Lett. **54** (1985) 753; **E673** collab., T.B.W. Kirk et al., Phys. Rev. Lett. **42** (1979) 619; **E369** collab., S.R. Hahn et al., Phys Rev. **D30** (1984) 671; **E705** collab., L. Antoniazzi et al., Phys. Rev. Lett. **70** (1993) 383; **E672/E706** collab., V. Koreshev et al., Phys. Rev. Lett **77** (1996) 4294
24. **CDF** collab., F. Abe et al., Phys. Rev. Lett. **79** (1997) 578

Index

Lecture Notes in Physics

For information about Vols. 1–602
please contact your bookseller or Springer-Verlag
LNP Online archive: springerlink.com

Vol.603: C. Noce, A. Vecchione, M. Cuoco, A. Romano (Eds.), Ruthenate and Rutheno-Cuprate Materials. Superconductivity, Magnetism and Quantum Phase.

Vol.604: J. Frauendiener, H. Friedrich (Eds.), The Conformal Structure of Space-Time: Geometry, Analysis, Numerics.

Vol.605: G. Ciccotti, M. Mareschal, P. Nielaba (Eds.), Bridging Time Scales: Molecular Simulations for the Next Decade.

Vol.606: J.-U. Sommer, G. Reiter (Eds.), Polymer Crystallization. Obervations, Concepts and Interpretations.

Vol.607: R. Guzzi (Ed.), Exploring the Atmosphere by Remote Sensing Techniques.

Vol.608: F. Courbin, D. Minniti (Eds.), Gravitational Lensing:An Astrophysical Tool.

Vol.609: T. Henning (Ed.), Astromineralogy.

Vol.610: M. Ristig, K. Gernoth (Eds.), Particle Scattering, X-Ray Diffraction, and Microstructure of Solids and Liquids.

Vol.611: A. Buchleitner, K. Hornberger (Eds.), Coherent Evolution in Noisy Environments.

Vol.612: L. Klein, (Ed.), Energy Conversion and Particle Acceleration in the Solar Corona.

Vol.613: K. Porsezian, V.C. Kuriakose (Eds.), Optical Solitons. Theoretical and Experimental Challenges.

Vol.614: E. Falgarone, T. Passot (Eds.), Turbulence and Magnetic Fields in Astrophysics.

Vol.615: J. Büchner, C.T. Dum, M. Scholer (Eds.), Space Plasma Simulation.

Vol.616: J. Trampetic, J. Wess (Eds.), Particle Physics in the New Millenium.

Vol.617: L. Fernández-Jambrina, L. M. González-Romero (Eds.), Current Trends in Relativistic Astrophysics, Theoretical, Numerical, Observational

Vol.618: M.D. Esposti, S. Graffi (Eds.), The Mathematical Aspects of Quantum Maps

Vol.619: H.M. Antia, A. Bhatnagar, P. Ulmschneider (Eds.), Lectures on Solar Physics

Vol.620: C. Fiolhais, F. Nogueira, M. Marques (Eds.), A Primer in Density Functional Theory

Vol.621: G. Rangarajan, M. Ding (Eds.), Processes with Long-Range Correlations

Vol.622: F. Benatti, R. Floreanini (Eds.), Irreversible Quantum Dynamics

Vol.623: M. Falcke, D. Malchow (Eds.), Understanding Calcium Dynamics, Experiments and Theory

Vol.624: T. Pöschel (Ed.), Granular Gas Dynamics

Vol.626: G. Contopoulos, N. Voglis (Eds.), Galaxies and Chaos

Vol.627: S.G. Karshenboim, V.B. Smirnov (Eds.), Precision Physics of Simple Atomic Systems

Vol.628: R. Narayanan, D. Schwabe (Eds.), Interfacial Fluid Dynamics and Transport Processes

Vol.630: T. Brandes, S. Kettemann (Eds.), Anderson Localization and Its Ramifications

Vol.631: D. J. W. Giulini, C. Kiefer, C. Lämmerzahl (Eds.), Quantum Gravity, From Theory to Experimental Search

Vol.632: A. M. Greco (Ed.), Direct and Inverse Methods in Nonlinear Evolution Equations

Vol.633: H.-T. Elze (Ed.), Decoherence and Entropy in Complex Systems, Based on Selected Lectures from DICE 2002

Vol.634: R. Haberlandt, D. Michel, A. Pöppl, R. Stannarius (Eds.), Molecules in Interaction with Surfaces and Interfaces

Vol.635: D. Alloin, W. Gieren (Eds.), Stellar Candles for the Extragalactic Distance Scale

Vol.636: R. Livi, A. Vulpiani (Eds.), The Kolmogorov Legacy in Physics, A Century of Turbulence and Complexity

Vol.637: I. Müller, P. Strehlow, Rubber and Rubber Balloons, Paradigms of Thermodynamics

Vol.638: Y. Kosmann-Schwarzbach, B. Grammaticos, K.M. Tamizhmani (Eds.), Integrability of Nonlinear Systems

Vol.639: G. Ripka, Dual Superconductor Models of Color Confinement

Vol.640: M. Karttunen, I. Vattulainen, A. Lukkarinen (Eds.), Novel Methods in Soft Matter Simulations

Vol.641: A. Lalazissis, P. Ring, D. Vretenar (Eds.), Extended Density Functionals in Nuclear Structure Physics

Vol.642: W. Hergert, A. Ernst, M. Däne (Eds.), Computational Materials Science

Vol.643: F. Strocchi, Symmetry Breaking

Vol.644: B. Grammaticos, Y. Kosmann-Schwarzbach, T. Tamizhmani (Eds.) Discrete Integrable Systems

Vol.645: U. Schollwöck, J. Richter, D.J.J. Farnell, R.F. Bishop (Eds.), Quantum Magnetism

Vol.646: N. Bretón, J. L. Cervantes-Cota, M. Salgado (Eds.), The Early Universe and Observational Cosmology

Vol.647: D. Blaschke, M. A. Ivanov, T. Mannel (Eds.), Heavy Quark Physics

Printing: Strauss GmbH, Mörlenbach
Binding: Schäffer, Grünstadt